Marine Propulsors

Marine Propulsors

Special Issue Editors

Sverre Steen
Kourosh Koushan

MDPI • Basel • Beijing • Wuhan • Barcelona • Belgrade

MDPI

Special Issue Editors
Sverre Steen
Norwegian University of Science and Technology
Norway

Kourosh Koushan
INTEF Ocean and Norwegian University of Science and Technology
Norway

Editorial Office
MDPI
St. Alban-Anlage 66
Basel, Switzerland

This is a reprint of articles from the Special Issue published online in the open access journal *Journal of Marine Science and Engineering* (ISSN 2077-1312) in 2018 (available at: http://www.mdpi.com/journal/jmse/special_issues/jz_marine_propulsors)

For citation purposes, cite each article independently as indicated on the article page online and as indicated below:

LastName, A.A.; LastName, B.B.; LastName, C.C. Article Title. *Journal Name* **Year**, *Article Number*, Page Range.

ISBN 978-3-03897-202-0 (Pbk)
ISBN 978-3-03897-203-7 (PDF)

Cover image courtesy of Sverre Steen.

Contents

About the Special Issue Editors

Sverre Steen specializes in marine hydrodynamics, focusing on propulsion and experimental methods. He has been a professor at the Department of Marine Technology at the Norwegian University of Science and Technology since 2004, and Head of Department at the same institute since 2016. He is also director of the Rolls-Royce University Technology Centre "Performance in a Seaway". In 2009, he founded the Symposium of Marine Propulsors together with Prof. Kourosh Koushan.

Kourosh Koushan is Senior Adviser at SINTEF Ocean (formerly MARINTEK) . He is also Adjunct Professor at the Norwegian University of Science and Technology (NTNU). He received his PhD from the Technical University of Berlin (1997). He joined SINTEF Ocean (in 1997) as research engineer, promoted to senior research engineer (2000), to principal research engineer (2005) and research director (2008). He is a member of the Executive Committee (since 2014) and Advisory Council (since 2011) of the International Towing Tank Conference (ITTC). Prof. Koushan together with Prof. Steen initiated smp (International Symposium on Marine Propulsors) in 2009.

Preface to "Marine Propulsors"

Marine propulsors are key components of the many thousands of ships operating in oceans, lakes, and rivers around the world. The performance of propulsors are vital for the efficiency, environmental impact and safety of the ships. Propulsor performance is also important for crew and passenger comfort. New types of propulsors, with electric drives, flexible blades, and multi-stage propellers require new knowledge and improved tools. Innovative main or auxiliary propulsor types, using renewable energy from waves or winds, are also being commercialized. The improvement of computers and computational fluid dynamics creates new opportunities for advanced design and performance prediction, and new instrumentation and data collection techniques enable more advanced experimental techniques.

This book is devoted to bringing the latest developments in research and technical developments regarding the hydrodynamic aspects of marine propulsors, to the benefit of both academics and industry. It consist of 16 peer-reviewed scientific papers previously published in the Journal of Marine Science and Engineering. Twelve of these papers are improved and extended versions of papers presented at the Fifth Symposium of Marine Propulsors in June 2017 in Espoo, Finland, while the remaining four are completely new submissions. The Symposia of Marine Propulsors are held bi-annually, the next one will take place in Rome in May 2019.

Sverre Steen, Kourosh Koushan
Special Issue Editors

Journal of
Marine Science and Engineering

MDPI

Editorial

Marine Propulsors

Sverre Steen [1] and Kourosh Koushan [2,*]

1 Department of Marine Technology, Faculty of Engineering Science and Technology, Norwegian University of Science and Technology, Otto Nielsens vei 10, N-7491 Trondheim, Norway; sverre.steen@ntnu.no
2 SINTEF Ocean and Department of Marine Technology, Faculty of Engineering Science and Technology, Norwegian University of Science and Technology, Otto Nielsens vei 10, N-7491 Trondheim, Norway
* Correspondence: Kourosh.Koushan@sintef.no; Tel.: +47-411-05-297

Received: 16 August 2018; Accepted: 16 August 2018; Published: 22 August 2018

Keywords: propellers; waterjets; unconventional propulsors (azimuthing, SPP, rim drive, etc.); cavitation; noise and vibration; numerical methods in propulsion; propulsor-ice interaction; propulsor dynamics; propulsion in seaways; propulsion in off-design conditions

---

This Special Issue is following up the success of the latest Symposium on Marine Propulsors (www.marinepropulsors.com, smp'17) by publishing extended or improved versions of the selected papers presented at the symposium. This issue also includes new original contributions. smp'17 was the fifth in a series of international symposiums dedicated to the hydrodynamics of all types of marine propulsors. The next symposium in this series will be held in Rome in May 2019. This Special Issue comprises 12 excellent papers originating from the symposium [1–12] and four outstanding new papers [13–16]. The papers disseminate state-of-the-art numerical and experimental research results on marine propulsors and marine renewable devices.

Marine propulsors are key components of the many thousands of ships operating in oceans, lakes, and rivers around the world. The performance of propulsors are vital for the efficiency, environmental impact, and safety of ships. Propulsor performance is also important for crew and passenger comfort. New types of propulsors with electric drives, flexible blades, and multi-stage propellers require new knowledge and improved tools. Innovative main or auxiliary propulsor types, using renewable energy from waves or winds, are also being commercialized. The improvement of computers and computational fluid dynamics creates new opportunities for advanced design and performance prediction, and new instrumentation and data collection techniques enable more advanced experimental techniques. This Special Issue of the *Journal of Marine Science and Engineering* is devoted to bringing the latest developments in research and technical developments regarding hydrodynamic aspects of marine propulsors, to the benefit of both academics and the industry.

Prof. Dr. Sverre Steen and Prof. Dr. Kourosh Koushan.

Guest Editors of "Marine Propulsors".

Conflicts of Interest: The authors declare no conflict of interest.

References

1. Kim, S.; Kinnas, S.A.; Du, W. Panel method for ducted propellers with sharp trailing edge duct with fully aligned wake on blade and duct. *J. Mar. Sci. Eng.* **2018**, *6*, 89. [CrossRef]
2. Helma, S.; Streckwall, H.; Richter, J. The effect of propeller scaling methodology on the performance prediction. *J. Mar. Sci. Eng.* **2018**, *6*, 60. [CrossRef]
3. Viitanen, V.M.; Hynninen, A.; Sipilä, T.; Siikonen, T. DDES of wetted and cavitating marine propeller for CHA underwater noise assessment. *J. Mar. Sci. Eng.* **2018**, *6*, 56. [CrossRef]
4. Salvatore, F.; Sarichloo, Z.; Calcagni, D. Marine turbine hydrodynamics by a boundary element method with viscous flow correction. *J. Mar. Sci. Eng.* **2018**, *6*, 53. [CrossRef]

5. Su, Y.; Kim, S.; Kinnas, S.A. Prediction of propeller-induced hull pressure fluctuations via a potential-based method: Study of the effects of different wake alignment methods and of the rudder. *J. Mar. Sci. Eng.* **2018**, *6*, 52. [CrossRef]
6. Maljaars, P.; Bronswijk, L.; Windt, J.; Grasso, N.; Kaminski, M. Experimental validation of fluid–structure interaction computations of flexible composite propellers in open water conditions using BEM-FEM and RANS-FEM methods. *J. Mar. Sci. Eng.* **2018**, *6*, 51. [CrossRef]
7. Bosschers, J. A semi-empirical prediction method for broadband hull-pressure fluctuations and underwater radiated noise by propeller tip vortex cavitation. *J. Mar. Sci. Eng.* **2018**, *6*, 49. [CrossRef]
8. Jones, M.C.; Paterson, E.G. Influence of propulsion type on the stratified near wake of an axisymmetric self-propelled body. *J. Mar. Sci. Eng.* **2018**, *6*, 46. [CrossRef]
9. Knight, B.; Freda, R.; Young, Y.L.; Maki, K. Coupling numerical methods and analytical models for ducted turbines to evaluate designs. *J. Mar. Sci. Eng.* **2018**, *6*, 43. [CrossRef]
10. Berchiche, N.; Krasilnikov, V.I.; Koushan, K. Numerical analysis of azimuth propulsor performance in seaways: Influence of oblique inflow and free surface. *J. Mar. Sci. Eng.* **2018**, *6*, 37. [CrossRef]
11. Regener, P.B.; Mirsadraee, Y.; Andersen, P. Nominal vs. effective wake fields and their influence on propeller cavitation performance. *J. Mar. Sci. Eng.* **2018**, *6*, 34. [CrossRef]
12. Baltazar, J.M.; Rijpkema, D.; de Campos, J.F.; Bosschers, J. Prediction of the open-water performance of ducted propellers with a panel method. *J. Mar. Sci. Eng.* **2018**, *6*, 27. [CrossRef]
13. Qiu, J.-T.; Yang, C.-J.; Dong, X.-Q.; Wang, Z.-L.; Li, W.; Noblesse, F. Numerical simulation and uncertainty analysis of an axial-flow waterjet pump. *J. Mar. Sci. Eng.* **2018**, *6*, 71. [CrossRef]
14. Maljaars, P.; Kaminski, M.; den Besten, H. Boundary element modelling aspects for the hydro-elastic analysis of flexible marine propellers. *J. Mar. Sci. Eng.* **2018**, *6*, 67. [CrossRef]
15. Ortolani, F.; Dubbioso, G.; Muscari, R.; Mauro, S.; Di Mascio, A. Experimental and numerical investigation of propeller loads in off-design conditions. *J. Mar. Sci. Eng.* **2018**, *6*, 45. [CrossRef]
16. Hally, D. Modelling a propeller using force and mass rate density fields. *J. Mar. Sci. Eng.* **2018**, *6*, 41. [CrossRef]

Journal of
Marine Science and Engineering

MDPI

Article

Panel Method for Ducted Propellers with Sharp Trailing Edge Duct with Fully Aligned Wake on Blade and Duct

Seungnam Kim *, Spyros A. Kinnas and Weikang Du

Ocean Engineering Group, CAEE, The University of Texas at Austin, Austin, TX 78712, USA; kinnas@mail.utexas.edu (S.A.K.); allendu1988@utexas.edu (W.D.)

* Correspondence: naoestar@utexas.edu; Tel.: +1-512-751-8829

Received: 5 April 2018; Accepted: 19 July 2018; Published: 23 July 2018

Abstract: A low-order panel method is used to predict the performance of ducted propellers. A full wake alignment (FWA) scheme, originally developed to determine the location of the force-free trailing wake of open propellers, is improved and extended to determine the location of the force-free trailing wakes of both the propeller blades and the duct, including the interaction with each other. The present method is applied on a ducted propeller with sharp trailing edge duct, and the predicted results over a wide range of advance ratios, with or without full alignment of the duct wake, are compared with each other, as well as with results from RANS simulations and with measurements from an experiment.

Keywords: ducted propeller; panel method; duct wake alignment; full wake alignment; force-free wake

1. Introduction

As a viable propulsion system, ducted propellers, which involve rotating blades inside a non-rotating nozzle, have been used widely in shipbuilding and offshore industries and beyond. Ducted propellers can provide more total thrust (due to blades and duct) with higher efficiency than open propellers, especially at low advance ratios. Additionally, they protect the propeller blades, even though they increase the risk of cavitation (in the case of accelerating ducts).

Accurate prediction of open or ducted propeller performance at on- and off-design conditions is very important. In recent years Reynolds-Averaged Navier–Stokes (RANS) simulations have been used to predict propeller performance over a wide range of operating conditions. However, due to the relatively long computation time and the often considerable effort to generate a proper grid, especially in the case of ducted propellers, RANS becomes a less viable tool in the early design stage. On the other hand, the boundary element method (BEM, or panel method) is a viable alternative numerical tool to RANS. BEM solves for the unknown perturbation potential on the boundary of the domain. Quantities inside the domain are determined in terms of the quantities on the boundary in BEM, while RANS discretizes the whole domain and solves for the unknown quantities at each cell. Compared to RANS, BEM has many fewer unknowns to solve for, and requires a significantly smaller effort to discretize the boundary instead of the whole domain.

Many types of panel methods have been proposed since the application of the surface source method to marine propellers by Hess and Valarezo [1] and the application of potential-based methods to open air propellers by Morino and Guo [2] and open or ducted marine propellers by Kerwin et al. [3]. Recently, Kinnas et al. [4] applied a perturbation potential based panel method to ducted propellers, in which they implemented the full wake alignment method of Tian and Kinnas [5] on the propeller blades, while they assumed a cylindrical duct wake. They showed that the panel method with the fully aligned blade wake predicted the performance of the ducted propeller more accurately than a simplified wake alignment model proposed by Greeley and Kerwin [6], which significantly overpredicted the forces in all operating conditions. Kinnas et al. [7] further improved the panel method for ducted propellers with an emphasis on the duct paneling and the blade wake alignment model. Baltazar et al. [8] developed a panel method in which they implemented a reduction in the pitch of the blade wake at its tip to account for the boundary layer over the duct inner surface.

In ducted propeller problems, the wake alignment model is critical because the blade trailing wake influences the loading distribution over the duct due to its proximity to the duct trailing edge. This influence becomes even more significant in zero-gap and square-tip ducted propeller cases. Previous studies [4,7] showed that the panel method with full wake alignment (FWA) applied to the blade wake predicted results in good correlation with the experiment and results from RANS. However, those studies applied the FWA scheme only to the blade wake, assuming that the duct wake has a cylindrical shape, with radius at the trailing edge of the duct, as shown in Figure 1a. Hence, the representation of the physical behavior of the vorticity behind the duct trailing edge has been neglected.

Although the assumption of cylindrical shape of the duct wake has a minor effect on the predicted propeller performance, due to its relatively large distance from the control points on propeller, the blade wake must be post-processed to prevent it from intersecting with the duct wake. Figure 1a depicts the intersection of the blade wake panels close to the tip with the duct wake, which eventually leads to a divergence of the alignment scheme. To avoid this issue, the locations of the blade wake panels that intersect the duct wake are adjusted radially so that they are always inside the duct wake, as shown in Figure 1b.

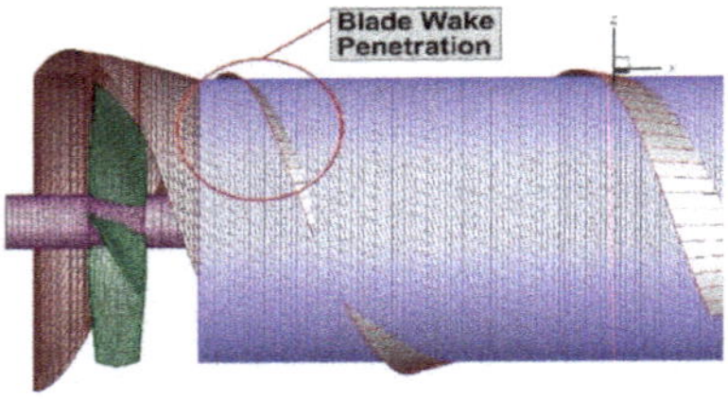

(a)

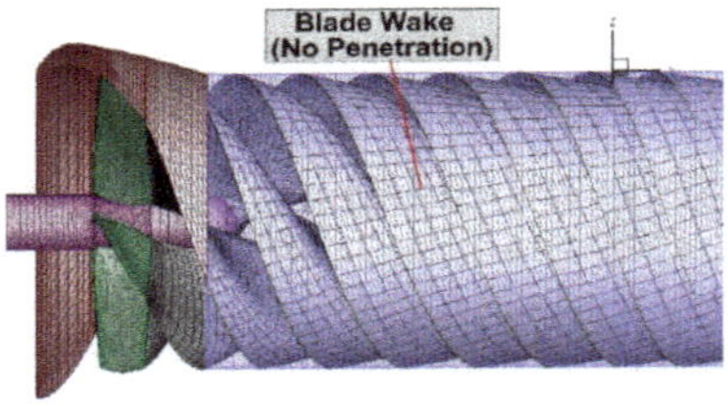

(b)

Figure 1. Penetration of the blade wake on duct wake before (**a**) and after (**b**) adjusting the location of the blade wake panels close to the tip. Only half of the duct geometry is shown.

The wake alignment scheme plays a crucial role in determining the location of the wake panels downstream, and subsequently it may significantly affect the predicted propeller performance. Considering that the influence coefficients from wake to propeller are calculated based on their relative locations, the predicted results could be incorrect if the wake panels are distorted during the alignment procedure. A typical example of such problem is the penetration of blade wake into duct wake, as shown in Figure 1a. The best way of avoiding this numerical problem is to apply the full wake alignment also on the duct wake such that the spatial location of the duct wake is determined based on the local flow, as in the case of the blade wake. To this end, the full wake alignment is applied to both the blade and duct wake within the framework of a low-order panel method. The detailed formulation of the full wake alignment scheme for the steady or unsteady performance of open/ducted propellers is provided in Kim [9], along with convergence and grid dependence studies. Therefore, this paper will focus on the application of the presented method to both the blade wake and the duct wake, in steady case, i.e., uniform inflow upstream of the ducted propeller.

In addition, repaneling on duct is employed to improve the convergence of the predicted forces. In the full wake alignment scheme the blade wake is being updated after each iteration until the pre-defined convergence criteria are satisfied. During the iterative process, the small distance between the blade wake and the duct inner side can result in slow convergence or even divergence of the results, especially for a square-tip blade case, which assumes a constant small gap between the blade tip and the duct, and high loading conditions. This is mainly due to the panel mismatch between the updated blade wake panels and the duct inner side panels, which results in singular behavior of the solution and/or of the induced velocities. To resolve this issue, a repaneling process is introduced, and consequently not only the convergence of the panel method but also the predicted pressure distribution on the duct is improved.

In the present method the complete interaction between the blade and the duct wake is included, by considering the induced velocities of one on the other. This interaction will cause the duct wake to curl around the blade wake due to the strong tip vortex of the blade. Results from the present method, with and without duct wake alignment, will be compared against each other, to those from RANS simulations, as well as to measurements from an experiment.

2. Methodology

2.1. Wake Alignment Procedure with Repaneling on Duct/Duct Wake

In the alignment procedure, the main contributor to the shape of the duct wake is the blade wake at the tip, due to its proximity to the duct wake. Hence, the duct wake alignment is conducted after the blade wake is updated at each iteration of the full wake alignment (FWA). Figure 2b describes the general flow chart of the FWA scheme, including the alignment of the duct wake. In Figure 2a, the alignment procedure without aligning the duct wake is also presented for comparison. Note that when the FWA is applied to the duct wake, the first duct wake panels (i.e., those closest to the duct trailing edge) are aligned with the last panels on the duct, but the rest of the duct wake panels are purely determined by the FWA scheme. This repaneling process is implemented on the duct panels using either of the following two repaneling options, which were first introduced in the case of cylindrical duct wake by Kinnas et al. [7]. Note at the end of each outer iteration loop the solution on the blade and duct is updated.

(a) Option 1: The duct panels are not modified, i.e., they are adapted to the initial blade and blade wake geometries, but are kept the same in subsequent iterations. However, the blade wake and duct wake panels keep changing throughout the iteration process.

(b) Option 2: At the beginning of each outer iteration, the panels on the duct surface are adapted to the blade wake panels, which are updated from the last inner iteration of the FWA.

In Option 2 the repaneling process is conducted at the beginning of each outer iteration based on the wake geometries from the previous iteration. With these wake geometries and the repaneled duct and duct wake, the panel method is implemented to solve for the updated potentials. Matching the panels on the duct with the blade wake at the tip can affect significantly the predicted loading distributions over the blade, especially toward the blade tip, as shown in Kim [9]. The local singular behavior in the vicinity of the duct inner side/blade wake intersection, is due to the panel mismatching, and can be effectively avoided by adjusting the panel arrangement on the duct surface to match that of the blade wake at the tip. The panel arrangements for the two options are shown in Figure 3.

Figure 4 shows the convergence history of the predicted thrust and torque coefficients using the two repaneling options (Options 1 and 2). The thrust and torque coefficients are defined as follows:

$$Thrust\ Coefficient\ (K_T) = \frac{T}{\rho n^2 D^4} \tag{1}$$

$$Torque\ Coefficient\ (K_Q) = \frac{Q}{\rho n^2 D^5}, \tag{2}$$

where T and Q are the propeller thrust and torque. ρ, n, and D are the fluid density, propeller rotational frequency (rev/s), and the diameter of propeller, respectively. The duct wake here is assumed to be cylindrical. Both options show stable convergence histories at high advance ratios, but only Option

2 remains stable even at the very high loading[1] condition, i.e., at low advance ratio: $J_s = \frac{V_s}{nD} = 0.2$, where V_s is the ship speed, or the uniform inflow speed upstream of the duct in the case of open water test. At this low advance ratio, the blade wake panels move significantly from their initial location (which is aligned with the duct inner side), and if Option 1 is implemented it leads to highly unmatched panels between the duct inner side and the outer edge of the blade wake, which eventually produces unstable results. On the other hand, the matched panels in Option 2 produce relatively stable convergence histories even at very high loading conditions.

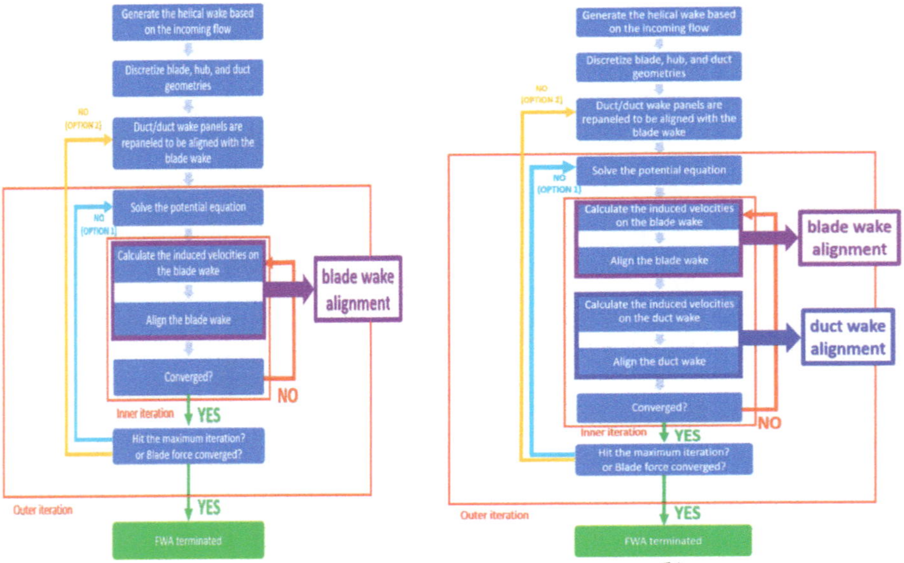

(**a**) (**b**)

Figure 2. Flowchart of the FWA without (**a**) and with (**b**) duct wake alignment.

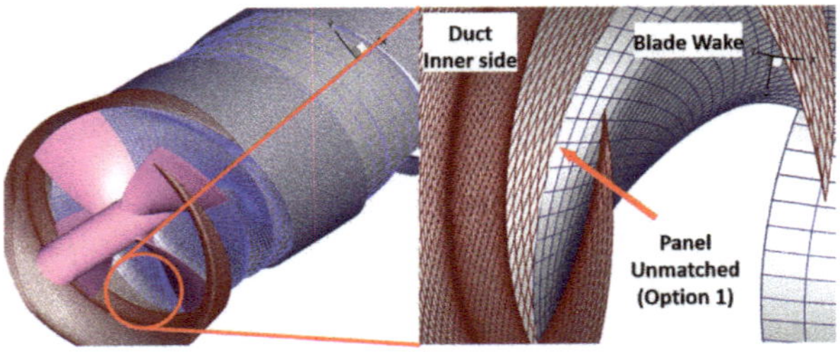

(**a**)

Figure 3. *Cont.*

[1] In the case of low advance ratios, the "local" thrust coefficient, C_T, based on the inflow to the propeller inside the duct, can be higher than 2.

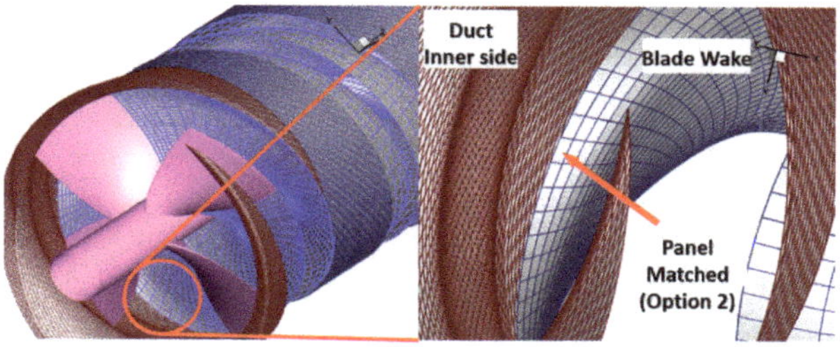

(**b**)

Figure 3. Unmatched (**a**) and matched (**b**) panels between the duct inner side and the outer edge of the blade wake using Option 1 and 2, respectively. Cylindrical duct wake is assumed with the penetration control on the blade wake. Only half the duct is shown for clarity.

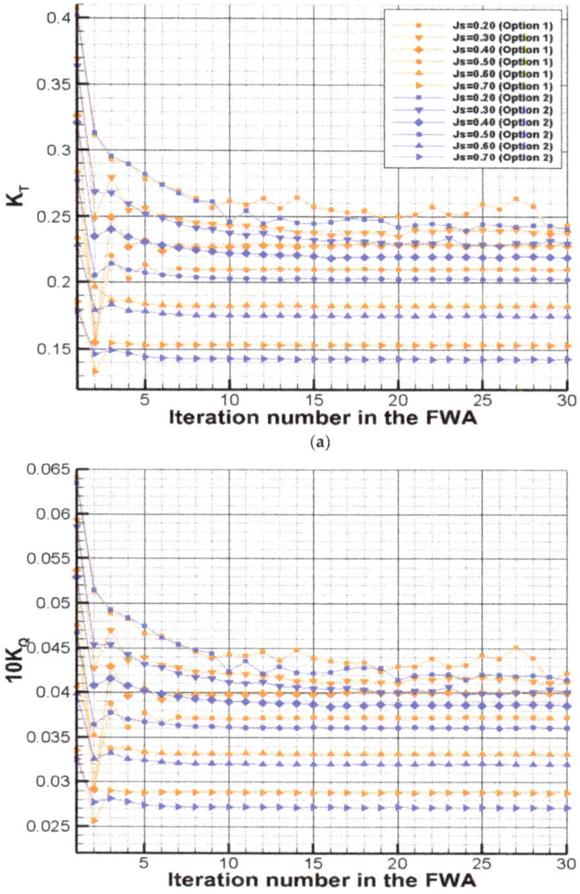

(**b**)

Figure 4. Convergence histories of the predicted K_T (**a**) and $10K_Q$ (**b**) on the blade using FWA with the two repaneling options. A cylindrical wake is assumed.

2.2. Full Wake Alignment Scheme on Duct Wake

As mentioned earlier, the duct wake is aligned based on the induced velocity from duct, blade, hub, and blade wake. The effects of the duct wake on itself (as in the case of blade wake) are also considered. The induced velocities are added to the inflow velocities to produce the total velocity, with which the four corners of the wake panels are aligned. The underlying duct wake alignment method is the same as that in the case of the blade wake, as introduced by Tian and Kinnas [5].

Consider a propeller subject to an axisymmetric effective wake, $\vec{U}_{eff}(\vec{x})$, which in this work is assumed to be uniform, i.e., $\vec{U}_{eff}(\vec{x}) = V_s$. $\vec{x} = (x, y, z)$ is a point in the flow-field, with x being along the propeller axis and positive downstream, y positive upwards (opposite to gravity), and z following the right hand rule. When the propeller rotates at a constant angular velocity, $\vec{\omega}$, the inflow velocity relative to a coordinate system that rotates with the propeller becomes (note in the case of a right handed propeller $\vec{\omega}$ is pointing towards $-x$):

$$\vec{U}_{in}(\vec{x}) = \vec{U}_{eff}(\vec{x}) - \vec{\omega} \times \vec{x}. \tag{3}$$

The potential, ϕ_p, which is independent of time in steady state, at an arbitrary point p on the discretized propeller geometry can be expressed using Green's third identity.

$$2\pi\phi_p = \iint_{S_B} \left[\phi_q \frac{\partial}{\partial n_q}\left(\frac{1}{R(p;q)}\right) - \frac{\partial \phi_q}{\partial n_q}\left(\frac{1}{R(p;q)}\right)\right] dS + \iint_{S_W} \Delta\phi_W(r_q) \frac{\partial}{\partial n_q}\left(\frac{1}{R(p;q)}\right) dS, \tag{4}$$

where ϕ_q denotes the potential at the variant point q, which has the distance $R(p;q)$ between the field point p. n_q denotes the unit normal vector at point q pointing out of the blades and duct propeller surface S_B, and $\Delta\phi_W$ is the potential jump (or, wake strength) across the blade/duct wake surface S_W.

Equation (4) states that the perturbation potential ϕ_p on the propeller surface is expressed as the superposition of the potentials due to source and dipole distribution on the propeller, and due to dipoles on the wake. By enforcing the kinematic boundary condition on the propeller and duct surface, the source strength can be determined as

$$\frac{\partial \phi_q}{\partial n_q} = -\vec{U}_{in}(\vec{x}) \cdot \vec{n}_q. \tag{5}$$

In steady state, note that the trailing wake strength $\Delta\phi_W$ on the blade wake is constant along the stream-wise direction, but changes in the span wise direction, while on the duct wake $\Delta\phi_W$ is also constant in the steam-wise direction, but changes in the circumferential direction. For both cases, $\Delta\phi_W$ is time-invariant.

Once the velocity potentials are determined by solving Equation (4), dipole strength ϕ_q on the propeller surface is known. The perturbation velocity $\vec{u}_i$ (or, inducted velocity) at the edge i of a vortex segment in the wake can be calculated from the Green's formula by taking the gradient of Equation (4) and after switching point p with point i. The relative locations of the blade/duct wake in the downstream are then determined by using this perturbation velocity induced from all blade and duct surfaces and their wakes:

$$\vec{u}_i = \frac{1}{4\pi} \iint_{S_B} \left[\phi_q \vec{\nabla} \frac{\partial}{\partial n_q}\left(\frac{1}{R(i;q)}\right) - \frac{\partial \phi_q}{\partial n_q} \vec{\nabla}\left(\frac{1}{R(i;q)}\right)\right] dS + \frac{1}{4\pi} \iint_{S_W} \Delta\phi_W(r_q) \vec{\nabla} \frac{\partial}{\partial n_q}\left(\frac{1}{R(i;q)}\right) dS. \tag{6}$$

The perturbation velocity is then decomposed into the two directions, along the inflow direction, $u_{i,s}$, and the other normal to the inflow direction, $\vec{u}_{i,n}$, as shown in Figure 5. The former component is added to the inflow velocity, since both are in the same direction.

$$u_{i,s} = \vec{u}_i \cdot \vec{s}_i \tag{7}$$

$$\vec{u}_{i,n} = \vec{u}_i - u_{i,s}\vec{s}_i, \tag{8}$$

where $\vec{s}_i = \vec{U}_{in,i}/|\vec{U}_{in,i}|$ is the unit vector in the direction of the unperturbed inflow velocity, $\vec{U}_{in,i}$, at the midpoint of each vortex segment spanning from point $\vec{X}_{i-1}$ to $\vec{X}_i$. The inflow velocity is re-evaluated at the updated location of the vortex segment, at each iteration step. The initial geometry of the trailing wake is placed along the direction of the unperturbed inflow, and the vector for each vortex segment is given as:

$$\Delta\vec{s}_i = \vec{U}_{in,i} * \Delta t = \begin{pmatrix} \frac{1}{2}(U_{in,i-1,x} + U_{in,i,x}) * \Delta t \\ \frac{1}{2}(U_{in,i-1,y} + U_{in,i,y}) * \Delta t \\ \frac{1}{2}(U_{in,i-1,z} + U_{in,i,z}) * \Delta t \end{pmatrix} \tag{9}$$

$$\vec{s}_i = \frac{\vec{U}_{in,i}}{|\vec{U}_{in,i}|} = \frac{\Delta\vec{s}_i}{|\Delta\vec{s}_i|}, \tag{10}$$

where $\Delta t = \Delta\theta/\omega$ is the time step size that corresponds to the initial geometry of the wake panels placed along the unperturbed flow direction, with $\Delta\theta$ being the constant increment in the direction of the blade angle. The essence of the full wake alignment (FWA) scheme is to fix the length of the projections of the vortex segments $|\Delta\vec{s}_i|$ in the inflow direction and then align the vortex segment with the updated velocity, as described in Tian and Kinnas [5]. In other to maintain the projected length of each segment we define an adjusted time step: $\Delta t_i^* = \frac{|\Delta\vec{s}_i|}{(|\vec{U}_{in,i}|+u_{i,s})}$. Please note that the provided expression for Δt_i^*. , has been corrected for a typo in the original formula presented in Tian and Kinnas [5]. The final equation to update the vertices of each vortex segment is as follows.

$$\vec{X}_i^{N+1} = \vec{u}_{i,n}\Delta t + \left(1 - \frac{\Delta t}{\Delta t_i^*}\right)\vec{X}_i^N + \frac{\Delta t}{\Delta t_i^*}\left(\vec{X}_{i-1}^N + \Delta\vec{s}_i\right), \tag{11}$$

where $\vec{X}_i^{N+1}$ denotes the coordinate of the aligned wake panel i. at $N + 1$ iteration step. The wake panels are repeatedly updated until the iteration number N reaches a pre-defined maximum iteration number, or a convergence criterion in terms of the wake geometry is satisfied. Equation (11) can be re-written as follows:

$$\vec{X}_i^{N+1} = \vec{X}_i^N + \varepsilon\left[(\vec{U}_{in,i} + \vec{u}_i)\Delta t - \frac{\Delta t}{\Delta t_i^*}\left(\vec{X}_i^N - \vec{X}_{i-1}^N\right)\right], \tag{12}$$

where ε is an under-relaxation factor, which at conditions close to design advance ratio is taken to be equal to 0.5, while at low-advance ratios (e.g., high loading conditions) is taken to be equal to 0.25.

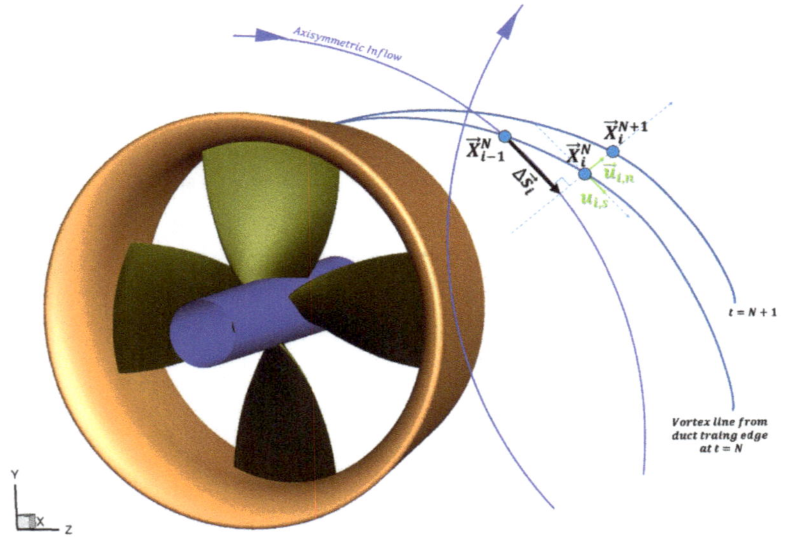

Figure 5. Schematic plot of a material line in the duct wake at iteration steps N and $N + 1$ in the full wake alignment (FWA) scheme.

2.3. Consideration of the Effects of Viscosity

To consider the effects of viscosity, the following two approaches can be used with the current panel method. One, named "viscous pitch correction", is by using an empirical correction to the pitch angle of the blade, and by applying a constant friction coefficient over the blades to account for the frictional forces, as described by Kerwin and Lee [10]. The other is by coupling the present panel method with a 2D boundary layer solver (XFOIL), modified to account for the effects of three dimensions, applied along each blade strip and its wake, with the effects of the other strips on the same and the other blades being included in an iterative manner, as described in Kinnas et al. [11]. In the latter approach the effects of viscosity on the pressure distribution on the blades, as well as the local friction coefficient over the blade surface, are evaluated by the combined panel method and the boundary layer solver. In the present work the effects of viscosity on the duct are evaluated via applying a uniform friction coefficient over its surface, and by ignoring the correction on its pitch angle in the case of ducts with sharp trailing edge. Ducts with blunt trailing edges are not considered in this paper. In that case the viscous effects on the duct loading are significant, and can be evaluated via coupling with a boundary layer solver applied over the circumferentially averaged flow, as presented in Kinnas et al. [7]. Alternatively, the effects of viscosity on the duct can be evaluated by coupling a vortex-lattice method (VLM) applied over the propeller blades, with an axisymmetric RANS solver applied over the duct, with the blades represented via body forces, as described in Tian et al. [12].

The two methods described above are applied on ducted propeller KA4-70 and the results are shown in Figure 6. The geometry of the ducted propeller is shown in Figure 7. The measured forces obtained by Bosschers and van der Veeken [13], are also included in Figure 6. Note that the present method, with full wake alignment applied on both blades and the duct, without the effects of viscosity predicts the propeller thrust quite well, but under predicts the propeller torque, and over predicts the duct thrust, significantly. When the effects of viscosity are applied, by using either one of the two methods presented above, the correlation with experiments improves significantly. It should be noted that if the simplified wake alignment approach of Greeley and Kerwin [6] were applied on the blade wakes, the propeller thrust and torque would be over predicted significantly, as presented in Kinnas et al. [4].

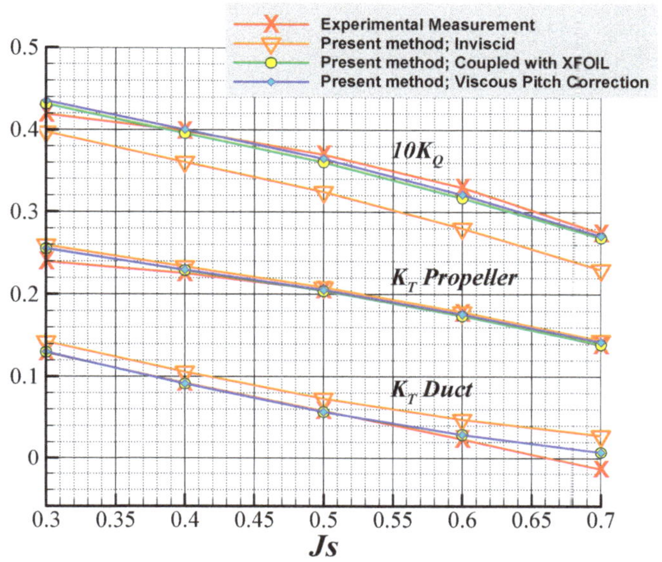

Figure 6. Correlations of the predicted force performance of the KA4-70 propeller with duct 19Am (with sharp trailing edge), as shown in Figure 7, from the presented method with/without viscous corrections, and experiment. Full Wake Alignment (FWA) is applied to the panel method.

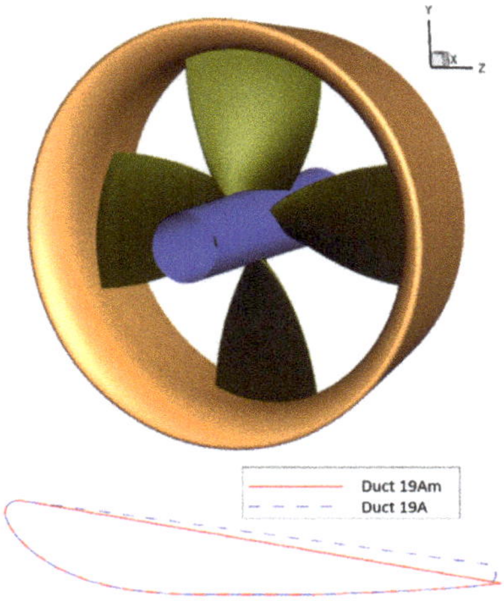

Figure 7. KA4-70 ducted propeller geometry (upper) and the duct cross section (lower). The modified Duct 19Am, with sharp trailing edge, is utilized in the calculations and the experiment at MARIN [13].

3. Results: Ducted Propeller with Square Tip and Sharp Trailing Edge Duct

FWA is applied to a square-tip ducted propeller to investigate the effects of the blade/duct wake on the predicted propeller performance. For this application, KA4-70 ducted propeller with Duct 19Am is adopted. The design advance ratio of this propeller is $Js = 0.5$, and zero gap is assumed between the duct inner side and the blade tip. The predicted force performance under different advance ratios are compared with the results from RANS (using ANSYS/Fluent) and measured values from the experiment, conducted by Bosschers and van der Veeken [13]. Pressure distributions over the blade surface and over the duct surface, as well blade and duct wake shapes, are also correlated with those predicted from RANS simulations.

3.1. Lower Order Panel Method

To discretize the propeller geometry, 60 × 20 (chordwise × spanwise) panels are used for each blade, and 160 × 20 (chordwise × circumferential) panels are used for the duct geometry between blades. The convergence history with number of iterations of the predicted blade forces with different number of panels on the blade and duct are shown in Figure 8. As shown, the converged forces are practically independent of panel numbers on the blade or duct. A similar conclusion can be drawn with number of panels on the blade and duct wakes, as long as a minimum of 100 panels in the streamwise direction are used, as shown in Kim [9]. In Figure 9, the predicted converged wake geometries are shown for various advance ratios. For all cases shown, FWA is applied to both the blade wake and the duct wake with 100 panels in the stream-wise direction.

The time required for the panel method calculations is highly dependent on the panel numbers rather than the geometrical operations, such as repaneling. Therefore, calculation time reduces significantly if the panel number is reduced. KA4-70 ducted propeller in this paper uses 9760 panels in total to discretize a quarter of the propeller geometry with 100 panels streamwise for each of the blade wake and duct wake. This large number of panels constitutes dense matrices, making the time required for solving the resulting system of equations relatively slow. Most time-consuming parts arise from the calculation of the influence coefficients, and the other is the calculation of the induced velocities on the wake panels. To resolve this numerical inefficiency, open multi-processing (OpenMP) parallel code is applied to the most time-consuming parts of the calculation. As a result, the total computing time is reduced to about 30 min on 8 Intel Xeon Platinum 8160 2.1 GHz cores (two hardware threads per core)[2]. In this calculation, the FWA is applied to both the blade/duct wake and goes through 30 iterations with repaneling Option 2. If the cylindrical duct wake is assumed with the FWA only applied to the blade wake as the least elaborated case, it only takes 15 min to finish the same iteration number under the same computing power.

For the lower advance ratios, especially below $Js = 0.5$, it is observed that the blade and duct wake are entangled with each other and that expand as convected downstream. Under high loading conditions, the strength of the vortices on the blade and duct gets stronger than in the case with high Js, leading to strong curling of both wakes. As the curling starts "early" and gets stronger with convection downstream, both the blade and duct wake panels get entangled, and subsequently get distorted due to the singular behavior. Another possible contribution to such numerical issue is the truncation of the wake surface. The wake panels around the truncated region, where the effect from propeller is minor due to the convected distance, are affected by the upstream wake in terms of the induced velocity. However, because of the lack of the same influence from the other side of the truncated region, the wake panels around the end expand under the one-sided wake-to-wake effect.

The expansion is more distinct with lower Js, at which the wake panels are closer to each other, leading to high influence among them. To ensure convergent solutions and accurate propeller forces,

[2] The simulation is performed at the Texas Advance Computing Center (TACC) at The University of Texas at Austin (Austin, TX 78703, USA). URL: http://www.tacc.utexas.edu.

therefore, the duct and blade wake need to be long enough with the end located away from the propeller, by thus using large number of panels in the streamwise direction. A lot of effort has been devoted to alleviate this numerical problem, but more research is needed.

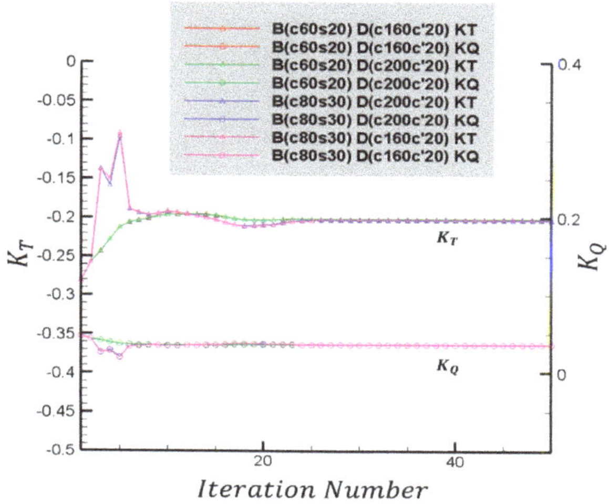

Figure 8. Convergence history of the predicted blade thrust and torque coefficients with number of panels on the blade and the duct. B(c60s20) and D(c160c'20), for example, represent 60 × 20 (chordwise × spanwise) panels on the blade and 160 × 20 (chordwise × circumferentially) panels on the duct. Full wake alignment is applied to both blade and duct wakes.

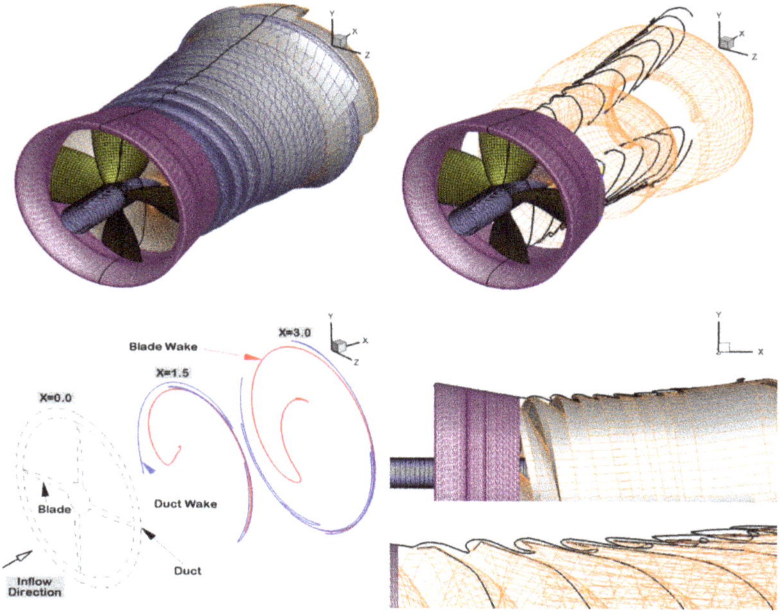

(a)

Figure 9. *Cont.*

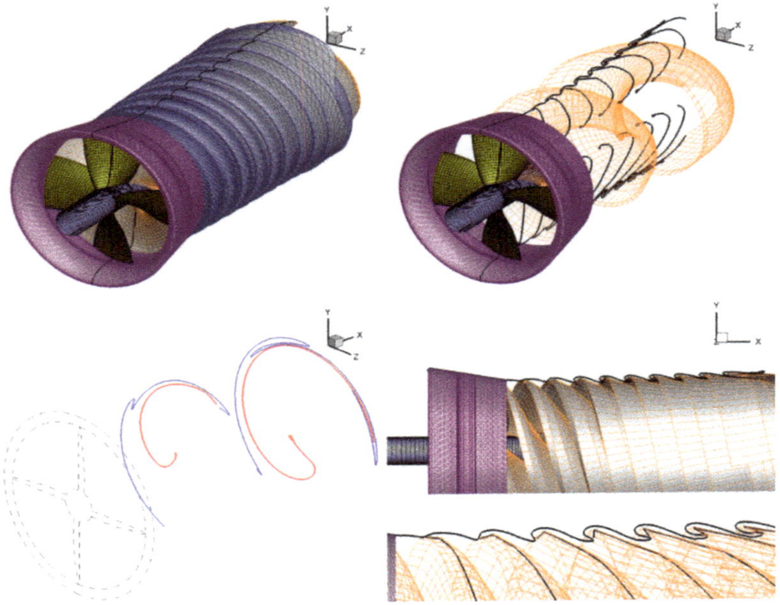

(**b**)

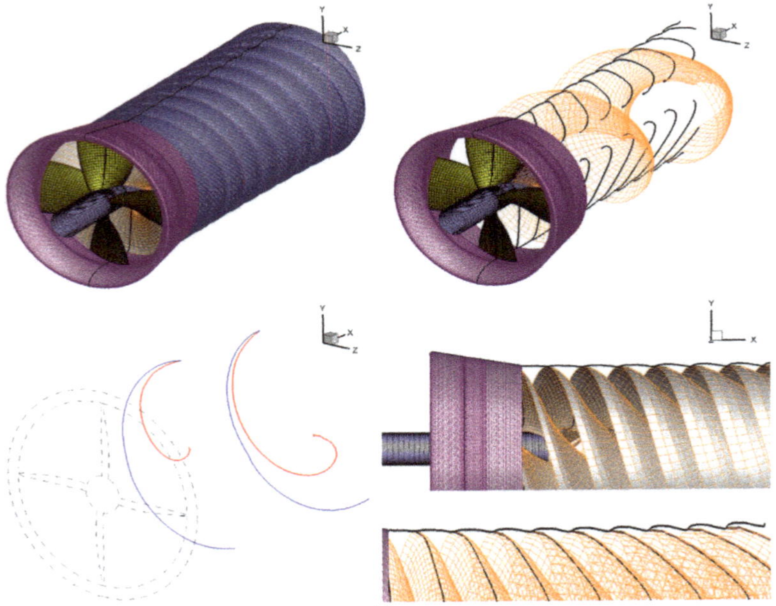

(**c**)

Figure 9. Converged blade and duct wake geometries from using full wake alignment on both, at (**a**) $J_S = 0.3$, (**b**) $J_S = 0.5$, (**c**) $J_S = 0.7$. Perspective views of the blade and duct wakes are shown together with their intersections with vertical planes through or perpendicular to the propeller axis.

Figure 10 shows the initial and the convergent fully aligned wake (FWA) shapes at the design advance ratio of 0.50. The wake shapes (black solid line) are plotted on a vertical plane, which passes through the center of the propeller geometry to show the details of the aligned wake. FWA starts its first iteration based on the helical and cylindrical shapes of the blade wake and duct wake, respectively, as shown in Figure 10a. Then, the solutions based on these initial geometries are used for the next iteration until the predicted thrust and torque coefficients converge. In Figure 10, the total velocity vectors on the blade and duct wake are plotted on both the blade wake and the duct wake together with the streamwise vortex elements. In addition, the angle between the total velocity vector and the corresponding vortex element are shown. These plots help visualize and verify the application of the force-free condition in the wake, which requires that the total velocity and the streamwise vorticity vectors be aligned with each other.

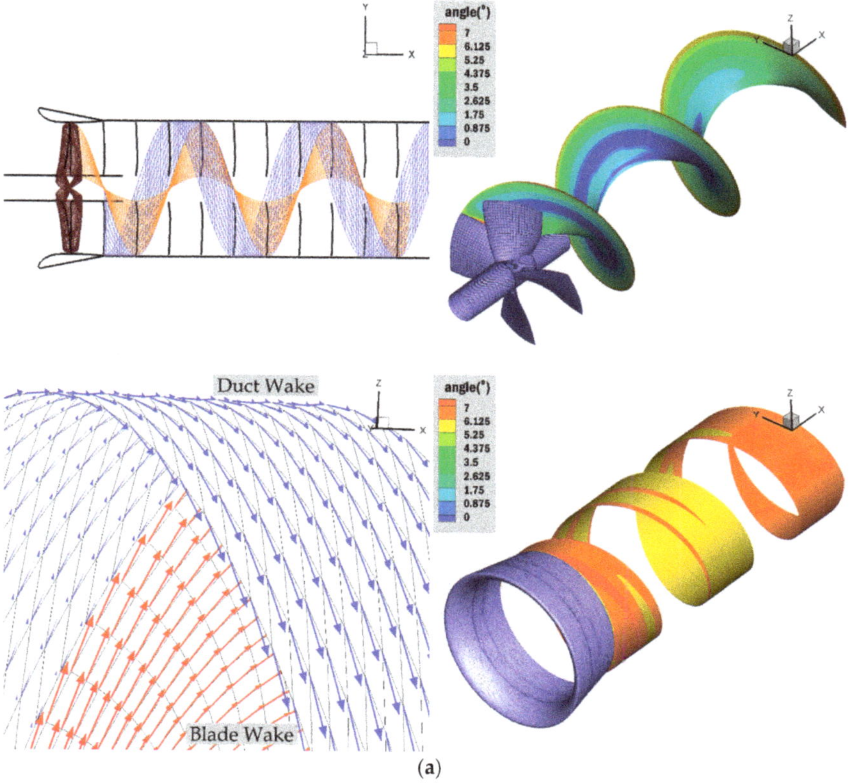

(**a**)

Figure 10. *Cont.*

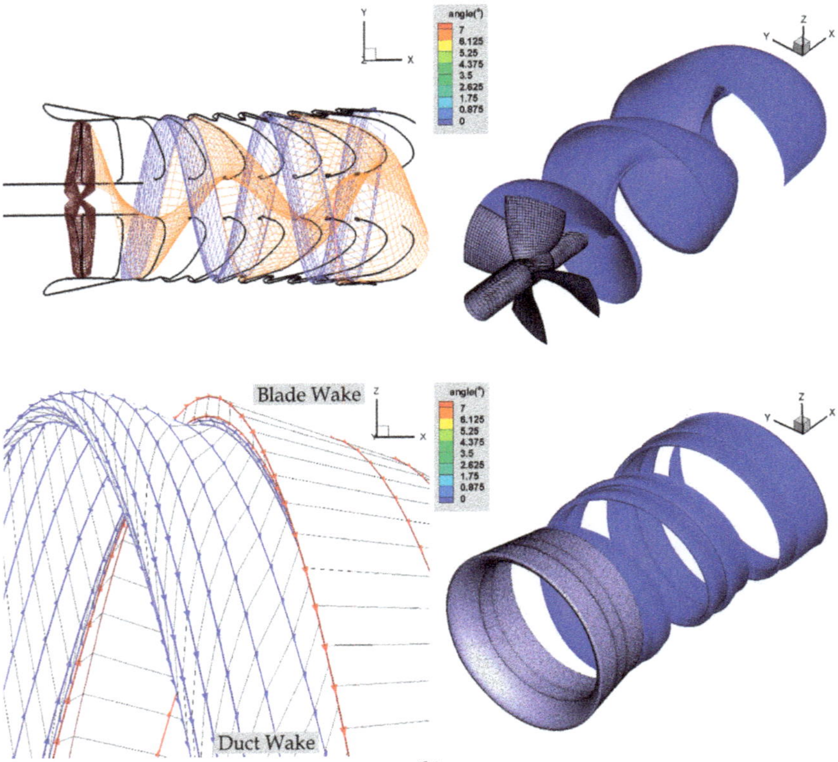

(**b**)

Figure 10. Blade and duct wake shapes at the first iteration (**a**), and converged last iteration, after full wake alignment on both blade and duct wakes (**b**). Shown are the blade and duct wake shapes, as intersected by a vertical plane through the propeller axis (**top left**), the total velocity vectors together with the streamwise vortex segments (**bottom left**), and contour plots of the angle of the total velocity vectors with the trailing vortex segments on the blade wakes (**top right**) and duct wakes (**bottom right**). The transverse vortex segments of zero strength are shown too. The total velocity vectors should be tangent to the streamwise vortex segments for force-free wake, as this can be verified by the contour plots in (**b**).

3.2. Reynolds Averaged Navier–Stokes Simulations

RANS simulations are conducted using ANSYS/Fluent (version 18.2) with the periodic interface, which requires only a quarter of the fluid domain to simulate the four-bladed ducted propeller in steady flow. For better resolution of the boundary layer along the propeller surface and to reduce the possible artificial diffusivity, structured meshing model is used for both the blade wake and duct wake. k–ω SST turbulent model is adopted with a Reynold's number of 1.0×10^6. QUICK scheme and SIMPLEC scheme are used for the spatial discretization and the pressure correction, respectively. Over 6 million polyhedral cells are used to discretize the domain with periodic boundary condition. It took over 2.75 h on 16 Intel Xeon Platinum 8160 2.1 GHz cores (two hardware threads per core)[3] to

[3] The simulation is performed at the Texas Advance Computing Center (TACC) at The University of Texas at Austin (Austin, TX 78703, USA). URL: http://www.tacc.utexas.edu.

achieve the converged blade thrust at $Js = 0.50$ after 7000 iterations, at which the continuity residual falls below 2.0×10^{-5}. At this level of convergence, the momentum and k–ω residuals are also less than 4.0×10^{-9} and 6.0×10^{-6}, respectively.

Figure 11 shows some views of the RANS mesh for the KA4-70 ducted propeller. A vertical x–y plane passing through the propeller axis is also shown on Figure 11. Contour plots of the predicted vorticity magnitude of points on that x–y plane will be compared with the wake shapes predicted by the present method. The same x–y plane is also shown in Figure 12 in the case of results from the present method.

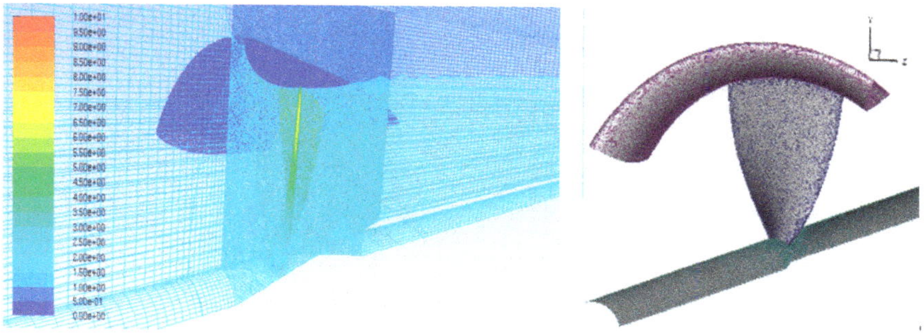

(**a**) (**b**)

Figure 11. Gridding of blade, hub, and duct surface in RANS simulation (**b**), and the x–y plane passing through the propeller axis (**a**). The scale shows the vorticity in s^{-1}.

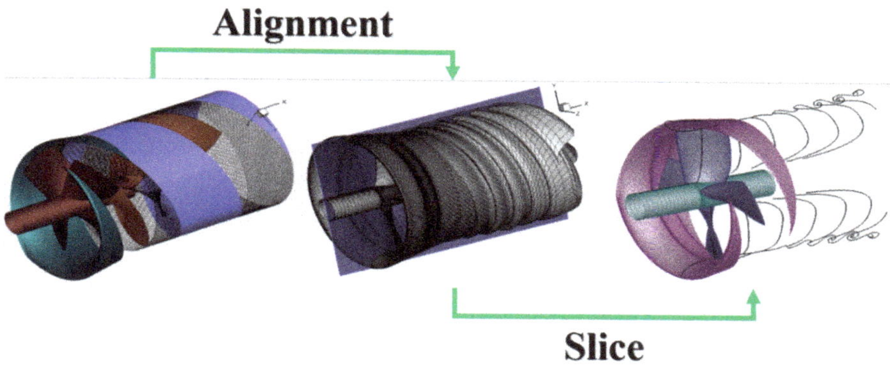

Figure 12. Perspective views of initial (**left**), and fully aligned duct wake shape (**middle**), predicted by the present method, and intersection of the duct wake shape with a x–y plane passing through the propeller axis (**right**).

3.3. Vorticity Predicted by RANS and Wake Shapes Predicted by the Present Method

Contour plots of the vorticity magnitude of points on the x–y plane, predicted from RANS simulations, overlaid with the wake shapes on the same plane, predicted by the present method, are shown in Figure 13 for three advance ratios. As shown, the lower the advance ratio is, the stronger the vorticity off the duct and blade trailing edge. The vorticity gradually diffuses as is convected downstream. The locations of the concentrated vorticity in the duct and blade wake, predicted by the present method, are in good agreement with the locations of the distributed vorticity predicted by RANS simulations. This good agreement is more evident near the duct or blade trailing edges.

Please note that the curling of the duct and tip vortex shapes, predicted by the present method, brings the vortices closer together, as can be seen in Figure 13, and that in RANS corresponds to regions of stronger distributed vorticity. This curling of the duct wake is due to the strong tip vortex at the blade tip, which locally forces the duct wake to wrap around it.

The forces predicted by the present method, using cylindrical duct wake or fully aligned duct wake, are shown in Figure 14, together with the results from RANS (ANSYS/Fluent), and the measured values from [13]. Overall, all methods seem to perform very well, especially around the design advance ratio of 0.5 (from 0.4–0.6).

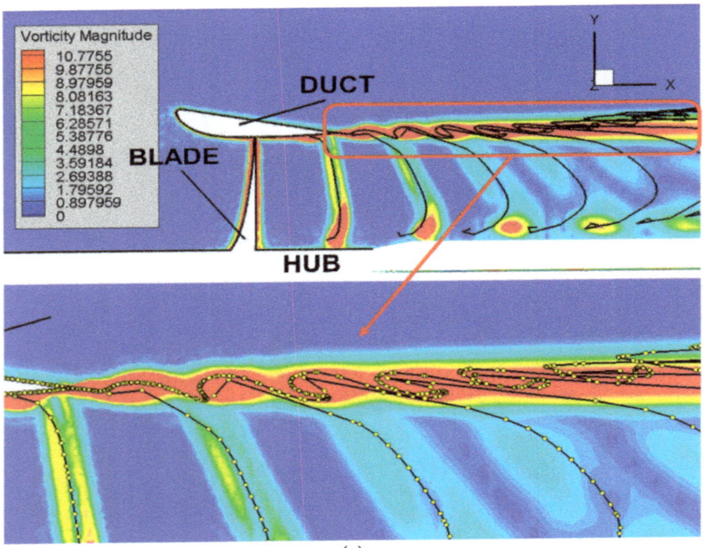

(**a**)

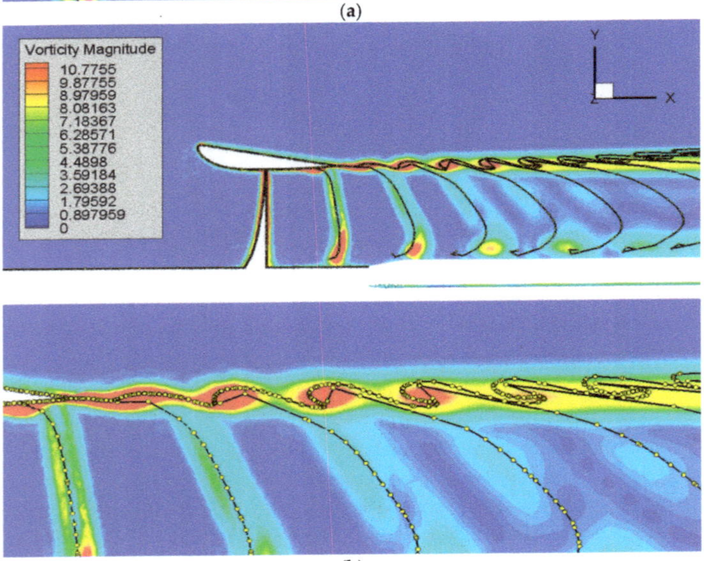

(**b**)

Figure 13. *Cont.*

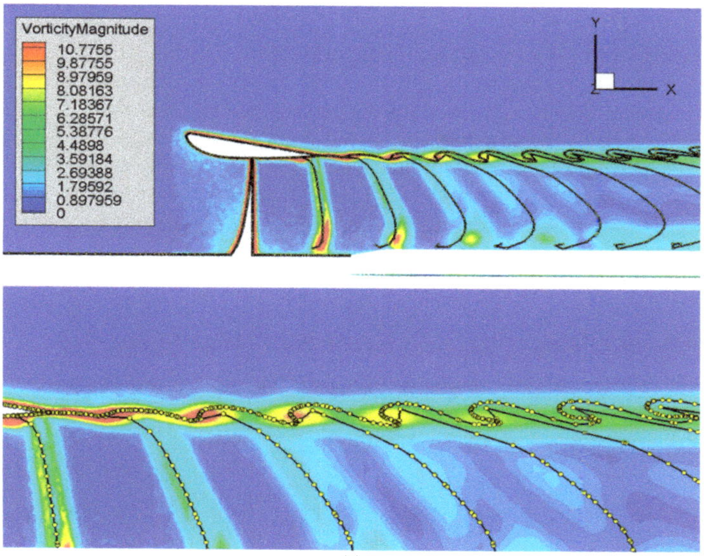

(c)

Figure 13. Comparison of the contour plots of vorticity in the duct and blade wake predicted from RANS, and the trailing wake shapes on the duct and blade predicted by the present method (shown with a black solid line) for advance ratios of 0.3 (**a**), 0.4 (**b**), and 0.5 (**c**). All shown quantities and wake shapes correspond to the intersection of a vertical plane through the axis of the ducted propeller.

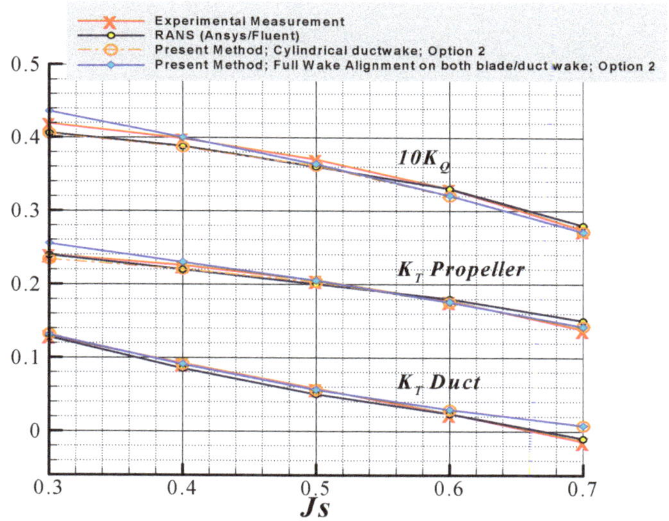

Figure 14. Predicted performance of the KA4-70 ducted propeller by the present method (with cylindrical duct wake and with full wake alignment on duct wake) and RANS, and compared with measured values.

3.4. Prediction of the Pressure Coefficients on the Blade and the Duct

Figure 15 shows the pressure coefficients along several blade sections at the design advance ratio of 0.5, as predicted by the present method, with various options of wake alignment, and by RANS. The predicted pressures are in general very good agreement with those from RANS simulations over most of the blade sections. The results from RANS at the section near the blade tip (especially at $r/R = 0.958$) seem to be non-smooth, and this may be attributed to the unstructured grid and the interpolation error in evaluating the pressures at points along each blade section. A structured grid on the blade in RANS could improve the accuracy of results at the blade tip. The pressure coefficient is defined as:

$$C_P = \frac{(P - P_0)}{(\frac{\rho}{2} n^2 D^2)} \tag{13}$$

where P_0 is the pressure far upstream.

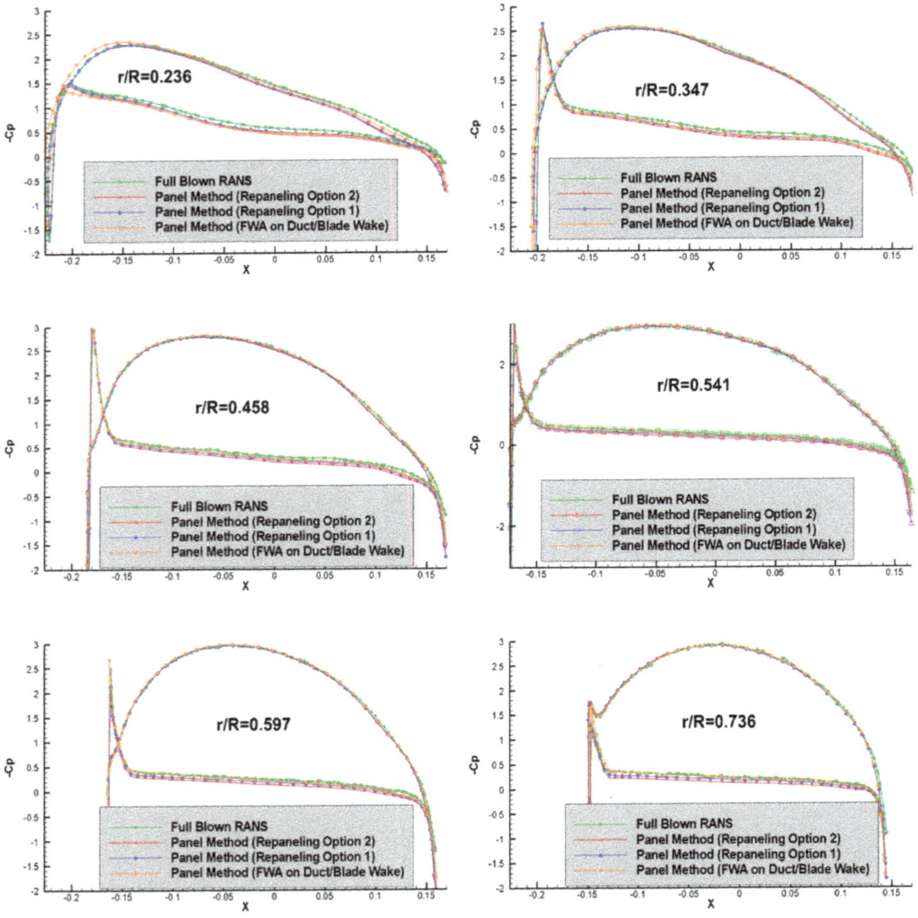

Figure 15. *Cont.*

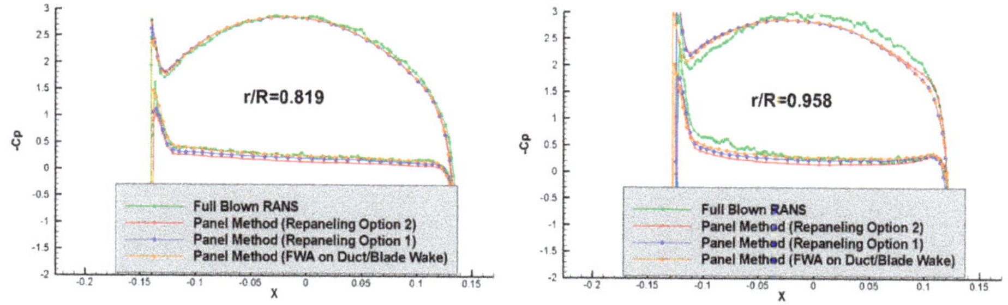

Figure 15. Correlation of the pressure coefficients predicted by the present method, with cylindrical duct wake, Options 1 or 2, and fully aligned duct wake (FWA), and from RANS at several blade sections. The radial location of each blade section is indicated in the figures.

The pressure distribution over the duct is evaluated by the present method and its circumferentially averaged value is shown in Figure 16, together with results from RANS, and from the RANS/VLM coupled method of Tian et al. [12]. As shown in Figure 16, the results from all methods seem to be in good agreement overall.

It is worthwhile explaining the discontinuity of the pressures predicted by the present method using Option 1. The different paneling regions over the inner side of the duct are shown in Figure 17, while the panel arrangement at the After Part (the blade wake/inner duct region) are shown in Figure 18. These unnatural pressure peaks in Option 1 are more distinct in Figure 19, which shows the pressure distributions along the several chordwise strips on duct. These singular pressures are due to the duct panels, which are intersected by the outer edge of the blade wake in Option 1, and are inevitably included in the evaluation of the duct pressures. However, if the duct panels are matched to the blade wake, as in the case of Option 2, the singular behavior is significantly reduced. This result clearly shows that the repaneling process (Option 2) should be included whenever an accurate evaluation of the duct pressures is required.

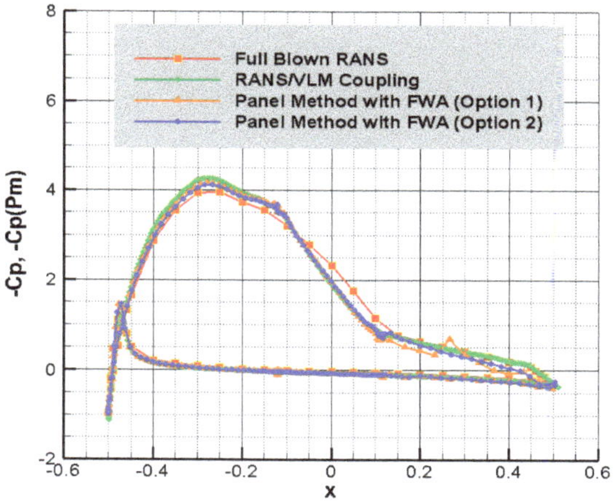

Figure 16. Circumferentially averaged pressure distribution on the duct, predicted by the present method (Option 1 or 2), RANS/VLM coupling method, and RANS simulation at the design advance ratio of 0.50.

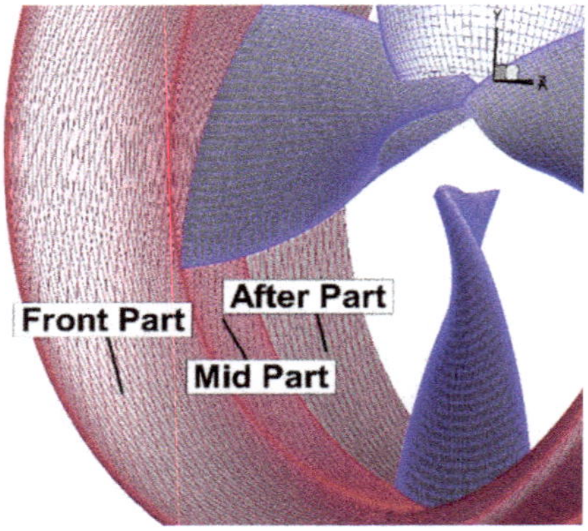

Figure 17. Paneling regions on the inner duct surface. The hub geometry is not included for clarity.

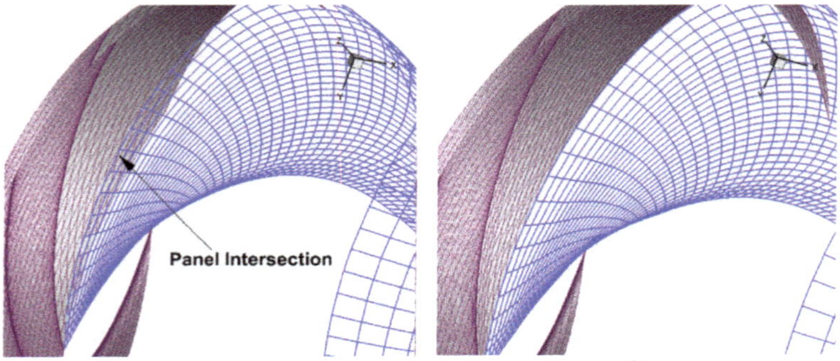

(a) (b)

Figure 18. Relative distribution of the blade wake and duct panels. Mismatched panel edges are shown at the blade wake/inner duct intersection (**a**), while adapted panels with matched edges are shown in Option 2 (**b**).

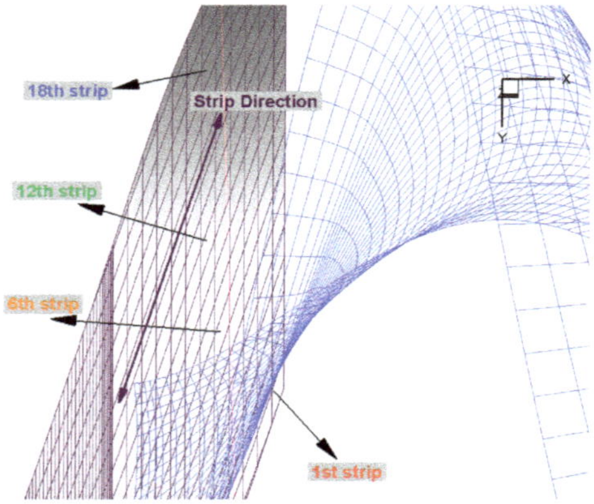

(a)

Figure 19. *Cont.*

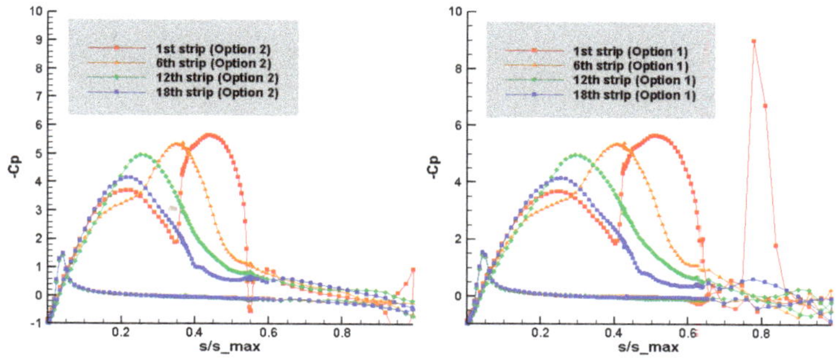

(b) (c)

Figure 19. Description of the strips that are adopted for the pressure plotting on the duct (**a**) and the corresponding pressure distributions on each strip using the present method with Option 2 (**b**) and Option 1 (**c**).

4. Conclusions and Future Work

To address the interaction between the propeller blades and duct, a full wake alignment (FWA) scheme has been applied on ducted propeller with a square tip blade and a sharp trailing edge duct. This paper describes the iterative algorithm to align both the blade and the duct wake, based on the local total velocity, thus avoiding using an oversimplified cylindrical duct wake. The viscous pitch correction or the viscous/inviscid interaction method are used to account for the effects of viscosity in the present panel method, and both methods, along with the FWA, are found to improve the correlation with the experimental measurements significantly.

By aligning both the blade and the duct wakes simultaneously, the present panel method can capture the behavior of the vorticity downstream of the blade and duct trailing edges. The location of the fully aligned duct wake was found to be in good agreement with that predicted from RANS simulations. Overall the present panel method with the FWA can predict the performance of ducted propellers with sharp trailing edge at high reliably over a wide range of advance ratios, even though the correlation with experiments worsens for lower advance ratios. Still, the discrepancy at low advance ratios is less significant, compared to the results from the panel method using the simplified wake model of Greeley and Kerwin [6]. Using Option 2, produces more stable results, especially at lower advance ratios, and improves the predicted pressure distributions on the duct. It was also found that using a cylindrical duct wake, with the provision of artificially restricting the blade wake from intersecting the duct wake, can produce equally reliable results to those from using full wake alignment on the duct, at about half the computing time. However, in the event details of the duct wake are needed (for example, in evaluating the performance of a rudder) the full wake alignment should be implemented.

Improving the prediction of the present method at even lower advance ratios should involve more careful wake alignment, by using more elements in the streamwise direction, thus allowing the low pitch blade and duct wake to extend further downstream. At the same time, extending the present panel method in the case of ducted propellers with blunt trailing edges, by employing the most recent method of Du and Kinnas [14], has been another objective of our research.

Author Contributions: S.K. participated in the improvement of the present method and the FWA scheme (45%). S.A.K. provided guidance throughout the research and participated in the improvement of the FWA scheme as well (35%). W.D. participated in the consideration of the viscous effects (20%).

Acknowledgments: Support for this research was provided by the U.S. Office of Naval Research (Grant Nos. N00014-14-1-0303 and N00014-18-1-2276; Ki-Han Kim) and by Phases VII and VIII of the "Consortium on Cavitation Performance of High Speed Propulsors".

Conflicts of Interest: The authors declare no conflicts of interest.

References

1. Hess, J.L.; Valarezo, W.O. Calculation of steady flow about propellers using a surface panel method. In Proceedings of the 23rd Aerospace Sciences Meeting, AIAA, Reno, NV, USA, 14–17 January 1985.
2. Morino, L.; Kuo, C.-C. Subsonic Potential Aerodynamics for Complex Configurations: A General Theory. *AIAA J.* **1974**, *12*, 191–197.
3. Kerwin, J.E.; Kinnas, S.A.; Lee, J.-T.; Shih, W.-Z. A Surface Panel Method for the Hydrodynamic Analysis of Ducted Propellers. *Trans. SNAME* **1987**, *95*, 93–122.
4. Kinnas, S.A.; Fan, H.; Tian, Y. A Panel Method with a Full Wake Alignment Model for the Prediction of the Performance of Ducted Propellers. *J. Ship Res.* **2015**, *59*, 249–257. [CrossRef]
5. Tian, Y.; Kinnas, S.A. A Wake Model for the Prediction of Propeller Performance at Low Advance Ratios. *Int. J. Rotat. Mach.* **2012**, *2012*, 372364. [CrossRef]
6. Greeley, D.S.; Kerwin, J.E. *Numerical Methods for Propeller Design and Analysis in Steady Flow*; Society of Naval Architects and Marine Engineers: Jersey City, NJ, USA, 1982; Volume 90, pp. 415–453.
7. Kinnas, S.A.; Su, Y.; Du, W.; Kim, S. A viscous/inviscid interactive method applied to ducted propellers with ducts of sharp or blunt trailing edge. In Proceedings of the 31st Symposium on Naval Hydrodynamics, Monterey, CA, USA, 11–16 September 2016.
8. Baltazar, J.; Falcão de Campos, J.A.C.; Bosschers, J. Open-water thrust and torque predictions of a ducted propeller system with a panel method. *Int. J. Rotat. Mach.* **2012**, *2012*, 474785. [CrossRef]
9. Kim, S. An Improved Full Wake Alignment Scheme for the Prediction of Open/Ducted Propeller Performance in Steady and Unsteady Flow. Master's Thesis, Ocean Engineering Group, The University of Texas, Austin, TX, USA, August 2017.
10. Kerwin, J.E.; Lee, C. *Prediction of Steady and Unsteady Marine Propeller Performance by Numerical Lifting-Surface Theory*; Trans. Society of Naval Architects and Marine Engineers: Jersey City, NJ, USA, 1978; Volume 86, pp. 218–256.
11. Kinnas, S.A.; Yu, X.; Tian, Y. Prediction of Propeller Performance under High Loading Conditions with Viscous/Inviscid Interaction and a New Wake Alignment Model. In Proceedings of the 29th Symposium on Naval Hydrodynamics, Gothenburg, Sweden, 26–31 August 2012.
12. Tian, Y.; Jeon, C.-H.; Kinnas, S.A. Effective Wake Calculation/Application to Ducted Propellers. *J. Ship Res.* **2014**, *58*, 1–13. [CrossRef]
13. Bosschers, J.; van der Veeken, R. *Open Water Tests for Propeller KA4-70 and Duct 19A with a Sharp Trailing Edge*; MARIN Report 224457-2-VT; Maritime Research Institute Netherlands (MARIN): Wageningen, The Netherlands, 2009.
14. Du, W.; Kinnas, S.A. A Flow Separation Model for Hydrofoil, Propeller, and Duct Sections with Blunt Trailing Edges. *J. Fluid Mech.* **2018**, under review.

Journal of
Marine Science and Engineering

MDPI

Article

Prediction of the Open-Water Performance of Ducted Propellers with a Panel Method

João Manuel Baltazar [1,*], Douwe Rijpkema [2], José Falcão de Campos [1] and Johan Bosschers [2]

1 Department of Mechanical Engineering, Instituto Superior Técnico, Av. Rovisco Pais 1, Universidade de Lisboa, 1049-001 Lisboa, Portugal; falcao.campos@tecnico.ulisboa.pt
2 Maritime Research Institute Netherlands, 2 Haagsteeg, 6708 PM Wageningen, The Netherlands; d.r.rijpkema@marin.nl (D.R.); j.bosschers@marin.nl (J.B.)
* Correspondence: joao.baltazar@tecnico.ulisboa.pt; Tel.: +351-218-419-289

Received: 16 January 2018; Accepted: 11 March 2018; Published: 19 March 2018

Abstract: In the present work, a comparison between the results obtained by a panel code with a Reynolds-averaged Navier-Stokes (RANS) code is made to obtain a better insight on the viscous effects of the ducted propeller and on the limitations of the inviscid flow model, especially near bollard pull conditions or low advance ratios, which are important in the design stage. The analysis is carried out for propeller Ka4-70 operating inside duct 19A. From the comparison, several modelling aspects are studied for improvement of the inviscid (potential) flow solution. Finally, the experimental open-water data is compared with the panel method and RANS solutions. A strong influence of the blade wake pitch, especially near the blade tip, on the ducted propeller force predictions is seen. A reduction of the pitch of the gap strip is proposed for improvement of the performance prediction at low advance ratios.

Keywords: ducted propeller; panel method; wake model; RANS comparison

1. Introduction

In recent years, substantial progress is being made in the computation of the flow around ducted propeller systems with Reynolds-averaged Navier-Stokes (RANS) equations by several research groups. For example, Sánchez-Caja et al. [1] used a RANS solver to simulate incompressible viscous flow around a propeller in the presence of a duct. Posteriorly, the influence of the rudder was also taken into account in the calculation of the viscous flow [2]. In addition, Abdel-Maksoud and Heinke [3], and Bhattacharyya et al. [4] have investigated the scale effects on ducted propellers numerically using RANS. Alternatively, Kim et al. [5] presented detailed RANS simulations for a ducted propeller system including verification and validation studies. Subvisual cavitation and acoustics modelling were also investigated in this work. However, the computational effort is still reasonably high due to the need for good numerical resolution in small flow regions dominated by strong viscous effects such as in the gap between the propeller blade tip and duct. This requirement poses considerable demands in the number of grid cells needed for accurate computations with associated long computational times, which still makes the method less useful for routine design studies.

On the other hand, in the past a number of methods based on inviscid potential flow theory have been proposed for the analysis of ducted propellers. Kerwin et al. [6] combined a panel method for the duct with a vortex lattice method for the propeller. This method was compared with experimental data available from open-water tests by Hughes et al. [7]. Later, Hughes [8] presented a complete three-dimensional panel method for both the propeller and duct, where a special procedure for modelling the gap flow is implemented. More recently, Lee and Kinnas [9] described a panel method for the unsteady flow analysis of ducted propeller with blade sheet cavitation. These methods are nowadays very efficient from the computational point of view which makes them particularly suited for design studies.

However, the inviscid methods have met serious limitations in their practical applications related to their inability to adequately model viscous effects occurring in a ducted propeller system, especially in the gap region and in the duct boundary layer leading to separation phenomenon. Due to the relative motion between the duct and the propeller blades, combined with the pressure difference across the gap, different viscous mechanisms occur simultaneously in this region, such as the tip-leakage vortex and the gap flow due to the duct and blade boundary layers. Schematics of this process are shown in Figure 1. The flow in the gap region is rather complex, since the duct and blade boundary layers influence the gap flow and consequently the characteristics of the tip-leakage vortex. The tip-leakage vortex is also responsible for the different loading of the ducted propeller in comparison with the open propeller and must be taken into account in the inviscid model. An extensive experimental investigation to examine the tip-leakage flow on ducted propulsors was carried out by Oweis et al. [10]. Another important viscous effect is the flow over the surface of the duct, where separation of the boundary layer creating a recirculation region may occur due to the thick blunt trailing-edge, which also affects the duct loading.

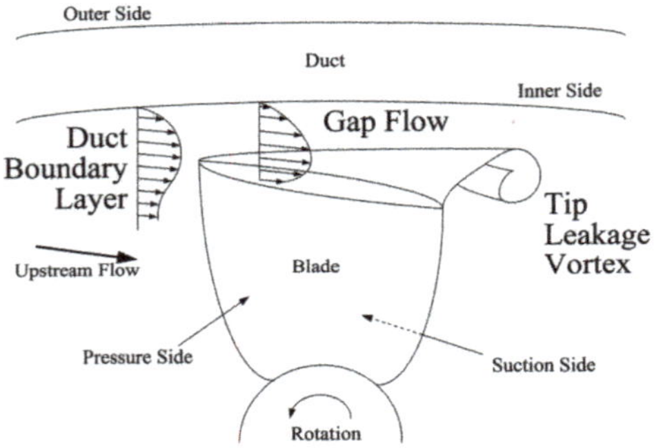

Figure 1. Schematic overview of the flow in the gap region between blade tip and duct inner side.

The calculation of the flow around a ducted propeller system in open-water with a panel code has been the subject of investigation by Instituto Superior Técnico (IST) and Maritime Research Institute Netherlands (MARIN), see [11–13]. In these studies, a low-order panel method has been used to predict the open-water diagram of a ducted propeller system, where several modelling aspects have been analysed for the improvement of the inviscid (potential) flow solution. The investigation comprehended the influence of the Kutta condition, gap flow model and wake model on the performance prediction. A similar method has also been implemented by MARIN in an in-house panel code [14,15].

In this investigation, an alternative Kutta condition has been proposed for the duct trailing-edge, which has a thick round geometry in comparison to the sharp trailing-edge of the propeller blades. In this new Kutta condition, the chordwise location for pressure equality on both sides of the duct trailing-edge has a strong influence on the propeller and duct force predictions.

To account for the gap flow between the blade tip and duct inner side, Hughes [8] proposed an iterative procedure where the gap flow is treated as a two-dimensional orifice. In his model, the gap between the blade tip and the duct inner surface is modelled as a rigid surface, named in this study as the gap strip. Then, transpiration velocities are computed from the pressure-difference in the gap strip, where an empirical discharge coefficient is used to take into account the loss of energy as the fluid passes through the gap. A similar gap model has been combined with a vortex-lattice method by Gu and Kinnas [16]. From previous comparative studies, see [11,13], similar results were found between the closed (sealed) gap model and the gap flow model with transpiration velocity based in the work of Hughes [8]. Therefore, the closed gap model is usually preferred in potential flow methods [13,15,17], since a negligible effect on the overall performance is obtained.

The influence of the blade wake geometry on the prediction of the ducted propeller performance with a panel method has been studied in detail, see [11,13]. In this work, the blade wake pitch is aligned with the local flow velocity using an Euler scheme, leading to an improvement in the prediction of the propeller forces. A similar model for alignment of the blade wake was presented by Kinnas et al. [17], and Kim et al. [18] extended the wake alignment scheme for both the blade wake and duct wake. However, from the work carried out by IST and MARIN [11], the loading predictions of the ducted propeller system were found to be critically dependent on the blade wake pitch especially at the tip. In this way, a simple model for the interaction between the blade wake and the boundary layer on the duct inner side was implemented in combination with the wake alignment model [11,13]. With this model, a reasonable to good agreement was obtained between the inviscid predictions and the experimental data from open-water tests [13]. However, significant differences were still seen in the open-water predictions at low advance ratios.

It is known that the prediction of the propeller performance at bollard pull conditions is important in the design stage. In the present work, a comparison between the results obtained by the panel method with RANS calculations is made to obtain a better insight on the viscous effects of the ducted propeller and on the limitations of the inviscid flow model. The comparison focuses mainly at the low advance ratios, where a new approach for the gap strip is proposed in the panel method. The paper is organised as follows: a description of the numerical methods is given in Section 2; the comparison of the inviscid predictions with the RANS calculations and the experimental open-water data is presented in Section 3; in Section 4 the main conclusions are drawn.

2. Numerical Methods

2.1. Problem Definition

Let us consider a propeller of radius R with a finite number of blades symmetrically distributed around an axisymmetric hub, rotating with constant angular velocity Ω inside a duct and advancing with constant axial speed U along its axis. The duct is also considered to be axisymmetric of inner radius at the propeller plane $R_d > R$ which defines a gap height $h = R_d - R$.

We introduce a Cartesian coordinate system (x, y, z) rotating with the propeller blades, with the x-axis positive downstream, the y-axis direction coincident with the propeller reference line, and the z-axis completing the right-hand-system. In this rotating reference frame the flow field is steady. We will use a cylindrical coordinate system (x, r, θ) which is related to the Cartesian system by the transformation

$$y = r\cos\theta, \ z = r\sin\theta. \tag{1}$$

The undisturbed inflow velocity in the rotating frame is

$$\vec{U}_\infty = U\vec{e}_x + \Omega r\vec{e}_\theta, \tag{2}$$

where $(\vec{e}_x, \vec{e}_r, \vec{e}_\theta)$ are the unit vectors of the cylindrical coordinate system. Figure 2 shows the coordinate systems used to describe the geometry and the fluid domain around the ducted propeller.

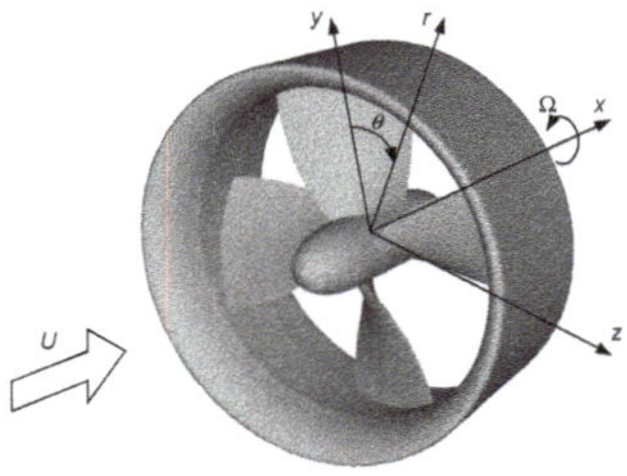

Figure 2. Propeller coordinate systems.

2.2. Panel Code PROPAN

Assuming an incompressible, ideal and irrotational flow at an infinity domain in all directions, the flow around the ducted propeller can be treated with a potential flow model. PROPAN is a IST in-house code which implements a low-order potential-based panel method for the calculation of the incompressible potential flow around marine propellers. The code has been widely used in the calculation of the three-dimensional potential flow around ducted propellers, [11,13,19].

Applying Green's second identity and using the so-called Morino formulation [20], which assumes that the perturbation potential is zero in interior region to the blade surfaces $\mathcal{S}_\mathcal{B}$, duct surface $\mathcal{S}_\mathcal{D}$ and hub surface $\mathcal{S}_\mathcal{H}$, we obtain the integral representation of the perturbation potential ϕ at a point p on the body surface,

$$2\pi\phi(p) = \iint_{\mathcal{S}_\mathcal{B}\cup\mathcal{S}_\mathcal{D}\cup\mathcal{S}_\mathcal{H}} \left[\phi(q)\frac{\partial}{\partial n_q}\left(\frac{1}{R(p,q)}\right) - \frac{\partial\phi}{\partial n_q}\frac{1}{R(p,q)}\right] dS + \iint_{\mathcal{S}_\mathcal{W}} \Delta\phi(q)\frac{\partial}{\partial n_q}\left(\frac{1}{R(p,q)}\right) dS, \quad p \in \mathcal{S}_\mathcal{B}\cup\mathcal{S}_\mathcal{D}\cup\mathcal{S}_\mathcal{H}, \tag{3}$$

where $R(p,q)$ is the distance between the field point p and the point q on the boundary $\mathcal{S}_\mathcal{B}\cup\mathcal{S}_\mathcal{D}\cup\mathcal{S}_\mathcal{H}\cup\mathcal{S}_\mathcal{W}$. The perturbation potential must satisfy the Neumann boundary condition at the body surface,

$$\frac{\partial\phi}{\partial n} \equiv \vec{n}\cdot\nabla\phi = -\vec{n}\cdot\vec{U}_\infty \quad \text{on } \mathcal{S}_\mathcal{B}\cup\mathcal{S}_\mathcal{D}\cup\mathcal{S}_\mathcal{H}, \tag{4}$$

where $\partial/\partial n$ denotes differentiation along the normal and $\vec{n}$ is the unit vector normal to the surface directed outward from the body. Equation (3) is a Fredholm integral equation of the second kind in the dipole distribution $\mu(q) = -\phi(q)$ on the surfaces $\mathcal{S}_\mathcal{B}$, $\mathcal{S}_\mathcal{D}$ and $\mathcal{S}_\mathcal{H}$. The Kutta condition yields the additional relationship to determine the dipole strength $\Delta\phi(q)$ in the wake surfaces $\mathcal{S}_\mathcal{W}$.

For the numerical solution of Equation (3), we discretise the body surfaces $\mathcal{S}_\mathcal{B}$, $\mathcal{S}_\mathcal{D}$ and $\mathcal{S}_\mathcal{H}$, and the wake surfaces $\mathcal{S}_\mathcal{W}$ in bi-linear quadrilateral elements which are defined by four points on the surface. We assume a constant strength of the dipole and source distributions on each element. In the numerical solution of the integral equation, Equation (3), the integrals over the body and wake surfaces are approximated by the summation of the integrals on the elements discretising the surfaces. The element integrals are calculated analytically following the formulation of Morino and Kuo [20].

To determine the dipole strength on the wake surfaces, an iterative pressure Kutta condition at the blade and duct trailing edges is applied. However, different forms of the Kutta condition are considered, since the blade trailing-edge has a sharp geometry, whereas the duct trailing-edge presents a blunt round geometry. For the blade trailing-edge, equal pressure on the collocation points of the panels adjacent to the trailing-edge on the upper and lower sides is considered. For the duct trailing-edge, the chordwise location of pressure equality on both sides is specified, which controls the strength of the shed vorticity. Due to the possible occurrence of flow separation, a constant pressure distribution downstream of the Kutta points is assumed in this model. Note that the potential flow solution at the duct trailing edge satisfies the integral equation, Equation (3), but the corresponding pressure distribution is disregarded aft the Kutta points. A schematic drawing of the pressure Kutta condition for a thick round trailing edge is shown in Figure 3. A detailed description of the iterative method for solution of the pressure Kutta condition may be found in [19].

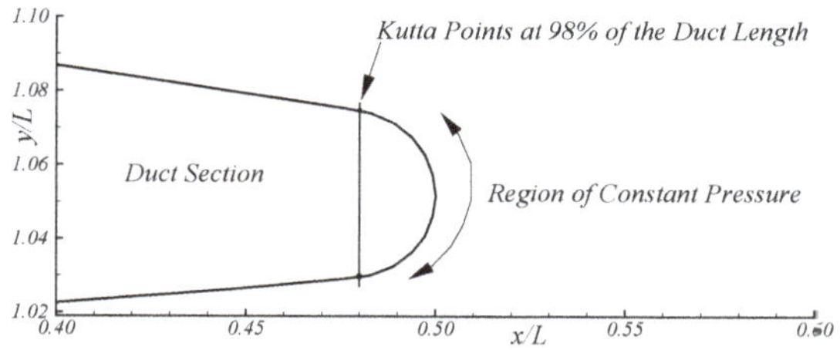

Figure 3. Pressure Kutta condition for the duct trailing edge with a thick round geometry and L denoting the duct length.

For the wake geometry, two wake models are considered: a rigid wake model (RWM) and a wake alignment model (WAM). In the rigid wake model, the geometry of the wake surfaces is specified empirically. For the blade wake, the pitch of the vortex lines is assumed constant along the axial direction and equal to the blade pitch. For the duct, the wake leaves the trailing edge at the bisector with constant radius. In the wake alignment model, the blade wake pitch is aligned with the local fluid velocity, while the radial position of the vortex lines is fully prescribed. The new axial $x_{i+1}^{(n+1)}$ and circumferential $\theta_{i+1}^{(n+1)}$ coordinates of the wake strip $i+1$ at the $(n+1)$th iteration are determined by using an Euler scheme:

$$\begin{cases} x_{i+1}^{(n+1)} = x_i^{(n)} + V_x\left(x_i^{(n)}, r_i^{(n)}, \theta_i^{(n)}\right)\Delta t \\ \theta_{i+1}^{(n+1)} = \theta_i^{(n)} + V_\theta\left(x_i^{(n)}, r_i^{(n)}, \theta_i^{(n)}\right)/r_i^{(n)}\Delta t \end{cases}, \tag{5}$$

where V_x and V_θ are the components of the mean vortex sheet velocity along the axial and circumferential directions, respectively, and Δt is the pseudo-time step for the Euler vortex convection scheme. The velocity components are calculated from the integral equation of the velocity, obtained by taking the gradient of Equation (3). For the time discretisation, an angular time step $\Delta\theta = \Omega\Delta t$ is introduced, which can also be expressed in terms of the number of time steps per propeller revolution $N_\theta = 2\pi/\Delta\theta$. To account for the interaction between the blade wake and the duct boundary layer on the inner side, see Figure 1, a correction to the blade wake pitch near the tip is introduced in the wake alignment model, whereas a reduction in the axial velocity is taken into account in the convection of the vorticity generated by the blade. Considering δ as the duct boundary layer thickness and assuming a power law distribution for the velocity profile, we have

$$\frac{V_x(R_d - r)}{V_x(\delta)} = \left(\frac{R_d - r}{\delta}\right)^{\frac{1}{n}}. \tag{6}$$

From Equation (6) the velocity deficit in the axial direction due to the duct boundary layer is taken into account in the wake alignment model, introducing a reduction in the pitch of the blade wake geometry near the tip. However, to avoid zero axial velocity at the duct inner surface in the wake alignment model, which would cause zero pitch for the tip vortex line and a mismatch with the duct grid, a linear variation of the axial velocity is considered in the gap region $(R \leq r \leq R_d)$. The corrected axial velocity at the duct surface $V_x(0)$ is obtained by linear extrapolation from the axial velocity at the gap $V_x(h)$, and at the edge of the duct boundary layer $V_x(\delta)$. This correction is applied along the entire length of the blade wake sheet.

The gap between the blade tip and the duct inner surface is modelled as a rigid surface, named in this study as the gap strip, and where the transpiration velocity is neglected. In this closed gap model, the boundary condition on the gap panels sets the source strength to cancel the normal component of the inflow velocity leading to the Neumann boundary condition, Equation (4). The gap strip extends from the blade tip to the duct inner side.

2.3. RANS Code ReFRESCO

Alternatively, in order to address the viscous effects a RANS solver can be used for the simulation of the incompressible viscous flow around ducted propellers. In the present study RANS code ReFRESCO version 2.1 is considered. ReFRESCO (www.refresco.org) is a community-based open-usage CFD code for the maritime world [21] and is currently being developed within a cooperation led by MARIN. It solves the steady incompressible RANS equations, complemented with turbulence models. The time-averaged continuity and momentum equations (RANS equations) written in the differential form and using the tensor notation are:

$$\begin{aligned} \frac{\partial V_i}{\partial x_i} &= 0, \\ \rho V_j \frac{\partial V_i}{\partial x_j} &= -\frac{\partial p}{\partial x_i} + \frac{\partial}{\partial x_j}\left[\mu\left(\frac{\partial V_i}{\partial x_j} + \frac{\partial V_j}{\partial x_i}\right)\right] + \frac{\partial \tau_{ij}}{\partial x_j}, \end{aligned} \tag{7}$$

where $x_i \equiv (x_1, x_2, x_3)$ are the coordinates of the reference system (x, y, z), V_i are the components of the mean velocity vector, ρ the fluid density, p the fluid static pressure, μ the fluid viscosity and τ_{ij} are the Reynolds stresses produced by the averaging process of the momentum equations. In this work, the selected turbulence model is based on the Boussinesq eddy-viscosity hypothesis that determines the Reynolds stresses from

$$\tau_{ij} = \mu_t \left(\frac{\partial V_i}{\partial x_j} + \frac{\partial V_j}{\partial x_i}\right) - \frac{2}{3}\rho k \delta_{ij}, \tag{8}$$

where μ_t is the eddy-viscosity, δ_{ij} is the Kronecker symbol and k is the turbulence kinetic energy. For all RANS calculations presented in this paper, the $k-\omega$ SST two-equation eddy-viscosity turbulence model proposed by Menter et al. [22] is used. In this model, two transport equations are solved: the turbulent kinetic energy k and the specific turbulent dissipation rate ω.

The equations are discretised using a face-based finite-volume approach with cell-centred collocation variables. A strong-conservation form and a pressure-correction equation based on the SIMPLE algorithm is used to ensure mass conservation. For open-water (steady) calculations the equations are solved in the body-fixed reference frame which is rotating with velocity Ω. A second-order convection scheme (QUICK) is used for the momentum equations and a first-order upwind scheme is used for the turbulence model equations.

3. Results

3.1. Grids and Numerical Set-Up

Results are presented for the propeller Ka4-70 with pitch-diameter ratio $P/D = 1.0$ inside the duct 19A operating in open-water conditions. The length-diameter ratio L/D of duct 19A is 0.5. The gap between the duct inner side and the blade tip is uniform and equal to 0.8% of the propeller radius. The geometry of the Ka-series and duct section is given by Kuiper [23].

The propeller operating conditions are defined by the advance coefficient $J = U/(nD)$, where $n = \Omega/2\pi$ is the rate of revolution and $D = 2R$ is the propeller diameter. The open-water characteristics are expressed in the propeller thrust coefficient K_{T_P}, the duct thrust coefficient K_{T_D}, the torque coefficient K_Q and the open-water efficiency η_0:

$$K_{T_P} = \frac{T_P}{\rho n^2 D^4}, \quad K_{T_D} = \frac{T_D}{\rho n^2 D^4}, \quad K_Q = \frac{Q}{\rho n^2 D^5}, \quad \eta_0 = \frac{U(T_P + T_D)}{2\pi n Q}, \tag{9}$$

where T_P is the propeller thrust, T_D the duct thrust and Q the propeller torque. Other used quantities are the vorticity $\vec{\omega} = \nabla \times \vec{V}$ and the pressure coefficient $Cp = (p - p_\infty)/(1/2\rho|\vec{U}_\infty|^2)$, where p_∞ is the undisturbed static pressure.

Convergence studies of the inviscid solution have been carried out for the panel method with the rigid wake model. The propeller blade discretisations ranged from 20×11 to 70×36, corresponding to the chordwise and spanwise radial directions, respectively. The number of panels on the duct, hub and wakes is modified according to the number of panels on the blade. The iterative pressure Kutta condition is applied which, in general, converged after three iterations to a precision of $|\Delta C_p| \leq 10^{-3}$ at the Kutta points.

The variation in thrust and torque for the different grid sizes obtained from the panel method computations at $J = 0.5$ is shown in Table 1. The variation of the open-water characteristics decrease with the grid refinement level. Differences lower than 1% are obtained for the 50×26 blade grid in comparison to the finest grid. In addition, the computational time for the 50×26 blade grid relative to the finest grid is in the order of 27%. Therefore, the 50×26 blade grid is used in the subsequent studies carried out with both the rigid wake and wake alignment models. For the wake alignment model, the wake geometry is obtained after five iterations using an angular step of $\Delta\theta = 4$ degrees. In combination with the wake alignment model, a duct boundary layer correction is applied with a thickness equal to $\delta/R = 4\%$ and a power law velocity profile with exponent equal to 1/7.

Table 1. Variation of open-water characteristics with different grid sizes compared to the finest grid. Panel method computations at $J = 0.5$.

Grid Size (Blade + Duct + Hub) [1]	ΔK_{T_P}	ΔK_{T_D}	ΔK_Q
20 × 11 + 100 × 40 + 39 × 32	−4.76%	−7.42%	−5.97%
30 × 16 + 130 × 80 + 51 × 48	−0.60%	−3.34%	−1.42%
40 × 21 + 160 × 120 + 63 × 64	−0.79%	−1.30%	−1.23%
50 × 26 + 190 × 160 + 75 × 80	−0.49%	−0.37%	−0.74%
60 × 31 + 220 × 200 + 87 × 96	−0.15%	−0.19%	−0.25%
70 × 36 + 250 × 240 + 98 × 112	–	–	–

[1] The grid size refers to the chordwise and spanwise radial directions for each propeller blade, and to the streamwise and circumferential directions for the duct and hub.

For the RANS simulations, the computational domain is defined as a cylindrical domain with a length of 5 propeller diameters in all directions. For this problem, five nearly-geometrically similar multi-block structured grids were generated using the commercial grid generation package GridPro (www.gridpro.com). The grids range from 1.1 to 26.8 million cells. A fine boundary-layer resolution is considered for all grids, where the maximum dimensionless distance to the wall of the first cell, known as y^+, is lower than 1. For the boundary conditions a uniform flow at the inlet and constant pressure at the outer boundary is applied. At the outlet, an outflow condition of zero downstream gradient is used. For the propeller blades, duct and hub, a non-slip boundary condition is set. A rotational velocity is prescribed to the blades and hub, while the duct does not rotate.

In Figure 4 an overview of the grids used for the calculations with panel code PROPAN and RANS code ReFRESCO is shown.

The variation in thrust and torque for the different grid sizes is listed in Table 2 at $J = 0.5$. A reduction in the variation of the open-water quantities with the increased number of cells is observed. Although the differences are less than 1% for the grid with 12.9 million cells, the finest grid (26.8 million cells) is used in the comparative study, since not only the propeller forces, but also local flow quantities are considered in the present analysis. For the prediction of the open-water performance, the ducted propeller is tested for a range of advance coefficients between 0.0 and 0.8

corresponding to Reynolds numbers from 3.5×10^5 to 3.8×10^5, where the Reynolds number is defined based on the propeller blade chord length at $0.7R$ and the resulting onset velocity at that radius.

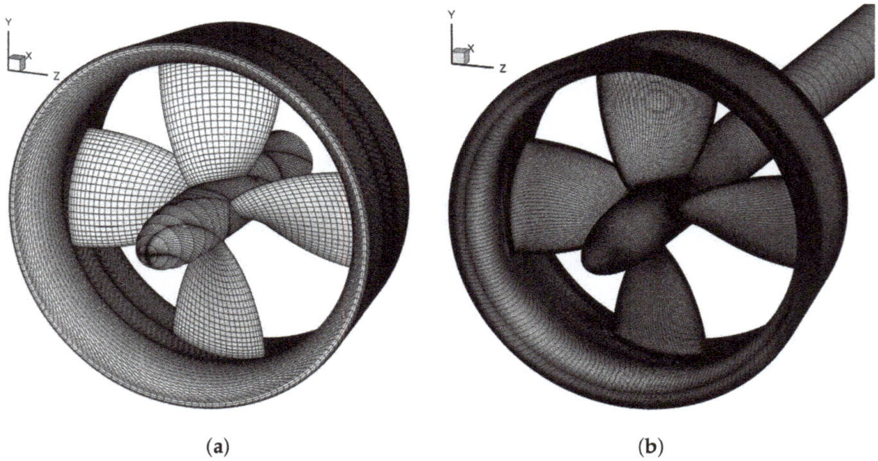

Figure 4. Overview of the surface grids used for the inviscid calculations with PROPAN (**a**) and RANS calculations with ReFRESCO (**b**).

Table 2. Variation of open-water characteristics with different grid sizes compared to the finest grid with M denoting million. RANS computations at $J = 0.5$.

Grid Size	ΔK_{T_P}	ΔK_{T_D}	ΔK_Q
1.1 M	3.25%	2.39%	3.55%
2.6 M	2.78%	2.78%	2.75%
7.7 M	0.67%	1.99%	0.69%
12.9 M	0.21%	0.99%	0.28%
26.8 M	–	–	–

3.2. Influence of the Wake Model

In this section the influence of the wake model is studied. The analysis is presented for three advance coefficients: 0.1, 0.2 and 0.5. The advance coefficients 0.1 and 0.2 refer to highly loaded conditions from where significant differences are still obtained with the present inviscid model, see Baltazar et al. [11,13].

The inviscid thrust and torque coefficients are compared with experimental open-water data in Table 3. Significant differences are seen in the propeller thrust and torque at low advance coefficients with the rigid wake model. By using the wake alignment model, which includes a correction in the axial velocity due to the interaction between the blade wake and the duct boundary-layer, a reduction in the propeller force coefficients is obtained and lower differences in comparison with the experimental data are observed. This effect has been studied before [11] and is related to the local reduction of the blade wake pitch near the tip which is responsible for lower incidence angles to the blade sections and as a consequence lower propeller forces. Figure 5 presents the blade wake geometries obtained with the rigid wake model (a) and wake alignment model (b) for $J = 0.2$. As we can see from Figure 5 the correction in the axial velocity due to the duct boundary-layer introduces a significant reduction in the vortex pitch at the blade wake tip.

Table 3. Inviscid thrust and torque coefficients for $J = 0.1$, 0.2 and 0.5 and comparison with experimental data [23]. Influence of the wake model.

Model	K_{T_P}	K_{T_D}	$10K_Q$
$J = 0.1$			
RWM	0.412	0.206	0.5882
WAM	0.313	0.231	0.4664
WAM with Reduced Gap Pitch	0.284	0.226	0.4228
Experiments	0.254	0.214	0.4387
$J = 0.2$			
RWM	0.383	0.160	0.5538
WAM	0.297	0.176	0.4456
WAM with Reduced Gap Pitch	0.273	0.171	0.4083
Experiments	0.248	0.166	0.4279
$J = 0.5$			
RWM	0.266	0.054	0.4041
WAM	0.208	0.057	0.3246
WAM with Reduced Gap Pitch	0.193	0.055	0.3010
Experiments	0.196	0.053	0.3506

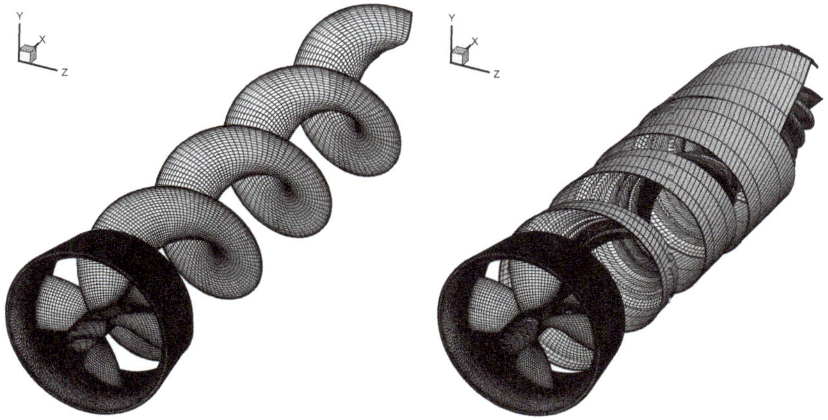

(**a**) (**b**)

Figure 5. Panel arrangement for propeller blades, duct, hub and blade wake at $J = 0.2$. Rigid wake model (**a**); Wake alignment model (**b**). Only one wake surface is shown.

Still, an over-prediction of the propeller forces is obtained with the panel method (Table 3). These differences suggest that larger corrections to the blade wake pitch are needed and a new wake model is considered, where the gap strip is rotated from the leading edge to reduce its pitch. In this study, the pitch of the gap strip is assumed to be equal to $P/D = 0.9$, whereas the blade pitch is constant and equal to $P/D = 1.0$. In Figure 6a detail of the gap strip and the obtained blade wake geometry are shown. We note that the gap strip is modelled as a rigid surface and is disconnected from the wake alignment model. The blade wake pitch near the tip may be controlled by the duct boundary-layer correction, Equation (6). However, for low advance ratios large corrections are needed and this has led to divergence of the Kutta condition and non-smooth surface grids. Therefore, the reduction of the gap strip pitch has proven to be a robust technique and can be applied at low advance ratios. As expected, a higher reduction in the propeller thrust and torque is obtained with the wake alignment model using a reduced pitch for the gap strip and approaches the results of the experimental data, see Table 3. The pitch angle of the blade wake β_v at the axial positions $x/R = 0.2$ and 0.4 downstream from the propeller is illustrated in Figure 7, where the local reduction of the wake pitch near the tip is visible.

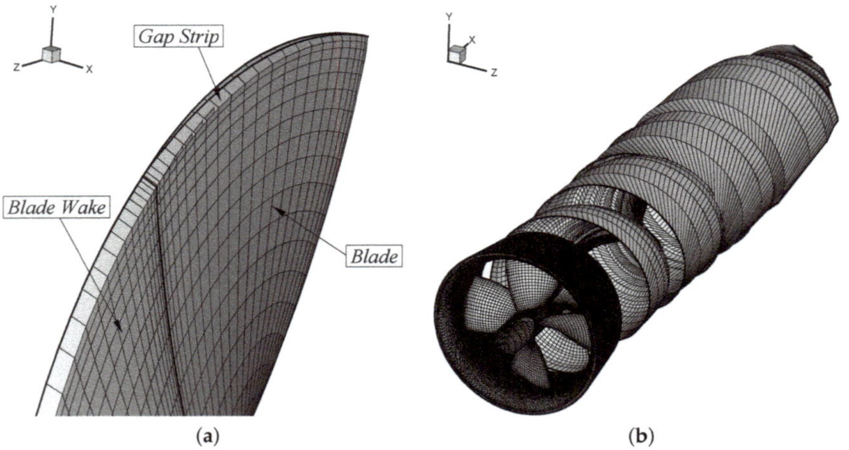

Figure 6. Wake alignment model with reduced gap pitch of $P/D = 0.9$. Detail of the gap strip and blade wake sheet (**a**); Overview of one blade wake geometry for $J = 0.2$ (**b**).

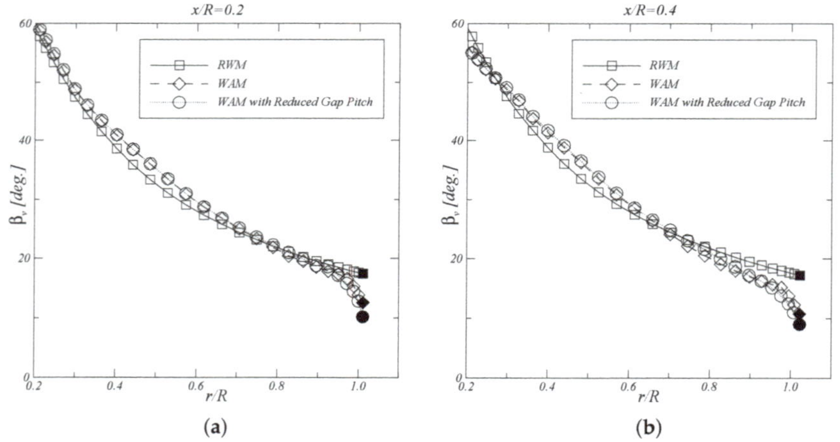

Figure 7. Vortex pitch β_v distribution at $x/R = 0.2$ (**a**) and $x/R = 0.4$ (**b**) for $J = 0.2$. Influence of the wake model. The filled symbols refer to the tip vortex.

3.3. Comparison Between PROPAN and ReFRESCO

In order to assess on the quality of the inviscid potential model, the results obtained with the panel code PROPAN and RANS code ReFRESCO are compared. The inviscid wake geometry obtained with the three wake models is compared with the vorticity field at the planes $x/R = 0.2$ and $x/R = 0.4$ downstream from the propeller, and the blade and duct pressure distributions are shown for the same advance ratios in Figures 8–13.

Once again, a reduction in the pitch of the tip vortex is seen when changing from the rigid wake model to the wake alignment model. However, the assessment of the correct location of the tip vortex core from the ReFRESCO calculations is difficult to make due to the interaction between the tip vortex and the duct boundary-layer, creating a viscous flow region at the duct inner side.

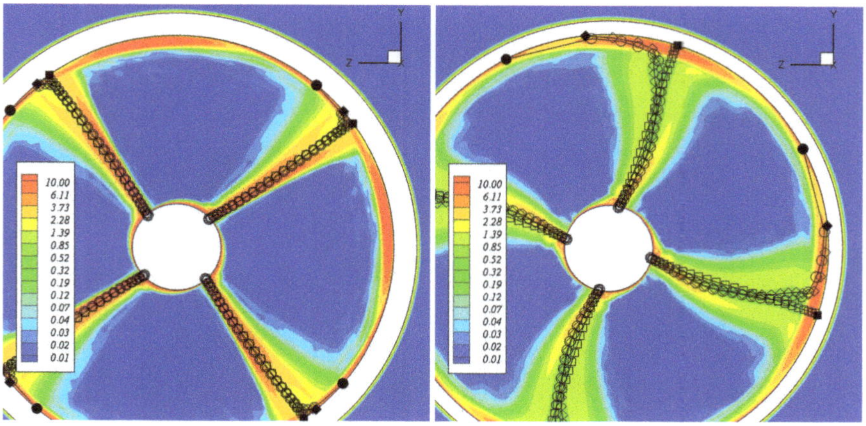

(a) (b)

Figure 8. Wake geometry at $x/R = 0.2$ (**a**) and $x/R = 0.4$ (**b**) for $J = 0.1$. The contours represent the ReFRESCO total vorticity field $|\vec{\omega}|/\Omega$. The symbols represent the PROPAN wake geometry: rigid wake model (squares), wake alignment model (diamonds) and wake alignment model with reduced gap pitch (circles). The filled symbols refer to the tip vortex.

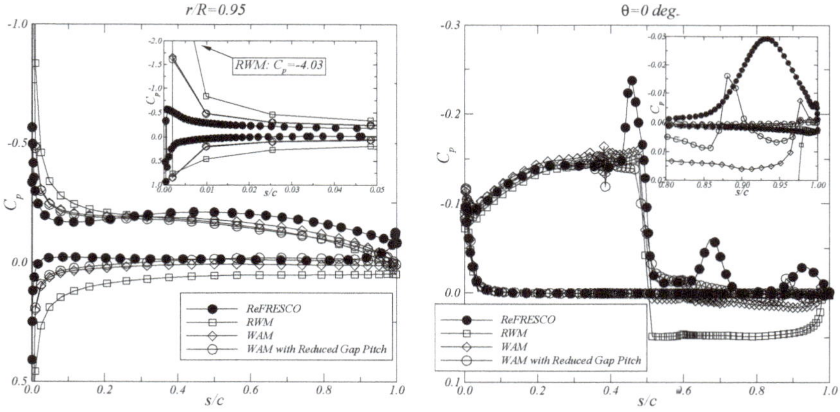

(a) (b)

Figure 9. Blade chordwise pressure distribution at $r/R = 0.95$ (**a**); Duct chordwise pressure distribution at $\theta = 0$ degrees (**b**). $J = 0.1$.

The comparison of the blade and duct pressure distributions is presented along the chordwise direction s/c at the radial section $r/R = 0.95$ and circumferential position $\theta = 0$ degrees, respectively. For the advance ratios 0.1 and 0.2, an improvement in the agreement between the inviscid and viscous pressure distributions is obtained when using the wake alignment model with reduced pitch for the gap strip. This comparison shows the influence of the tip vortex pitch on the prediction of the pressure distribution, especially on the duct inner side downstream of the propeller. A decrease in the duct pressure downstream of the propeller is obtained with the reduction of the tip vortex pitch, which is consistent with the viscous results. A reduction in the suction peak at the blade leading edge is also visible, since the blade wake sheet strongly affects the local flow direction to the blade sections. For the pressure distribution at $J = 0.5$, a good agreement in the blade suction peak and duct pressure distribution downstream from the propeller is achieved with the wake alignment model. In this case, no significant improvements are obtained when combining the wake alignment model with the reduced gap pitch. For this advance coefficient, the assumption of a duct thickness equal to $\delta/R = 4\%$ is sufficient for the correct prediction of the propeller and duct loads.

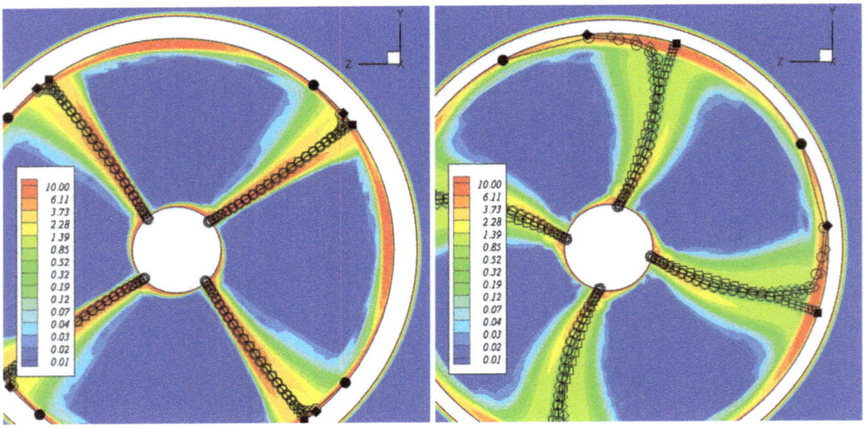

(**a**) (**b**)

Figure 10. Wake geometry at $x/R = 0.2$ (**a**) and $x/R = 0.4$ (**b**) for $J = 0.2$. The contours represent the ReFRESCO total vorticity field $|\vec{\omega}|/\Omega$. The symbols represent the PROPAN wake geometry: rigid wake model (squares), wake alignment model (diamonds) and wake alignment model with reduced gap pitch (circles). The filled symbols refer to the tip vortex.

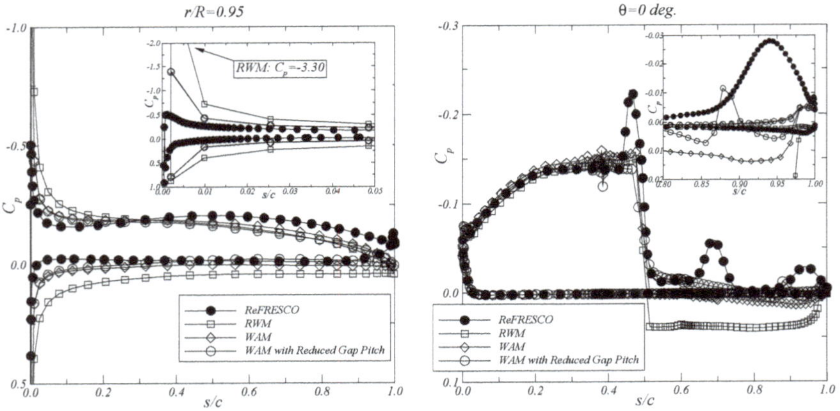

(**a**) (**b**)

Figure 11. Blade chordwise pressure distribution at $r/R = 0.95$ (**a**); Duct chordwise pressure distribution at $\theta = 0$ degrees (**b**). $J = 0.2$.

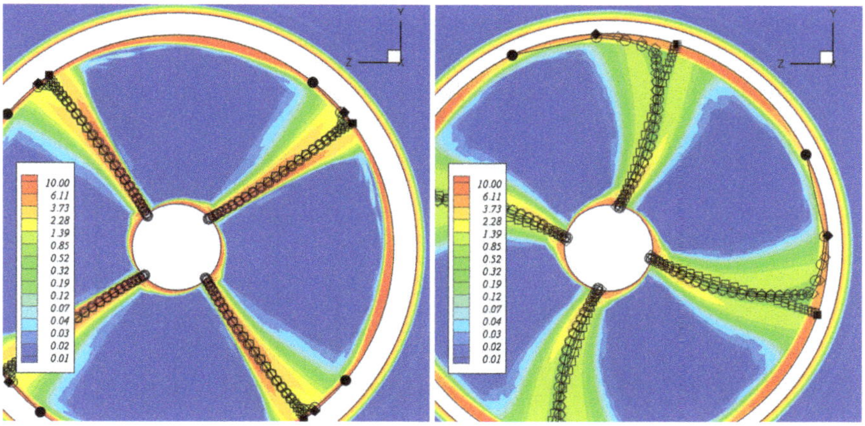

(a) (b)

Figure 12. Wake geometry at $x/R = 0.2$ (**a**) and $x/R = 0.4$ (**b**) for $J = 0.5$. The contours represent the ReFRESCO total vorticity field $|\vec{\omega}|/\Omega$. The symbols represent the PROPAN wake geometry: rigid wake model (squares), wake alignment model (diamonds) and wake alignment model with reduced gap pitch (circles). The filled symbols refer to the tip vortex.

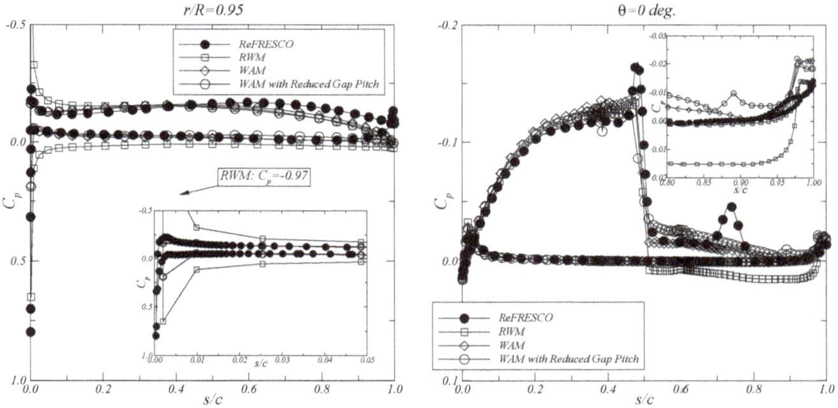

(a) (b)

Figure 13. Blade chordwise pressure distribution at $r/R = 0.95$ (**a**); Duct chordwise pressure distribution at $\theta = 0$ degrees (**b**). $J = 0.5$.

However, from this study two exceptions are observed in the comparison of the pressure distributions: at the blade leading edge near the tip and in the duct inner side For the blade pressure, a larger suction peak is obtained with the inviscid model. This suction peak decreases with the reduction of the blade wake pitch at the tip. For the duct pressure at the inner side, local pressure minima are observed in the viscous computations, which are related to the passage of the tip vortices from the different blades. This effect is not captured by the inviscid calculations due to the closed gap model, where the blade wake is attached to the duct inner side.

The correlation between the position of the tip vortex core and the peaks of low pressure is illustrated in Figure 14 for the plane $z = 0$ at $J = 0.2$, which corresponds to the circumferential position $\theta = 0$ degrees. In this figure the inviscid wake geometries are compared with the viscous total vorticity field along the longitudinal direction. A good agreement is obtained with the aligned wakes, except near the blade wake tip, where some differences are still observed.

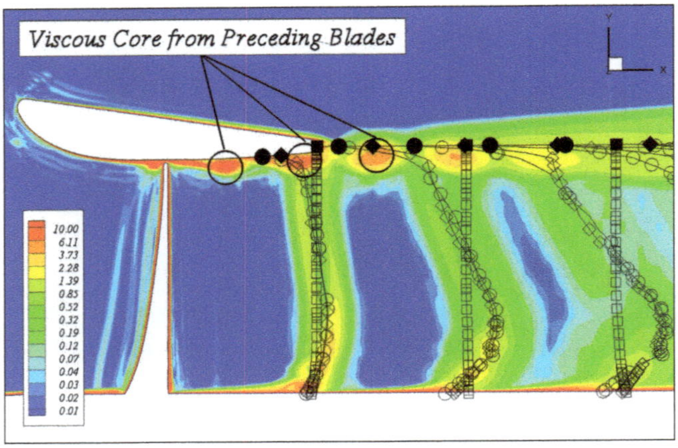

Figure 14. Wake geometry at $z = 0$ for $J = 0.2$. The contours represent the ReFRESCO total vorticity field $|\vec{\omega}|/\Omega$. The symbols represent the PROPAN wake geometry: rigid wake model (squares), wake alignment model (diamonds) and wake alignment model with reduced gap pitch (circles). The filled symbols refer to the tip vortex.

3.4. Prediction of the Open-Water Performance

In this section the predicted thrust and torque coefficients are compared with experimental data available from open-water tests [23]. In the wake alignment model the duct boundary-layer thickness ($\delta/R = 4\%$) is assumed independent of the inflow conditions for the entire open-water range. In addition, in the wake alignment model with reduced gap pitch a constant value of $P/D = 0.9$ is also considered for all advance coefficients. The ReFRESCO calculations are also included in the comparison. Figure 15 illustrates the comparison of the thrust and torque coefficients with the experiments. A section viscous drag coefficient of 0.007 and suppression of the chordwise component of the blade section lift are considered for all inviscid computations. This suppression models the effect of flow separation which eliminates the non-physical suction peaks at the leading edge in the potential flow theory. No viscous drag correction to the duct thrust has been applied.

As expected, a significant over-prediction of the propeller thrust and torque is obtained with the rigid wake model. This result shows that the prescribed wake geometry with constant pitch and equal to the blade pitch completely misses the propeller and duct loads. Alternatively, the propeller thrust and torque are well predicted for the advance ratios higher than 0.3 when using the wake alignment model without gap pitch correction. For the advance ratios lower than 0.3 a significant improvement in the comparison with the experiments is obtained with the wake alignment model using a reduced pitch for the gap strip.

Although the assumptions of constant duct boundary-layer thickness and gap strip pitch independent of the inflow conditions are questionable, a reasonable to good agreement of the propeller forces is obtained with the wake alignment model when compared with the experiments. For example, at $J = 0.1$ the differences between the measured and the predicted propeller thrust reduce from 20% to 7% by applying the gap pitch reduction. For the propeller torque, the differences decrease from 12% to 0.3% with the reduced gap pitch. A smaller influence of the gap pitch is observed for the higher advance coefficients, which is due to the decrease of the tip vortex strength. The duct thrust coefficient agrees well with the measurements for low advance coefficients. For high advance coefficients, an over-prediction of the duct thrust is seen, which is due to the occurrence of flow separation on the outer side of the duct and it is not modelled in the inviscid method.

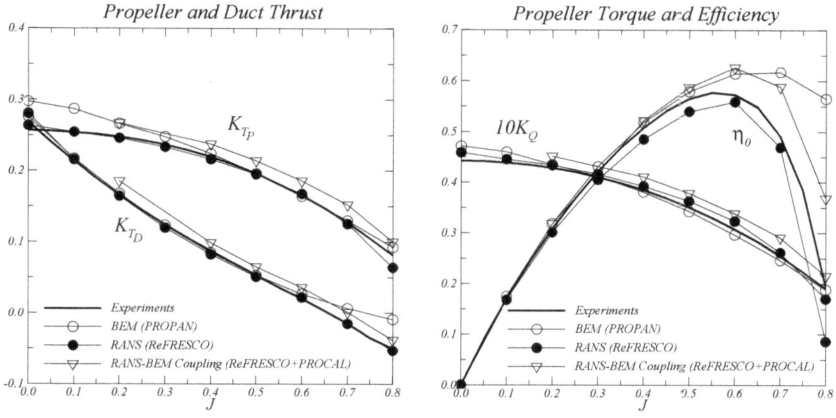

(a) (b)

Figure 15. Comparison between numerical and experimental data from open-water tests. Propeller and duct thrust (**a**); Propeller torque and open-water efficiency (**b**).

A good agreement of the propeller forces with the experimental data is obtained with the viscous calculations using code ReFRESCO for advance ratios up to 0.7. In this range the differences are in the order of 1%. This agreement legitimates the use of RANS simulations for the present comparison study.

4. Conclusions

The investigation presented in this paper focused on the improvement of the inviscid performance predictions with a panel method for ducted propellers. For this study, a comparison between the results obtained by a panel method with a RANS solver is made to obtain a better insight on the viscous effects of the ducted propeller and on the limitations of the inviscid potential flow model. Special attention is given near bollard pull conditions, which are important in the design of ducted propeller systems. Results show that an alignment model of the wake geometry with the local flow is essential for an accurate prediction of the propeller and duct loads. Due to the strong interaction between the blade wake and the boundary-layer on the duct inner side, a correction in the axial velocity is also taken into account in the wake alignment model. As a consequence, a local reduction of the blade wake pitch near the tip is obtained influencing the propeller loading. However, this mechanism is not sufficient to correctly predict the propeller forces near bollard pull and an additional correction is proposed for the gap strip. In the present work, the pitch of the gap strip is empirically prescribed in a first attempt to model the tip leakage vortex. In this way, a stronger reduction in the blade wake pitch is obtained and the agreement of the propeller forces with the experiments improves significantly over the entire open-water range. The wake alignment scheme combined with empirical corrections for the blade wake pitch near the tip and gap strip, to take into account the viscous effects of the gap flow in the potential flow model, has proven to be robust, efficient and to provide accurate predictions of the open-water performance of ducted propellers.

Acknowledgments: The authors acknowledge the Laboratory for Advanced Computing at University of Coimbra (www.lca.uc.pt) for providing computing resources that have contributed to part of the results reported in this paper.

Author Contributions: J.M.B. and J.F.C. developed the inviscid panel method with the contribution from J.B. in the implementation of the wake alignment model. D.R. generated the grids and conduct the viscous calculations with the RANS solver. J.M.B. ran the inviscid code, analysed the output data and compared the inviscid and viscous results. J.B. provided scientific advice and supervision throughout. All authors discussed the results and commented on the manuscript.

Conflicts of Interest: The authors declare no conflict of interest.

Abbreviations

The following abbreviations are used in this manuscript:

IST	Instituto Superior Técnico
MARIN	Maritime Research Institute Netherlands
QUICK	Quadratic upwind interpolation for convective kinematics
RANS	Reynolds-averaged Navier-Stokes
RWM	Rigid wake model
SIMPLE	Semi-implicit method for pressure linked equations
SST	Shear stress transport
WAM	Wake alignment model

References

1. Sánchez-Caja, A.; Rautaheimo, P.; Siikonen, T. Simulation of Incompressible Viscous Flow Around a Ducted Propeller Using a RANS Equation Solver. In Proceedings of the Twenty-Third Symposium on Naval Hydrodynamics, Val de Reuil, France, 17–22 September 2000; The National Academies Press: Washington, DC, USA, 2001; pp. 527–539.
2. Sánchez-Caja, A.; Pylkkänen, J.V.; Sipilä, T.P. Simulation of the Incompressible Viscous Flow Around Ducted Propellers With Rudders Using a RANSE solver. In Proceedings of the 27th Symposium on Naval Hydrodynamics, Seoul, Korea, 5–10 October 2008; Curran Associates, Inc.: Red Hook, NY, USA, 2010; pp. 968–982.
3. Abdel-Maksoud, M.; Heinke, H.-J. Scale Effects on Ducted Propellers. In Proceedings of the Twenty-Fourth Symposium on Naval Hydrodynamics, Fukuoka, Japan, 8–13 July 2002; The National Academies Press: Washington, DC, USA, 2003; pp. 744–759.
4. Bhattacharyya, A.; Krasilnikov, V.; Steen, S. Scale effects on open water characteristics of a controllable pitch propeller working within different duct designs. *Ocean Eng.* **2016**, *112*, 226–242.
5. Kim, J.; Paterson, E.G.; Stern, F. RANS simulation of ducted marine propulsor flow including subvisual cavitation and acoustic modeling. *ASME J. Fluids Eng.* **2006**, *128*, 799–810.
6. Kerwin, J.E.; Kinnas, S.A.; Lee, J.-T.; Shih, W.-Z. A Surface Panel Method for the Hydrodynamic Analysis of Ducted Propellers. In *Transactions of Society of Naval Architects and Marine Engineers*; Society of Naval Architects and Marine Engineers: New York, NY, USA, 1987; p. 4.
7. Hughes, M.J.; Kinnas, S.A.; Kerwin, J.E. Experimental validation of a ducted propeller analysis method. *ASME J. Fluids Eng.* **1992**, *114*, 214–219.
8. Hughes, M.J. Implementation of a Special Procedure for Modeling the Tip Clearance Flow in a Panel Method for Ducted Propulsors. In Proceedings of the Propellers & Shafting '97 Symposium, Virginia Beach, VA, USA, 23–24 September 1997; Society Naval Architects and Marine Engineers: Alexandria, VA, USA, 1997; Paper Number 17.
9. Lee, H.; Kinnas, S.A. Prediction of Cavitating Performance of Ducted Propellers. In Proceedings of the Sixth International Symposium on Cavitation, Wageningen, The Netherlands, 11–15 September 2006.
10. Oweis, G.F.; Fry, D.; Jessup, S.D.; Ceccio, S.L. Development of a tip-leakage flow—Part 1: The flow over a range of Reynolds numbers. *ASME J. Fluid Eng.* **2006**, *128*, 751–764.
11. Baltazar, J.; Falcão de Campos, J.A.C.; Bosschers, J. Open-water thrust and torque predictions of a ducted propeller system with a panel method. *Int. J. Rotating Mach.* **2012**, *2012*, 474785, doi:10.1155/2012/474785. Available online: https://www.hindawi.com/journals/ijrm/2012/474785/ (accessed on 16 January 2018).
12. Baltazar, J.; Rijpkema, D.; Falcão de Campos, J.A.C. A Comparison of Panel Method and RANS Calculations for a Ducted Propeller System in Open-Water. In Proceedings of the Third International Symposium on Marine Propulsors, Launceston, Australia, 5–8 May 2013; Binns, J., Brown, R., Bose, N., Eds.; Australian Maritime College, University of Tasmania: Newnham, TAS, Australia, 2013; pp. 338–346.

13. Baltazar, J.; Falcão de Campos, J.A.C.; Bosschers, J. Potential Flow Modelling of Ducted Propellers with a Panel Method. In Proceedings of the Fourth International Symposium on Marine Propulsors, Austin, TX, USA, 31 May–4 June 2015; Kinnas, S.A., Ed.; The University of Texas at Austin: Austin, TX, USA, 2015; pp. 184–192.
14. Vaz, G.; Bosschers, J. Modelling Three Dimensional Sheet Cavitation on Marine Propellers Using a Boundary Element Method. In Proceedings of the Sixth International Symposium on Cavitation, Wageningen, The Netherlands, 11–15 September 2006.
15. Bosschers, J.; Willemsen, C.; Peddle, A.; Rijpkema, D. Analysis of Ducted Propellers by Combining Potential Flow and RANS Methods. In Proceedings of the Fourth International Symposium on Marine Propulsors, Austin, TX, USA, 31 May–4 June 2015; Kinnas, S.A., Ed.; The University of Texas at Austin: Austin, TX, USA, 2015; pp. 639–648.
16. Gu, H.; Kinnas, S.A. Modeling of Contra-Rotating and Ducted Propellers via Coupling of a Vortex-Lattice with a Finite Volume Method. In Proceedings of the Propellers & Shafting 2003 Symposium, Virginia Beach, VA, USA, 17–18 September 2003; Society Naval Architects and Marine Engineers: Alexandria, VA, USA, 2003.
17. Kinnas, S.A; Fan, H.; Tian, Y. A Panel Method with a Full Wake Alignment Model for the Prediction of the Performance of Ducted Propellers. *J. Ship Res.* **2015**, *59*, 246–257.
18. Kim, S.; Du, W.; Kinnas, S.A. Panel Method for Ducted Propellers with Sharp and Round Trailing Edge Duct With Fully Aligned Wake on Blade and Duct. In Proceedings of the Fifth International Symposium on Marine Propulsors, Espoo, Finland, 12–15 June 2017; Sánchez-Caja, A., Ed.; VTT Technical Research Center of Finland Ltd.: Espoo, Finland, 2017; pp. 627–636.
19. Baltazar, J. On the Modelling of the Potential Flow About Wings and Marine Propellers Using a Boundary Element Method. Ph.D. Thesis, Instituto Superior Técnico, Lisbon, Portugal, 30 September 2008.
20. Morino, L.; Kuo, C.-C. Subsonic potential aerodynamics for complex configurations: A general theory. *AIAA J.* **1974**, *12*, 191–197.
21. Vaz, G.; Jaouen, F.; Hoekstra, M. Free-Surface Viscous Flow Computations. Validation of URANS Code FreSCo. In Proceedings of ASME 28th International Conference on Ocean, Offshore and Arctic Engineering, OMAE2009-79398, Honolulu, HI, USA, 31 May–5 June 2009; pp. 425–437.
22. Menter, F.; Kuntz, M.; Langtry, R. Ten Years of Industrial Experience With the SST Turbulence Model. In Proceedings of the Fourth International Symposium on Turbulence, Heat and Mass Transfer, Antalya, Turkey, 12–17 October 2003; Hanjalić, K., Nagano, Y., Tummers, M.J., Eds.; Begell House: Danbury, CT, USA, 2003; pp. 625–632.
23. Kuiper, G. *The Wageningen Propeller Series*; Maritime Research Institute Netherlands: Wageningen, The Netherlands, 1992; ISBN 9090072470.

Journal of
Marine Science and Engineering

Article

Boundary Element Modelling Aspects for the Hydro-Elastic Analysis of Flexible Marine Propellers

Pieter Maljaars *, Mirek Kaminski and Henk den Besten

Maritime & Transport Technology Department, Delft University of Technology, Mekelweg 2, 2628 CD Delft, The Netherlands; m.l.kaminski@tudelft.nl (M.K.); henk.denbesten@tudelft.nl (H.d.B.)
* Correspondence: p.j.maljaars@tudelft.nl; Tel.: +31-15-278-5923

Received: 10 April 2018; Accepted: 1 June 2018; Published: 5 June 2018

Abstract: Boundary element methods (BEM) have been used for propeller hydrodynamic calculations since the 1990s. More recently, these methods are being used in combination with finite element methods (FEM) in order to calculate flexible propeller fluid–structure interaction (FSI) response. The main advantage of using BEM for flexible propeller FSI calculations is the relatively low computational demand in comparison with higher fidelity methods. However, the BEM modelling of flexible propellers is not straightforward and requires several important modelling decisions. The consequences of such modelling choices depend significantly on propeller structural behaviour and flow condition. The two dimensionless quantities that characterise structural behaviour and flow condition are the structural frequency ratio (the ratio between the lowest excitation frequency and the fundamental wet blade natural frequency) and the reduced frequency. For both, general expressions have been derived for (flexible) marine propellers. This work shows that these expressions can be effectively used to estimate the dry and wet fundamental blade frequencies and the structural frequency ratio. This last parameter and the reduced frequency of vibrating blade flows is independent of the geometrical blade scale as shown in this work. Regarding the BEM-FEM coupled analyses, it is shown that a quasi-static FEM modelling does not suffice, particularly due to the fluid-added mass and hydrodynamic damping contributions that are not negligible. It is demonstrated that approximating the hydro-elastic blade response by using closed form expressions for the fluid added mass and hydrodynamic damping terms provides reasonable results, since the structural response of flexible propellers is stiffness dominated, meaning that the importance of modelling errors in fluid added mass and hydrodynamic damping is small. Finally, it is shown that the significance of recalculating the hydrodynamic influence coefficients is relatively small. This fact might be utilized, possibly in combination with the use of the closed form expressions for fluid added mass and hydrodynamic damping contributions, to significantly reduce the computation time of flexible propeller FSI calculations.

Keywords: flexible (composite) propellers; BEM modelling; fluid–structure interaction

1. Introduction

Over the last two decades, an increased interest in flexible propellers can be noticed given the growing list of publications on the hydro-elastic analysis of flexible marine propellers. Several publications present a methodology for the numerical analysis of these types of propellers in steady and unsteady inflow conditions. These methods typically involve partitioned fluid–structure interaction (FSI) computations, meaning that the fluid and structural problem are separately solved and coupling iterations are required to converge to the fully coupled solution. For steady propeller FSI computations mainly Reynolds-averaged Navier–Stokes (RANS) methods [1–4] and boundary element methods (BEM) [5–12] have been used for solving the fluid part of the coupled problem.

The fundamental difference between RANS and BEM is that, in the latter, a potential flow is assumed, meaning that phenomena as flow separation, flow transition, boundary layers and vorticity dynamics are not modelled. Results of validation studies on flexible propellers in open water conditions with BEM-FEM (finite element method) and RANS-FEM methods have been presented in [13]. Despite the limitations of a BEM and the complicated flow characteristics considered in that work, a fairly good estimate of the propeller hydro-elastic response was obtained with the BEM-FEM approach. Given the relatively low computational demand of BEM in comparison to a RANS method, BEM is an attractive method for the FSI analysis of flexible propellers.

The BEM modelling of flexible propellers in steady and unsteady flow requires several modelling choices—for instance, how to include propeller deformations in the BEM analysis. Is a full geometry update necessary at every time step or can an accurate solution be obtained more efficiently by partly updating the propeller geometry every time step? How can fluid added mass and hydrodynamic damping be included: explicitly with closed form expressions [7] or implicitly in the BEM calculation? Answers to these questions do not seem to be available in literature. The main purpose of this work is to fill this knowledge gap and therefore several modelling choices have been evaluated.

The BEM models for steady and unsteady flexible propeller analyses as proposed in this work look similar to the model as presented in [7]. However, on the following aspects, these models are different. In [7], the BEM modelling relies on the assumption of negligible small blade deformations. The most extensive model presented in this work is applicable for large elastic blade deformations, since blade deformation and vibration effects are implicitly included in the BEM calculation by updating the blade geometry and the body boundary impermeability condition at each computation step. Another difference is that the BEM modelling approach as presented in [7] is based on an explicit treatment of fluid added mass and hydrodynamic damping forces, while, in this work, both an implicit and explicit modelling of fluid added mass and hydrodynamic damping effects have been considered.

The consequences of BEM modelling choices depend significantly on flow condition and structural behaviour. The two dimensionless quantities that characterise propeller structural behaviour and flow condition are the structural frequency ratio (the ratio between the lowest excitation frequency and the fundamental wet blade natural frequency) and the reduced frequency. For both, general expressions have been derived for (flexible) marine propellers.

This paper is structured as follows. In Section 2, general expressions for reduced frequencies and structural frequency ratios of flexible composite propellers are derived. Section 3 describes the BEM modelling of flexible propellers. Section 4 presents the derivation of closed form expressions for fluid added mass and hydrodynamic damping. In Section 5, the hydrodynamic loads on a plunging hydrofoil are investigated and important conclusions are drawn with respect to the frequency dependence of fluid added mass and hydrodynamic damping. Section 6 presents different BEM models for steady and unsteady flexible propeller calculations and include the results of a comparative study. Section 7 contains the conclusions.

2. Flow and Structural Response Characterisation

In this section, expressions will be derived to estimate the characteristics of fluid and structural response for flexible marine propellers. Section 2.1 provides expressions for the fundamental blade natural frequency in air and water and the structural frequency ratio. Section 2.2 shows a typical value for the propeller flow reduced frequency. In Section 2.3, estimated natural frequencies in air and water obtained for flexible versions of the highly skewed Seiun–Maru propeller have been compared to FEM calculation results.

2.1. Structural Frequency Ratio

The structural response of a one degree-of-freedom (1DOF) linear mass-spring-damper system can be assigned to one of the following regimes, depending on the ratio between excitation frequency ω and natural frequency ω_0:

- $\omega << \omega_0$; quasi-static regime, structural response dominated by stiffness.
- $\omega = \omega_0$; resonance regime, structural response dominated by damping.
- $\omega >> \omega_0$; dynamic regime, structural response dominated by mass.

The fundamental natural frequency of a propeller blade is the frequency of the first bending mode. The blade natural frequencies depend on the stiffness and mass and can be computed using solution techniques available in FEM software. For a quick approximation of the first two natural frequencies in air and water, formulas are provided in [14,15]. The fundamental dry natural frequency, ω_0^{dry}, in rad/s can be estimated for moderately skewed propellers with

$$\omega_0^{dry} = \frac{2\pi}{3.28\,(R - r_h)^2}\sqrt{\left(\frac{E}{\rho}\right)\left(\frac{t_m}{c_m}\right)c_r t_r}, \tag{1}$$

where R is the propeller radius, r_h is the hub radius, E the Young's modulus ρ the blade material density, t_m the mean blade thickness, t_r the root blade thickness, c_m the mean chord length and c_r the root chord length. The fundamental wet natural frequency of nickel aluminium bronze (NAB) propellers is generally 62–64% of the value in air [16], meaning that, for the first mode, the modal fluid added mass is approximately 2.5 times the NAB blade modal mass. Thus, the fundamental wet natural frequency, ω_0^{wet}, for any propeller material is given by

$$\omega_0^{wet} = \frac{2\pi}{3.28\,(R - r_h)^2}\sqrt{\left(\frac{E}{\rho + 2.5\rho_{NAB}}\right)\left(\frac{t_m}{c_m}\right)c_r t_r}, \tag{2}$$

where ρ_{NAB} is the density of the NAB material.

From Equations (1) and (2), it can be concluded that, when NAB blade material with a typical Young's modulus and density of 110 GPa and 7600 kgm^{-3} is replaced by a glass–epoxy composite material with a typical Young's modulus and density of 20 GPa and 1700 kgm^{-3}, the dry natural frequencies will slightly decrease. When a carbon–epoxy material is considered with a typical Young's modulus and density of 75 GPa and 1600 kgm^{-3}, the dry natural frequencies will be significantly higher than for the NAB equivalent. The fundamental wet natural frequency of a glass–epoxy blade will be considerably smaller than its NAB equivalent, while for a carbon–epoxy blade, it is approximately the same. Due to the lower material density of fibre reinforced plastics, the fluid added mass has a more pronounced effect on the wet natural frequencies than in the case of a NAB propeller. From Equations (1) and (2), it can be concluded that the dry and wet blade frequencies scale inversely proportional with the geometrical blade scale.

Analogous to the structural frequency ratio of a 1DOF linear mass-spring-damper system, the structural response behaviour of propeller blades can be characterised by the ratio of the frequency corresponding to the dominant blade mode and the excitation frequency. As will be shown latter on, the response of flexible propellers is stiffness dominated, then, the first blade mode will dominate the structural response. Hence, the wet fundamental blade frequency is the typical frequency used for the structural frequency ratio. The lowest excitation frequency is the shaft rotation frequency. Since the shaft rotation speed, n, is related to the blade radius by the tip speed $v_{tip} = 2\pi n R$, the following expression for the ratio between excitation frequency $\omega = 2\pi n$ and wet fundamental blade frequency, i.e., the structural frequency ratio, $\frac{\omega}{\omega_0^{wet}}$, can be derived

$$\frac{\omega}{\omega_0^{wet}} = \frac{3.28 v_{tip}\,(R - r_h)^2}{2\pi R\sqrt{\left(\frac{E}{\rho + 2.5\rho_{NAB}}\right)\left(\frac{t_m}{c_m}\right)c_r t_r}}. \tag{3}$$

In order to avoid cavitation, a typical value for the maximum allowable tip speed is around 35 m/s. Therefore, the maximum tip speed in ship propeller design is a constant rather than a variable.

By proportionally scaling the propeller dimensions R, r_h, t_m, c_m, c_r and t_r, the structural frequency ratio does not change. Hence, the structural frequency ratio is independent of the geometrical propeller scale.

2.2. Propeller Flow Reduced Frequency

For fluids around lifting bodies, a dimensionless number exists that describes the unsteadiness of the flow and is called the reduced frequency. In the expression for the reduced frequency, k, an oscillation frequency is related to the flow speed:

$$k = \frac{\omega c}{2v_0}, \tag{4}$$

where c is the chord length and v_0 the undisturbed flow velocity. For small reduced frequencies, the unsteadiness of the flow is negligible and a quasi-steady approach can be justified. For higher reduced frequencies, the circulatory lift reduces, which is called the lift deficiency and a phase lag between the circulatory part of the lift and the body motion exists due to the wake vorticity. Lift deficiency and phase lag functions for a flat plate foil in small amplitude unsteady motion were derived by Theodorsen [17], see Figure 1.

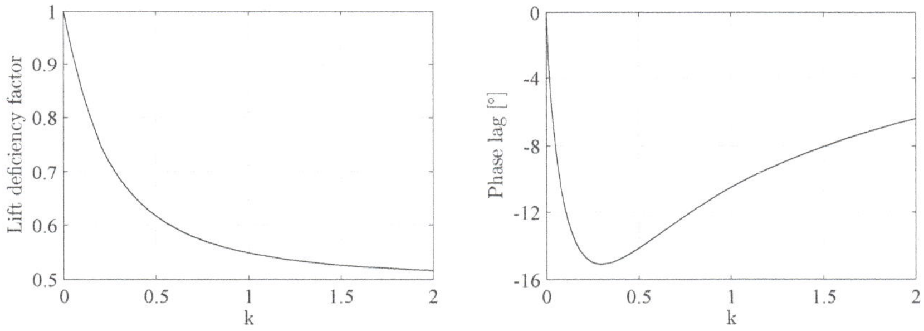

Figure 1. Graphical description of Theodorsen's lift deficiency and phase lag functions for small amplitude unsteady motions of a flat plate foil.

The propeller flow reduced frequency at 70% of the blade radius is obtained from the shaft rotation speed and is given by

$$k_{0.7} = \frac{2\pi n c_{0.7}}{2v_{0,0.7}}, \tag{5}$$

where $c_{0.7}$ is the chord length at $0.7R$ and is typically 40% of the blade diameter, D. $v_{0,0.7}$ is the undisturbed flow speed at $0.7R$. $v_{0,0.7}$ is approximately $0.7\pi nD$. Hence, for conventional propellers, a typical value for the reduced frequency at $0.7R$ is 0.57. This value can be considered as highly unsteady according to Figure 1 and appears to be independent of the geometrical propeller scale.

2.3. Seiun–Maru Propeller Frequencies

The expressions for the fundamental natural frequencies have been verified by comparing estimated dry and wet fundamental blade frequencies to FEM calculated frequencies for the highly skewed Seiun–Maru propeller. The Seiun–Maru is a Japanese training vessel. The geometry of the propeller is in the public domain [18] and its main particulars are summarized in Table 1.

Table 1. Main particulars of the highly skewed Seiun–Maru propeller.

Diameter	3600 mm
Pitch Ratio (mean)	0.92
Expanded Area Ratio	0.7
Number of Blades	5
Blade Thickness Ratio	0.0496
Boss Ratio	0.1972
Total Skew Angle	45°
Rake Angle	−3.03°
Blade Section	Modified SRI-B

Figure 2 shows the dry and wet fundamental blade frequencies calculated with the FEM and estimated with Equations (1) and (2). For the various materials, the properties as mentioned before have been used. For the FEM calculations, a Poisson ratio of 0.3 has been adopted and the fluid added mass has been calculated with the approach described in Section 4. The differences between the frequencies obtained with FEM and the estimation formulas are smaller than 20% for all the cases. It is expected that, for propellers with less skew, the differences will be smaller, since Equation (1) was proposed for moderately skewed propellers [14,15]. Nevertheless, the estimated natural frequencies are accurate enough to identify the structural response regime and Equation (3) is proposed to attribute the structural behaviour to stiffness, damping or mass dominated response. For a maximum allowable tip speed of 35 m/s, the structural frequency ratios as given in Figure 3 have been obtained for the various Seiun–Maru propellers. It can be concluded that the structural response of composite propeller blades is expected to be predominantly quasi-static and probably a quasi-static structural approach might give a good approximation of the blade response. This will be further evaluated in Section 6.

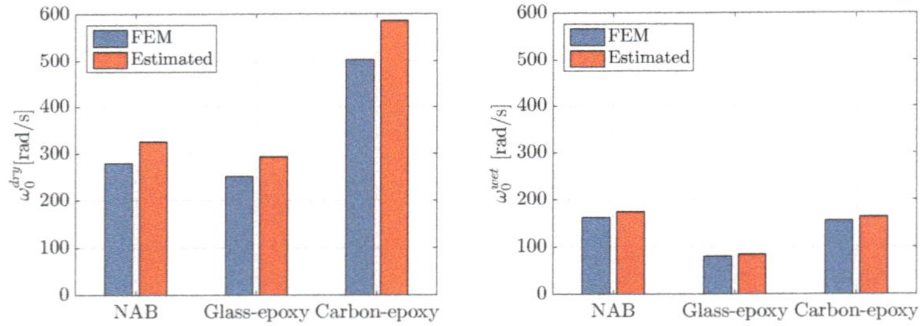

Figure 2. Dry and wet fundamental blade frequencies of the Seiun–Maru propeller for different blade materials obtained with FEM and estimated with Equations (1) and (2).

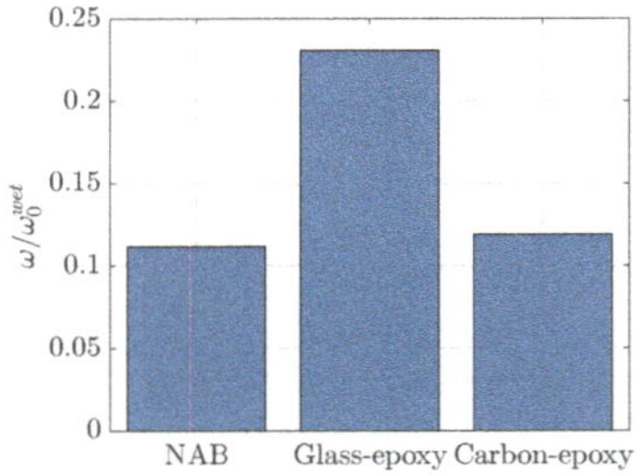

Figure 3. Structural frequency ratios for the various Seiun–Maru propellers.

3. Hydrodynamic Method for Propeller Forces

3.1. Potential Flow Theory

In the present study, the BEM PROCAL has been used for the hydrodynamic calculations. PROCAL is a panel method developed by the Maritime Research Institute Netherlands (MARIN) for the Cooperative Research Ships [19,20]. PROCAL solves the integral equation for the velocity potential in a fluid domain based on the Morino formulation [21]. In this integral formulation, the propeller induced velocity disturbances are considered irrotational. Then, by defining the scalar variable ϕ as the disturbance velocity potential, the total velocity, $\mathbf{v}$, relative to the operating propeller becomes

$$\mathbf{v}(\mathbf{x},t) = \mathbf{v}_0(\mathbf{x},t) + \nabla\phi(\mathbf{x},t), \tag{6}$$

where t is the time and $\mathbf{x}$ is the position vector in a Cartesian coordinate system. The undisturbed velocity, $\mathbf{v}_0$, can be written as the sum of the ship's effective wake field velocity and the effect of the propeller angular velocity, with the wake field velocity, $\mathbf{v}_w$, and with the propeller angular velocity, $\boldsymbol{\Omega}$,

$$\mathbf{v}_0(\mathbf{x},t) = \mathbf{v}_w(\mathbf{x},t) - \boldsymbol{\Omega} \times \mathbf{x}. \tag{7}$$

The flow is assumed to be incompressible and has a constant density. Therefore, Laplace's equation applies to the disturbance velocity potential,

$$\nabla^2\phi(\mathbf{x},t) = 0. \tag{8}$$

Then, the fluid pressures p are related to the total velocity and the disturbance velocity potential according to Bernoulli's law,

$$\frac{\partial\phi}{\partial t} + \frac{1}{2}|\mathbf{v}|^2 + \frac{p}{\rho} + gz = \frac{p_{ref}}{\rho} + \frac{1}{2}|\mathbf{v}_0|^2. \tag{9}$$

For a propeller, p_{ref} is the pressure far upstream (along the shaft axis) and it obeys the hydrostatic law, $p_{ref} = p_{atm} + \rho g z_{shaft}$ being p_{atm} the atmospheric pressure at the free surface, and submergence, z, at the shaft as, z_{shaft}. In order to solve Equation (8), boundary conditions have to be imposed on the propeller surface (S_B) and wake sheet (S_W), which contains the shed vorticity. On the propeller surface, an impermeability condition is imposed,

$$\nabla\phi \cdot \mathbf{n} = -\mathbf{v}_0 \cdot \mathbf{n}, \tag{10}$$

where $\mathbf{n}$ is the surface normal. For the wake sheet, a kinematic and dynamic boundary condition can be formulated. The kinematic boundary condition prescribes that the wake sheet is a stream-surface of the flow,

$$\mathbf{v} \cdot \mathbf{n} = 0. \tag{11}$$

The wake sheet itself is an imaginary surface in the flow with zero thickness. The wake sheet cannot support a pressure difference between its upper and lower side. This dynamic boundary condition can be written as

$$\Delta p = p^+ - p^- = 0, \tag{12}$$

where $^+$ and $^-$ denote the upper and lower side of the wake sheet. At the blade trailing edge, this is the so-called Kutta condition.

3.2. Integral Formulation for Disturbance Potential

A relation between the potential at any point in the fluid domain and the source strengths (normal component of the disturbance velocity at the body boundary) exists and is given by the integral equation following from the third Green's identity. Using Morino's formulation [21], this can be written as

$$\epsilon(\mathbf{a})\phi(\mathbf{a},t) = \int_{S_B} \left[\phi(\mathbf{b},t)\frac{\partial G(\mathbf{a},\mathbf{b})}{\partial \mathbf{n}_b} - G(\mathbf{a},\mathbf{b})\frac{\partial \phi(\mathbf{b},t)}{\partial \mathbf{n}_b} \right] dS + \int_{S_W} \left[\Delta\phi(\mathbf{b},t)\frac{\partial G(\mathbf{a},\mathbf{b})}{\partial \mathbf{n}_b} - G(\mathbf{a},\mathbf{b})\Delta\left(\frac{\partial \phi(\mathbf{b},t)}{\partial \mathbf{n}_b}\right) \right] dS, \tag{13}$$

where **a** is a point in the fluid domain, and **b** is a point on the fluid domain boundary surface. G is the Green's function for the Laplace equation defined as

$$G(\mathbf{a},\mathbf{b}) = \frac{1}{r(\mathbf{a},\mathbf{b})} \qquad r(\mathbf{a},\mathbf{b}) = |\mathbf{r}| = |\mathbf{a}-\mathbf{b}|\,. \tag{14}$$

$\mathbf{n_b}$ is the outward normal at **b**. ϵ is a constant that depends on the field point **a** and is 2π if **a** is on the fluid boundary surface. With the dynamic boundary condition for the wake sheet, the following integral equation is obtained:

$$2\pi\phi(\mathbf{a},t) = \int_{S_B} \left[\phi(\mathbf{b},t)\frac{\partial G(\mathbf{a},\mathbf{b})}{\partial \mathbf{n}_b} - \frac{\partial \phi(\mathbf{b},t)}{\partial \mathbf{n}_b} G(\mathbf{a},\mathbf{b}) \right] dS + \int_{S_W} \Delta\phi(\mathbf{b},t)\frac{\partial G(\mathbf{a},\mathbf{b})}{\partial \mathbf{n}_b} dS. \tag{15}$$

In this integral equation, the geometry of the body and wake sheet are constant. This means that the points **b** on the fluid domain boundary surfaces and the surface areas are time invariant, which is obviously not the case for a deformable body. In that case, the integral formulation of Equation (15) transforms to the unsteady flow and time-variant body integral equation,

$$2\pi\phi(\mathbf{a}(t),t) = \int_{S_{B(t)}} \left[\phi(\mathbf{b}(t),t)\frac{\partial G(\mathbf{a}(t),\mathbf{b}(t))}{\partial \mathbf{n}_b(t)} - \frac{\partial \phi(\mathbf{b}(t),t)}{\partial \mathbf{n}_b(t)} G(\mathbf{a}(t),\mathbf{b}(t)) \right] dS + \int_{S_{W(t)}} \Delta\phi(\mathbf{b}(t),t)\frac{\partial G(\mathbf{a}(t),\mathbf{b}(t))}{\partial \mathbf{n}_b(t)} dS. \tag{16}$$

3.3. Numerical Formulation

The integral equations of Equations (15) and (16) are solved in PROCAL by approximating the surfaces S_B and S_W by N_{total} number of panels. On each panel, a collocation point is defined where the integral equation is applied. Finally, a system of equations is obtained, unknown in the strengths of the source and dipole elements. With the imposed boundary conditions, the system of equations can be solved and the potential at the boundaries is obtained. This subsection briefly describes and presents the system of equations for the non-cavitating propeller calculations considered in this work. For the formulation of the discretised problems, the discretisation parameters and equations as presented in [19] have been used. For a more thorough derivation, one is referred to that publication.

3.3.1. Geometry Discretisation

In PROCAL, the entire fluid domain boundary surface, consisting of body surface S_B and wake surface S_W, is decomposed in a key part containing one blade with corresponding hub section and wake sheet and a symmetry part including the other blades, hub sections and wake sheets. The motivation for this subdivision in key- and symmetry surfaces is to obtain a smaller system of equations by utilizing the symmetry properties of ship propellers. The symmetry of flexible propellers in an unsteady flow can be questioned due to time varying deformations and will be discussed further in Section 6.

Figure 4 shows the discretised geometry of a propeller and wake. The discretisation parameters displayed in this figure have the following meaning:

- N_{surfs}: Number of surfaces, N_{surfs} is 2; the blade surfaces and the hub surfaces.
- N_{sym}: Number of symmetries for each surface. In general, N_{sym} is equal to the number of propeller blades.
- N_i: Number of panels in the streamwise direction.
- N_j: Number of panels in the radial direction.
- N_{wi}: Number of panels on the wake sheet in the streamwise direction.
- N_{wj}: Number of panels on the wake sheet in the radial direction.

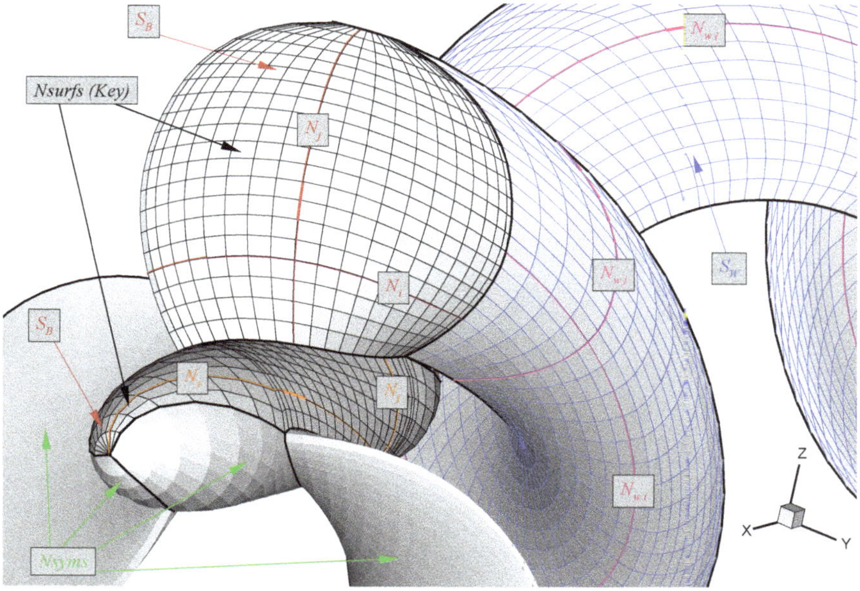

Figure 4. Geometry discretisation parameters for an unsteady propeller calculation. (Image republished from [19] with permission of the author.)

3.3.2. Steady Flow and Rigid Propeller Formulation

For a rigid propeller in a steady flow, Equation (15) can be discretized as follows:

$$\sum_{isurf=1}^{N_{surfs}} \sum_{isym=1}^{N_{syms}} \left[\sum_{j=1}^{N_j} \sum_{i=1}^{N_i} (D_{nij}\phi_{ij} - S_{nij}\sigma_{ij}) + \sum_{j=1}^{N_{wj}} \sum_{i=1}^{N_{wi}} W_{nij}\Delta\phi_{ij} \right] = 0 \qquad n = 1, \ldots, N_{total}, \tag{17}$$

where D_{nij}, S_{nij} and W_{nij} are the hydrodynamic influence coeficients for the body dipoles, body sources and wake dipoles, respectively. They are defined as

$$D_{nij} = \begin{cases} -\dfrac{1}{2\pi}\displaystyle\int_{S_{B_{ij}}} \dfrac{\mathbf{n}_{ij}\cdot\mathbf{r}_{nij}}{r^3_{nij}} dS_{B_{ij}}, & \text{if } n \text{ does no refer to element } i,j, \\ 1-\dfrac{1}{2\pi}\displaystyle\int_{S_{B_{ij}}} \dfrac{\mathbf{n}_{ij}\cdot\mathbf{r}_{nij}}{r^3_{nij}} dS_{B_{ij}}, & \text{if } n \text{ refers to element } i,j, \end{cases} \tag{18}$$

and

$$S_{nij} = \frac{1}{2\pi}\int_{S_{B_{ij}}} \frac{1}{r_{nij}} dS_{B_{ij}} \qquad W_{nij} = -\frac{1}{2\pi}\int_{S_{W_{ij}}} \frac{\mathbf{n}_{ij}\cdot\mathbf{r}_{nij}}{r^3_{nij}} dS_{W_{ij}}. \tag{19}$$

σ_{ij} and ϕ_{ij} are the source and dipole strengths respectively at collocation point i,j on the body surface. $\Delta\phi_{ij}$ is the dipole strength at collocation point i,j on the wake surface. N_{total} is the total number of panels for which the flow quantities have to be solved: $N_{total} = \sum_{isurf=1}^{N_{surfs}} N_i N_j$.

In case of a steady flow calculation, the symmetry surfaces can be taken into account by adding the influence coefficients of key- and symmetry surfaces together; for instance, for the dipole influence coefficients,

$$D^*_{nij} = D_{nij} + \sum_{isym=2}^{N_{syms}} D_{nij}. \tag{20}$$

In general, the strength of a vortex shed by a lifting body is time invariant. This follows from the dynamic and kinematic boundary condition on the wake surface (Equations (11) and (12)); by using Bernoulli's theorem, this yields that the dipole strengths are time invariant following a wake particle. Furthermore, for steady flow conditions the dipole strengths $\Delta\phi_{ij}$ on the wake sheet are the same at fixed j but different i. Accordingly, the dipole influence coefficients of each j strip can be added,

$$W^*_{nj} = \sum_{i=1}^{N_{wi}} W^*_{nij}. \tag{21}$$

Making use of the symmetry properties of the problem and the summation of the wake dipole influence coefficients, Equation (17) can be written as

$$\sum_{isurf=1}^{N_{surfs}} \left[\sum_{j=1}^{N_j}\sum_{i=1}^{N_i} D^*_{nij}\phi_{ij} + \sum_{j=1}^{N_{wj}} W^*_{nj}\Delta\phi_j \right] = \sum_{isurf=1}^{N_{surfs}} S^*_{nij}\sigma_{nij} \qquad n = 1,\ldots,N_{total}. \tag{22}$$

The body source strengths are known from the impermeability boundary condition (Equation (10)) and moved to the right-hand-side of the equation. The matrix defined at the left-hand-side of Equation (22) is a full square matrix with dimension $\left(N_{total} + \sum_{isurf=1}^{Nsurfs} N_{wj}\right)^2$. However, the number of known source strengths is equal to N_{total}, meaning that an additional set of $\sum_{isurf=1}^{Nsurfs} N_{wj}$ number of equations have to be defined in order to close the system of equations. These additional equations can be deduced from the Kutta condition imposing that the tangential velocities at the blade trailing edge, computed on the upper and lower side of the body surface, are equal. The discretised translation of the Kutta condition is the Morino Kutta condition [21] expressing that the difference of the potential values between the upper and lower side of the body trailing edge is equal to the wake dipole strength,

$$\Delta\phi_j = \phi_{N_i j} - \phi_{1j}. \tag{23}$$

This equation holds for every strip j of the propeller, resulting in a number of N_{wj} additional equations per surface. Then, the system of equations becomes

$$\begin{bmatrix} \mathbf{D}^* & \mathbf{W}^* \\ -\mathbf{I}_u + \mathbf{I}_l & \mathbf{I} \end{bmatrix} \begin{bmatrix} \phi \\ \Delta\phi \end{bmatrix} = \begin{bmatrix} \mathbf{S}^* & \mathbf{0} \\ \mathbf{0} & \mathbf{0} \end{bmatrix} \begin{bmatrix} \sigma \\ 0 \end{bmatrix}, \tag{24}$$

where $\mathbf{D}^*$, $\mathbf{W}^*$ and $\mathbf{S}^*$ are the body dipole, wake dipole and body source influence coefficients matrices, respectively. The matrices $\mathbf{I}_u$, $\mathbf{I}_l$ and $\mathbf{I}$ contain only ones and zeros to account for the Morino Kutta condition. Equation (24) can be written as

$$[\mathbf{B}^*]\{\phi\} = [\mathbf{S}^*]\{\sigma\}, \tag{25}$$

where $[\mathbf{B}^*] = [\mathbf{D}^*] + [\mathbf{W}^*][\mathbf{I}_u - \mathbf{I}_l]$ is the total dipole influence coefficients matrix.

3.3.3. Unsteady Flow and Rigid Propeller Formulation

For an unsteady flow and rigid propeller, the inflow conditions and, consequently, the source strengths depend on the blade position.

In PROCAL, an unsteady problem is solved by performing k steady flow like calculations, where k is the number of revolutions, N_{revs}, required for convergence times the number of timesteps during one revolution, N_t, i.e., $k = N_{revs}N_t$.

There is an important difference between a steady computation and the k-th calculation of an unsteady computation. For steady calculations, the symmetry properties of the propeller inflow and geometry are exploited to obtain a smaller system of equations. In case of an unsteady flow, the inflow is generally non-symmetric. To avoid simultaneously solving the system of equations for all key and symmetry surfaces, an iterative procedure is applied.

In this iterative procedure, the distinction between the key and symmetry surfaces is maintained. The system of equations is solved for the key surfaces only, while, for the symmetry surfaces, a previous solution of the key surfaces is taken from the timestep k_{isym} when the key surfaces were at the same spatial position as the symmetry surfaces at the current timestep. This means that the contribution of the symmetry surfaces can be moved to the right-hand-side of the equation. Then, the solution is obtained by solving the systems of equations every timestep and a multiple number of revolutions until convergence. In this way, the discretised system of equations are given by

$$\begin{aligned}
\sum_{isurf=1}^{N_{surfs}} \left[\sum_{j=1}^{N_j}\sum_{i=1}^{N_i} D_{nij}\phi_{ij}^k + \sum_{j=1}^{N_{wj}} W_{n1j}\Delta\phi_{1j}^k \right] &= \sum_{isurf=1}^{N_{surfs}} \left[\sum_{j=1}^{N_j}\sum_{i=1}^{N_i} S_{nij}\sigma_{ij}^k + \sum_{j=1}^{N_{wj}}\sum_{i=2}^{N_{wi}} W_{nij}\Delta\phi_{ij}^k \right] \\
&\quad - \sum_{isurf=1}^{N_{surfs}} \sum_{isym=2}^{N_{syms}} \left[\sum_{j=1}^{N_j}\sum_{i=1}^{N_i} D_{nij}\phi_{ij}^{k_{isym}} \right. \\
&\quad \left. + \sum_{j=1}^{N_j}\sum_{i=1}^{N_i} S_{nij}\sigma_{ij}^{k_{isym}} - \sum_{j=1}^{N_{wj}}\sum_{i=1}^{N_{wi}} W_{nij}\Delta\phi_{ij}^{k_{isym}} \right], \\
&\quad n = 1, \ldots, N_{total}, \\
&\quad k = 1, \ldots, N_{revs}N_t.
\end{aligned} \tag{26}$$

3.3.4. Unsteady Flow and Flexible Propeller Formulation

In case of an unsteady flow and flexible propeller, all the influence coefficients are time dependent, which results in

$$
\begin{aligned}
\sum_{isurf=1}^{N_{surfs}} \left[\sum_{j=1}^{N_j} \sum_{i=1}^{N_i} D_{nij}^{k} \phi_{ij}^{k} + \sum_{j=1}^{N_{wj}} W_{n1j}^{k} \Delta\phi_{1j}^{k} \right] = & \sum_{isurf=1}^{N_{surfs}} \left[\sum_{j=1}^{N_j} \sum_{i=1}^{N_i} S_{nij}^{k} \sigma_{ij}^{k} + \sum_{j=1}^{N_{wj}} \sum_{i=2}^{N_{wi}} W_{nij}^{k} \Delta\phi_{ij}^{k} \right] \\
& - \sum_{isurf=1}^{N_{surfs}} \sum_{isym=2}^{N_{syms}} \left[\sum_{j=1}^{N_j} \sum_{i=1}^{N_i} D_{nij}^{k_{isym}} \phi_{ij}^{k_{isym}} \right. \\
& \left. + \sum_{j=1}^{N_j} \sum_{i=1}^{N_i} S_{nij}^{k_{isym}} \sigma_{ij}^{k_{isym}} - \sum_{j=1}^{N_{wj}} \sum_{i=1}^{N_{wi}} W_{nij}^{k_{isym}} \Delta\phi_{ij}^{k_{isym}} \right], \\
& n = 1, \ldots, N_{total}, \\
& k = 1, \ldots, N_{revs} N_t.
\end{aligned}
\quad (27)
$$

Equation (27) shows that the symmetry influence coefficients are time dependent as well and are equal to the influence coefficients of the symmetry surfaces on the key surfaces at timestep k_{isym} when the key surfaces were at the same spatial position as the symmetry surfaces at the current timestep. This means that the symmetry surface influence coefficients have to be stored in memory N_t times increasing the required computer memory significantly. Therefore, in the BEM model for unsteady flexible propeller calculations, it has been decided to use the symmetry surface influence coefficients at time step k for the symmetry surface influence coefficients of time step k_{isym}. This modelling choice, together with the importance of the recalculation of the key blade influence coefficients, will be discussed further in Section 6.

3.3.5. The Kutta Condition for Flexible Propellers

It has to be discussed whether the Kutta condition can be applied in the unsteady flexible propeller BEM model because the Kutta condition was proposed solely for steady flows [22]. Later, the Kutta condition has been modified for unsteady models, which had been questioned [23]. Some guidelines for the applicability of the Kutta condition for unsteady flows have been given in [24]. Given the relatively small deformations of flexible propellers and that the unsteady blade forces mainly originate from the non-uniform wakefield rather than from the blade vibrations, it has been assumed that the Kutta condition is valid for flexible propeller calculations.

3.3.6. The Wake Geometry of Flexible Propellers

In case of a vibrating blade, the wake geometry will follow the blade movements and therefore it has to be explained how the wake geometry has been modelled in the unsteady flexible propeller BEM model.

The effect of the wake geometry of a plunging airfoil for reduced frequencies up to 8 has been studied in the past [25]. Two different wake models were investigated. A free wake model in which the vortices shed from the trailing edge move in accordance with the induced velocities of all other vortices, in addition to the free-stream velocity and a prescribed wake model in which case the vortices are convected downstream with the free-stream velocity. The results show that the wake model has little effect on the forces generated by the airfoil. Therefore, in the unsteady flexible propeller BEM model a prescribed wake geometry is used. The wake geometry depends on the blade geometry, which means that, for every calculation step in which the blade influence coefficients are recalculated, the prescribed wake geometry is redefined and the wake influence coefficients are recomputed as well.

4. Propeller Fluid Added Mass and Hydrodynamic Damping

4.1. Decomposition of Total Pressure Field

For the derivation of the fluid added mass and hydrodynamic damping matrices, the total disturbance potential, Φ, is decomposed in two parts. Without any assumption, Φ can be decomposed into a disturbance potential φ due to only the vibration velocities of the blade in the non-uniform wakefield and a disturbance potential ϕ for the flexible propeller in the non-uniform wakefield excluding the blade vibration velocity contribution,

$$\Phi = \phi + \varphi. \tag{28}$$

Note that this decomposition of the total disturbance potential looks similar to what has been presented in [7], but differs on the following aspects. In [7], φ is the disturbance potential of the flexible blade in a uniform wakefield, here φ is the disturbance potential due to only the vibration velocities of the blade in the non-uniform wakefield. Here, ϕ denotes the remaining part of the disturbance potential, including the disturbance potential due to the rigid blades in a non-uniform wakefield and the disturbance potential due to the blade deformations in that wakefield. In [7], formally, the later part is included in φ, but has been neglected by assuming small blade deformations, which is expressed in the kinematic boundary condition for φ.

With $\mathbf{v}_0$ as inflow velocity, the total velocity $\mathbf{v}$ is equal to

$$\mathbf{v} = \mathbf{v}_0 + \nabla\phi + \nabla\varphi. \tag{29}$$

ϕ and φ can both be solved with the hydrodynamic method described before. The kinematic boundary condition for ϕ is the impermeability condition as given in Equation (10), but for a deforming blade,

$$\nabla\phi \cdot \mathbf{n}(\mathbf{x},\delta) = -\mathbf{v}_0 \cdot \mathbf{n}(\mathbf{x},\delta), \tag{30}$$

where the surface normal vector is also a function of blade deformation, δ, than only of $\mathbf{x}$. The kinematic boundary condition for φ is

$$\frac{\partial\varphi}{\partial n} = \frac{\partial\delta}{\partial t} \cdot \mathbf{n}(\mathbf{x},\delta). \tag{31}$$

This boundary condition means that the flow on the deforming blade should have the same velocity as the vibration velocity of the deformed blade itself. The term $\frac{\partial\delta}{\partial t} \cdot \mathbf{n}(\mathbf{x},\delta)$ denotes the normal component of the blade vibration velocity.

The pressures on the blade follow from the unsteady Bernoulli equation,

$$p = p_{ref} - \rho\left(\frac{\partial\phi}{\partial t} + \frac{\partial\varphi}{\partial t} + \frac{1}{2}|\mathbf{v}|^2 - \frac{1}{2}|\mathbf{v}_0|^2\right). \tag{32}$$

A pressure contribution due to vibration velocities only can be obtained by decomposing p in p_ϕ and p_φ, where p_φ denotes the pressure contribution due to blade vibrations and p_ϕ is the remaining force contribution,

$$\begin{aligned} p &= p_\phi + p_\varphi, \\ p_\phi &= p_{ref} - \rho\left(\frac{\partial\phi}{\partial t} + \mathbf{v}_0 \cdot \nabla\phi + \nabla\phi \cdot \nabla\varphi + \frac{1}{2}|\nabla\phi|^2\right), \\ p_\varphi &= \rho\left(-\frac{\partial\varphi}{\partial t} - \mathbf{v}_0 \cdot \nabla\varphi + \frac{1}{2}|\nabla\varphi|^2\right). \end{aligned} \tag{33}$$

For a total decomposition of the pressure contribution due to ϕ and φ, the term $\nabla\phi \cdot \nabla\varphi$ in Equation (33) has been neglected [7]. In addition, the second order term $\frac{1}{2}|\nabla\varphi|^2$ is excluded by assuming that φ is small compared to ϕ [7].

4.2. Fluid Added Mass and Hydrodynamic Damping Matrices

To obtain closed form expressions for added mass and hydrodynamic damping, the assumptions regarding the pressure contributions of $\nabla\phi \cdot \nabla\varphi$ and $\frac{1}{2}|\nabla\varphi|^2$ are adopted, which results in,

$$p_\phi = p_{ref} - \rho\left(\frac{\partial\phi}{\partial t} + \mathbf{v}_0 \cdot \nabla\phi + \frac{1}{2}|\nabla\phi|^2\right), \tag{34}$$

$$p_\varphi = \rho\left(-\frac{\partial\varphi}{\partial t} - \mathbf{v}_0 \cdot \nabla\varphi\right). \tag{35}$$

The pressure p_ϕ can be obtained by solving ϕ with the hydrodynamic method described in Section 3.3.4. The pressure contribution p_φ can be related to blade vibration velocities and accelerations. Similar to a rigid blade problem, the vibration velocity induced potential φ can be obtained by applying Green's identities. Hence, in discrete form,

$$\{\varphi\} = \begin{cases} [\mathbf{B}^*]^{-1}[\mathbf{S}^*]\left\{\dfrac{\partial\varphi}{\partial n}\right\}, & \text{with a constant strength vortex wake sheet,} \\ [\mathbf{D}^*]^{-1}[\mathbf{S}^*]\left\{\dfrac{\partial\varphi}{\partial n}\right\}, & \text{without a vortex wake sheet.} \end{cases} \tag{36}$$

The dipole and source influence coefficient matrices $[\mathbf{B}^*]$, $[\mathbf{D}^*]$ and $[\mathbf{S}^*]$ are taken time-invariant, assuming that the change in influence coefficients with time is negligible, which is assumed valid for small blade deformations [7]. By using the matrices with summed key- and symmetry influence coefficients, the implicit assumption is that the blade vibration problem is symmetric. The two cases of a constant strength vortex wake sheet and without a vortex wake sheet (i.e., not satisfying the Kutta condition) are introduced here because it will be shown in Section 5.1 that these two cases provide the upper and lower limit of the frequency dependent fluid added mass and hydrodynamic damping.

According to Equation (31), $\frac{\partial\varphi}{\partial n}$ is related to the blade vibration velocities. When $\dot{\mathbf{u}}$ are the blade nodal velocities obtained from the structural calculation, the kinematic boundary condition for φ can be written as

$$\left\{\frac{\partial\varphi}{\partial n}\right\} = [\mathbf{T}][\mathbf{N}]\{\dot{\mathbf{u}}\}, \tag{37}$$

where $[\mathbf{N}]$ is the transformation matrix to obtain the surface normal velocities from the 3D nodal velocities. According to Equation (31), this transformation matrix is a function of the blade deformation. To obtain closed form expressions for added mass and hydrodynamic damping, this has to be neglected, which is assumed valid for the case of small blade deformations [7]. This means that, for the closed form expressions, the kinematic boundary condition as given in [7] is applied

$$\frac{\partial\varphi}{\partial n} = \frac{\partial\delta}{\partial t} \cdot \mathbf{n}(\mathbf{x}), \tag{38}$$

where $\mathbf{n}$ is for the undeformed blade geometry. The transformation matrix $[\mathbf{T}]$ relates the normal velocities at the nodes to the collocation points in the BEM analysis. In case of a constant strength vortex wake sheet, the pressures, $p_{\mathbf{v}_0\nabla\varphi}$, due to $\mathbf{v}_0\nabla\varphi$ at the BEM collocation points are given by

$$\{p_{\mathbf{v}_0\nabla\varphi}\} = -\rho[\mathbf{V}_0][\boldsymbol{\nabla}][\mathbf{B}^*]^{-1}[\mathbf{S}^*][\mathbf{T}][\mathbf{N}]\{\dot{\mathbf{u}}\}, \tag{39}$$

where $[\mathbf{V}_0]$ is the matrix with inflow velocities and $[\boldsymbol{\nabla}]$ the discrete form of the gradient operator. Then, the forces at the BEM collocation points, $f_{\mathbf{v}_0\nabla\varphi}$, are obtained after multiplying the pressures with the BEM panel areas:

$$\{f_{\mathbf{v}_0\nabla\varphi}\} = -\rho\,[\mathbf{A}]\,[\mathbf{V}_0]\,[\boldsymbol{\nabla}]\,[\mathbf{B}^*]^{-1}\,[\mathbf{S}^*]\,[\mathbf{T}]\,[\mathbf{N}]\,\{\dot{\mathbf{u}}\}\,, \tag{40}$$

where $[\mathbf{A}]$ is the matrix with the BEM panel areas on its diagonal. The 3D force components at the nodes are equal to

$$\{\mathbf{f}_{\mathbf{v}_0\nabla\varphi}\} = -\rho\,[\mathbf{T}]^T\,[\mathbf{N}]^T\,[\mathbf{A}]\,[\mathbf{V}_0]\,[\boldsymbol{\nabla}]\,[\mathbf{B}^*]^{-1}\,[\mathbf{S}^*]\,[\mathbf{T}]\,[\mathbf{N}]\,\{\dot{\mathbf{u}}\} \tag{41}$$

from which it can be concluded that the hydrodynamic damping matrix, $[\mathbf{C}_h]$, is equal to

$$[\mathbf{C}_h] = -\rho\,[\mathbf{T}]^T\,[\mathbf{N}]^T\,[\mathbf{A}]\,[\mathbf{V}_0]\,[\boldsymbol{\nabla}]\,[\mathbf{B}^*]^{-1}\,[\mathbf{S}^*]\,[\mathbf{T}]\,[\mathbf{N}]\,. \tag{42}$$

Due to $\mathbf{v}_0$, the hydrodynamic damping matrix depends on the inflow velocity. Hence, in case of a propeller operating in a non-uniform flow, the hydrodynamic damping matrix will change in accordance with the blade rotation angle. To arrive at a constant hydrodynamic damping matrix, the authors propose to decompose the non-uniform propeller inflow velocity into a circumferentially averaged constant velocity field in which the free-stream velocity only depends on the radial position and a disturbing flow field, denoted with $\bar{\mathbf{v}}_0$ and $\tilde{\mathbf{v}}_0$. Then, the total inflow velocity is

$$\mathbf{v}_0\,(\mathbf{x},t) = \bar{\mathbf{v}}_0\,(\mathbf{x}) + \tilde{\mathbf{v}}_0\,(\mathbf{x},t)\,. \tag{43}$$

The constant hydrodynamic damping matrix is then equal to

$$[\mathbf{C}_h] = -\rho\,[\mathbf{T}]^T\,[\mathbf{N}]^T\,[\mathbf{A}]\,[\bar{\mathbf{V}}_0]\,[\boldsymbol{\nabla}]\,[\mathbf{B}^*]^{-1}\,[\mathbf{S}^*]\,[\mathbf{T}]\,[\mathbf{N}] \tag{44}$$

in case of a constant strength vortex wake sheet. Without a vortex wake sheet, the constant hydrodynamic damping matrix is

$$[\mathbf{C}_h] = -\rho\,[\mathbf{T}]^T\,[\mathbf{N}]^T\,[\mathbf{A}]\,[\bar{\mathbf{V}}_0]\,[\boldsymbol{\nabla}]\,[\mathbf{D}^*]^{-1}\,[\mathbf{S}^*]\,[\mathbf{T}]\,[\mathbf{N}]\,. \tag{45}$$

In a similar way, the fluid added mass matrix can be derived. Analogous to Equation (37), it can be written

$$\left\{\frac{\partial^2\varphi}{\partial t\partial n}\right\} = [\mathbf{T}]\,[\mathbf{N}]\,\{\ddot{\mathbf{u}}\}\,. \tag{46}$$

In case of a constant strength vortex wake sheet, the pressures due to $\frac{\partial\varphi}{\partial t}$, $p_{\frac{\partial\varphi}{\partial t}}$, are given by

$$\left\{p_{\frac{\partial\varphi}{\partial t}}\right\} = -\rho\,[\mathbf{B}^*]^{-1}\,[\mathbf{S}^*]\,[\mathbf{T}]\,[\mathbf{N}]\,\{\ddot{\mathbf{u}}\}\,. \tag{47}$$

Then, the forces at the BEM collocation points, $f_{\frac{\partial\varphi}{\partial t}}$, are

$$\left\{f_{\frac{\partial\varphi}{\partial t}}\right\} = -\rho\,[\mathbf{A}]\,[\mathbf{B}^*]^{-1}\,[\mathbf{S}^*]\,[\mathbf{T}]\,[\mathbf{N}]\,\{\ddot{\mathbf{u}}\}\,. \tag{48}$$

The 3D force components at the nodes are equal to

$$\left\{\mathbf{f}_{\frac{\partial\varphi}{\partial t}}\right\} = -\rho\,[\mathbf{T}]^T\,[\mathbf{N}]^T\,[\mathbf{A}]\,[\mathbf{B}^*]^{-1}\,[\mathbf{S}^*]\,[\mathbf{T}]\,[\mathbf{N}]\,\{\ddot{\mathbf{u}}\} \tag{49}$$

from which it can be concluded that the fluid added mass matrix, $[\mathbf{M}_h]$, is equal to

$$[\mathbf{M}_h] = -\rho\,[\mathbf{T}]^T\,[\mathbf{N}]^T\,[\mathbf{A}]\,[\mathbf{B}^*]^{-1}\,[\mathbf{S}^*]\,[\mathbf{T}]\,[\mathbf{N}] \tag{50}$$

in case of a constant strength vortex wake sheet. Without a vortex wake sheet, the fluid added mass matrix is

$$[\mathbf{M}_h] = -\rho\,[\mathbf{T}]^T\,[\mathbf{N}]^T\,[\mathbf{A}]\,[\mathbf{D}^*]^{-1}\,[\mathbf{S}^*]\,[\mathbf{T}]\,[\mathbf{N}]\,. \tag{51}$$

In the derivation of the closed form expressions for added mass and hydrodynamic damping, the following assumptions have been made:

(1) $\frac{1}{2}\,|\nabla\varphi|^2$ and $\nabla\phi\cdot\nabla\varphi$ are negligible.
(2) The influence coefficient matrices can be taken time-invariant.
(3) The summed key- and symmetry influence coefficients matrices can be used.
(4) The blade deformation can be neglected in the kinematic boundary condition.
(5) The fluctuating part of the hydrodynamic damping in case of an non-uniform wakefield is negligible.

It will be shown in the next section that the first assumption is only valid for high reduced frequencies for which conditions the hydrodynamic damping force is small anyway and the added mass force dominates. Assumptions 2 and 5 will be evaluated in Section 6 by comparing results of fully coupled FSI analyses of the Seiun–Maru propeller obtained from different BEM modelling approaches.

4.3. Fluid Added Mass Validation

In order to validate the fluid added mass matrix calculation, the natural frequencies of the four Boswell propellers [26] have been calculated and compared to results as presented in [7]. The four propellers were designed to study the influence of the skew angle on propeller performance and cavitation behaviour and have skew angles of 0°, 36°, 72° and 108°, identified by number 4381, 4382, 4383 and 4384, respectively. The natural frequencies of the propellers have been calculated with the commercial FEM package MSC MARC/Mentat using the following material properties: fluid density 1000 kgm^{-3}, blade material density 2800 kgm^{-3}, Young's modulus 75 GPa and Poisson ratio 0.33. The FEM models consist of one propeller blade without the hub part. The stiffness contribution of the hub has been modelled by a full clamping of the propeller blade at the blade–hub interface. The models were discretised by quadratic solid elements using a $29 \times 30 \times 4$ element distribution, meaning that 29, 30 and 4 elements are distributed in chord-wise, radial and through-thickness direction, respectively [13,27]. The added mass matrices have been computed from Equation (51) and diagonalised with a Hinton–Rock–Zienkiewicz lumping technique, which means that the diagonal entries of the full added mass matrix are scaled with the ratio of the sum of the entries that contribute to the motion in the same direction over the sum of the diagonal entries that contribute to the motion in that direction [7].

Figure 5 shows the dry and wet natural frequencies for the four propeller blades from Young [7] and present work. By comparing the results, it can be concluded that natural frequencies from present work are consistent with the results as presented by Young [7]. The dry natural frequencies of [7] and present work are close together. The biggest differences are in the wet natural frequencies, probably due to a slightly different added mass calculation—for instance, by calculating the added mass from the influence coefficients of one blade only instead of including the influence coefficients of the symmetry blades in the added mass calculation.

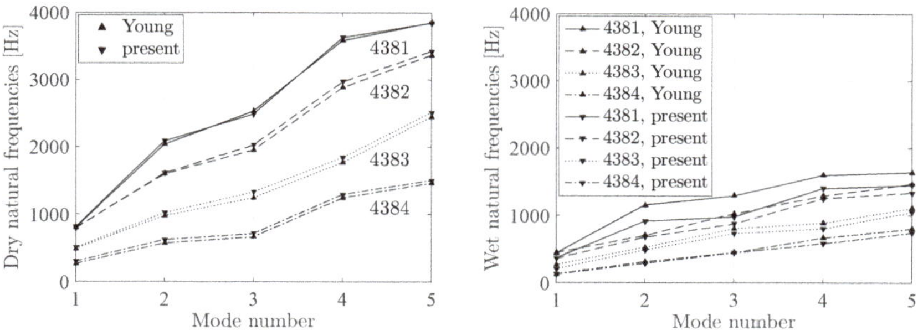

Figure 5. Comparison of dry (**left**) and wet (**right**) natural frequencies of the Boswell propeller blades from Young [7] and present work.

5. Hydrodynamic Loads on a Plunging Hydrofoil

In this section, the hydrodynamic loads on a plunging hydrofoil are investigated in order to evaluate the modelling of fluid added mass and hydrodynamic damping with the closed form expressions. The problem considered here concerns the hydrodynamic loads on a prismatic hydrofoil with a span of 20 m, chord of 1 m, a NACA 0012 cross section profile and a zero angle of attack for a prescribed sinusoidal plunge motion in water (density ρ is 1000 kg/m^{-3}). The flow velocities, $\mathbf{v}_0$, in (m/s) are characterized by the following three velocity components:

$$\mathbf{v}_0 = \begin{bmatrix} v_x & v_y & v_z \end{bmatrix}^T = \begin{bmatrix} 1 & 0.1 \cdot sin(\omega t) & 0 \end{bmatrix}^T, \tag{52}$$

where v_x and v_z are the chord- and spanwise velocity components, respectively. The plunge velocity is denoted by v_y and ω is the plunge frequency.

5.1. Fluid Added Mass and Hydrodynamic Damping of a Plunging Hydrofoil

The fluid added mass and hydrodynamic damping of the plunging hydrofoil have been obtained in two different ways. The first approach is from the closed form expressions for fluid added mass and hydrodynamic damping (Equations (44), (45) and (50), (51)) and taking the sum of matrix elements that contribute to the plunge motion. The second approach is by obtaining the hydrodynamic loads on a plunging hydrofoil from an unsteady PROCAL calculation.

The values obtained from the closed form expressions for the fluid added mass and hydrodynamic damping depend on the wake model. The two limit cases are a constant strength vortex wake sheet like in steady flow calculations and without a vortex wake sheet. Without a vortex wake sheet, the fluid added mass of the hydrofoil in water as obtained from Equation (51) is approximately 16 $\times$ 10^3 kg, (note that this corresponds to the added mass of flat plat with length 20 m and width 1 m). For that case, the hydrodynamic damping obtained with Equation (45) is zero, since without a vortex wake sheet the circulation is zero.

By satisfying the Kutta condition and assuming a constant strength vortex wake sheet, the added mass of the hydrofoil calculated from Equation (50) is much bigger: 46 $\times$ 10^3 kg. A hydrodynamic damping of 52 $\times$ 10^3 kg/s has been obtained from Equation (44) in that case. As a sanity check, this value has been compared to the lift force of a NACA 0012 profile, for instance presented in [28]. For the maximum plunge velocity of 0.1 m/s, the angle of attack is approximately 0.1 rad. For this small angle, the lift force is approximately equal to the force in plunge direction, which is 0.1 $\times$ 52 $\times$ 10^3 = 52 $\times$ 10^2 N; then, the lift coefficient is equal to 0.52, which corresponds to [28].

In the second approach, the prescribed plunge velocity is imposed on a BEM model of the hydrofoil and the forces are calculated with an unsteady PROCAL calculation. The hydrodynamic forces in plunge direction are subdivided in a circulatory and non-circulatory part. The non-circulatory part is the force contribution from the acceleration potential in the Bernoulli equation and is due to the body acceleration. The circulatory part is the remaining force contribution and contains the hydrodynamic damping part.

The non-circulatory plunge force is subdivided into two force contributions, one 90° out of phase with the body acceleration and the other in antiphase with the body accelerations. The added mass is obtained by dividing the latter part by the body acceleration. In the same way, the hydrodynamic damping is determined: the hydrodynamic damping is the circulatory plunge force in antiphase with the body velocity, divided by the body velocity.

Figure 6 shows the results obtained for the fluid added mass and hydrodynamic damping. The results show that fluid added mass and damping depend on the reduced frequency of the plunging motion. The low frequency limits are obtained for a constant strength vortex wake sheet (Equations (44) and (50)). The high frequency limits are obtained from the closed form expressions without taking into account the vortex wake sheet (Equations (45) and(51)). This shows that the assumption of a constant added mass [29] and hydrodynamic damping is only valid for sufficiently low or high reduced frequencies.

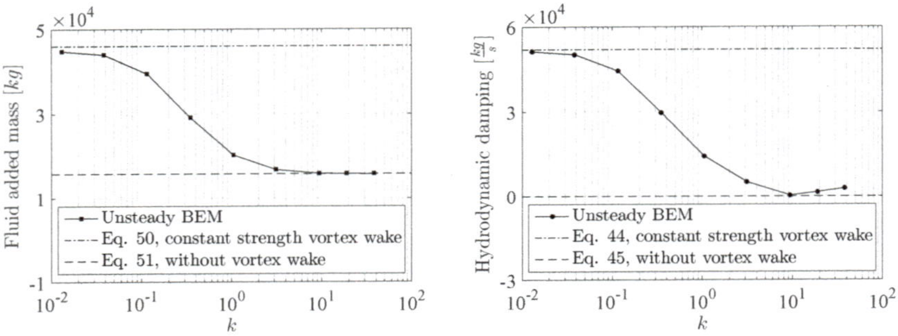

Figure 6. Added mass and hydrodynamic damping of the hydrofoil for different plunge frequencies.

5.2. Circulatory and Non-Circulatory Forces on a Plunging Hydrofoil

Figure 7 shows for which reduced frequencies the circulatory and non-circulatory forces in plunge direction can be correctly predicted with the closed form expressions for fluid added mass and hydrodynamic damping. The left graph of Figure 7 presents the amplitude of the non-circulatory force in plunge direction obtained with the unsteady BEM as a function of the reduced frequency, together with the force amplitudes which are obtained from the high and low frequency limit of the fluid added mass force. The right graph shows something similar but then for the circulatory force in plunge direction. Figure 8 presents the phase shift between the circulatory plunge force and the body velocity, and non-circulatory plunge force and body acceleration. The most important results of Figures 7 and 8 can be summarized as follows:

- For small reduced frequency, the amplitude of the circulatory plunge force is underestimated with the low frequency limit of the hydrodynamic damping force as obtained from Equation (44). This is a result of neglecting the pressure term due to $\frac{1}{2}\nabla\varphi^2$ in the derivation of the closed form expressions. However, the circulatory plunge force is well in antiphase with the body velocity.
- For small reduced frequency, the amplitude of the non-circulatory plunge force agrees well with the low frequency limit of the added mass force as obtained from Equation (50). However, the non-circulatory plunge force is not perfectly in antiphase with the body acceleration. This phase lag is due to unsteady wake effects.

- For high reduced frequency, the amplitude of the non-circulatory plunge force agrees well with the high frequency limit of the added mass force as obtained from Equation (51). Furthermore, the non-circulatory plunge force is well in antiphase with the body acceleration.
- For high reduced frequency, the amplitude of the circulatory plunge force approaches the high frequency limit of the hydrodynamic damping force as obtained from Equations (45).

This means that, only for high reduced frequencies ($k > 4$), the plunge forces can be correctly estimated with the closed form expressions. As revealed in Section 2.2, the reduced frequency for propeller blade vibrations excited by the first shaft harmonic is around 0.5. Hence, the question that remains is how accurately the hydro-elastic response of flexible propellers is predicted by modelling the fluid added mass and hydrodynamic damping effects with the closed form expressions. This will be discussed further in Section 6.

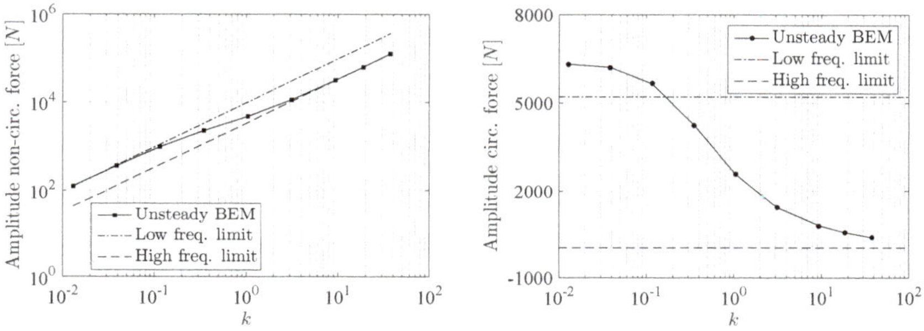

Figure 7. Amplitude of non-circulatory and circulatory forces in plunge direction for different plunge frequencies.

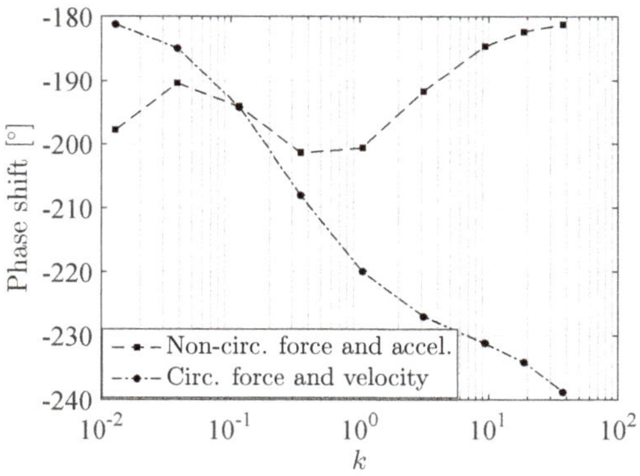

Figure 8. Phase shift between non-circulatory plunge force relative to acceleration and circulatory plunge force relative to velocity for different plunge frequencies.

6. Steady and Unsteady Flexible Propeller Calculations with Different BEM-FEM Coupled Approaches

The following questions regarding the FSI modelling of flexible propellers with a BEM-FEM coupled approach could be raised based on what has been discussed:

(1) How important is the re-calculation of key and symmetry body and wake influence coefficients in the BEM modelling?

(2) Is it valid to use for the symmetry surface influence coefficients at time step k_{isym} the symmetry influence coefficients of time step k, in order to reduce computer memory?
(3) Can the hydro-elastic response of flexible propellers be accurately predicted by modelling the fluid added mass and hydrodynamic damping effects with the closed form expressions and what has been taken for the the fluid added mass and hydrodynamic damping matrix, i.e., the low frequency limit, high frequency limit or something in between?
(4) As the structural response is stiffness dominated, would a quasi-static FEM calculation be sufficient?

In order to give answers to these modelling questions, different BEM-FEM coupling approaches were implemented and the hydro-elastic responses of quasi-isotropic glass–epoxy and carbon–epoxy Seiun–Maru propellers in uniform and non-uniform flows have been compared. In order to judge from these results which modelling approach is acceptable and which is not, criteria on accuracy levels have to be given. The purpose of the BEM-FEM coupled calculations is to predict correctly the hydro-elastic behaviour of flexible propellers. A measure for the hydro-elastic behaviour from a propeller performance perspective is the blade thrust change due to flexibility. An accuracy level of 10% on the blade thrust change is considered as acceptable. This seems significant but means that an accuracy of 1% in blade thrust is required in case the blade thrust change is 10% of the blade thrust itself. Additionally, in order to assess the structural response results, an accuracy level of 5% is considered as acceptable.

6.1. BEM Models for Steady and Unsteady Flexible Propeller Calculations

In contrast to standard PROCAL applications, for flexible propellers, the blades deform and for unsteady cases induce fluid velocities and accelerations due to blade vibrations as well. As a result of blade deformations:

- panel normal vectors become time-dependent, which is reflected in the panel source strengths.
- pressures have to be evaluated from the computed velocity potentials on a modified grid.
- influence coefficients of key- and symmetry blades and wake surfaces become time-dependent.

The blade vibration velocity and acceleration hydrodynamic effects can be included in the FSI calculations in two ways: either implicitly by imposing the kinematic boundary condition of Equation (31) in the PROCAL calculations or explicitly by including the fluid added mass and hydrodynamic damping effects in the FSI calculations with the closed form expressions of Equations (44), (45) and (50), (51). Based on this, the following three BEM models have been proposed:

- Fully geometry dependent BEM model with fluid added mass and hydrodynamic damping effects implicitly included, denoted by FGDI-BEM.
- Fully geometry dependent BEM model with fluid added mass and hydrodynamic damping effects explicitly included, denoted by FGDE-BEM.
- Partially geometry dependent BEM model with fluid added mass and hydrodynamic damping effects implicitly included, denoted by PGDI-BEM.

Uniform flow calculations can be considered as a special case of an unsteady calculation without time effects, blade velocities and accelerations, which means the implicit or explicit modelling of fluid added mass and hydrodynamic damping is irrelevant and the different BEM models are denoted by FGD-BEM and PGD-BEM.

6.1.1. FGDI-BEM Model

In the FGDI-BEM model, all the geometry dependent items in a PROCAL calculation, including source strengths, blade and wake influence coefficients, are recalculated based on the deformed blade geometry and the pressures are evaluated on the modified BEM model. Note that, for the symmetry surface influence coefficients of time step k_{isym}, the symmetry influence coefficients of time step k have been used in order to save computer memory. The blade vibration velocity and acceleration effects are implicitly included in the BEM calculation by imposing Equation (31) as an additional boundary condition supplementary to the boundary condition of Equation (30).

6.1.2. FGDE-BEM Model

The FGDE-BEM model is the same as the FGDI-BEM model except that PROCAL solves only the disturbance velocity potential without the contribution of blade vibration velocities, ϕ, from the kinematic boundary condition of Equation (30). The blade vibration velocity and acceleration effects are included in the FSI analyses by means of the closed form expressions for fluid added mass and hydrodynamic damping. The main advantage of this approach is that the fluid added mass and hydrodynamic damping can be explicitly included in the structural computation, which stabilizes the FSI solution and reduces CPU time [30].

The question that remains is which fluid added mass and hydrodynamic damping matrix have to be taken. The reduced frequency for propeller blade vibration flows is around 0.5 for an oscillation frequency equal to the shaft rotation rate. For higher harmonics, the reduced frequency is obviously a multiple of the fundamental reduced frequency. From the left graph of Figure 6, it can be concluded that, for such high reduced frequencies, the fluid added mass is approaching the high frequency limit. Therefore, the added mass matrix has been computed without taking the vortex wake sheet into account, i.e., Equation (51).

Based on the right graph of Figure 6, 50% of the low frequency limit of the hydrodynamic damping term, i.e., Equation (44), has been taken for the hydrodynamic damping matrix.

6.1.3. PGDI-BEM Model

In the PGDI-BEM model, blade vibration and velocity effects have been incorporated in the same way as in the FGDI-BEM model. The difference is to which level the blade geometry update is included in the BEM calculations. In the PGDI-BEM model for each time step, source strengths are recalculated from the deformed blade geometry and pressures and forces are evaluated on the modified BEM model. However, the blade and wake influence coefficients are kept constant throughout the analyses. This can reduce the CPU time significantly. The blade and wake influence coefficients used in the unsteady calculations with the PGDI-BEM model are obtained for the averaged deformed blade geometry. The average deformed blade geometry is calculated from a steady BEM-FEM computation with the circumferentially averaged wakefield $\bar{\mathbf{v}}_0$ for the inflow velocities. The steady calculations with the PGD-BEM model are conducted with the hydrodynamic influence coefficients of the undeformed propeller blade geometry.

6.2. *Steady and Unsteady BEM-FEM Coupling*

In [13], the BEM-FEM coupling for the steady analyses has been presented, including a validation study. The coupling is based on a partitioned approach in which fluid and structure problem are solved in separate codes. For the fluid part, any of the PROCAL BEM models as described above has been used. For the structural modelling, the FEM package MSC MARC/Mentat has been used. The blade FEM modelling has been extensively described in [13,27]. In summary, the FEM models consist of one propeller blade without the hub part. The stiffness contribution of the hub has been modelled by a full clamping of the propeller blade at the blade–hub interface. The models are discretised by quadratic solid elements using a $29 \times 30 \times 4$ element distribution, meaning that 29, 30 and 4 elements are distributed in chord-wise, radial and through-thickness direction, respectively. The unsteady FEM calculations have been conducted in modal space and a model reduction is applied by using only the first 40 mode shapes.

Since a partitioned FSI approach has been adopted, coupling iterations between BEM and FEM solver are required to converge to the monolithic solution. For steady analyses, these coupling iterations converge in any case. However, without a dedicated coupling algorithm instabilities occur in unsteady analyses due to strong fluid added mass effects. Special attention has been paid to the coupling algorithm for the unsteady BEM-FEM analyses. A detailed description of this BEM-FEM coupling is out of the scope of this work and will be revealed in a future publication. In short, in the unsteady BEM-FEM coupling, the periodicity of the problems is utilized by coupling the BEM and FEM solver on period level instead per time step. This time periodic coupling [31] is the most obvious approach to couple PROCAL with the structural problem because PROCAL is completely periodical in nature. Think of the vortex wake shedding and the use of symmetry blades and therefore a non-periodic solution could not be treated in PROCAL. Furthermore, the time periodic approach allows for first letting PROCAL converge before new disturbances enter the equations. After all, the PROCAL computations itself are already iterative and, in strongly coupled FSI problems, it is advantageous to have the inner loop converge first. In order to stabilize the coupling iterations on period level, a Quasi-Newton inverse least squares partitioned solution method has been implemented [32].

The unsteady BEM-FEM coupling procedure has been verified with the FSI solutions of a pitching hydrofoil in water with a span of 1 m, chord of 0.1 m, a NACA 0009 cross section profile and a zero initial angle of attack subjected to a controlled external moment, M_0, equal to $34.9 sin(\omega t)$ Nm [33]. The inflow velocity, v_0, is equal to 5 m/s. For this problem, the equation for the pitch motion, θ, reads:

$$I_\theta \ddot{\theta} + C_\theta \dot{\theta} + K_\theta \theta = M_0 + M_{fluid}, \tag{53}$$

where I_θ is the structural moment of inertia equal to 1.429×10^{-3} kgm^2, C_θ is the structural damping parameter equal to 0.096 kgms^{-1}, K_θ is the structural torsional stiffness equal to 1000 Nm. M_{fluid} is the moment exerted by fluid forces. M_{fluid} has been subdivided in fluid added mass, fluid damping and fluid stiffness forces defined by the following approximation formulas [29]:

$$\begin{aligned}
M_{fluid} &= -M_f \ddot{\theta} - C_f \dot{\theta} - K_f \theta, \\
M_f &= \frac{\pi}{128} \rho c^4, \\
C_f &= \begin{cases} 0.105 k^{-0.4} \rho v_0 c^3, & \text{for k} \leq 4, \\ 0.010 k^{0.6} \rho v_0 c^3, & \text{for k} \geq 12, \end{cases} \\
K_f &= \begin{cases} (0.090k - 0.80)\, \rho v_0^2 c^2, & \text{for k} \leq 4, \\ (0.065k - 0.90)\, \rho v_0^2 c^2, & \text{for k} \geq 12. \end{cases}
\end{aligned} \tag{54}$$

The monolithic solution is obtained by moving the fluid forces from the right-hand-side of the equation of motion to left-hand-side and then the exact solution for θ can be computed. The coupling procedure solution is obtained by keeping the fluid forces on the right-hand-side as unknowns and iterations are required to converge to the monolithic solution. Figure 9 shows the monolithic solution and the time periodic coupling solution for two of the cases with fluid added mass to structural mass ratio of 1.72 as also presented in [33]. For both cases, the time period coupling procedure converges to the monolithic solution, for the $\omega = 4.83$ rad/s case in four coupling iterations and for the case with $\omega = 483$ rad/s in seven coupling iterations. This shows that the coupling procedure is appropriate for periodic FSI problems with strong fluid added mass effects, like in flexible propeller FSI analyses.

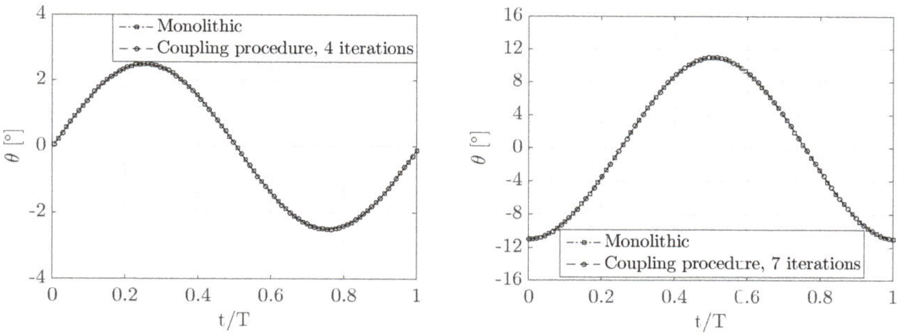

Figure 9. Comparison of monolithic and coupling procedure solution of the predicted pitch motion, θ, as a function of the normalized time, t/T, for $\omega = 4.83$ rad/s (**left**) and $\omega = 483$ rad/s (**right**) graph.

6.3. Steady Analyses with FGD-BEM Model and PGD-BEM Model

In order to obtain an answer to the first modelling question raised in the beginning of this section, calculations were performed with the FGD-BEM and PGD-BEM model on the Seiun–Maru propeller for various uniform flow conditions. All steady BEM-FEM coupled calculations are performed for a quasi-isotropic glass–epoxy material, with material properties as presented before. The advance ratio J, free stream velocity V, propeller rotation rate and blade thrust T_0 in undeformed configuration are given in Table 2. The rotation rate and advance speeds were selected in such a way that the undeformed propeller thrust is more or less the same for all the flow conditions and the power required for the advance ratio of 0.7 corresponds roughly to the maximum continuous rating power of the Seiun–Maru vessel [34].

Table 2. Flow conditions for steady flexible propeller calculations.

J	V (m/s)	(rpm)	T_0 (kN)
0.3	2.98	165	96.6
0.5	5.72	191	96.0
0.7	9.97	237	96.2
0.9	19.6	364	97.3

Figure 10 shows the results that are obtained with the FGD- and PGD-BEM model. In the left graph, the reduction in blade thrust due to blade flexibility calculated with the two models is given as a percentage of T_0, together with the percentage difference between the blade thrust reductions obtained with both models. It can be seen that with the PGD-BEM model the blade thrust reduction is smaller for all advance ratios. The relative difference between the blade thrust reductions obtained with both models reduces with increasing advance ratio. The explanation for this is that for increasing advance ratio, the relative contribution of the disturbance velocity potential, calculated with the PGD-BEM model in a simplified way, reduces.

The right graph shows the tip displacements obtained with both models. Since the blade thrust calculated with both models is close, the differences in tip displacements are relatively small as well. From the results, it can be concluded that without recalculation of the blade and wake influence coefficients, the inaccuracy in thrust change for the two lowest advance ratios is slightly larger than the 10%, but satisfies the 5% deformation inaccuracy. The PGD-BEM results for the two highest advance coefficients satisfy the accuracy criteria, which means that updating of the influence coefficients is not necessary for these conditions.

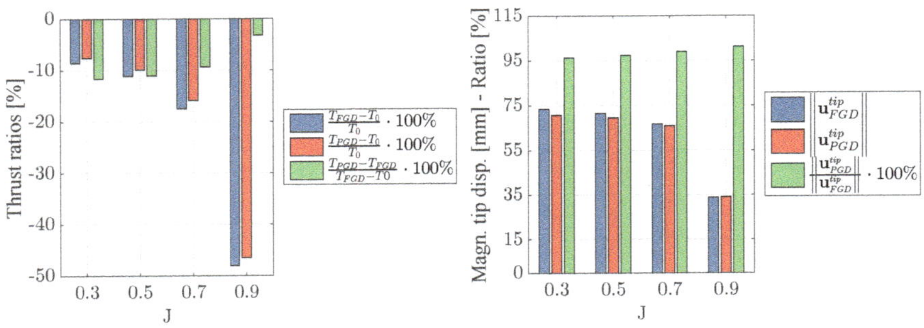

Figure 10. Blade thrust (**left**) and tip displacement (**right**) results obtained with the FGD- and PGD-BEM model for various advance ratios. The subscripts $_{FGD}$ and $_{PGD}$ indicate with which BEM model the results were obtained.

6.4. Unsteady Analyses

This section presents results obtained for various non-uniform flows, fully BEM-FEM coupled calculations on the Seiun–Maru propeller and its ship effective wakefield [35] as shown in Figure 11. All the calculations are performed for quasi-isotropic glass–epoxy or carbon–epoxy material, with material properties as presented before. All the calculations are performed for a ship speed of 21 knots and a rotation rate of 240 rpm, which is roughly the rotation rate resulting in the maximum continuous engine power of 7723 kW [34], according to the graph showing the power against propeller revolution rate as presented in [18].

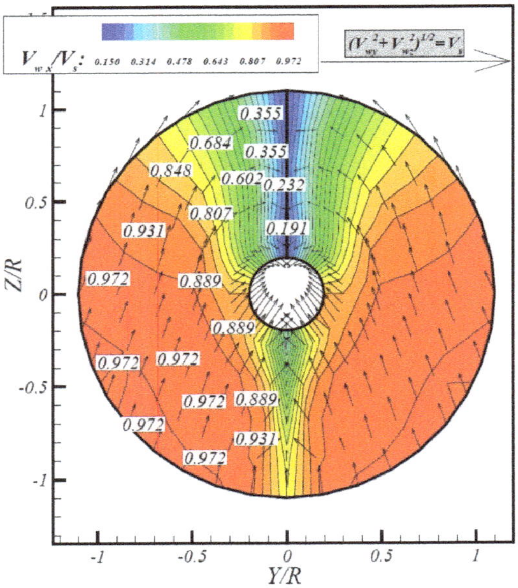

Figure 11. Seiun–Maru ship effective wakefield. (Image republished from [19] with permission of the author.)

First, several results obtained with the FGDE-BEM model are presented together with a result, which is obtained from a static calculation of the blade structural response. In addition, the contributions of fluid added mass and hydrodynamic damping forces are revealed. Secondly, results obtained with the FGDI-, FGDE- and PGDI-BEM model are compared, results obtained from a static blade structural response calculation are included in this comparison. Finally, the results of two different PGDI-BEM model calculations are compared.

6.4.1. Unsteady Analyses with FGDI-BEM, PGDI-BEM and FGDE-BEM Model

In this subsection, results obtained with the FGDI-BEM, PGDI-BEM and FGDE-BEM models have been compared in order to find answers to modelling question 2 and 3 and to answer modelling question 1 from an unsteady problem perspective.

Figures 12 and 13 show the blade thrust for one revolution calculated for the glass–epoxy and carbon–epoxy propeller with the three methods, together with the blade thrust obtained for an unsteady PROCAL analysis with a completely rigid propeller.

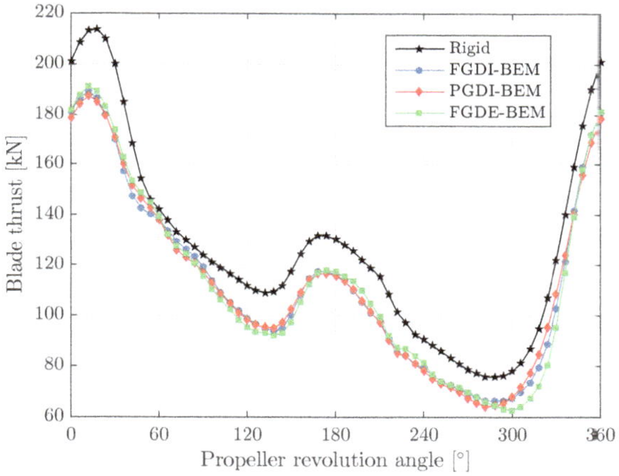

Figure 12. Blade thrust computed with different BEM-FEM couplings for the glass–epoxy Seiun-Maru propeller.

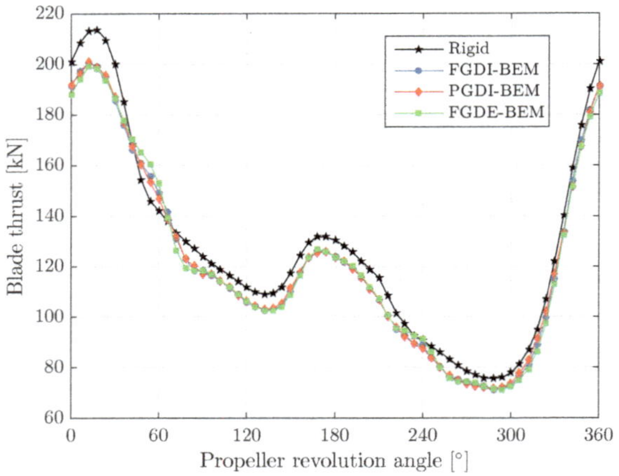

Figure 13. Blade thrust computed with different BEM-FEM couplings for the carbon–epoxy Seiun–Maru propeller.

Figures 14 and 15 show the modal participation factors for the first three dry blade modes as obtained from these analyses. Figure 16 summarizes the reduction of minimum, maximum and average blade thrust due to blade flexibility as obtained from these calculations. An interesting outcome of these calculations is that the reduction of the peak thrust due to blade flexibility is significantly larger than the reduction of the average thrust. This demonstrates the potential advantage of flexible propellers to improve cavitation inception speeds and cavitation noise.

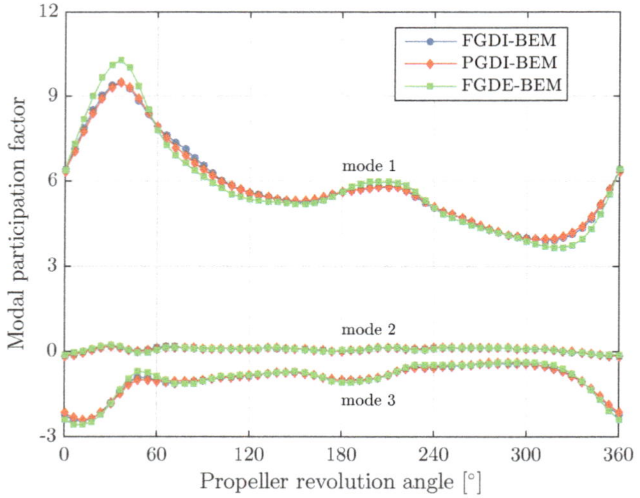

Figure 14. Modal participation factors for the first three modes computed with different BEM-FEM couplings for the glass–epoxy Seiun–Maru propeller.

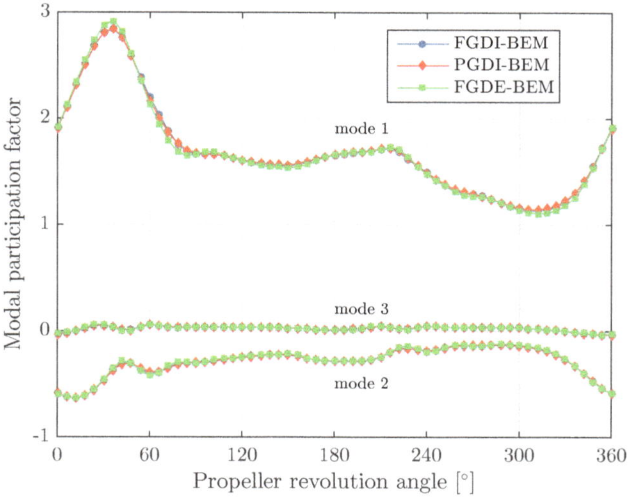

Figure 15. Modal participation factors for the first three modes computed with different BEM-FEM couplings for the carbon–epoxy Seiun–Maru propeller.

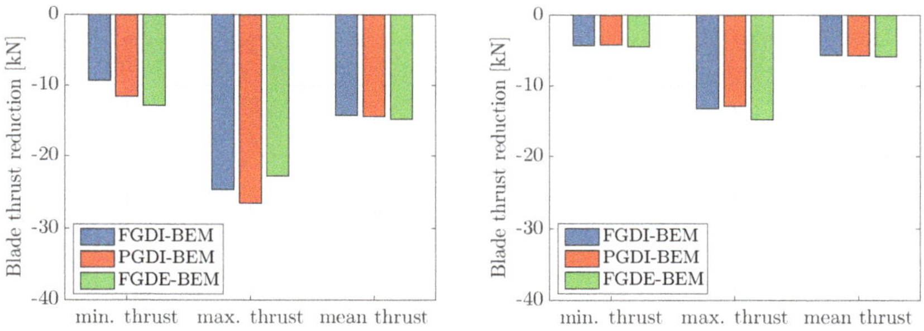

Figure 16. Thrust reduction for minimum, maximum and average blade thrust calculated for the glass–epoxy (**left graph**) and the carbon–epoxy (**right graph**) propeller as obtained from the BEM-FEM coupled calculations.

The FGDI-BEM model is the most extensive and complete BEM modelling of the flexible blades; therefore, all the other results will be compared to results obtained with this model. First, it can be seen that the differences between the results obtained with the FGDI-BEM and the PGDI-BEM model are small, meaning that the computationally intensive process of recalculating all the influence coefficients at every time step is relatively unimportant. This is in line with the results obtained with the FGD-BEM and PGD-BEM model for the steady cases. In answer to modelling question 1, it can be concluded that updating of the influence coefficients can be omitted for these unsteady flexible propeller calculations. Based on this conclusion, an answer can be given to the second modelling question as well. Since updating of the influence coefficients makes such a small contribution to the final results, it can be concluded that the symmetry surface influence coefficients of time step k instead of k_{isym} can be safely used for the symmetry surfaces because the symmetry surfaces are in the far-field with respect to the key blade and therefore the error in thrust reduction from this modelling choice will be much smaller than the modelling error introduced by not updating the key surface influence coefficients.

An answer to the third modelling question can be based on a comparison of the results obtained with the FGDE-BEM model to the results obtained with the FGDI-BEM calculations. It can be concluded that the hydro-elastic response prediction of the glass–epoxy flexible Seiun–Maru propeller with the FGDE-BEM model does not satisfy the accuracy criterion regarding the change in blade thrust. However, it can be seen that the hydro-elastic responses are fairly well predicted by modelling the fluid added mass and hydrodynamic damping effects with the closed form expressions, more specifically by using the high frequency limit for the fluid added mass matrix and for the hydrodynamic damping matrix by taking 50% of the low frequency limit. The main reason that the results obtained with the FGDE-BEM model are fairly close to results obtained with the FGDI-BEM model is that the structural response of flexible propellers is dominated by stiffness and therefore the consequences of modelling errors in the fluid added mass and hydrodynamic damping contributions are relatively small.

6.4.2. Unsteady Analyses with FGDE-BEM Model

The purpose of this subsection is to give an answer to the fourth modelling question. It will be shown whether a quasi-static approach for the blade structural response calculation gives acceptable results (i.e., inaccuracy in thrust change <10% and inaccuracy in structural response <5%) and what the contribution is of the fluid added mass and hydrodynamic damping forces. Therefore, several computations with the unsteady FGDE-BEM-FEM coupling were conducted. The FGDE-BEM model has been taken, since the fluid added mass and hydrodynamic damping forces are explicitly calculated and their contribution to the total force can be easily presented.

Figures 17 and 18 show the blade thrust for one revolution computed with four different FGDE-BEM-FEM coupled calculations for the glass–epoxy and carbon–epoxy propeller, including fluid added mass and hydrodynamic damping, without hydrodynamic damping, without fluid added mass and a quasi-static analysis of the structural response. In these figures, the force contributions due to fluid added mass and hydrodynamic damping are included as well.

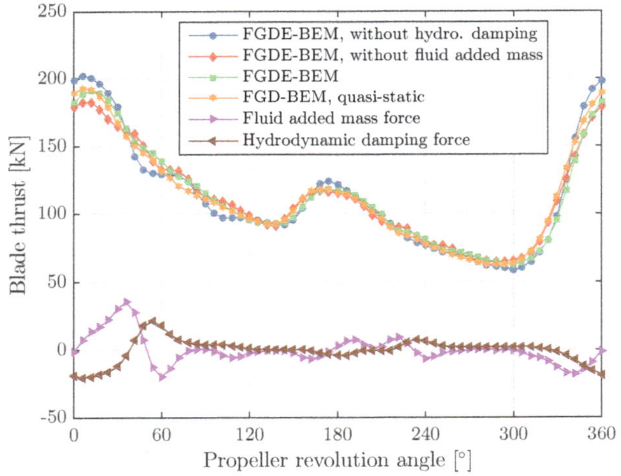

Figure 17. Blade thrust, fluid added mass and hydrodynamic damping force computed with the FGDE-BEM-FEM coupling with and without including the fluid added mass and hydrodynamic damping in the calculations, for the glass–epoxy Seiun–Maru propeller.

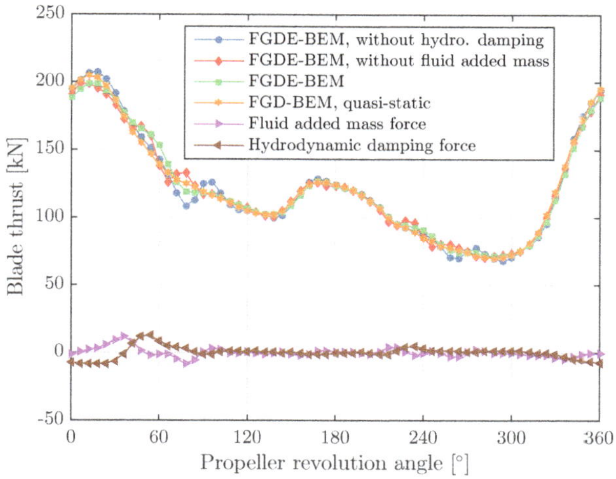

Figure 18. Blade thrust, fluid added mass and hydrodynamic damping force computed with the FGDE-BEM-FEM coupling with and without including the fluid added mass and hydrodynamic damping in the calculations, for the carbon–epoxy Seiun–Maru propeller.

Figures 19 and 20 show the modal participation factors for the first three dry blade modes as obtained from the three different analyses. Figure 21 summarizes the reduction of minimum, maximum and average blade thrust due to blade flexibility.

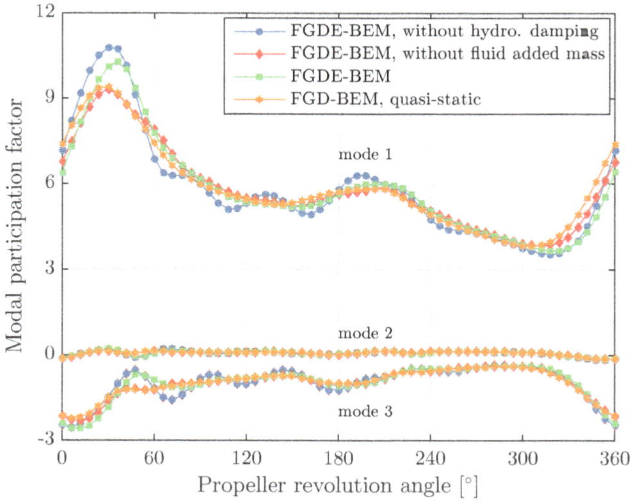

Figure 19. Modal participation for first three modes computed with the FGDE-BEM-FEM coupling with and without including the fluid added mass and hydrodynamic damping in the calculations, for the glass–epoxy Seiun–Maru propeller.

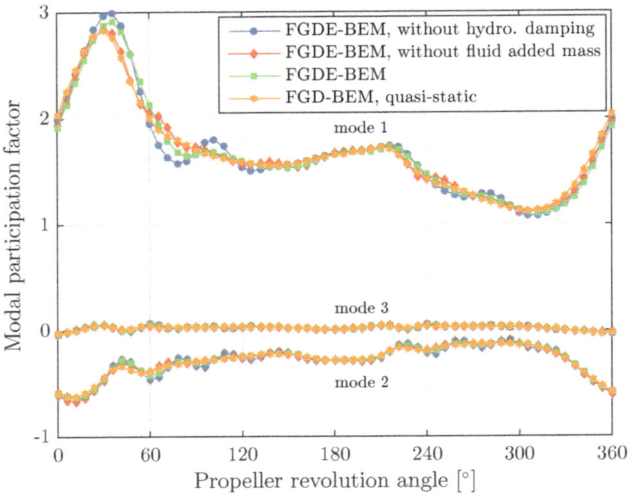

Figure 20. Modal participation for first three modes computed with the FGDE-BEM-FEM coupling with and without including the fluid added mass and hydrodynamic damping in the calculations, for the carbon–epoxy Seiun–Maru propeller.

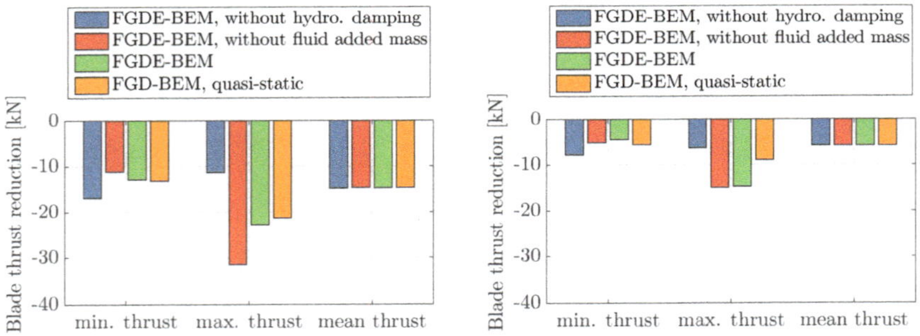

Figure 21. Thrust reduction for minimum, maximum and average blade thrust calculated for glass–epoxy (**left**) and carbon–epoxy (**right**) propeller as obtained from the various FGDE-BEM calculations.

These results show that a quasi-static FEM modelling cannot be used for an accurate prediction of the hydro-elastic response for these cases. This can be illustrated for instance with the blade thrust reduction results for the carbon–epoxy calculation. It can be expected that a quasi-static FEM approach will work the best for this case, since the structural frequency ratio for the carbon–epoxy propeller is significantly smaller than for the glass–epoxy propeller. However, a significant difference between reduction of the maximum thrust value calculated with the FGDE-BEM and the quasi-static approach is obtained for this case. Therefore, it is concluded that a quasi-static FEM modelling of flexible propellers should be applied with great care.

The importance of including fluid added mass and hydrodynamic damping in the calculations has been illustrated with two other calculations: one, BEM-FEM coupled calculation with the FGDE-BEM model without hydrodynamic damping included and one without fluid added mass included. These results show that, by neglecting fluid added mass or hydrodynamic damping, the average thrust corresponds very well with the average thrust obtained from a calculation in which both terms are included. However, significant differences can be seen in the calculated maximum blade thrust values and the shape of the blade thrust curves during a full revolution. The same conclusion can be drawn from the structural responses, as the differences in results obtained from the different calculations are too big to justify neglecting fluid added mass and/or hydrodynamic damping.

7. Conclusions

This work presents expressions to characterize the flow and structural response of flexible marine propellers. From these formulas, the conclusion can be drawn that the structural frequency ratio of flexible blades and the reduced frequency of vibrating blade flows is independent of the geometrical propeller scale.

The formulas were used to estimate the dry and wet fundamental blade frequencies of the highly skewed Seiun–Maru propeller for different blade materials. These frequencies were compared to results obtained from FEM calculations, from which can be concluded that reasonable results are obtained by using the estimation formulas. From the prediction of the structural frequency ratio for a quasi-isotropic glass–epoxy and carbon–epoxy Seiun–Maru propeller, it can be expected that the structural response is stiffness dominated. It can be discussed what the structural behaviour of flexible propellers is in general. It is the authors' opinion that the structural response of flexible propellers is dominated by the stiffness in general because the structural frequency ratio is independent of the geometrical propeller scale. Secondly, the Seiun–Maru propeller has relatively low blade frequencies due to the heavily skewed blade, and, therefore, in the case of other blade geometries, the structural frequency ratio is probably smaller.

A large part of this publication is used to explain possible BEM modelling approaches of flexible propellers. A derivation is presented on how fluid added mass and hydrodynamic damping matrices can be obtained from the hydrodynamic influence coefficients. To obtain closed form expressions for the added mass and hydrodynamic damping, several assumptions were made. The validity of using the closed form expressions for added mass and hydrodynamic damping forces has been studied by investigating the hydrodynamic loads on a plunging hydrofoil. The results of this study show that fluid added mass and hydrodynamic damping forces are frequency dependent due to flow memory effects as a consequence of the vortex wake sheet. The high frequency limit of the added mass and hydrodynamic damping forces is obtained without considering a vortex wake sheet. The low frequency limit is obtained for a constant strength vortex wake sheet. For sufficiently high reduced frequencies ($k > 4$), the added mass force converges to the high frequency limit and the hydrodynamic damping approaches the high frequency limit for the damping force, which is zero. From the results obtained for the hydrofoil only it cannot be concluded that the closed form expressions can be used for correctly modelling fluid added mass and hydrodynamic damping effects in case of flexible propellers. Three other modelling questions regarding the hydro-elastic response modelling were raised and answers to these questions have been obtained by comparing results of different BEM-FEM coupled analyses of the Seiun–Maru propeller in steady and unsteady flows. In order to assess the different results, criteria on accuracy levels has been adopted. As acceptable accuracy level for the blade thrust change 10% has been used. That seems significant but means, in the case of 10% blade thrust change due to flexibility, an accuracy of 1% on the blade thrust itself. For the structural response, an accuracy level of 5% in blade deformation response has been considered.

From the study with the different BEM-FEM calculations, the following conclusions can be drawn. Firstly, the recalculation of influence coefficients seems to be of minor importance and can be omitted to reduce CPU time.

Secondly, by modelling explicitly the blade vibration and acceleration hydrodynamic effects by using the high frequency limit for the fluid added mass term and by taking 50% of the low frequency limit for the hydrodynamic damping term, the hydro-elastic response was reasonably well predicted, although the glass–epoxy propeller case did not satisfy the accuracy criteria. The main reason for these reasonable results is that the structural response of flexible propellers is dominated by the stiffness and therefore the consequences of modelling errors in the fluid added mass and hydrodynamic damping contributions are relatively small. For cases with structural frequencies around 0.2, the application of this modelling approach needs to be considered carefully and is not recommended for cases with higher structural frequency ratios as considered in this work.

Regarding a quasi-static FEM modelling of the structural response of flexible propellers, it can be concluded that this is not recommended for two reasons. First of all, the fluid added mass and hydrodynamic damping contributions are relatively small but not negligible. Secondly, the structural response of flexible propellers is dominated by stiffness, but already the results for the carbon–epoxy propeller with a structural frequency ratio of around 0.15 show that dynamic effects have to be included. This would be even more important for cases were the structural frequency ratio is higher or the higher harmonics of the blade force are more significant.

Author Contributions: P.M. performed all the computations and he mainly wrote the paper. M.K. and H.d.B. contributed to the writing of the paper.

Acknowledgments: The authors gratefully acknowledge the Defence Materiel Organisation, Maritime Research Institute Netherlands, Wärtsilä Netherlands B.V., Solico B.V. and NWO, domain Toegepaste en Technische Wetenschappen (project number 13278) for their financial support. The authors are also grateful to the Cooperative Research Ships (CRS) for making their BEM software PROCAL available for this work. The authors also gratefully acknowledge Maritime Research Institute Netherlands for making their propeller geometry toolbox PROPART available for this work.

Conflicts of Interest: The authors declare no conflict of interest.

References

1. Mulcahy, N.; Prusty, B.; Gardiner, C. Hydroelastic tailoring of flexible composite propellers. *Ship Offshore Struct.* **2010**, *5*, 359–370. [CrossRef]
2. He, X.; Hong, Y.; Wang, R. Hydroelastic optimisation of a composite marine propeller in a non-uniform wake. *Ocean Eng.* **2012**, *39*, 14–23. [CrossRef]
3. Taketani, T.; Kimura, K.; Ando, S.; Yamamoto, K. Study on performance of a ship propeller using a composite material. In Proceedings of the Third International Symposium on Marine Propulsors, Launceston, Tasmania, Australia, 5–8 May 2013.
4. Solomon Raj, S.; Ravinder Reddy, P. Bend-twist coupling and its effect on cavitation inception of composite marine propeller. *Int. J. Mech. Eng. Technol.* **2014**, *5*, 306–314.
5. Kuo, J.; Vorus, W. Propeller blade dynamic stress. In Proceedings of the Tenth Ship Technology and Research (STAR) Symposium, Norfolk, VA, USA, 21–24 May 1985; pp. 39–69.
6. Georgiev, D.; Ikehata, M. Hydro-elastic effects on propeller blades in steady flow. *J. Soc. Naval Archit. Jpn.* **1998**, *184*, 1–14. [CrossRef]
7. Young, Y. Time-dependent hydro-elastic analysis of cavitating propulsors. *J. Fluids Struct.* **2007**, *23*, 269–295. [CrossRef]
8. Blasques, J.; Berggreen, C.; Andersen, P. Hydro-elastic analysis and optimization of a composite marine propeller. *Mar. Struct.* **2010**, *23*, 22–38. [CrossRef]
9. Ghassemi, H.; Ghassabzadeh, M.; Saryazdi, M. Influence of the skew angle on the hydro-elastic behaviour of a composite marine propeller. *J. Eng. Marit. Environ.* **2012**, *226*, 346–359.
10. Sun, H.; Xiong, Y. Fluid-structure interaction analysis of flexible marine propellers. *Appl. Mech. Mater.* **2012**, *226–228*, 479–482. [CrossRef]
11. Lee, H.; Song, M.; Suh, J.; Chang, B.J. Hydro-elastic analysis of marine propellers based on a BEM-FEM coupled FSI algorithm. *Int. J. Naval Archit. Ocean Eng.* **2014**, *6*, 562–577. [CrossRef]
12. Maljaars, P.; Dekker, J. Hydro-elastic analysis of flexible marine propellers. In *Maritime Technology and Engineering*; Guedes Soares, C., Santos, T., Eds.; CRC Press-Taylor & Francis Group, London, UK, 2014; pp. 705–715.
13. Maljaars, P.; Bronswijk, L.; Windt, J.; Grasso, N.; Kaminski, M. Experimental validation of fluid–structure interaction computations of flexible composite propellers in open water conditions using BEM-FEM and RANS-FEM methods. *J. Mar. Sci. Eng.* **2018**, *6*. [CrossRef]
14. Baker, G. Vibration patterns of propeller blades. *Trans. NECIES* **1940**, *57*, 1040–1041.

15. Fischer, R. Singing propellers—Solution and case histories. *Mar. Technol.* **2008**, *45*, 221–227.
16. Carlton, J. *Marine Propellers and Propulsion*; Butterworth Heinemann: Oxford, UK, 1994.
17. Theodorsen, T. *General Theory of Aerodynamic Instability and the Mechanism of Flutter*; NACA Report 496; 1935.
18. Ukon, Y.; Yuasa, H. Pressure distribution and blade stress on a higly skewed propeller. In Proceedings of the 19th Symposium of Naval Hydrodynamics, Seoul, Korea, 23–28 August 1992; pp. 793–814.
19. Vaz, G. Modelling of Sheet Cavitation on Hydrofoils and Marine Propellers Using Boundary Element Methods. Ph.D Thesis, Instituto Superior Técnico, Lisbon, Portugal, 2005.
20. Vaz, G.; Bosschers, J. Modelling of three dimensional sheet cavitation on marine propellers using a boundary element method. In Proceedings of the Sixth International Symposium on Cavitation, Wageningen, The Netherlands, 11–15 September 2006.
21. Morino, L.; Kuo, C. Subsonic potential aerodynamics for complex configurations: A general theory. *AIAA J.* **1974**, *12*, 191–197.
22. La Mantia, M.; Dabnichki, P. Influence of the wake model on the thrust of oscillating foil. *Eng. Anal. Bound. Elem.* **2011**, *35*, 404–414. [CrossRef]
23. La Mantia, M.; Dabnichki, P. Unsteady panel method for flapping foil. *Eng. Anal. Bound. Elem.* **2009**, *33*, 572–580. [CrossRef]
24. Katz, J.; Plotkin, A. *Low-Speed Aerodynamics*; Camebridge University Press: New York, NY, USA, 2001.
25. Young, J. Numerical Simulation of the Unsteady Aerodynamics of Flapping Airfoils. Ph.D Thesis, The University of New South Wales, Sydney, Australia, 2005.
26. Boswell, R. *Design, Cavitation Performance and Open-Water Performance of a Series of Research Skewed Propellers*; Report 3339, DTNSRDC; 1971.
27. Maljaars, P.; Kaminski, M.; den Besten, J. Finite element modelling and model updating of small scale composite propellers. *Compos. Struct.* **2017**, *176*, 154–163. [CrossRef]
28. Abbott, I.; von Doenhoff, A. *Theory of Wing Sections*; Dover Publications: Mineola, NY, USA, 1959.
29. Muñch, C.; Ausoni, P.; Braun, O.; Farhat, M.; Avellan, F. Fluid–structure coupling for an oscillating hydrofoil. *J. Fluids Struct.* **2010**, *26*, 1018–1033. [CrossRef]
30. Maljaars, P.; Kaminski, M.; den Besten, J. A new algorithm for computing the steady state fluid–structure interaction response of periodic problems. *J. Fluids Struct.* **2018**, submitted.
31. Beulen, B.; Rutten, M.; van de Vosse, F. A time-periodic approach for fluid–structure interaction in distensible vessels. *J. Fluids Struct.* **2009**, *25*, 954–966. [CrossRef]
32. Degroote, J.; Haelterman, R.; Annerel, S.; Bruggeman, P.; Vierendeels, J. Performance of partitioned procedures in fluid–structure interaction. *Comput. Struct.* **2010**, *88*, 446–457. [CrossRef]
33. Young, Y.; Chae, E.; Akcabay, D. Hybrid algorithm for modeling of fluid–structure interaction in incompressible, viscous flows. *Acta Mech. Sin.* **2012**, *28*, 1030–1041. [CrossRef]
34. Kato, H.; Kodama, Y. Microbubbles as a skin friction reduction device—A midterm review of the research. In Proceedings of the 4th Symposium on Smart Control of Turbulence, Tokyo, Japan, 2–4 March 2003.
35. Gindroz, B.; Hoshino, T.; Pylkkanen, J. 22nd ITTC Propulsion Committee Propeller RANS/Panel Method Workshop. In Proceedings of the 22nd ITTC, Grenoble, France, 5–6 April 1998.

Journal of
Marine Science and Engineering

MDPI

Article

Experimental Validation of Fluid–Structure Interaction Computations of Flexible Composite Propellers in Open Water Conditions Using BEM-FEM and RANS-FEM Methods

Pieter Maljaars [1,*], Laurette Bronswijk [1,2], Jaap Windt [3], Nicola Grasso [3] and Mirek Kaminski [1]

1 Department of Maritime & Transport Technology, Delft University of Technology, Mekelweg 2, 2628 CD Delft, The Netherlands; L.Bronswijk@academy.marin.nl (L.B.); M.L.Kaminski@tudelft.nl (M.K.)
2 Maritime Research Institute Netherlands, MARIN Academy, Haagsteeg 2, 6708 PM Wageningen, The Netherlands
3 Maritime Research Institute Netherlands, Haagsteeg 2, 6708 PM Wageningen, The Netherlands; J.Windt@marin.nl (J.W.); N.Grasso@marin.nl (N.G.)
* Correspondence: P.J.Maljaars@tudelft.nl; Tel.: +31-15-27-85923

Received: 29 March 2018; Accepted: 24 April 2018; Published: 7 May 2018

Abstract: In the past several decades, many papers have been published on fluid–structure coupled calculations to analyse the hydro-elastic response of flexible (composite) propellers. The flow is usually modelled either by the Navier–Stokes equations or as a potential flow, by assuming an irrotational flow. Phenomena as separation of the flow, flow transition, boundary layer build-up and vorticity dynamics are not captured in a non-viscous potential flow. Nevertheless, potential flow based methods have been shown to be powerful methods to resolve the hydrodynamics of propellers. With the upcoming interest in flexible (composite) propellers, a valid question is what the consequences of the potential flow simplifications are with regard to the coupled fluid–structure analyses of these types of propellers. This question has been addressed in the following way: calculations and experiments were conducted for uniform flows only, with a propeller geometry that challenges the potential flow model due to its sensitivity to leading edge vortex separation. Calculations were performed on the undeformed propeller geometry with a Reynolds-averaged-Navier–Stokes (RANS) solver and a boundary element method (BEM). These calculations show some typical differences between the RANS and BEM results. The flexible propeller responses were predicted by coupled calculations between BEM and finite element method (FEM) and RANS and FEM. The applied methodologies are briefly described. Results obtained from both calculation methods have been compared to experimental results obtained from blade deformation measurements in a cavitation tunnel. The results show that, even for the extreme cases, promising results have been obtained with the BEM-FEM coupling. The BEM-FEM calculated responses are consistent with the RANS-FEM results.

Keywords: flexible composite propellers; BEM-FEM coupling; RANS-FEM coupling; propeller deformation measurements

1. Introduction

In the last several decades, many papers have been published on the hydro-elastic analysis of flexible (composite) propellers. The majority of the studies were limited to steady inflow conditions. In these studies, mainly three different approaches were used for the hydrodynamic calculations of flexible propellers viz. Reynolds averaged Navier–Stokes methods (RANS), boundary element methods

(BEM) and vortex lattice methods (VLM). In BEM and VLM computations, the flow is assumed to be a potential flow.

RANS solvers were used in the flexible propeller FSI computations presented in [1–4]. BEM solvers were applied in the hydro-elastic propeller analysis as presented in [5–12]. In the hydro-elastic coupling, procedures presented in [13,14] VLM solvers were applied.

The fundamental difference between RANS computations and potential flow based calculations is that phenomena as separation of the flow, flow transition, boundary layer build-up and vorticity dynamics are not modelled in the latter one. For rigid propellers, the consequences of these flow simplifications have been studied in the past. Results of such studies were presented in the proceedings of the first and fourth Symposium on Marine Propulsors, for instance. These proceedings include papers with results of comparative studies with different RANS and BEM calculations together with experimental results for rigid propellers [15,16]. Around the optimum efficiency, a good agreement between BEM, RANS and experimental results is obtained. An increasing inaccuracy of the BEM results can be expected for larger skew angles and smaller advance coefficients [12]. This is explained by the increased leading edge vortex strength for decreasing advance ratios and increasing skew angles. This indicates that, for blade integrated values like thrust and torque, vorticity phenomena may not be negligible for high loading conditions and propellers with considerable skew. However, comparative studies between RANS and potential flow based hydro-elastic calculations for flexible propellers are still lacking.

Recently, the influence of viscous effects on the hydro-elastic response of hydrofoils has been investigated [17,18]. It has been shown that the flow-induced bend-twist coupling effects of a flexible hydrofoil in fully turbulent and attached flow conditions are well predicted with an inviscid fluid–structure interaction (FSI) method [17]. However, when the stability boundaries are reached (i.e., static/dynamic divergence or flutter velocity boundaries), viscous FSI methods were recommended to predict the dynamic response, especially for solid-to-fluid added mass ratios smaller than four [18]. Therefore, it is expected that, for uniform flow conditions, in which dynamic instabilities are irrelevant, an accurate prediction of the FSI response of marine propellers can be obtained with an inviscid method when the flow is fully turbulent and attached.

However, both conditions are not met for the small scale propellers considered in this work. Measurements and calculations have been performed at a transitional flow regime, partly laminar and partly turbulent. Secondly, experiments have been conducted at relatively high angles of attack resulting in a leading edge vortex and flow separation at the trailing edge. Finally, due to the finite blade size, a vortex is generated at the blade tips. The influence of these viscous effects and vorticity phenomena on the hydro-elastic response prediction is not known. Therefore, experimental and RANS-FEM results have been used to validate a BEM-FEM coupled calculation for uniform flows.

The main purpose of this paper is to validate the RANS-FEM and BEM-FEM coupled calculations with experimental results and to show what the consequences of the potential flow simplifications are with regard to the coupled fluid–structure response. This work is structured as follows: first, the different propellers are described in Section 2. Section 3 provides information about the structural modelling. In Section 4, the fluid models are explained. The BEM-FEM and RANS-FEM coupling are described in Section 5. Section 6 presents a comparison of open water measurements and BEM and RANS calculations for a rigid propeller. In Section 7, the cavitation tunnel experiments are explained. Section 8 presents an uncertainty estimation for measurements and RANS-FEM and BEM-FEM calculations. In Section 9, the experimental results obtained for the flexible propellers are compared to results obtained with the BEM-FEM and RANS-FEM coupling. Conclusions and recommendations are given in Section 10.

2. General Description of Propellers and Flow Conditions

2.1. Propellers

Four propellers are considered. The propellers have a diameter (D) of 0.34 m and are similar in size and geometry, but differ with respect to blade material. Two propellers are made out of glass fibre reinforced epoxy with different laminate orientations. The other two propellers are made of isotropic material: one bronze propeller and one made of epoxy. The propellers have been labelled as follows:

- Propeller bronze: this propeller is assumed completely rigid.
- Propeller epoxy: this propeller is the most flexible one.
- Propeller 45: [+45°/−45°] laminate lay-up.
- Propeller 90: [0°/90°] laminate lay-up.

The 0°-direction of the laminae is parallel to the z-axis of the propeller blade coordinate system, where positive x-, y- and z-axis are pointing forward, portside and upwards, respectively. All the results are presented according to this coordinate system.

2.2. Material Properties

The Young's moduli E, Poisson ratios v and the shear moduli G of the epoxy material and the composite laminate are given in Tables 1 and 2.

Table 1. Elastic properties of epoxy.

E (GPa)	v (-)	G (GPa)
3.60	0.300	1.39

Table 2. Composite [0°/90°] laminate elastic properties.

E_{11} (GPa)	E_{22} (GPa)	E_{33} (GPa)	v_{12} (-)	v_{13} (-)	v_{23} (-)	G_{12} (GPa)	G_{13} (GPa)	G_{23} (GPa)
18.8	18.8	8.00	0.132	0.275	0.275	2.81	2.49	2.49

2.3. Flow Conditions

For the open water calculations, computations have been performed for a constant rotational speed (n) of 1170 rpm and varying advance speed, keeping the Reynolds number as constant as possible for different flow conditions.

The flexible propeller calculations have been performed for the measured conditions with advance ratios around 0.37, 0.64 and 0.85 as given in Table 3. All calculations have been performed for uniform inflow conditions. The last column of Table 3 presents the Reynolds number, Re, based on the chord length (C) and flow velocity at 0.7 of the propeller radius (r),

$$Re_{0.7r} = \frac{\rho C_{0.7r}\sqrt{v^2 + (0.7\pi n D)^2}}{\mu}, \quad (1)$$

with ρ the density of water taken as 1000 kgm^{-3} and μ the dynamic viscosity equal to $1.01 \cdot 10^{-3}$ Pa·s.

Table 3. Flow conditions for the flexible propeller calculations and measurements.

	$J = \frac{60v}{nD}$	v (m/s)	n (rpm)	$Re_{0.7r} \times 10^6$
Prop. Epoxy, blade 1	0.37	1.88	900	1.28
Prop. Epoxy, blade 1	0.64	3.97	1098	1.60
Prop. Epoxy, blade 1	0.85	6.73	1392	2.10
Prop. 45, blade 1	0.38	1.92	900	1.28
Prop. 45, blade 1	0.64	3.99	1098	1.60
Prop. 45, blade 1	0.85	6.75	1398	2.10
Prop. 90, blade 2	0.38	1.98	900	1.28
Prop. 90, blade 2	0.66	4.09	1098	1.60
Prop. 90, blade 2	0.85	6.74	1398	2.10

3. Structure Model

For FEM modelling and computations, MSC Marc/Mentat has been used. The FEM models consist of one propeller blade without the hub part. The stiffness contribution of the hub has been modelled by a full clamping of the propeller blade at the blade–hub interface. The models were discretised by quadratic solid elements. A mesh convergence study has been conducted in order to ensure a mesh independent solution for the calculated displacements. Structured FEM meshes have been used, identified with the following three discretisation parameters: N_c the number of elements in chordwise direction, N_r the number of elements in radial direction and N_n the number of elements in through-thickness direction. Table 4 presents the displacements of the tip for a static load case obtained with FEM models for different meshes.

Table 4. Results of finite element method mesh convergence study.

N_c	N_r	N_n	Tip Displ. (mm)
116	120	4	16.76
58	60	4	16.76
58	60	8	16.76
29	30	4	16.74

These results show that, for the $29 \times 30 \times 4$ element distribution, the tip displacement differs approximately 0.1% from the grid independent solution. Therefore, the $29 \times 30 \times 4$ element distribution has been used for all the FEM calculations.

In the FEM modelling, special attention has been given to the establishment of the material orientations in composite blades. In [19,20], the importance of a proper material orientation for doubly curved structures has been described. Standard commercial FEM software packages are usually not able to define unambiguously the material orientations in complex geometries [19]. In [20], an approach has been presented to determine the element dependent material orientations in doubly curved structures. In this method, the through thickness direction and the projection of the transverse laminate (90°)-direction on the element surface is used to establish the material orientation per element. A more detailed description of this approach and the blade FEM modelling can be found in that paper.

4. Fluid Models

4.1. BEM Model

For this work, the BEM PROCAL, developed by the Maritime Research Institute Netherlands (MARIN) [21,22] was used. In BEM, only the surfaces of the object have to be discretized. From a grid sensitivity study, it was concluded that a 29×30 element distribution is sufficient to obtain a grid independent solution. In the previous section, it was revealed that, in the FEM model,

29 and 30 elements are distributed along the blade chord and radius, respectively. This means that the BEM and FEM solvers require a similar distribution of panels or elements on the propeller surface. This property has been used by applying identical mesh distributions in the FEM and the BEM model in order to avoid the need for an interpolation of pressure and structural response between the two grids.

Two corrections are applied on the PROCAL pressures before they are imposed on the FEM model. The first correction is to rectify for an overestimation of the pressures at the propeller tip. In general, with a potential flow solver, a pressure difference between blade face and back at the tip will be computed, while, in reality, this pressure difference will be zero. To correct for this, the calculated pressures from a certain propeller radius are smoothed to a zero pressure coefficient at the tip. The default from where this tip pressure correction is applied is 95% of the propeller radius. This value has been used throughout this work. The pressure smoothing radius has an important influence on the propeller blade deformations at the tip, since the tip is relatively flexible and therefore the tip deformations are sensitive to small changes in pressure distribution [23].

The second correction is a viscous correction to include frictional losses. The viscous shear stresses have been computed using the maximum skin friction coefficient calculated from the Blasius formulation for laminar flow, ITTC formulation for transition to turbulent flow [24] and the Prandtl–Schlichting formulation. The skin friction coefficients have been used to calculate the blade tangential forces from the total velocities and element areas per element. These forces have been imposed on the FEM model.

The correction for the minimum allowable pressure coefficient as explained in Section 6.1.1 has not been used in the BEM-FEM calculations because this correction applies only for the lowest advance ratios of 0.37 and 0.38, but hardly affects the hydro-elastic response for these conditions.

4.2. RANS Model

RANS calculations have been conducted with the CFD software ReFRESCO developed at MARIN. ReFRESCO [25] is a community based open-usage/open-source CFD code for the maritime world. It solves multiphase (unsteady) incompressible viscous flows using the Navier–Stokes equations, complemented with turbulence models, cavitation models and volume-fraction transport equations for different phases. The equations are discretised using a finite-volume approach with cell-centered collocated variables, in strong-conservation form and a pressure-correction equation based on the SIMPLE algorithm is used to ensure mass conservation [26,27].

Since open water conditions are considered, the computations can be done using a body fitted, rotating reference system. For this, the propeller has been modelled in a rotating circular domain with diameter and length three and five times the propeller diameter, respectively. The propeller is located in the middle of this domain. The computational domain consists of a structured multi-block grid built with GridPro, using a standard block topology developed at MARIN. The topology is applied quite often and gives good quality grids, most of the time. However, for the propeller considered here, grid generation was cumbersome due to the relatively high skew of the propeller blades, the relatively thin leading edge and propeller section thickness's. The quality of the grid was not very high, which leads to convergence problems of the solver when using higher order discretisations of the convective fluxes (QUICK scheme). Therefore, for all calculations, a blending between central and upwind discretisation was used. Figures of the domain, grid and propeller are depicted in Figure 1. For all the calculations, the $k - \sqrt{k}L$ turbulence model has been used as described in [28]. The $k - \sqrt{k}L$ turbulence model gives very similar results as the $k - \omega$ SST model, but shows in general an improved iterative convergence. This has been shown explicitly for propeller flows [29].

The boundary conditions applied on the domain and propeller are given in Table 5. At the inlet, the velocity is imposed. At the outlet, a combination of an outflow and pressure condition is imposed. Behind the propeller, the velocity derivatives are imposed to be zero; at the remainder of the outlet, the pressure is imposed. The propeller, hub and shaft have a no-slip boundary condition. In the calculations, no wall functions have been used, i.e., in all calculations, the non-dimensional wall distance y+ was below 1. At the outer surface of the circular domain, a free-slip boundary condition is applied, i.e., the velocities normal to the surface are zero and the tangential velocities are free.

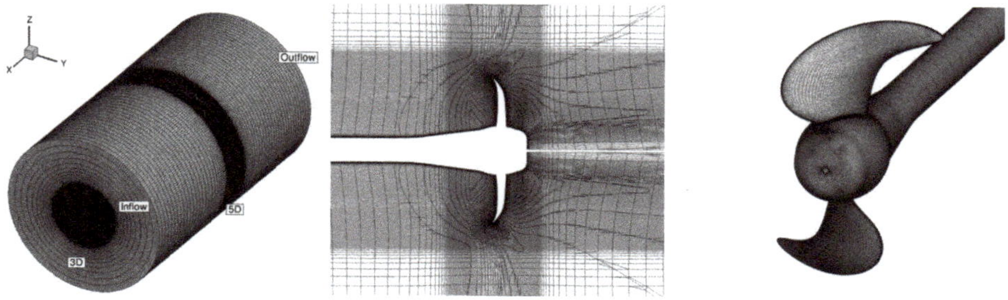

Figure 1. Domain and grid for Reynolds-averaged-Navier–Stokes computations.

Table 5. Boundary conditions for Reynolds-averaged-Navier–Stokes calculations.

	Boundary Condition
Inlet	Prescribed inflow velocity.
Inner outlet	Neumann boundary condition on velocity and pressure.
Outer outlet	Dirichlet boundary condition on pressure, Neumann on velocity.
Propeller, hub and shaft surfaces	Velocity is zero, no wall functions applied (y+ should be <1).
Outer surface flow domain	Normal and tangential velocity are zero and free, respectively.

Thrust and Torque RANS Discretisation Uncertainties

The discretisation errors in the thrust and torque values obtained with the RANS calculations have been estimated with the method described in [30]. In this approach, the numerical uncertainty is obtained using solutions on systematically refined grids. The error is estimated with power series expansions as a function of the typical cell size. The expansions are fitted to the data in the least-squares sense. For the present uncertainty estimation, four grids with cell densities described in Table 6 have been used. The third row gives the relative step size. This parameter identifies the representative grid cell size and is the cubic root of the ratio between the amount of grids cells for finest grid and the considered grid.

Table 6. Amount of cells and relative step size for different grids.

Grid	Amount of Cells	Relative Step Size
A	917,000	2.18
B	2,390,000	1.58
C	3,790,000	1.36
D	9,460,000	1.00

Figure 2 shows the results of the discretisation uncertainty quantification for the thrust and torque values computed for the three advance coefficients. The figures show that the order of accuracy p is in the range 0.5 to 1.7. $p = *1,2$ means that a fit was made using first and second order exponents. The order of accuracy is smaller than the typical value of two that would have been obtained when a quadratic upwind differencing (QUICK) scheme was used for the convective flux discretisation. However, with a QUICK scheme, the computations did not converge.

The results show that the discretisation uncertainties are small ($<3\%$) for the advance ratios 0.37 and 0.64. The uncertainties are the highest for the largest advance coefficient, 7.7% and 11.2% for the thrust and torque, respectively. These values are higher than generally accepted. Therefore, for the advance ratio of 0.85, a computation has been performed with a further refined grid. Unfortunately, this calculation required a smaller blending factor to converge and therefore it was decided to stay with grid D.

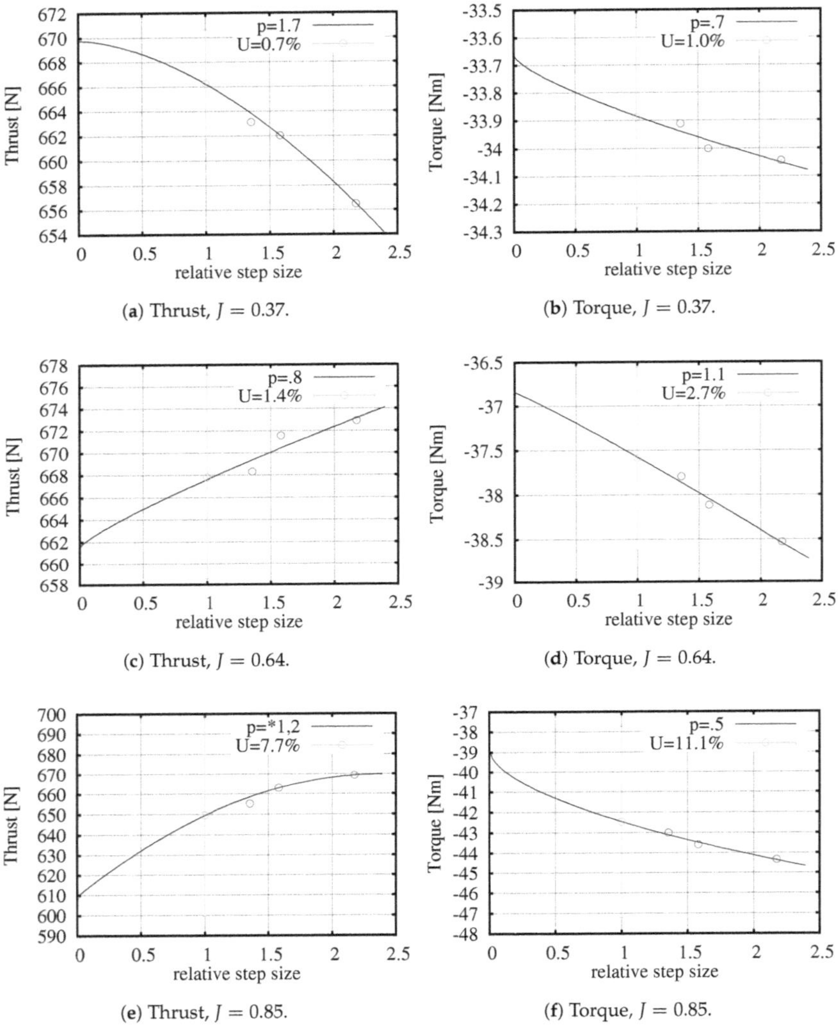

(**a**) Thrust, $J = 0.37$. (**b**) Torque, $J = 0.37$.

(**c**) Thrust, $J = 0.64$. (**d**) Torque, $J = 0.64$.

(**e**) Thrust, $J = 0.85$. (**f**) Torque, $J = 0.85$.

Figure 2. Numerical uncertainty of thrust and torque, U and order of accuracy p for various advance ratios.

5. Fluid–Structure Coupling

5.1. BEM-FEM Coupling

Figure 3 shows the coupling procedure between the BEM solver PROCAL and the FEM software MSC Marc. In this coupling, the following nonlinear equation is solved:

$$\mathbf{Ku} = \mathbf{f}_{BEM}\left(\mathbf{u}\right) + \mathbf{f}_{viscous}\left(\mathbf{u}\right) + \mathbf{f}_{centrifugal}, \tag{2}$$

where $\mathbf{f}_{BEM}$, $\mathbf{f}_{viscous}$ and $\mathbf{f}_{centrifugal}$ denote potential flow force, viscous forces and centrifugal forces, respectively. $\mathbf{K}$ is the stiffness matrix. Essentially, all the variables in Equation (2) are a function of the deformations $\mathbf{u}$. Since the blade deformations are relatively small, geometric linear elastic analyses are performed with the FEM solver and therefore the stiffness matrix and centrifugal force vector are not functions of $\mathbf{u}$.

The first step in the coupled BEM-FEM calculation is a PROCAL calculation on the undeformed geometry. The tip pressure correction is applied on the pressures obtained with PROCAL as explained above. Next, the viscous forces are calculated to account for frictional losses. Subsequently, the calculated pressures and viscous forces are used to calculate the structural response of the propeller blade. The structural deformations are used to construct a new propeller geometry from which new fluid pressures are calculated. When not converged, the iteration loop starts again.

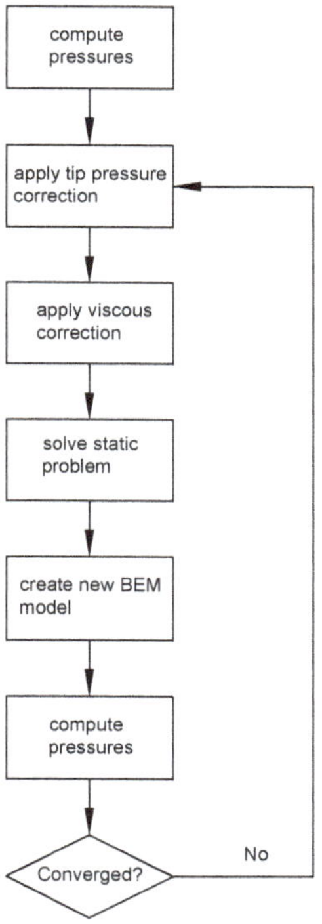

Figure 3. Flow chart of BEM-FEM coupling.

5.2. RANS-FEM Coupling

Recently, MARIN has developed an FSI module in ReFRESCO, the implementation and a verification study have been presented in [31]. The method is a partitioned strong coupling approach, meaning that the fluid and structural problem are solved separately and coupling iterations are performed each timestep to obtain the coupled solution. To stabilize this procedure under-relaxation is applied by means of the Aitken adaptive under-relaxation method. For transfer of information across the fluid–structure interface, a radial basis-function (RBF) interpolation is used.

In [31], a verification study was presented for the unsteady problem of the flow around a rigid cylinder with a flexible flap (the Turek benchmark [32]). Results obtained with the FSI module of ReFRESCO were in good agreement with results presented in literature.

The FSI module has been developed for unsteady FSI problems. For steady FSI problems, like the flow around a flexible propeller in open water conditions, the FSI module can be used by performing unsteady simulations until equilibrium is obtained. This was applied in this work, since only open water conditions have been considered. For steady FSI problems the equilibrium solution should be irrespective of the time step in the RANS and FEM calculation, which was the case for all the computations. The most convenient was to apply the same time step in RANS and FEM calculation.

Figure 4 shows the flow chart for one time step with the RANS-FEM coupling. Each time step starts with a RANS computation. Then, the pressures are transferred to the FEM calculation by using an RBF interpolation. With FEM, the structural response is computed. Subsequently, the structural response is transferred to the RANS calculation by again using an RBF interpolation. Based on the blade deformations, the RANS grid is adapted and the pressures are recomputed until convergence is obtained. Then, the following time step is resolved in the same way. For the steady problems considered in this work, the RANS-FEM calculation runs through a number of time steps until the steady state solution is obtained.

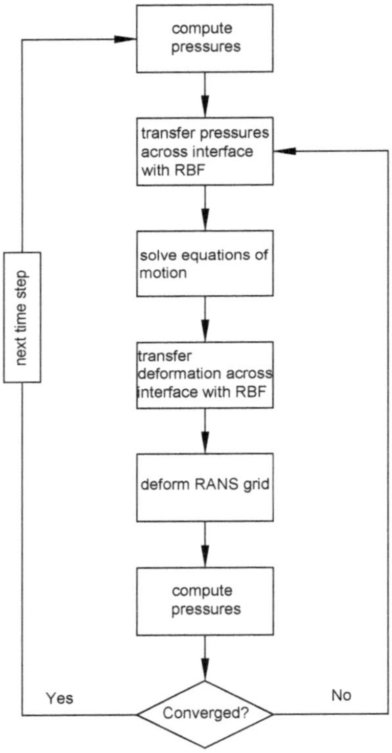

Figure 4. Flow chart of one time step with RANS-FEM coupling.

6. Comparison of Experimental, BEM and RANS Results for the Bronze Propeller

6.1. Open Water Diagram Bronze Propeller

The open water diagram of the bronze propeller has been measured and compared to calculated open water diagrams with PROCAL and ReFRESCO. Open water measurements have been performed in the deepwater towing tank of MARIN at a constant rotational speed of 1170 rpm and varying advance speed. The Reynolds number based on velocity and chord length at 0.7 of the propeller radius then varies between 1.64×10^6 and 1.90×10^6. A total measurement uncertainty (precision and bias) of 3% on thrust and torque can be adopted for these measurements.

6.1.1. Comparison of Experimental and BEM Results

For the BEM calculations, a discretisation error of 0.5% in thrust and torque is estimated by taking the percentage difference between the mesh independent thrust and torque values and the values obtained for the considered panel distribution. The measured and calculated open water curves are depicted in Figure 5 with their uncertainty bands. Particularly for $J < 0.4$ and $J > 1$, a relatively large discrepancy between measured and calculated curves is seen, especially for the torque coefficient. For $J > 1$, viscous effects play an increasingly important role and therefore the results of the PROCAL calculations diverge from the measured results. The discrepancy for low advance coefficients is attributed to the relatively sharp leading edge the potential flow modelling. This results, for heavy loading conditions, in unrealistic high flow velocities and consequently low pressures at the leading edge, since the flow does not separate in the BEM calculation. In reality, the flow will separate, which will limit the suction pressure. The torque is mainly affected by the unrealistic low pressures at the leading edge because the surface normals at the leading edge point mainly in the direction of the blade nose-tail line and therefore the low leading edge pressures reduce the drag rather than the lift. A correction on the suction pressures can be applied by restricting the minimum pressure coefficient, C_p, obtained with the BEM calculation by replacing lower pressures with that value. By properly selecting the minimum allowable pressure coefficient for the different advance ratios, the dotted KQ curve of Figure 5 is obtained. For $J > 0.4$, no minimum pressure correction has been applied.

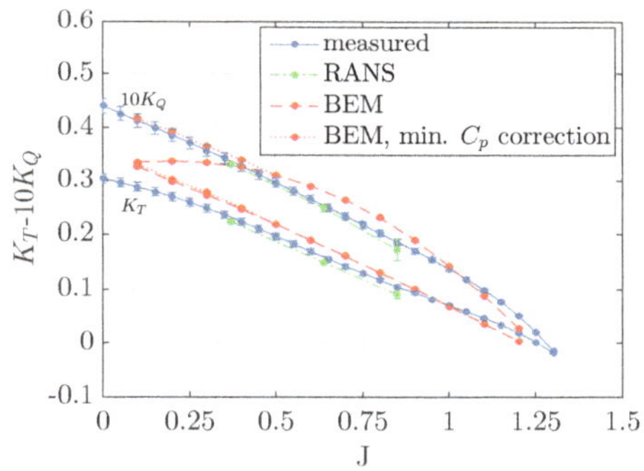

Figure 5. Measured and calculated open water diagrams of the bronze (rigid) propeller.

6.1.2. Comparison of Experimental and RANS Results

The calculated open water coefficients with ReFRESCO together with the discretisation uncertainty bandwidths are depicted in Figure 5 as well. The RANS computed thrust and torque coefficients are smaller than the measured values. The uncertainty bars show that the differences can be explained from uncertainties in the RANS computations and the measurements, except for the highest advance coefficient. That means that, with the present RANS calculation, the thrust and therefore the pressure distribution cannot be correctly resolved for this propeller at $J = 0.85$. A plausible explanation for this is the modelling error originating from the turbulence model, while a transition model might be more appropriate. This modelling uncertainty might be significant for this condition, since, for an increasing advance ratio, the lift force decrease and viscous forces become more important.

6.2. Comparison of BEM and RANS Pressure Distributions

Figure 5 shows significant differences between BEM and RANS results. Not surprisingly, the RANS results seem more accurate than the BEM results, since the RANS calculation includes viscous flow modelling and a vorticity phenomena. To investigate the differences between BEM and RANS results in more detail, pressure distributions obtained with both methods for the first three flow conditions of Table 3 are compared. For these flow conditions, the limiting streamlines on the propeller suction side are depicted in Figure 6. As indicated by the contraction of the streamlines, three flow separation areas can be distinguished: separation at the leading edge resulting in a leading edge vortex, a separation area at the trailing edge and flow separation at the propeller tip. The least flow separation is obtained for the highest advance coefficient.

The pressure distributions for the three BEM and RANS computations are presented in Figures 7–12. In these figures, the pressure is made dimensionless with the fluid density, propeller rotation rate and diameter. These results show that the suction side pressure is generally lower for the BEM computations. Overall, the BEM computed pressures on the pressure side are higher than those obtained from the RANS computations. These results correspond with the differences in thrust and torque between the RANS and BEM computations as presented in Table 7.

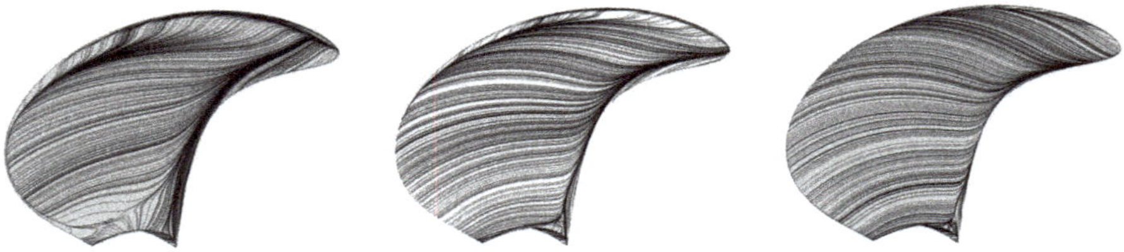

Figure 6. Limiting streamlines on propeller suction side for $J = 0.37$, $J = 0.64$ and $J = 0.85$.

Table 7. K_T and $10K_Q$ for rigid propeller Reynolds-averaged-Navier–Stokes and boundary element method calculations.

	K_T (BEM)	K_T (RANS)	%	$10K_Q$ (BEM)	$10K_Q$ (RANS)	%
$J = 0.37$	0.255	0.222	−13%	0.344	0.331	−4%
$J = 0.64$	0.177	0.149	−16%	0.278	0.247	−11%
$J = 0.85$	0.114	0.089	−22%	0.208	0.172	−17%

For the conditions $J = 0.37$ and $J = 0.64$, the angle of attack on the blades is high, leading to a strong suction peak in the pressure distribution. In the RANS computations, the flow separates as shown in Figure 6. At the blade tip, the differences in pressure distribution obtained from RANS and BEM computations are evident: the BEM calculations show unrealistic pressures due to a non-physical modelling of the flow. In the BEM-FEM coupled calculations, the tip pressure correction is applied to correct for that.

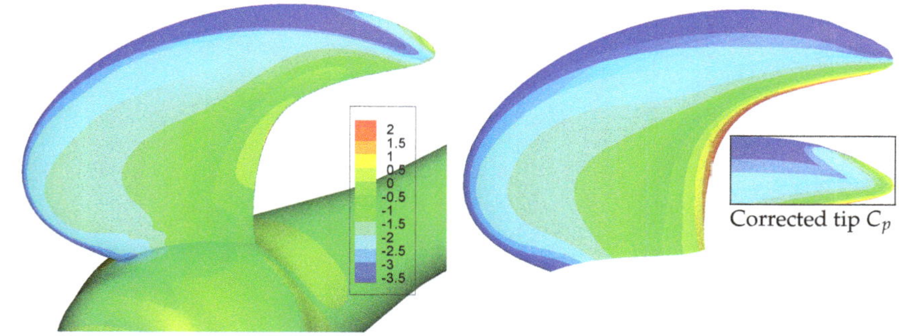

Figure 7. Pressure coefficient, C_p, on the suction side for $J = 0.37$, RANS (**left**) BEM (**right**). The insert figure shows the BEM pressure distribution after the tip pressure correction.

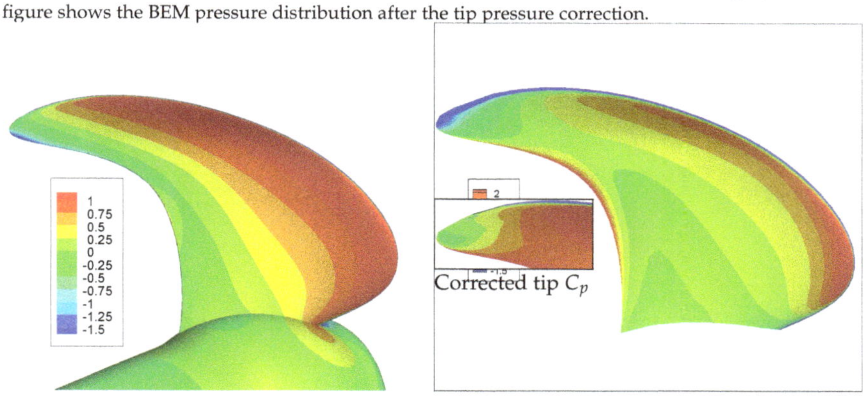

Figure 8. Pressure coefficient, C_p, on the pressure side for $J = 0.37$, RANS (**left**) BEM (**right**). The insert figure shows the BEM pressure distribution after the tip pressure correction.

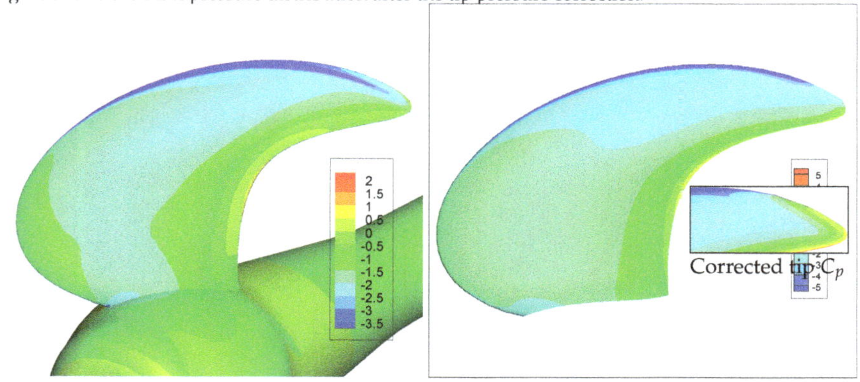

Figure 9. Pressure coefficient, C_p, on the suction side for $J = 0.64$, RANS (**left**) BEM (**right**). The insert figure shows the BEM pressure distribution after the tip pressure correction.

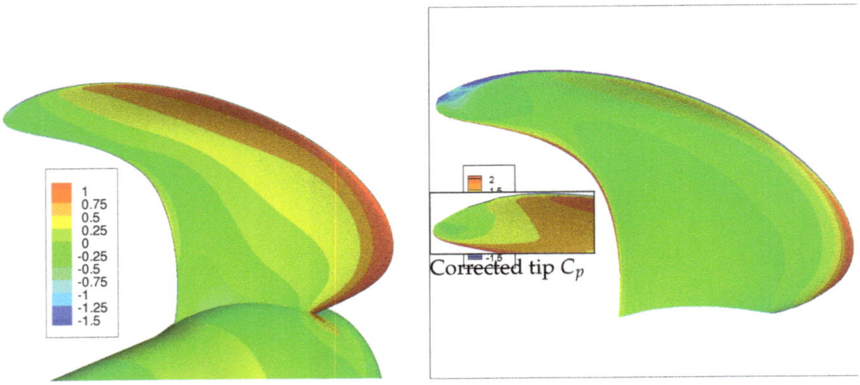

Figure 10. Pressure coefficient, C_p, on the pressure side for $J = 0.64$, RANS (**left**) BEM (**right**). The insert figure shows the BEM pressure distribution after the tip pressure correction.

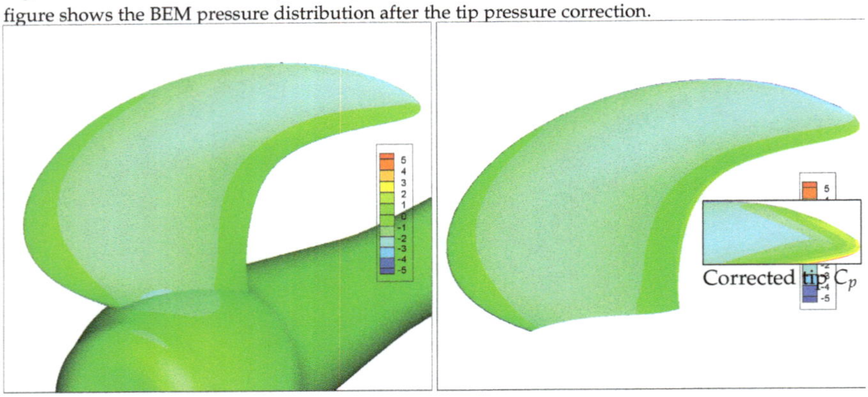

Figure 11. Pressure coefficient, C_p, on the suction side for $J = 0.85$, RANS (**left**) BEM (**right**). The insert figure shows the BEM pressure distribution after the tip pressure correction.

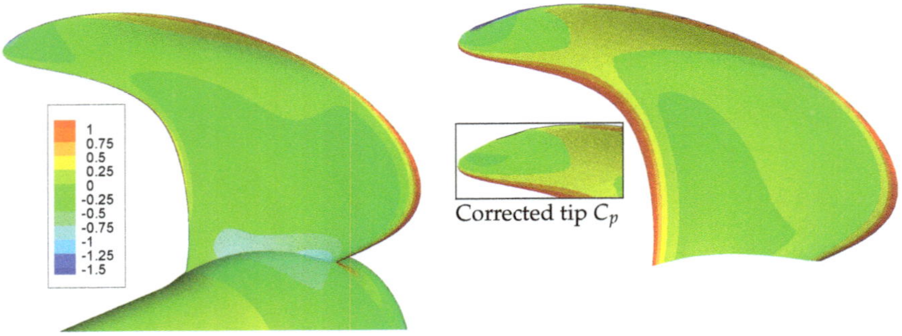

Figure 12. Pressure coefficient, C_p, on the pressure side for $J = 0.85$, RANS (**left**) BEM (**right**). The insert figure shows the BEM pressure distribution after the tip pressure correction.

7. Flexible Propeller Cavitation Tunnel Experiments

Usually, thrust, torque and blade deformations are measured and used for validation of the hydro-elastic propeller calculations. In this work, the focus is on the blade deformations rather than thrust and torque values for several reasons: first of all, the deformation field provides spatially distributed information about the propeller response in contrast to the blade integrated values like thrust and torque. Secondly, the uncertainties in the measured blade deformations are smaller than the uncertainties in thrust and torque changes due to blade flexibility. The reason is that thrust and torque changes can be easily affected by unintended small deviations in propeller geometry introduced in the complicated manufacturing of flexible propellers. Stereo-photography with a Digital Image Correlation (DIC) technique was selected to measure the propeller blade deformations. With this system, a very accurate recording of the complete 3D blade deformation field was achieved.

7.1. Test Set-Up

The measurements were conducted by MARIN using their cavitation tunnel test facility. The propeller is mounted on the tunnel shaft, which is connected to an encoder. This encoder sends impulse signals used for the triggering of the strobe lights and the cameras. The test set-up used in the cavitation tunnel is explained in Figure 13. It consists of the following elements:

- Two synchronized and calibrated cameras with FireWire interface; resolution: 1388 × 1038 pixels; maximum frame rate 16 fps at full resolution.
- Stroboscopic lights with flash duration in the micro second range. Flash duration is kept as short as possible to avoid motion blur at the blade tip.
- The shaft encoder mounted on the shaft, provides 360 pulses per revolution.
- A pulse selector is able to select one of these 360 pulses as a trigger, which is sent to the stroboscopes and the cameras. Therefore, a trigger can be supplied, with a resolution of one degree for every blade position. The cameras and the strobe are synchronized such that the strobe flash falls within the time frame that the camera shutter is open.

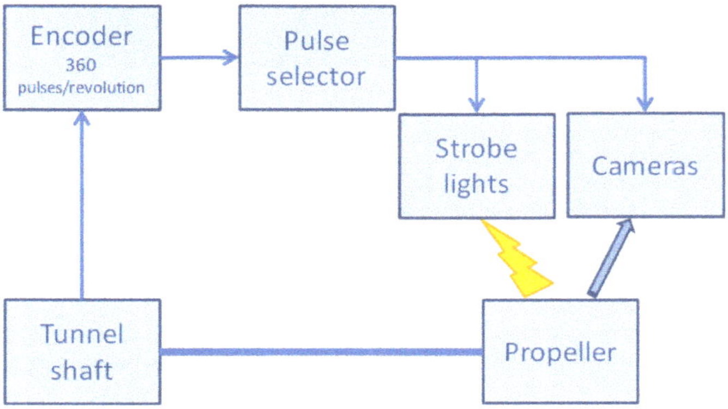

Figure 13. Cavitation tunnel set-up diagram.

Figure 14 shows the initially proposed camera set-up. Two purpose-built windows were mounted in place of the cavitation tunnel lateral windows to have an optimal camera view. Figure 15 shows a picture of the realized test set-up. During the experiments, one of the windows was moved to the bottom of the tunnel to further improve the view on the blade surfaces. In addition, the cameras were mounted on a vibration damping structure to ensure their isolation from vibrations.

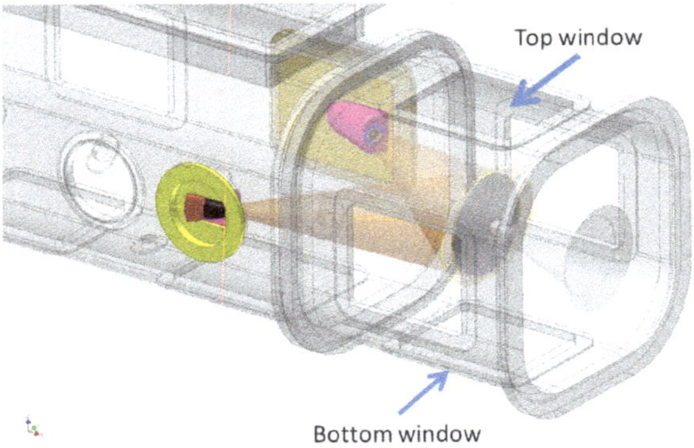

Figure 14. Proposed camera set-up.

Figure 15. Picture of the cavitation tunnel test set-up.

7.2. Measurement Technique

The image data acquired with the calibrated stereo camera system have been used to compute the blade deformations by means of DIC. DIC is a full-field image analysis method, based on grey value digital images, that finds the displacements and deformations of an object in three-dimensional space [33]. During the blade deformation, the method tracks the grey value pattern from which the deformations of the object are calculated. This method can be used in several applications. In particular, it has been successfully applied for blade deformation measurements both in uniform flow in the cavitation tunnel and in behind ship model condition in the towing tank [34].

In order to use the DIC technique, the surface of the measured object must have a random speckle pattern with no preferred orientation and sufficiently high contrast. The size of the features in the pattern should be large enough to be distinguished. If the material does not naturally show a usable speckle pattern, this must be applied through printing or painting. With this technique, very accurate measurements of the blade response were achieved. Several images for each blade position were acquired and image averaging was applied to filter out displacements resulting from high frequency vibrations of the propeller blade, and to remove possible bubbles or particles in the water. The results were further post-processed and a procedure was applied to correct for rigid body motions induced by vibrations and movements of the shaft.

8. Experimental, Modelling and Discretisation Uncertainties Flexible Propeller Cases

8.1. Experimental Uncertainties

Regarding the experimental uncertainties, two types of errors can be distinguished: precision errors and systematic errors. Precision errors are due to the statistical variability in the measured data set and can be reduced by repeating the test a number of times until the true average value of the measurement distribution is obtained. To reduce the precision errors in the measurements, averaging of the results was applied. Thrust, torque, flow speed and rotation rate are time averaged values. To obtain the true average displacements, image averaging with 30 images was applied. Therefore, it is expected that the precision errors are small compared to the systematic errors. Only for the DIC measurements was a precision error determined because this precision error is not reduced with the image averaging, since the results are averaged before the cross-correlation. In order to estimate this precision error, a test was performed. A rigid flat plate, with a speckle pattern, was mounted on the cavitation tunnel shaft. The displacements of the plate in the axial direction due to different tunnel speeds (from 0 to 6 m/s) were measured. Given the stiffness of the plate, bend and shear deformations of the plate can be neglected. Therefore, the measured displacements are due to the compression of the tunnel shaft and are assumed constant over the plate area. The distribution of the rigid displacement is an indicator of the precision error. A 95% confidence interval of around 20 μm was obtained. From the results, it was also concluded that this precision error is independent of the displacement magnitude or tunnel flow speed (see also [23]).

A DIC systematic error of 30 μm has been assumed. This value is based on the measured blade response adjacent to the hub where a zero blade displacements can be expected, but which is generally not the case. The total uncertainty for the DIC measurements becomes 50 μm by simply adding the precision and systematic error together.

8.2. Modelling Uncertainties

Modelling uncertainties originate from different sources. There are modelling uncertainties due to simplifications in the mathematical models. For instance, in the BEM calculations, by assuming a potential flow and in the RANS calculations by adopting the $k-\sqrt{k}L$ turbulence model instead of a transition model, which would be able to predict the transition from laminar to turbulent flow.

There are also modelling uncertainties in the RANS-FEM and BEM-FEM calculations due to not exactly modelling the properties and conditions as appearing in the cavitation tunnel experiments. For this class of modelling errors, uncertainty levels have been estimated. The following modelling errors have been judged to be the most important, modelling errors due to:

1. using the design propeller geometry instead of the as-built geometry,
2. using the design propeller stiffness instead of the actual propeller stiffness,
3. neglecting of the cavitation tunnel walls.

The first modelling error is due to using the design propeller stiffness instead of the actual propeller stiffness. Based on measured and computed blade natural frequencies, a stiffness systematic error of ±5% and ± 10% has been assumed for epoxy propeller and composite propellers, respectively.

An important modelling error is introduced by performing the calculations with the design geometry instead of the as-built geometry. The influence of the difference between as-built and design geometry on the blade forces was investigated by comparing results of BEM calculations obtained for the different geometries. This investigation indicated that, depending on the propeller blade and flow condition, a significant difference in thrust force due to inaccuracies in the blade geometry can be assumed (see also [23]).

The modelling error due to neglecting the tunnel walls is only relevant for the BEM-FEM calculations, since the RANS-FEM calculations were performed in a bounded circular domain. This error has been estimated with Glauert's correction for tunnel wall effects [35]. With Glauert's correction, the unbounded flow velocity is replaced by an equivalent mean tunnel flow speed resulting in the same thrust.

8.3. Discretisation Uncertainties

The discretisation uncertainties in the BEM-FEM calculations are smaller than 0.5% and assumed negligible compared to the modelling errors as described above. The RANS grid discretisation uncertainties in bend and twist deformations have been estimated with the same method and grids as used for the thrust and torque uncertainty estimation in Section 4.2. This means that, in total, twelve RANS-FEM calculations have been performed—for every flow condition, four calculations with the four RANS grids. Then, for each radial station, the uncertainties in mid-chord bend deformation and twist deformation were calculated from the solutions of the four systematically refined grids.

8.4. Total Uncertainties

Tables 8 and 9 present the uncertainties at the propeller tip for the RANS-FEM calculations. The separate uncertainty contributions are are added in quadrature, resulting in the total uncertainties as given in the last columns. From these results, it can be concluded that the discretisation uncertainty dominates in the total bend and twist deformation uncertainty of the lowest and highest advance ratio. From this, it can be concluded that convergence of thrust and torque (as shown in Section 4.2) does not automatically mean that the blade structural response is converged as well.

The separate contributions of the modelling uncertainties for the composite propellers are not shown here. However, for propeller 90, the stiffness uncertainty has the largest contribution. For propeller 45, the blade geometry uncertainty has more or less the same magnitude as the stiffness uncertainty. The contribution in uncertainty due to the presence of the tunnel walls for the epoxy propeller BEM-FEM calculation was relatively large, since the stiffness uncertainty is smaller than for the composite propellers.

Table 8. Uncertainties (lower and upper levels) in mid-chord bend deformation at the tip in (mm).

	RANS-FEM Result	Modelling Uncert. 1	Modelling Uncert. 2	Discretisation Uncert.	Total Uncert.
$J = 0.37$	4.34	−0.157; 0.0	−0.212; 0.211	−0.386; 0.386	−0.468; 0.440
$J = 0.64$	4.11	−0.184; 0.0	−0.200; 0.201	−0.025; 0.025	−0.273; 0.202
$J = 0.85$	4.20	−0.233; 0.0	−0.185; 0.183	−0.290; 0.290	−0.415; 0.343

Table 9. Uncertainties (lower and upper levels) in twist deformation at the tip in (°).

	RANS-FEM Result	Modelling Uncert. 1	Modelling Uncert. 2	Discretisation Uncert.	Total Uncert.
$J = 0.37$	−2.29	0.0; 0.066	−0.089; 0.090	−0.508; 0.508	−0.515; 0.520
$J = 0.64$	−1.95	0.0; 0.083	−0.093; 0.091	−0.092; 0.092	−0.131; 0.154
$J = 0.85$	−2.58	0.0;0.128	−0.101; 0.102	−0.335; 0.335	−0.350; 0.373

9. Comparison of Experimental, BEM-FEM and RANS-FEM Results

Figures 16–18 show the uncertainty intervals for measured and calculated bend and twist deformations against the radial position on the blades. By investigating the results obtained for the epoxy propeller, in general, the measured bending responses are well predicted with RANS-FEM calculations. The uncertainties bandwidths for the RANS-FEM calculated twist deformations of the lowest and highest advance ratio are relatively large, mainly caused by the discretisation uncertainty, but do not partially overlap with the measurements. The differences between the twist deformations obtained from the RANS-FEM calculation and measurements for $J = 0.85$ is explainable given the deviation between RANS computed thrust and the rigid propeller open water measurement as presented in Section 6.1.2 for this condition. It was pointed out that these differences might originate from modelling errors in the turbulence modelling. A plausible explanation for the differences at

$J = 0.37$ is the severe flow separation that might be not correctly resolved with a RANS model and the turbulence model. It can be concluded that the best agreement between measurements and the RANS-FEM calculations is obtained for the advance ratio of $J = 0.64$, in which leading edge vortex separation is present but limited compared to the lowest advance ratio, and viscous forces might be less dominating than for the highest advance coefficient.

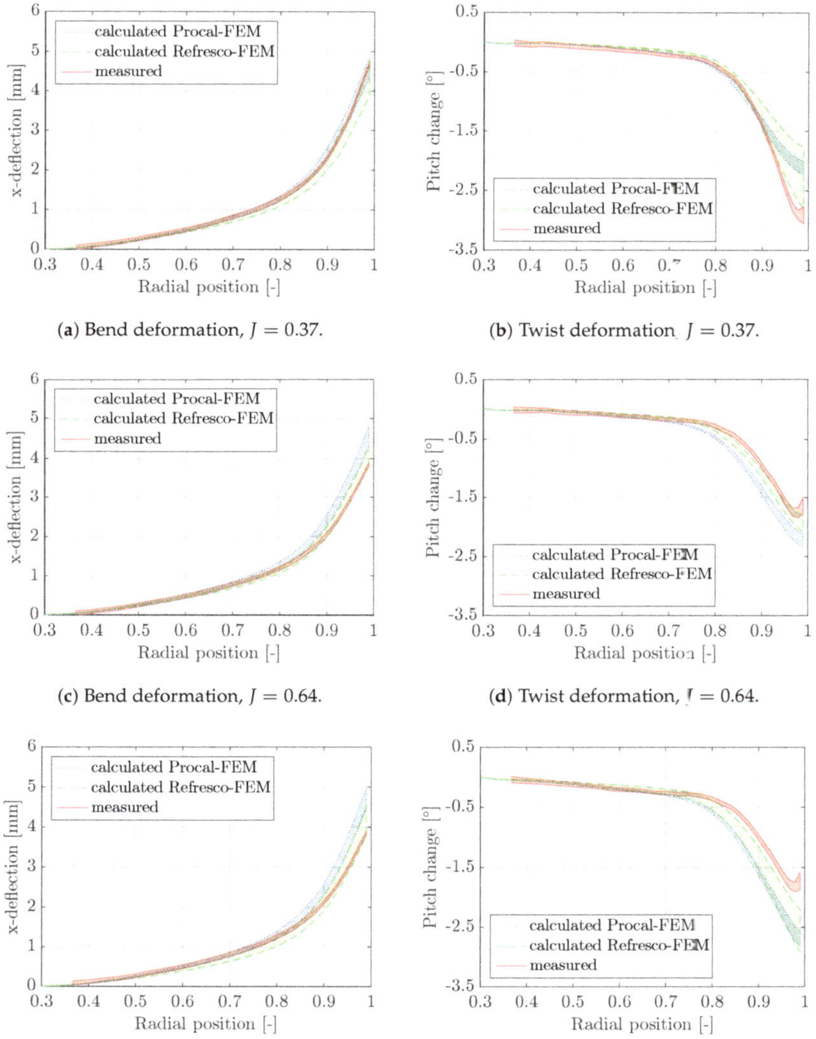

(**a**) Bend deformation, $J = 0.37$.
(**b**) Twist deformation, $J = 0.37$.
(**c**) Bend deformation, $J = 0.64$.
(**d**) Twist deformation, $J = 0.64$.
(**e**) Bend deformation, $J = 0.85$.
(**f**) Twist deformation, $J = 0.85$.

Figure 16. Uncertainty intervals for bend (**left**) and twist (**right**) deformations of mid-chord points of epoxy propeller blade 1 against the radial position on the blade, for the measured and calculated responses.

In general, the responses calculated with the BEM-FEM coupling are larger and differ more from the measured deformations than those obtained with RANS-FEM calculations. This is in line with the differences in thrust and torque computed with BEM and RANS compared to the measured open water diagram as presented in Figure 5. By looking at the BEM-FEM calculated response of the composite blades, it can be concluded that, for all the blades and loading conditions, the uncertainty intervals for measured and BEM-FEM calculated responses overlap up to 0.7 of the propeller radius. Overall, the best resemblance between measured and BEM-FEM calculated responses is obtained for the epoxy propeller. This is explained by the less complicated modelling of the epoxy material than that of composite material.

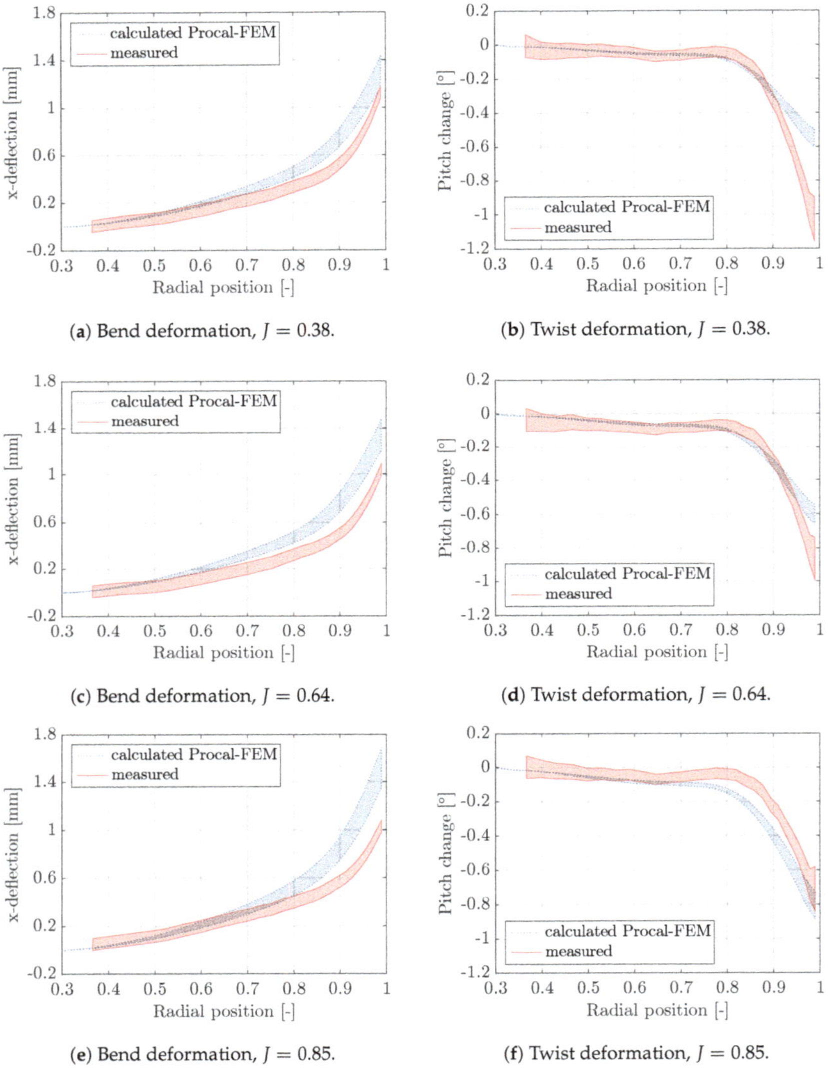

(**a**) Bend deformation, $J = 0.38$.

(**b**) Twist deformation, $J = 0.38$.

(**c**) Bend deformation, $J = 0.64$.

(**d**) Twist deformation, $J = 0.64$.

(**e**) Bend deformation, $J = 0.85$.

(**f**) Twist deformation, $J = 0.85$.

Figure 17. Uncertainty intervals for bend (**left**) and twist (**right**) deformations of mid-chord points of propeller 45 blade 1 against the radial position on the blade, for the measured and calculated responses.

In general, the results obtained for epoxy and composite propellers are consistent: bend deformations are always over-predicted with the BEM-FEM coupling. For the twist deformations, it depends on the loading condition: for the lowest and the highest advance coefficient, the twist deformations are under-predicted and over-predicted, respectively. For the intermediate advance coefficient, the best agreement between measured and calculated twist deformations is obtained.

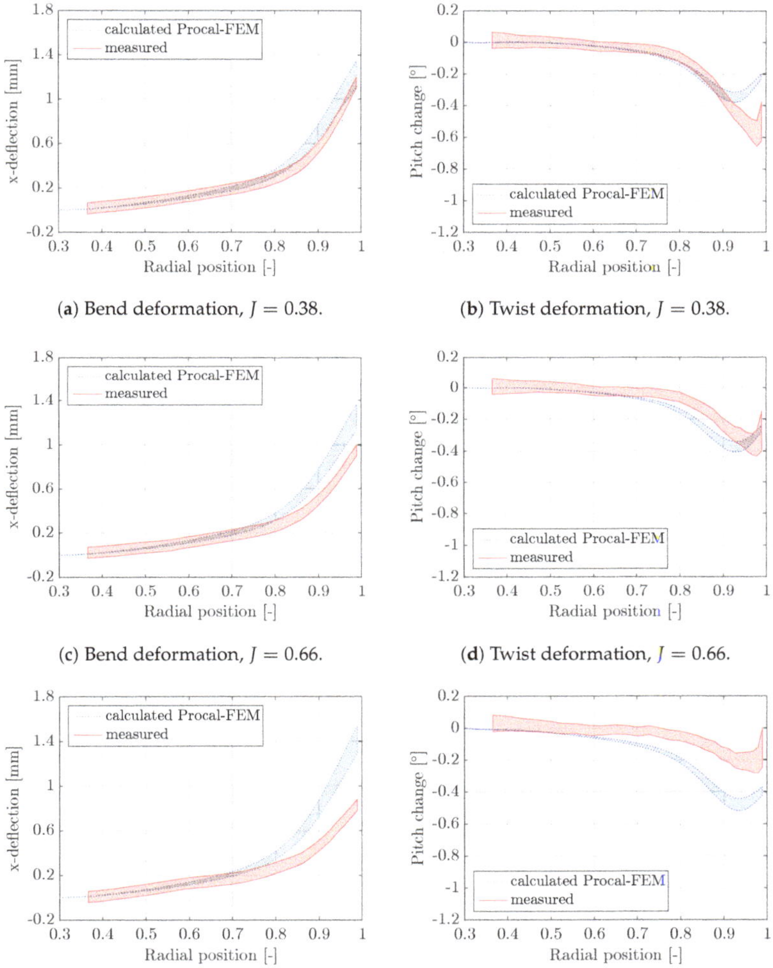

(**a**) Bend deformation, $J = 0.38$. (**b**) Twist deformation, $J = 0.38$.

(**c**) Bend deformation, $J = 0.66$. (**d**) Twist deformation, $J = 0.66$.

(**e**) Bend deformation, $J = 0.85$. (**f**) Twist deformation, $J = 0.85$.

Figure 18. Uncertainty intervals for bend (**left**) and twist (**right**) deformations of mid-chord points of propeller 90 blade 2 against the radial position on the blade, for the measured and calculated responses.

Since no RANS-FEM computations are performed for the composite propellers, it is difficult to examine whether the differences between calculated and measured responses for these propellers can be explained from the BEM modelling uncertainty. For propeller 45, it is assumed that this is the case, since qualitatively the predicted and measured response for this propeller is very similar to the response of the epoxy propeller. For propeller 90, the differences between measured and calculated response are larger, especially for the advance ratio of 0.85. Furthermore, the twist deformation of this propeller is completely different from the other two propellers, but qualitatively well-predicted with the BEM-FEM coupling.

10. Conclusions, Recommendations and Further Work

In this work, a BEM-FEM coupling is presented for analysing the hydro-elastic behaviour of flexible propellers in uniform flows. This code has been validated with small scale experiments and compared to the results of RANS-FEM calculations.

From the comparison between the measured open water diagram and the open water curves calculated with BEM and RANS, it can be concluded that, for the two lowest advance ratios considered in this work, the resemblance between RANS predicted and measured K_T and K_Q values is acceptable. The differences that were found for the highest advance ratio might originate from the turbulence modelling. A transition model might be more appropriate, especially for the highest advance coefficient, but was considered out of the current scope. It can be concluded that the open water curves calculated with PROCAL clearly show the limitations of BEM. For high advance coefficients ($J > 1$), the viscous effects play an increasing role and therefore the PROCAL results become inaccurate. For $J < 0.4$, a strong leading edge vortex is generated in the RANS results, and it is hypothesized that this is the reason that the results of the PROCAL calculations diverge from measured results for low advance ratios.

Interesting results are obtained from the uncertainty analysis with the RANS grids. However, the thrust and torque values were converged, and it has been shown that this does not automatically mean that the blade structural response is converged as well. Therefore, it is recommended to include a criteria on the convergence of blade bending and twisting in grid refinement studies for these types of calculations.

From the comparison of RANS-FEM and BEM-FEM results to the results of the blade deformation measurements on the epoxy propeller, it can be concluded that the bending response is well predicted with the RANS-FEM and BEM-FEM simulations. For the BEM-FEM coupling, this is despite the limitations of BEM and the complicated flow characteristics, and, therefore, it is expected that, for many other propeller geometries, the BEM-FEM coupling can correctly predict the bending response. Depending on the advance ratio, a fair to poor prediction of the twist deformations of the epoxy propeller is obtained with the RANS-FEM and BEM-FEM coupling. The best agreement between measured and calculated twist deformations is obtained for the RANS-FEM results for an advance ratio of 0.64, in which leading edge vortex separation is present but limited compared to the lowest advance ratio and viscous forces might be less dominating than for the highest advance coefficient.

The results of this work show that the differences between measured and predicted responses of the composite propellers are larger than for the epoxy propeller. This is attributed to the more complicated FEM modelling of the composite material and most likely there is a bigger spread between design and actual material properties than for the epoxy material. Therefore, it is recommended to validate these types of fluid–structure analyses with propellers made out of isotropic and flexible material. When composite materials are used, it is recommended to do the experiments on a larger scale or with very flexible composite blades so that the measurement uncertainty becomes less dominant.

Regarding the consequences of the potential flow simplifications on the FSI response of flexible propeller, it can be concluded that, for the lowest and the highest advance coefficient, the uncertainty bandwidths of the twist deformation curves obtained with BEM-FEM and RANS-FEM overlap. This is due to the large RANS-FEM discretisation uncertainty in these calculations, rather than a correct prediction of the twist deformations with the BEM-FEM coupling, since, for these conditions, the BEM-FEM predicted twist responses deviate significantly from the measurements. The differences between measured and BEM-FEM calculated twist deformations are, for the low advance ratio, explained by the strong leading edge vortex separation, which is not computed in the BEM. For the highest advance ratio, viscous effects play an increasingly important role. It is obvious that the best resemblance between measurements, RANS-FEM and BEM-FEM results, is obtained for the intermediate advance ratio for which flow separation and viscous effects play a less dominant role. Hence, it can be concluded that, with a BEM-FEM approach, the bending response can be well predicted; however, in case of extreme flow separation and viscous effects, a BEM-FEM approach for computing the twist deformations is not recommended.

Author Contributions: Pieter Maljaars developed the BEM-FEM coupling and performed the computations with this coupling approach; furthermore, he compared all the results and he mainly wrote the paper. Laurette Bronswijk executed the RANS and RANS-FEM computations during her MSc. graduation internship at the MARIN Academy. Jaap Windt is co-developer of the RANS-FEM coupling and supervised Laurette Bronswijk. Nicola Grasso performed the experiments. Mirek Kaminski contributed by managing the project.

Acknowledgments: The authors gratefully acknowledge the Defence Materiel Organisation, Maritime Research Institute Netherlands, Wärtsilä Netherlands B.V., Solico B.V. and NWO, domain Toegepaste en Technische Wetenschappen (project number 13278) for their financial support. The authors gratefully thank the Cooperative Research Ships (CRS) for making their BEM software PROCAL available for this work. The authors gratefully acknowledge Maritime Research Institute Netherlands for making their propeller geometry toolbox PROPART available for this work. The authors gratefully acknowledge Airborne for the manufacturing of the epoxy and composite propellers.

Conflicts of Interest: The authors declare no conflict of interest.

References

1. Mulcahy, N.; Prusty, B.; Gardiner, C. Hydroelastic tailoring of flexible composite propellers. *Ship Offshore Struct.* **2010**, *5*, 359–370.
2. He, X.; Hong, Y.; Wang, R. Hydroelastic optimisation of a composite marine propeller in a non-uniform wake. *Ocean Eng.* **2012**, *39*, 14–23.
3. Taketani, T.; Kimura, K.; Ando, S.; Yamamoto, K. Study on performance of a ship propeller using a composite material. In Proceedings of the Third International Symposium on Marine Propulsors, Launceston, Australia, 5–8 May 2013.
4. Solomon Raj, S.; Ravinder Reddy, P. Bend-twist coupling and its effect on cavitation inception of composite marine propeller. *Int. J. Mech. Eng. Technol.* **2014**, *5*, 306–314.
5. Kuo, J.; Vorus, W. Propeller blade dynamic stress. In Proceedings of the Tenth Ship Technology and Research (STAR) Symposium, Norfolk, VA, USA, 21–24 May 1985; pp. 39–69.
6. Georgiev, D.; Ikehata, M. Hydro-elastic effects on propeller blades in steady flow. *J. S. Nav. Archit. Jpn.* **1998**, *184*, 1–14.
7. Young, Y. Time-dependent hydro-elastic analysis of cavitating propulsors. *J. Fluids Struct.* **2007**, *23*, 269–295.
8. Blasques, J.; Berggreen, C.; Andersen, P. Hydro-elastic analysis and optimization of a composite marine propeller. *Mar. Struct.* **2010**, *23*, 22–38.
9. Ghassemi, H.; Ghassabzadeh, M.; Saryazdi, M. Influence of the skew angle on the hydro-elastic behaviour of a composite marine propeller. *J. Eng. Marit. Environ.* **2012**, *226*, 346–359.
10. Sun, H.; Xiong, Y. Fluid-structure interaction analysis of flexible marine propellers. *Appl. Mech. Mater.* **2012**, *226–228*, 479–482.
11. Lee, H.; Song, M.; Suh, J.; Chang, B. Hydro-elastic analysis of marine propellers based on a BEM-FEM coupled FSI algorithm. *Int. J. Nav. Archit. Ocean Eng.* **2014**, *6*, 562–577.
12. Maljaars, P.; Dekker, J. Hydro-elastic analysis of flexible marine propellers. In *Maritime Technology and Engineering*; Guedes Soares, C., Santos, T., Eds.; CRC Press-Taylor & Francis Group: London, UK, 2014; pp. 705–715.
13. Atkinson, P.; Glover, E. *Propeller Hydro-Elastic Effects; Propellers '88 Symposium*; SNAME: Virginia Beach, VA, USA, 1988.
14. Lin, H.; Lin, J. Nonlinear hydroelastic behavior of propellers using a finite element method and lifting surface theory. *J. Mar. Sci. Technol.* **1996**, *1*, 114–124.
15. Salvatore, F.; Streckwall, H.; van Terwisga, T. Propeller cavitation modelling by CFD-Results from the VIRTUE 2008 Rome workshop. In Proceedings of the First International Symposium on Marine Propulsors, Trondheim, Norway, 22–24 June 2009.

16. Vaz, G.; Hally, D.; Huuva, T.; Bulten, N.; Muller, P.; Becchi, P.; Herrer, J.; Whitworth, S.; Macé, R.; Korsström, A. Cavitating flow calculations for the E779A propeller in open water and behind conditions: Code comparison and solution validation. In Proceedings of the Fourth International Symposium on Marine Propulsors, Austin, TX, USA, 31 May–4 June 2015.
17. Chae, E.; Akcabay, D.; Young, Y. Influence of flow-induced bend-twist coupling on the natural vibration responses of flexible hydrofoils. *J. Fluids Struct.* **2017**, *69*, 323–340.
18. Chae, E.; Akcabay, D.; Lelong, A.; Astolfi, J.; Young, Y. Numerical and experimental investigation of natural flow-induced vibrations of flexible hydrofoils. *Phys. Fluids* **2016**, *28*, 075102.
19. Chen, J.; Hallet, S.; Wisnom, M. Modelling complex geometry using solid finite element meshes with correct composite material orientations. *Comput. Struct.* **2010**, *88*, 602–609.
20. Maljaars, P.; Kaminski, M.; den Besten, J. Finite element modelling and model updating of small scale composite propellers. *Compos. Struct.* **2017**, *176*, 154–163.
21. Vaz, G. Modelling of Sheet Cavitation on Hydrofoils and Marine Propellers Using Boundary Element Methods. Ph.D. Thesis, Instituto Superior Técnico, Lisbon, Portugal, 2005.
22. Vaz, G.; Bosschers, J. Modelling of three dimensional sheet cavitation on marine propellers using a boundary element method. In Proceedings of the Sixth International Symposium on Cavitation, Wageningen, The Netherlands, 11–15 September 2006.
23. Maljaars, P.; Grasso, N.; Kaminski, M.; Lafeber, W. Validation of a steady BEM-FEM coupled simulation with experiments on flexible small scale propellers. In Proceedings of the Fifth International Symposium on Marine Propulsors, Espoo, Finland, 12–15 June 2017.
24. ITTC 1978. Report of performance committee. In Proceedings of the 15th International Towing Tank Conference, The Hague, The Netherlands, 3–10 September 1978.
25. ReFRESCO. ReFRESCO Community Site. Available online: http://www.refresco.org (accessed on 12 March 2018).
26. Vaz, G.; Jaouen, F.; Hoekstra, M. Free-surface viscous flow computations. Validation of URANS code FRESCO. In Proceedings of the ASME 2009 28th International Conference on Ocean, Offshore and Arctic Engineering, Honolulu, HI, USA, 31 May–5 June 2009.
27. Klaij, C.; Vuik, C. Simple-type preconditioners for cell-centered, colocated finite volume discretization of incompressible Reynolds-averaged Navier–Stokes equations. *Int. J. Numer. Methods Fluids* **2013**, *71*, 830–849.
28. Menter, F.; Egorov, Y.; Rusch, D. Steady and unsteady flow modelling using the k-skl model. In Proceedings of the Turbulence Heat and Mass Transfer, Dubrovnik, Croatia, 25–29 September 2006.
29. Rijpkema, D.; Baltazar, J.; Falcao de Campos, J. Viscous flow simulations of propellers in different Reynolds number regimes. In Proceedings of the Fourth International Symposium on Marine Propulsors, Austin, TX, USA, 31 May–4 June 2015.
30. Eça, L.; Hoekstra, M. A procedure for the estimation of the numerical uncertainty of CFD calculations based on grid refinement studies. *J. Comput. Phys.* **2014**, *262*, 104–130.
31. Jongsma, S.; van der Weide, E.; Windt, J. Implementation and verification of a partitioned strong coupled fluid–structure interaction approach in a finite volume method. In Proceedings of the International Conference in Hydrodynamics, Egmond aan Zee, The Netherlands, 18–23 September 2016.
32. Turek, S.; Hron, J. Proposal for numerical benchmarking of fluid–structure interaction between an elastic object and laminar incompressible flow. In *Fluid–Structure Interaction-Modelling, Simulation, Optimization*; Number 53; Springer: Berlin/Heidelberg, Germany, 2006; pp. 371–385.
33. Sutton, J.; Ortue, J.; Schreier, H. *Image Correlation for Shape, Motion and Deformation Measurements: Basic Concepts, Theory and Applications*; Springer: New York, NY, USA, 2009.
34. Zondervan, G.; Grasso, N.; Lafeber, W. Hydrodynamic design and model testing techniques for composite ship propellers. In Proceedings of the Fifth International Symposium on Marine Propulsion, Espoo, Finland, 12–15 June 2017.
35. Glauert, H. *The Elements of Aerofoil and Airscrew Theory*, 2nd ed.; Cambridge University Press: New York, NY, USA, 1947.

Journal of
Marine Science and Engineering

MDPI

Article

Nominal vs. Effective Wake Fields and Their Influence on Propeller Cavitation Performance

Pelle Bo Regener [1,*], Yasaman Mirsadraee [1,2] and Poul Andersen [1]

[1] Department of Mechanical Engineering, Technical University of Denmark, Nils Koppels Allé 403, 2800 Kgs. Lyngby, Denmark; yasmir@mek.dtu.dk (Y.M.); pa@mek.dtu.dk (P.A.)
[2] MAN Diesel & Turbo, Engineering and R&D Propulsion, 2450 København SV, Denmark
* Correspondence: pboreg@mek.dtu.dk

Received: 27 February 2018; Accepted: 29 March 2018; Published: 5 April 2018

Abstract: Propeller designers often need to base their design on the nominal model scale wake distribution because the effective full scale distribution is not available. The effects of such incomplete design data on cavitation performance are examined in this paper. The behind-ship cavitation performance of two propellers is evaluated, where the cases considered include propellers operating in the nominal model and full scale wake distributions and in the effective wake distribution, also in the model and full scale. The method for the analyses is a combination of RANS for the ship hull and a panel method for the propeller flow, with a coupling of the two for the interaction of ship and propeller flows. The effect on sheet cavitation due to the different wake distributions is examined for a typical full-form ship. Results show considerable differences in cavitation extent, volume, and hull pressure pulses.

Keywords: propeller cavitation; wake scaling; effective wake; RANS-BEM coupling

1. Introduction

1.1. Motivation

Propeller cavitation is strongly influenced by the non-uniform inflow to the propeller. As the ship wake is dominated by viscous effects, it is subject to major scale effects. Still, propeller designers are generally only provided with the nominal wake field measured at model scale. This then needs to be scaled to match the estimated full scale effective wake fraction from a self-propulsion test to ensure that the average axial velocity in the propeller disk corresponds to the average effective inflow. However, scaling the field uniformly will not result in the right velocity distribution. In addition to the scale effects, the actual inflow field to the propeller is not the nominal field, but the effective wake field, including hull–propeller interaction. As a result, the propeller designer might base design decisions on insufficiently accurate information.

A successful propeller design is a trade-off between propeller cavitation performance and total propeller efficiency. The ability to predict cavitation performance at early design stages will eliminate the need for overly conservative designs. This paper intends to highlight the role of the wake field in the design, analysis, and optimization of a conventional ship propeller.

1.2. Background

Due to the higher Reynolds number at full scale compared to model scale, the boundary layer around the ship hull changes and hence the velocity distribution near the hull is altered. This difference results in a narrower wake peak and a lower wake fraction in full scale. The presence of the propeller behind the ship adds to the complexity of the problem because the propeller–hull interaction modifies the inflow field to the propeller as well.

Single-screw ship wake fields are usually characterized by a strongly non-uniform distribution of velocities with a wake peak at the 12 o'clock position, where the axial velocities are particularly low. This means that the blade sections of a propeller operating behind the ship experience strong variations in angle of attack. As the hydrostatic pressure acting on the blade reaches its minimum at the same time as the blade experiences high angles of attack while passing through the wake peak, this region of the wake field is particularly critical in terms of cavitation.

Analyzing the different factors influencing propeller cavitation and related erosion and vibration issues, an ITTC propulsion committee [1] pointed out that, for large container ships with highly-loaded propellers, the wake field characteristics—and not propeller geometry details—are the key to achieving decent propeller cavitation performance.

Especially the depth of the wake peak, i.e., the difference between the lowest axial velocity occurring there and the maximum velocity in the propeller disk, is of decisive importance for the cavitation performance of a propeller behind the ship. When uniformly scaling the nominal wake velocities to match the effective wake fraction, the width and depth of the wake peak are unlikely to be represented properly.

As this has been known for many years, different methods exist for estimating the full scale wake field of a ship, covering a rather wide range of complexity and sophistication. Usually the nominal wake field, measured at model scale, serves as input for these methods. A review of the most commonly used scaling methods was carried out some years ago by an ITTC specialist committee on wake field scaling [2]. That report mentions the simplest form of wake scaling, where one only scales the wake field by changing the magnitude of the velocities uniformly to match a target wake fraction, as already described above. In that case, the shape of the isolines of the input field (usually the measured nominal wake field) remains unchanged. Therefore, even calling this procedure a "scaling method" is questionable. While the shortcomings of this approach are well-known, it still appears to be commonly used for its simplicity.

Adding complexity while still only requiring very limited effort, the semi-empirical scaling method described by Sasajima and Tanaka [3] decomposes the wake into a frictional and potential component and contracts the frictional wake based on horizontal velocity profiles. This still-popular method was recommended (with warnings) by the above-mentioned ITTC committee in 2011 for the case when full scale wake data are not available.

The main challenges related to scale effects on the effective wake field and the corresponding effect on propeller cavitation are outlined in a review by van Terwisga et al. [4]. The work by Bosschers et al. [5] is one of the relatively few examples that explicitly show and discuss differences in propeller cavitation due to the effect of hull–propeller interaction and local changes in the wake distribution. That paper focuses on scale effects, offers comparisons with experiments at model and full scale, and uses similar methods as the present work for computation of effective wake fields and sheet cavitation prediction.

Gaggero et al. [6] compared the cavitation performance of conventional propellers in nominal full scale wake fields from CFD calculations to the performance in nominal full scale wake fields obtained by applying an empirical wake scaling method to measured nominal fields at model scale, and observed noticeable differences.

The present work focuses on the differences in predicted sheet cavitation behavior purely due to the wake distribution, using computed nominal and effective wake fields at both model and full scale. Section 4 presents analysis results for one propeller and different wake distributions. Section 5 introduces another propeller designed for the same ship, showing the difference in cavitation behavior between propellers designed for different wake distributions.

2. Methods

2.1. Boundary Element Method for Propeller Analysis

A potential-based boundary element method ("panel code") serves as the main tool for propeller analysis in the present work. As for all potential flow-based methods, inviscid, incompressible, and irrotational flow is assumed. Sheet cavitation is modeled in a partially nonlinear way. The present implementation's approach to cavitation modeling—whose basic formulation is reproduced below—follows the approach initially introduced by Kinnas and Fine [7] and is able to predict unsteady sheet cavitation in inhomogeneous inflow, including supercavitation.

Mathematical Formulation

The velocity potential must satisfy the Laplace equation:

$$\nabla^2\Phi = 0. \tag{1}$$

Given the linearity of Equation (1), the total velocity vector $\boldsymbol{U}_{\text{Total}} = \nabla\Phi$ can be split into a known onset part $\boldsymbol{U}_{\text{Onset}}$ (dependent on the wake field with the local velocity $\boldsymbol{U}_{\text{Wake}}$ and the propeller rotation) and the gradient of a propeller geometry-dependent perturbation potential ϕ that is to be determined:

$$\boldsymbol{U}_{\text{Total}} = \boldsymbol{U}_{\text{Onset}} + \nabla\phi. \tag{2}$$

For a domain bound by the blade surface S_B (with a continuous distribution of sources and dipoles) and the force-free wake surface S_W (with a continuous distribution of dipoles), application of Green's second identity gives the potential at a field point p, when the integration point q lies on the domain boundary. G is defined as the inverse of the distance R between these two points, $G = \frac{1}{R}$. The term $\Delta\phi_q$ corresponds to the potential jump across the wake sheet at an integration point q on S_W. An additional term appears in the presence of supercavitation, as additional sources—with the strength $\Delta\frac{\partial\phi_q}{\partial n}$—are placed on the cavitating part of the wake, $S_{CW} \subset S_W$.

If the field point p lies on the blade surface, the potential ϕ_p is found from

$$2\pi\phi_p = \int_{S_B}\left[\phi_q\frac{\partial G}{\partial n} - G\frac{\partial\phi_q}{\partial n}\right]dS - \int_{S_{CW}}\left[G\Delta\frac{\partial\phi_q}{\partial n}\right]dS + \int_{S_W}\left[\Delta\phi_q\frac{\partial G}{\partial n}\right]dS. \tag{3}$$

As the surface S_W in principle consists of two surfaces collapsed into one infinitely thin wake sheet, the integral equation reads slightly differently if the field point lies on S_W. The potential ϕ_p for a field point on the wake surface is

$$4\pi\phi_p = 2\pi\Delta\phi_q + \int_{S_B}\left[\phi_q\frac{\partial G}{\partial n} - G\frac{\partial\phi_q}{\partial n}\right]dS - \int_{S_{CW}}\left[G\Delta\frac{\partial\phi_q}{\partial n}\right]dS + \int_{S_W}\left[\Delta\phi_q\frac{\partial G}{\partial n}\right]dS. \tag{4}$$

Equations (3) and (4) are then discretized using flat quadrilateral panels arranged in a structured mesh. Introducing influence coefficient matrices $\boldsymbol{A}$ through $\boldsymbol{H}$ that describe the influence from unit strength singularities located on panel j on the control point of panel i, a system of $J_B + J_{CW}$ equations and unknowns results. For each panel on the blade, one equation of the following form exists:

$$2\pi\phi_i = \sum_{J_B}(-\phi_j\boldsymbol{A}_{ij}) + \sum_{J_B}(\sigma_j\boldsymbol{B}_{ij}) + \sum_{J_{CW}}(\sigma_j\boldsymbol{C}_{ij}) - \sum_{J_W}(\Delta\phi_j\boldsymbol{G}_{ij}), \tag{5}$$

and, for each cavitating wake panel, there is an additional equation of the form

$$4\pi\phi_i = \sum_{J_B}(-\phi_j\boldsymbol{D}_{ij}) + \sum_{J_B}(\sigma_j\boldsymbol{E}_{ij}) + \sum_{J_{CW}}(\sigma_j\boldsymbol{F}_{ij}) - \sum_{J_W}(\Delta\phi_j\boldsymbol{H}_{ij}) + 2\pi\Delta\phi_j. \tag{6}$$

On the wetted part of the blade, the source strengths $\sigma_i = \frac{\partial \phi}{\partial n}$ are known from the kinematic boundary condition, Equation (8), and the dipole strength is the unknown:

$$\nabla \Phi \cdot \boldsymbol{n} = \boldsymbol{U}_{\text{Onset}} \cdot \boldsymbol{n} + \frac{\partial \phi}{\partial n} = 0, \tag{7}$$

$$\frac{\partial \phi}{\partial n} = -\boldsymbol{U}_{\text{Onset}} \cdot \boldsymbol{n}. \tag{8}$$

On the cavitating part of the blade and wake surfaces, a dynamic boundary condition is applied, prescribing the pressure to correspond to the given cavitation number σ_n. To achieve this, the corresponding local "cavity velocity" needs to be found. For convenience, this part is formulated in curvilinear coordinates aligned with the panel edges. The v-direction is pointing outwards in the spanwise direction and the s-direction is the chordwise direction, positive towards the trailing edge on the suction side of the blade. The angle between the $\hat{s}$ and $\hat{v}$ vectors of a panel is designated θ and is usually close to 90°. With U_s and U_v as the s- and v-components of the onset velocity vector and z as the vertical distance from the propeller shaft, the chordwise cavity velocity corresponding to the cavitation number σ_n at shaft depth is

$$\frac{\partial \phi}{\partial s} = -U_s + \cos(\theta) \left(\frac{\partial \phi}{\partial v} + U_v \right) + \sin(\theta) \sqrt{f}, \tag{9}$$

where

$$f = (nD)^2 \sigma_n + |\boldsymbol{U}_{\text{Onset}}|^2 - \left(\frac{\partial \phi}{\partial v} + U_v \right)^2 - 2\frac{\partial \phi}{\partial t} - 2gz. \tag{10}$$

To be able to provide a Dirichlet boundary condition on the potential, Equation (9) is integrated in chordwise direction and added to the potential at the chordwise detachment point ϕ_0, which is assumed known and practically expressed by extrapolation from the wetted part ahead of the cavity:

$$\phi = \phi_0 + \int_0^{s_p} \frac{\partial \phi}{\partial s} \mathrm{d}s. \tag{11}$$

The cavity extent on the blade (and wake) needs to be found iteratively. After an initial guess based on the non-cavitating pressure distribution and cavitation number, the cavity thickness is computed. The cavity extent is then changed until the cavity thickness is sufficiently close to zero at the edges of the cavity sheet, so it detaches from the blade and closes on the blade or wake. For the cavity closure, the pressure recovery model described in [7] is used.

The present implementation also includes additional features described in [8], such as the split panel technique for faster convergence and more flexibility in terms of mesh and timestep size. However, the present implementation (the panel code "ESPPRO", developed at the Technical University of Denmark), uses lower-order extrapolation schemes throughout for increased numerical stability. In addition, spatial derivatives in the cavity height equation are discretized using lower-order finite differences.

The results obtained with the "ESPPRO" code agree well with the results reported for similar methods in validation studies for ship propellers, such as [9,10].

2.2. RANS-BEM Coupling

In recent years, viscous flow simulations around the hull coupled with potential flow-based propeller models have become a popular choice for numerical self-propulsion simulations. Usually, field methods solving the Reynolds-averaged Navier–Stokes (RANS) equations are used for the hull part, and panel methods (boundary element methods, BEM) are a common choice for the propeller calculations, as they allow for a good representation of the flow physics while only requiring limited computational effort. Computational approaches using this combination of tools are commonly called "RANS-BEM Coupling".

In such an approach, the exact propeller geometry is not resolved in CFD, but the propeller is accounted for by modeling its effect on the flow by introducing body forces, provided by the propeller model. This is an iterative process: the total velocity field in the coupling plane (an approximation of the propeller plane) is passed from the RANS solver to the propeller model, which then returns the propeller forces that correspond to this inflow field. The key part here is that the inflow field to the propeller model is not the total wake field as extracted from the global CFD simulation, but rather the effective wake field, i.e., with the propeller-induced velocities subtracted from the total wake field. The induced velocity field is approximated by using the values from the previous coupling iteration. Thereby, the effective wake field is not only a byproduct but also an inherent part of this iterative coupling.

By being able to compute not only the effective wake *fraction* but also the *distribution* of the effective wake velocities in the propeller disk, the RANS-BEM coupling approach provides a major advantage over all CFD simulations beyond substantially reducing the computational effort.

In the present work, the RANS-BEM coupling approach is used to determine effective wake fields at model and full scale to later investigate differences in propeller cavitation. On the RANS side, the XCHAP solver from the commercial SHIPFLOW package is used. XCHAP solves the steady-state RANS equations on overlapping, structured grids using the finite volume method and employs the EASM (Explicit Algebraic Stress Model) turbulence model. Nominal wake fields are found using the same RANS solver and identical grids, but with the propeller model switched off.

The DTU-developed panel code ESPPRO (whose basic formulation is described in the previous section) serves as the propeller model. The unsteady propeller forces are time-averaged over one revolution before being passed to XCHAP. In line with that, the induced velocity field is also time-averaged to compute the effective wake field in the subsequent coupling iteration.

When modeling the propeller in RANS by a body force distribution only, the blade blockage effect is present in the BEM, but not transferred to the RANS side of the computation. The implications of this are shown in [11], which also discusses two possibilities to address this problem in a consistent manner: either by accounting for the blockage on the RANS side by mass sources, or by removing the blockage from the BEM side. In the present work, the latter option—the approach initially described in [12]—is followed, excluding the contribution of all sources when computing forces and velocities used as part of the RANS-BEM coupling.

To reduce computational effort, the non-cavitating condition is assumed in self-propulsion and the cavitation model described previously remains disabled. This is not expected to have a noticeable effect on the computed effective wake distributions, as in the present model a sheet cavity is primarily represented by the presence of additional sources. As mentioned above, sources do not contribute to relevant quantities used in the RANS-BEM coupling. The remaining relevant effect of the sheet cavity on the computed effective wake field, the change in the dipole strength distribution on the blade, is considered negligibly small for the expected cavitation extent.

Previous work on computing effective wake fields using RANS-BEM coupling by Rijpkema et al. [12] highlighted the importance and influence of the location of the coupling plane on self-propulsion results. As singularities are placed on the propeller blade surfaces in the panel method, the induced velocities can not be computed in the propeller plane directly. To avoid evaluating induced velocities too close to the singularities, the coupling plane is usually chosen to be upstream of the propeller. Extrapolating the effective wake to the propeller plane linearly from two upstream planes was found [12] to give the best results in terms of predicting the self-propulsion point (effective wake fraction in self-propulsion).

For the present work, the distribution of effective velocities in the coupling plane is of higher interest than the mean velocity, i.e., the absolute value of the wake fraction. Therefore, and to remove any potential extrapolation artifacts that affect the distribution, the effective wake is computed on a single curved surface that follows the blade leading edge contour closely (at an axial distance of 2% of the propeller diameter), essentially resulting in a curved coupling plane just upstream of the propeller.

3. Case Study

All calculations are carried out for a state-of-the-art handysize bulk carrier, representing a modern single-screw full hull form. The block coefficient is 0.82 and the aftbody is of pram-with-gondola-type. The simple conventional propeller used for the analyses was specifically designed for this ship and the present work (see Section 5 for more details on the design method).

In this work, nominal wake fields at model and full scale for this ship are obtained by running steady-state RANS-based CFD simulations. Effective wake fields at the self-propulsion point, also at both model and full scale, are computed using the hybrid RANS-BEM method described in the previous section. Additionally, Sasajima's wake scaling method is applied to the computed model scale nominal wake field for comparison with the computed full scale fields.

The effect of the wake distribution on propeller cavitation performance, including cavitation extent, cavitation volume, and hull pressure pulses, is examined by using the panel code with the sheet cavitation model described in Section 2.1.

For the cavitation analyses, the axial components of all five wake fields (nominal and effective at model and full scale, plus the result of Sasajima's scaling method) are then uniformly scaled to the same axial wake fraction, so the propeller is running at the same operating point in all wake fields. Any differences in results are then due to the different velocity *distributions* in the propeller disk and different in-plane velocity components.

All cavitation simulations are carried out at a cavitation number (based on the propeller speed n) of $\sigma_n = 1.8$, corresponding to the full scale condition. The propeller for this ship is moderately loaded ($C_{TH} = 1.4$) at the operating point considered.

Note that approach, case, and all simulation conditions are identical to those used in the earlier version of the present work, presented and published by the authors at the Fifth International Symposium on Marine Propulsors (smp'17) [13]. Nominal and effective wake fields are unchanged, while there are differences in cavitation simulation results. These differences are due to changes made to the ESPPRO code and the corresponding preprocessor. The improvements allow for higher spatial resolution and higher-quality meshes, and an improved cavity shape-finding mechanism that reduces the openness tolerance by one order of magnitude compared to previous versions. Apart from increasing accuracy, the latter also affects the rate at which the cavity can grow and shrink. The present results thereby reflect progress made between January 2017 and January 2018. The authors are not aware of major problems in either of the two versions, and the results primarily differ in magnitude, showing the same trends and leading to the same conclusions.

4. Results and Discussion

4.1. Wake Fields

Figure 1a–d show the nominal and effective wake distributions based on the RANS and RANS-BEM results. Figure 1e shows the full scale wake field after applying Sasajima's scaling method, based on the computed nominal wake at model scale and the potential wake (see Figure 2), computed using the panel code SHIPFLOW-XPAN.

Note, however, that all fields shown have been "scaled" or "corrected" to match the same axial wake fraction of 0.25. This value corresponds to the full scale prognosis for the effective wake fraction based on model tests with a stock propeller for a ship very similar to the case considered here. Apart from this number, no further model test results have been used in any part of this work. For an original wake fraction w_0 and a target value of w_t, the "corrected" non-dimensional axial velocity u at any point is $u = 1 - f + u_0 f$, where u_0 is the original non-dimensional axial velocity at that point and $f = \frac{w_t}{w_0}$.

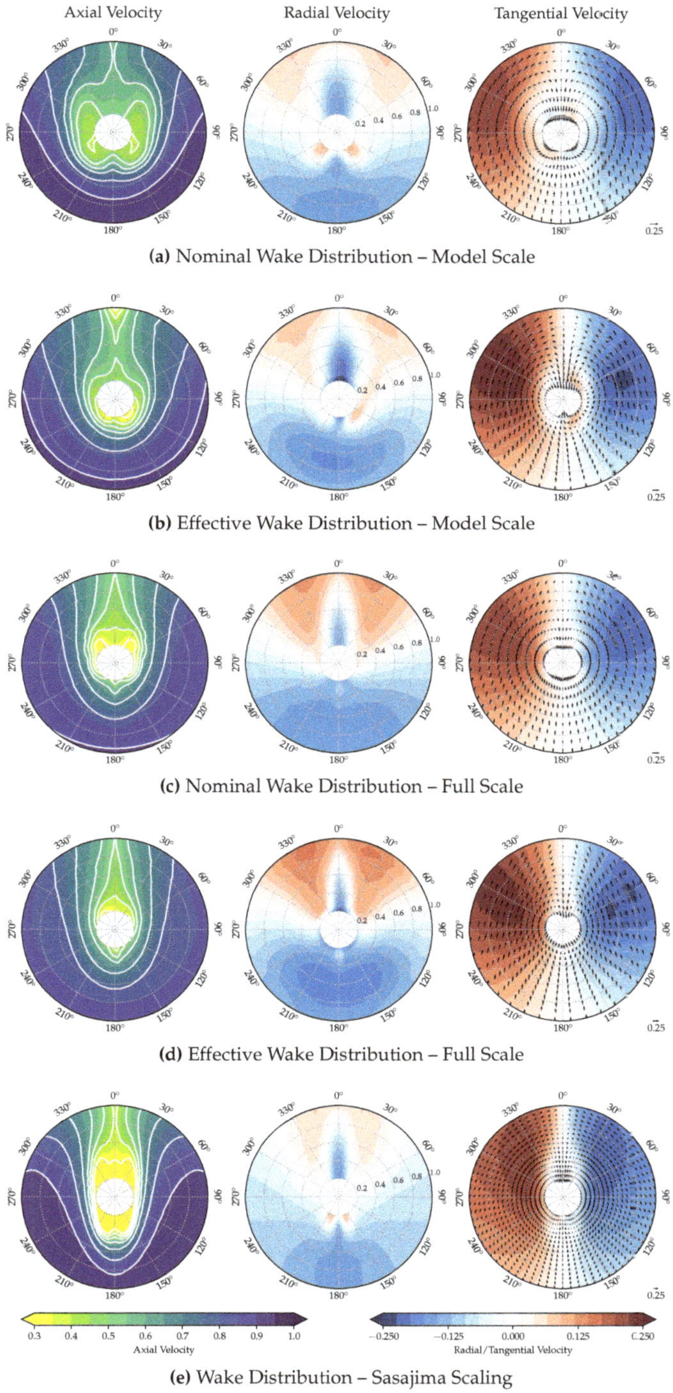

Figure 1. Wake fields for cavitation analysis, all scaled to $w = 0.25$.

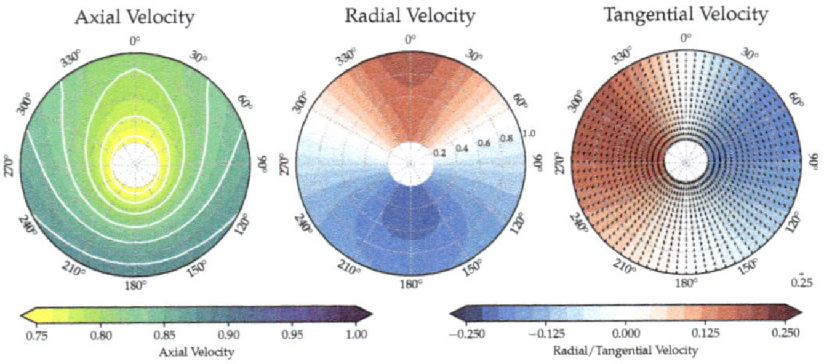

Figure 2. Potential wake field.

The uncorrected wake fractions are listed in Table 1. As expected, the wake fractions of the model scale fields are substantially higher than the fractions in full scale, and the computed effective wake fractions are lower than their nominal counterparts. Applying Sasajima's semi-empirical scaling method to the computed nominal model scale wake field results in a wake fraction that is surprisingly close to the computed effective wake fraction in full scale. Still, while the axial wake fractions are similar, the differences in axial and radial wake distribution are substantial, as can be seen from Figure 1d,e.

Table 1. Wake fractions before scaling.

Wake Field	Axial Wake Fraction
Nominal Model Scale	0.360
Effective Model Scale	0.300
Nominal Full Scale	0.240
Effective Full Scale	0.237
Sasajima Scaling	0.236
Target Value	0.250

A strong bilge vortex can be seen in Figure 1a, which results in low axial velocities in the region between the hub and 40% of the propeller radius. It should be noted that, in this radial range of this particular wake field, the axial velocities in the usual "wake peak" region between 330° and 30° are actually higher than in the lower half of the field.

Looking at the effective wake distribution resulting from the self-propulsion simulation at model scale, Figure 1b, the bilge vortex appears substantially weaker and closer to the centerline. This is also obvious from the radial and tangential velocity components, which are otherwise of similar magnitude as in Figure 1a. The effective field shown in Figure 1b exhibits a much more defined and pronounced wake peak at 12 o'clock, the axial velocities reaching consistently low values in this region.

The asymmetric flow at the innermost radii seen in Figure 1b is attributed to the lack of the propeller hub in both the RANS and the BEM part of the simulation. No propeller shaft, hub, or even stern tube was part of the RANS grids. The lack of the hub in the propeller panel code results in an unrealistic flow around the open blade root. Consequently, secondary flow structures emerge at low Reynolds numbers. While undesirable, the effect is local and is not expected to influence the cavitation behavior. As can be seen in Figure 1d, this is of less concern at full scale.

Moving to full scale, the nominal wake distribution (Figure 1c) changes significantly compared to model scale, as expected. The bilge vortex is remarkably less dominant, and the isolines of the axial wake distribution are more U-shaped. With a thinner boundary layer and a weaker bilge vortex, the in-plane velocity components change as well. In both computed full scale fields, the radial and tangential components indicate a less vortical and more upwards-directed flow. Except for the remainders of the bilge vortex, the in-plane velocity distribution approaches the potential one (Figure 2).

The trend towards a more defined and narrower wake peak continues moving on to the effective wake distribution at full scale, shown in Figure 1d. A bilge vortex can hardly be observed anymore.

Applying the method by Sasajima and Tanaka [3] leads to a very different wake distribution. The method works by contracting the nominal model scale wake field horizontally while also scaling the axial velocities, depending on the relationship of frictional and potential wake. For this particular case—with the above-described flow features in the nominal wake field at model scale—this results in a box-shaped region of very low axial velocities, visible in Figure 1e. Given that all fields shown in Figure 1 are uniformly scaled to the same wake fraction, the velocities in the outer and lower regions are very high, compensating for the large low-velocity region described previously.

4.2. Sheet Cavitation

The unsteady sheet cavitation behavior of a conventional propeller was analyzed in the five wake fields shown in Figure 1 using ESPPRO, the panel code for propeller analysis described in Section 2.1. As mentioned before, the input to the simulations differed only in the wake distribution. Wake fraction, cavitation number, scale, and all other parameters are identical across all cases presented below. All results discussed in this section refer to Propeller "M". A second propeller called Propeller "F" will be introduced and discussed in Section 5.

Figure 3a gives an overview over the global differences in sheet cavitation over one revolution. The lines indicate the radial extent of sheet cavitation for all blade angles in the different wakes. The lower cutoff threshold for these plots is a cavity thickness of 5% of the blade section thickness at $r/R = 0.7$.

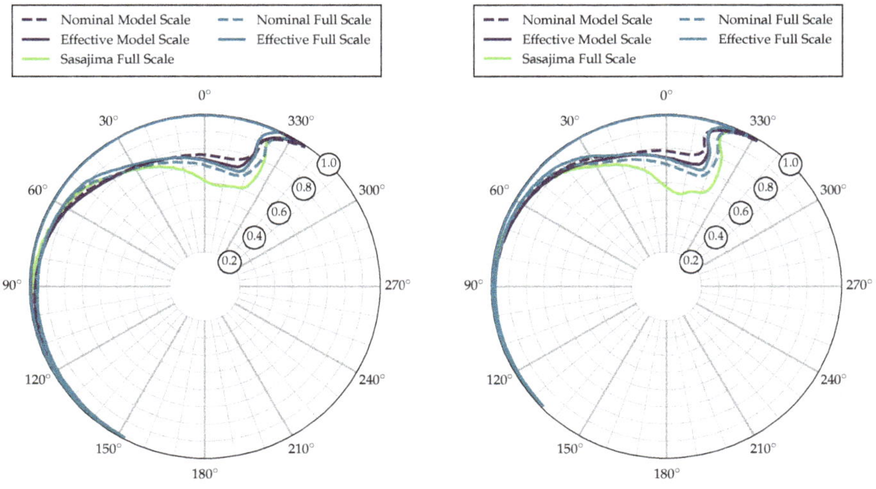

(a) Propeller "M" **(b)** Propeller "F"

Figure 3. Radial cavitation extent.

As can be seen from Figure 3a, there is little variation in cavitation inception angle (around 330°) yet noticable variation in terms of radial extent between the five wake fields over a relatively wide range of blade angles. The nominal wake distribution at model scale clearly results in the smallest cavitation extent, underpredicting the extent seen in the full scale effective field, especially at the blade angles with maximum cavitation extent. Using the full scale distribution obtained by Sasajima's method, the extent is overpredicted significantly, which is not surprising given the wake field seen in Figure 1e. The other three curves—representing the effective distribution at model scale and the two full scale fields—result in remarkably similar extents. There also appear to be differences in the closing behavior of the cavity, as can be seen from Figure 3a in the range 30–90°.

The time-variation in sheet cavitation is more easily quantified by looking at the cavity volume on one blade (non-dimensionalized by D^3), as shown in Figure 4a. Confirming the general trends between the wakes visible in Figure 3a, the small differences in inception and closure angle and large differences in maximum cavitation extent appear more clearly from Figure 4a. The plot also already indicates that the time-derivatives of the cavity volume are rather different between wake distributions when the cavity is shrinking between approx. 10–70° blade angle, hinting at major differences in associated hull pressure pulses. The second derivative of a B-Spline interpolation of the cavity volume with respect to blade angle (or time) is shown in Figure 5a.

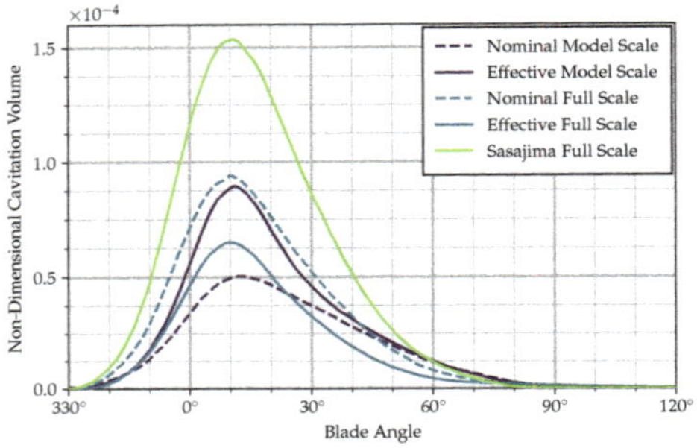

(a) Propeller "M"

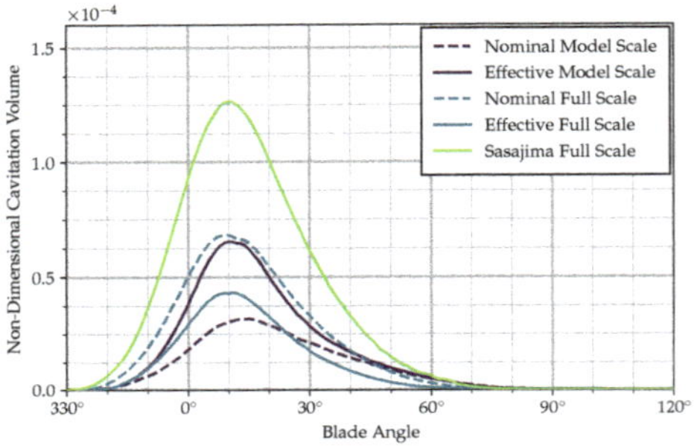

(b) Propeller "F"

Figure 4. Cavitation volume.

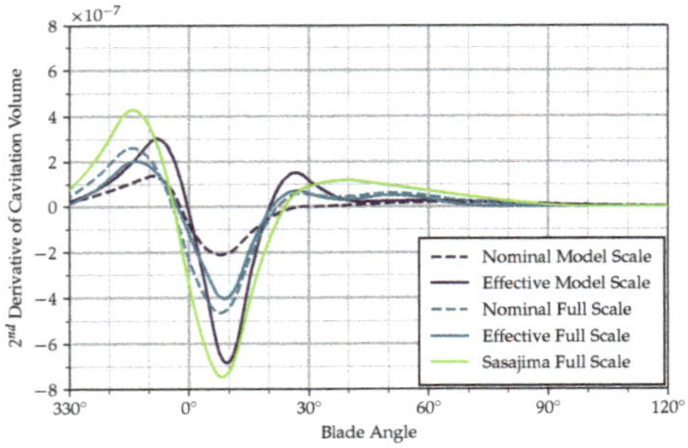

(a) Propeller "M"

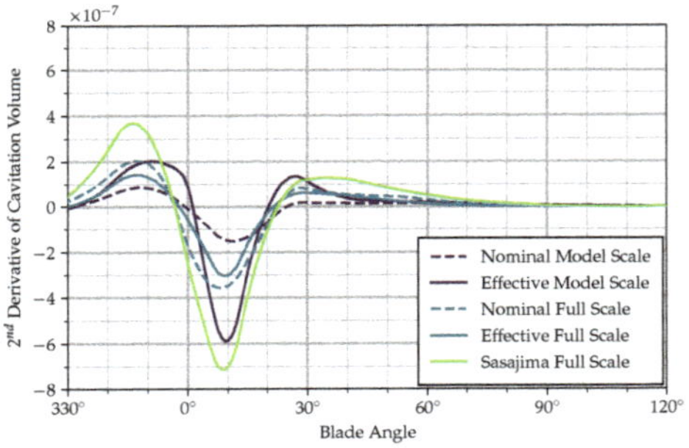

(b) Propeller "F"

Figure 5. Second derivatives of cavitation volume.

Disregarding the curve corresponding to the field scaled by Sasajima's method, particularly large differences exist between the results based on the nominal model scale distribution and the other computed curves. In that inflow field, the maximum cavity volume is more than 20% smaller than for the same propeller in the full scale effective wake distribution. In addition, second derivatives of the cavity volume (see Figure 5a) appear to be considerably different, the nominal model scale distribution again resulting in the smallest amplitudes.

Hull pressure pulses are evaluated in a single point, located on the centerline, 17% of the propeller diameter above the propeller plane. These calculations were done in the BEM part of the simulation, applying the Bernoulli equation at an offbody point and Fourier-transforming the time signal. The results are given in Figure 6, which provides the amplitudes at first and second harmonic of the blade frequency. The pressure pulse results also reflect the findings described previously. For example, in the full scale effective wake distribution, the value for the first harmonic is 17% larger than in the nominal distribution at model scale. The differences are even larger for the second harmonic. Higher harmonics have not been evaluated for this paper, as the driving factors, such as tip vortex cavitation, are not captured or modeled in the present propeller analysis method.

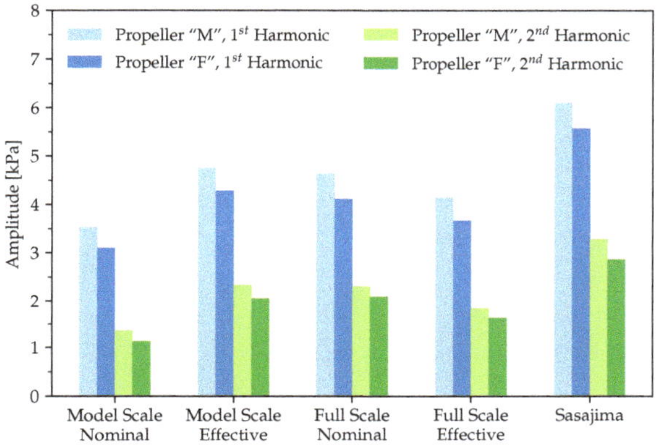

Figure 6. Pressure pulse amplitudes.

5. Results for Alternative Propeller Design

For the purpose of the analyses described in this paper, two simple conventional propellers have been designed for the bulk carrier case mentioned in Section 3. Resistance and thrust deduction values from model tests (which were carried out for the hull used in the present computations, but with minor aftbody modifications) establish the thrust requirement for the propeller design. For the design point, the predicted full scale effective wake fraction (based on self-propulsion model tests with a stock propeller) of $w = 0.25$ is used.

For the propeller design, a lifting line-based propeller design tool is employed that finds the optimum radial circulation (load) distribution for a circumferentially averaged wake field and required thrust. Radial pitch and camber distributions can then be found from the circulation distribution, assuming a standard NACA66 profile. Based on the designer's experience, the propeller was chosen to be 3-bladed with moderate skew and no rake. The expanded blade area ratio was selected as $A_E/A_0 = 0.3$.

The nominal wake field obtained from SHIPFLOW-XCHAP (see Figure 1a) is circumferentially averaged and scaled to the target effective wake fraction. This circumferentially averaged and scaled nominal wake field at model scale (which still varies radially, see Figure 7) is then used as input for the design tool. The optimum radial distribution of circulation found from lifting line theory for this case is shown in Figure 8. The corresponding propeller—which is the basis for all cavitation analysis results discussed so far—is referred to as Propeller "M" (as it is based on the model scale nominal wake).

Using this propeller, numerical self-propulsion simulations have been carried out to find the effective wake fields at model and full scale shown in Figure 1b,d, using the method described in Section 2.2.

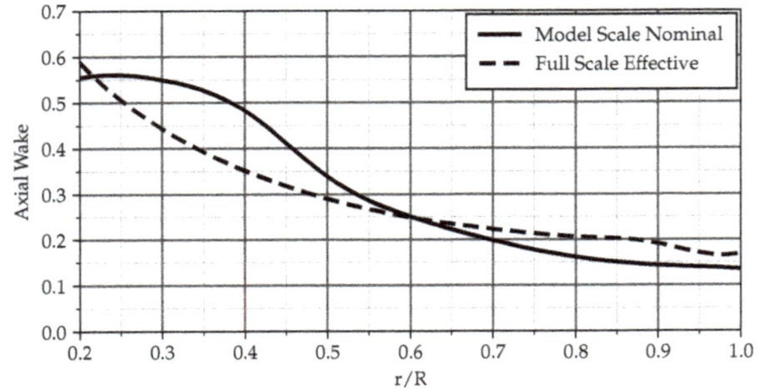

Figure 7. Circumferentially averaged axial velocities used for propeller design.

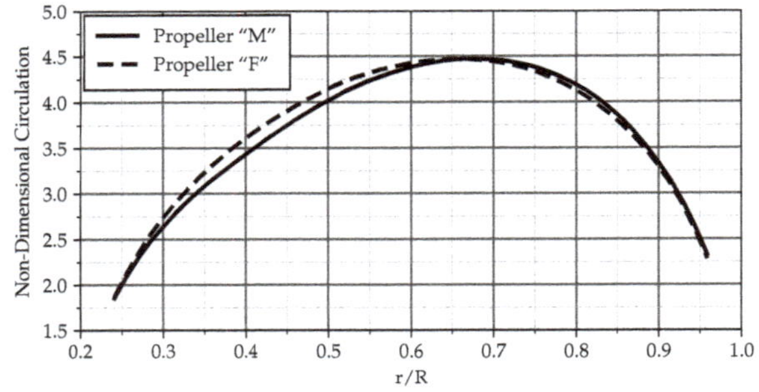

Figure 8. Radial circulation distributions.

In order to study the effect of nominal vs. effective wake on propeller design and resulting differences in propeller cavitation performance, the same design process is repeated with the full scale effective wake distribution as input, yielding Propeller "F". Other input parameters, such as design wake *fraction*, advance ratio, number of blades, and blade area ratio, remain unchanged. The circumferentially averaged axial inflow for this case and the resulting radial circulation distribution are shown as dashed lines in Figures 7 and 8. An inward shift of the loading compared to Propeller "M"—corresponding to the difference in inflow—can be seen from the latter.

The differences in effective wake distribution due to differences in geometry and radial load distribution between Propellers "M" and "F" are assumed to be negligible.

It can be seen from Figure 3b as well as Figure 4b that the characteristics of sheet cavitation extent and behavior of Propeller "F" are generally similar to Propeller "M". The magnitudes of all values, however, are significantly lower. The cavitation volume in the full scale effective distribution is reduced by more than 30%, compared to Propeller "M". Similar trends can be seen for the second cavity volume derivatives in Figure 5. Corresponding reductions in pressure pulses appear from Figure 6.

6. Conclusions

For the examined case of a modern and representative full-form ship, using the model scale nominal wake distribution for propeller design and cavitation analysis leads to a significant underprediction of cavitation extent, volume, and pressure pulses, compared to the behavior of the same propeller in the full scale effective wake field. Therefore, a conservative design is required if the propeller designer only has access to the measured nominal wake field. Otherwise, the expected extent of cavitation and acceptable levels of pressure pulses might be exceeded in full scale.

Compared to the propeller designed on the basis of the nominal wake in the model scale, the propeller designed for the effective full scale wake distribution performs better in all cavitation criteria considered. This highlights the importance of accurate wake data and the benefits of those—or hull geometry information—being available to the propeller designer. Knowing the effective wake distribution at full scale allows for a more realistic cavitation prediction in the propeller design process, enabling more efficient propeller designs.

Acknowledgments: The study was primarily funded by the Department of Mechanical Engineering at the Technical University of Denmark. The early development and implementation of the methods described in this paper were partly funded by Innovation Fund Denmark through the "Major Retrofitting Technologies for Containerships" project. Yasaman Mirsadraee's PhD project is supported by Innovation Fund Denmark under the Industrial PhD programme.

Author Contributions: P.B.R. and Y.M. developed and implemented the sheet cavitation model in the panel code, continuing earlier work by P.A. and others. P.B.R. developed and implemented the RANS-BEM coupling. Y.M. designed the propellers used for this study. P.B.R. ran the RANS-BEM simulations and the subsequent BEM simulations. The results were further analyzed and discussed by all authors. P.B.R. wrote the paper with continuous input and comments from Y.M. and P.A. P.A. initiated and supervised the project.

Conflicts of Interest: The authors declare no conflict of interest.

References

1. Jessup, S.; Mewis, F.; Bose, N.; Dugue, C.; Esposito, P.G.; Holtrop, J.; Lee, J.T.; Poustoshny, A.; Salvatore, F.; Shirose, Y. Final Report of the Propulsion Committee. In Proceedings of the 23rd International Towing Tank Conference, Venice, Italy, 8–14 September 2002; Volume I, pp. 89–151.
2. Fu, T.C.; Takinaci, A.C.; Bobo, M.J.; Gorski, W.; Johannsen, C.; Heinke, H.J.; Kawakita, C.; Wang, J.B. Final Report of The Specialist Committee on Scaling of Wake Field. In Proceedings of the 26th International Towing Tank Conference, Rio de Janeiro, Brazil, 28 August–3 September 2011; Volume II, pp. 379–417.
3. Sasajima, H.; Tanaka, I. Report of the Performance Committee, Appendix X: On the Estimation of Wake of Ships. In Proceedings of the 11th International Towing Tank Conference, Tokyo, Japan, 11–20 October 1966; pp. 140–144.
4. Van Terwisga, T.; van Wijngaarden, E.; Bosschers, J.; Kuiper, G. Achievements and Challenges in Cavitation Research on Ship Propellers. *Int. Shipbuild. Prog.* **2007**, *54*, 165–187.
5. Bosschers, J.; Vaz, G.; Starke, B.; van Wijngaarden, E. Computational analysis of propeller sheet cavitation and propeller-ship interaction. In Proceedings of the RINA Conference "MARINE CFD2008", Southampton, UK, 26–27 March 2008.
6. Gaggero, S.; Villa, D.; Viviani, M.; Rizzuto, E. Ship wake scaling and effect on propeller performances. In *Developments in Maritime Transportation and Exploitation of Sea Resources*; CRC Press: London, UK, 2014; pp. 13–21.
7. Kinnas, S.A.; Fine, N.E. A Numerical Nonlinear Analysis of the Flow Around Two- and Three-dimensional Partially Cavitating Hydrofoils. *J. Fluid Mech.* **1993**, *254*, 151–181.
8. Fine, N.E. Nonlinear Analysis of Cavitating Propellers in Nonuniform Flow. Ph.D. Thesis, Massachusetts Institute of Technology, Cambridge, MA, USA, 1992.
9. Vaz, G.; Bosschers, J. Modelling Three Dimensional Sheet Cavitation on Marine Propellers Using a Boundary Element Method. In Proceedings of the 6th International Symposium on Cavitation (CAV2006), Wageningen, The Netherlands, 11–15 September 2006.
10. Vaz, G.; Hally, D.; Huuva, T.; Bulten, N.; Muller, P.; Becchi, P.; Herrer, J.L.R.; Whitworth, S.; Mace, R.; Korsström, A. Cavitating Flow Calculations for the E779A Propeller in Open Water and Behind Conditions: Code Comparison and Solution Validation. In Proceedings of the 4th International Symposium on Marine Propulsors (smp'15), Austin, TX, USA, 31 May–4 June 2015; pp. 330–345.
11. Hally, D. Propeller Analysis Using RANS/BEM Coupling Accounting for Blade Blockage. In Proceedings of the 4th International Symposium on Marine Propulsors (smp'15), Austin, TX, USA, 31 May–4 June 2015; pp. 297–304.

12. Rijpkema, D.; Starke, B.; Bosschers, J. Numerical simulation of propeller–hull interaction and determination of the effective wake field using a hybrid RANS-BEM approach. In Proceedings of the 3rd International Symposium on Marine Propulsors (smp'13), Tasmania, Australia, 5–7 May 2013; pp. 421–429.
13. Regener, P.B.; Mirsadraee, Y.; Andersen, P. Nominal vs. Effective Wake Fields and their Influence on Propeller Cavitation Performance. In Proceedings of the 5th International Symposium on Marine Propulsors (smp'17), Helsinki, Finland, 12–15 June 2017; pp. 331–337.

Journal of
Marine Science and Engineering

MDPI

Article

Prediction of Propeller-Induced Hull Pressure Fluctuations via a Potential-Based Method: Study of the Effects of Different Wake Alignment Methods and of the Rudder

Yiran Su *, Seungnam Kim and Spyros A. Kinnas

Ocean Engineering Group, Department of Civil, Architectural and Environmental Engineering, The University of Texas at Austin, Austin, TX 78712, USA; naoestar@utexas.edu (S.K.); kinnas@mail.utexas.edu (S.A.K.)

* Correspondence: yiransu@utexas.edu; Tel.: +1-512-706-5771

Received: 4 April 2018; Accepted: 2 May 2018; Published: 8 May 2018

Abstract: In order to predict ship hull pressure fluctuations induced by marine propellers, a combination of several numerical schemes is used. The propeller perturbation flow is solved by the boundary element method (BEM), while the coupling between a BEM solver and a Reynolds-averaged Navier-Stokes (RANS) solver can efficiently predict the effective wake. Based on the BEM solution under the predicted effective wake, the propeller-induced potential on the ship hull can be evaluated. Then, a pressure-BEM solver is used to solve the diffraction pressure on the hull in order to obtain the solid boundary factor which leads to the total hull pressure. This paper briefly introduces the schemes and numerical models. To avoid numerical instability, several simplifications need to be made. The effects of these simplifications are studied, including the rudder effect and the wake alignment model effect.

Keywords: hull pressure; pressure fluctuation; diffraction potential; wake alignment scheme; boundary element method; cavitation

1. Introduction

Propeller-induced noise and vibration is one of the major issues that threatens onboard comfort, causes mechanical failures, and potentially impacts marine animals. To resolve these issues during the design stage, it is important to have a reliable prediction of the propeller-induced hull pressures. Research has been done on predicting the hull pressure both experimentally and numerically. Most of the experimental studies are performed in the model-scale with multiple pressure transducers mounted on the hull surface above the propeller to monitor the hull pressure [1,2]. According to the level of simplification that can be made, the numerical approaches for underwater noise simulations can be divided into compressible Navier–Stokes equation-based approaches, Lighthill equation-based approaches, Ffowcs–Williams–Hawkings equation-based approaches [3,4], Helmholtz equation-based approaches [5], and hybrids of any two. These equations are usually implemented by either a finite volume method in which the propeller is modelled by a rotating boundary [4,6] or by the boundary element method (BEM) [3,7] in which the propeller is represented by sources and dipoles on the boundary surface.

To numerically predict the propeller-induced hull pressure, the BEM method is often used because the propeller induced pressure field is a small-amplitude high-frequency field which can be decoupled from the background flow. Most of these BEM applications focus on the effect of cavitation because it is the main contributor of high-level noise and vibrations [7,8]. For a cavitating propeller, the cavitation source is the major excitation in the acoustic BEM model so that the influence of the propeller lifting force, the blade thickness, and the blade trailing wake can all be neglected. However, the continuous low-level noise and vibration induced by non-cavitating propellers or marginally-cavitating propellers can also be a problem. In this case, the lifting surface, the blade thickness, and the wake all have a comparable influence on the induced pressure.

This paper focuses on a streamlined procedure of predicting the ship hull pressures. A BEM/RANS[1] interactive scheme is first used to predict the effective wake and the propeller-induced pressure field under this effective wake [9,10]. Then, a pressure-BEM solver is used to predict the diffraction pressure which leads to the solid boundary factor. The method considers the lifting surface effect, the blade thickness effect, cavitation source effect, and trailing wake effect so that it can predict the hull pressure induced by either a wetted propeller or a cavitating propeller.

2. Methodology

2.1. Boundary Element Method

The BEM can be used to solve various types of partial differential equations. In this application, the control Equation (1) can be obtained by inserting the Laplace equation into the Green's third identity. G is Green's function; S_B represents all the boundary surfaces in the fluid domain; n_q is the normal vector at the point q pointing into the flow field.

$$\frac{\phi_p}{2} = \int_{S_B} \left[\frac{\partial \phi_q}{\partial n_q} G(p,q) - \phi_q \frac{\partial G(p,q)}{\partial n_q} \right] ds \tag{1}$$

Equation (1) states that the value of the ϕ_p at any point on the boundary surface S_B depends only on the values of ϕ_q and $\frac{\partial \phi_q}{\partial n_q}$ at any point q on the boundary of the body B. Moreover, the value of ϕ_p can be expressed as the superposition of the potentials due to distribution of sources and normal dipoles on the boundary of the body of strengths $\frac{\partial \phi_q}{\partial n_q}$ and $-\phi_q$ repectively.

Special care should be given to the branch wake surface S_W behind the hydrofoil of propeller. By considering the wake surface, Equation (1) renders:

$$\frac{\phi_p}{2} = \int_{S_B} \left[\frac{\partial \phi_q}{\partial n_q} G(p,q) - \phi_q \frac{\partial G(p,q)}{\partial n_q} \right] ds - \int_{S_W} \Delta\phi_W \frac{\partial G(p,q)}{\partial n^+} ds \tag{2}$$

After enforcing the kinematic boundary condition on the strength of the source potential $\frac{\partial \phi_q}{\partial n_q}$ and the Morino's condition on the wake strength $\Delta\phi_W$, Green's foumula finally becomes:

$$\frac{\phi_p}{2} = \int_{S_B} \left[\left(-\vec{U}_{in} \cdot n_q \right) G(p,q) - \phi_q \frac{\partial G(p,q)}{\partial n_q} \right] ds - \int_{S_W} \Delta\phi_W \frac{\partial G(p,q)}{\partial n^+} ds \tag{3}$$

where $\vec{U}_{in}$ is the inflow velocity. Equation (3) is a Fredholm integral equation of the second kind for the unknown ϕ. This analytic formulation will be solved for the unknown quantity, ϕ_p by using numerical implementation.

To predict the propeller performance, the total flow $\vec{U}_T$ can be docomposed to the known inflow $\vec{U}_{IN}$ (or background flow) and the unknown propeller induced flow $\vec{U}_P$, which can be treated as a potential flow. Therefore, the propeller perturbation potential is governed by the Laplace equation and can be solved via the boundary element method. In the current boundary element solver, constant strength panels are placed on the surface of the propeller and on the propeller trailing wake surfaces.

1 RANS is the abbreviation of the Reynolds-averaged Navier–Stokes method.

Equation (3) can then be discretized into a linear matrix system which can be solved by either a direct method or an iterative method.

2.2. Wake Alignment Model

Assuming a propeller blade is a lifting body and the blade trailing wake is a material surface on which the potential field is not continuous, the trailing wake surfaces can be treated as a boundary in the BEM model. The strength of the trailing wake is calculated by the Kutta condition and the location of the wake sheet in the downstream is determined by the wake alignment model, which ensures both sides of the wake surface have the same pressure (force free condition).

The basic philosophy behind the wake alignment model is that the wake sheet is a material surface that needs to convect with the local stream. Two types of wake alignment models are available in the current BEM solver depending the time dependency: the steady wake alignment model and the unsteady wake alignment model. Whether the time variation of the incoming flow is considered (unsteady) or not (steady) is the major difference when implementing the alignment procedure. Brief description of each alignment scheme is as follows:

(a) Steady wake alignment: the full wake alignment (FWA) scheme [11,12] considers only the zeroth harmonic of the inflow velocity, therefore the higher harmonic components are neglected in the alignment process. Axisymmetric variation of the incoming flow in radial direction is the most general case, and consequently the propeller performance becomes invariant to the angular position of the blade. The shape of the blade wake is the same for all blades. Figure 1a shows propellers under three different axisymmetric inflows. The induced velocity (or, perturbation velocity) $\vec{U}_P$ in Figure 1b is evaluated based on the effects from the blade, hub, duct (in case duct geometry is included), and the wake itself. Along with the axisymmetric inflow, the induced velocity constitutes the total velocity prior to implementing the alignment procedure in the FWA.

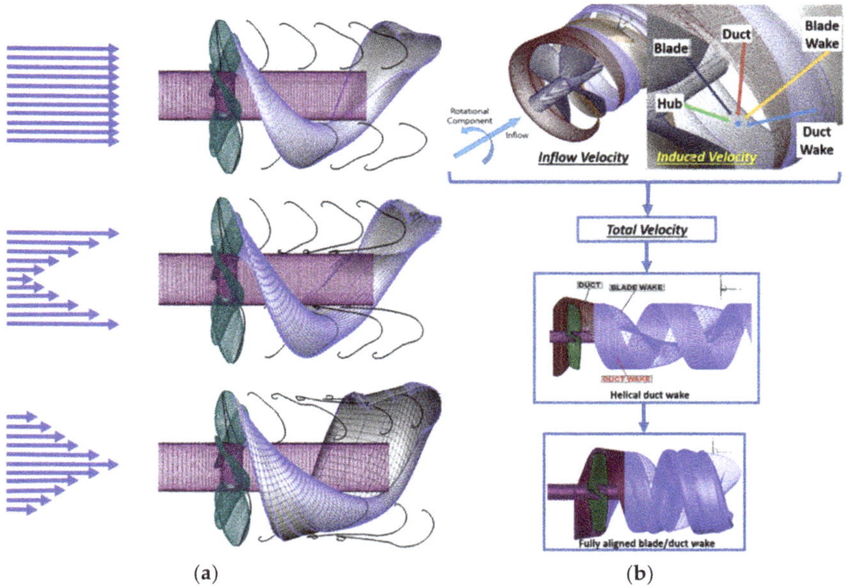

Figure 1. (**a**) Propeller under the axisymmetric inflow; (**b**) Contributors of the perturbation velocity on the wake surface, taken from Kim [12].

(b) Unsteady wake alignment: it is similar to the steady case, but the consideration of time variation of the incoming flow is included. Full harmonics in the inflow are considered to evaluate the velocity components on the wake panels depending on their locations. The alignmen starts with the FWA in steady state ($t = 0$) and procedes into the unsteady alignment model ($t > 0$) using the Euler-explicit sceheme with revolutions in a progressive manner.

$$
\begin{aligned}
X_{i+1}^{n+1} &= X_i^n + X\vec{U}_{TX_i}^{n}\Delta t \\
Y_{i+1}^{n+1} &= Y_i^n + Y\vec{U}_{TY_i}^{n}\Delta t \\
Z_{i+1}^{n+1} &= Z_i^n + Z\vec{U}_{TZ_i}^{n}\Delta t
\end{aligned}
\tag{4}
$$

In the above equations, i and n denote the ith node point at the nth time step, $\vec{U}_T$ is the total velocity, and Δt is the time step size. X, Y, and Z are the coordinates of the nodal points on the wake surfaces. Due to the time consideration of the total flow, each blade has a different wake geometry for different blade angles. At each time step, only the key wake is aligned, updated, and then saved before proceeding to the next time step. After that, the saved key wake can be used for other blades when they reach the same blade angle in the future time steps. This procedure will be repeated for several revolutions until the converged unsteady force performances are obtained.

In the current panel model, some elements of the steady wake alignment scheme introduced by Tian & Kinnas [11] are coupled with the unsteady wake alignment model by Lee [13] to improve the convergence of the alignment procedure. Different from the Lee's unsteady scheme, which evaluates the total velocity at the panel center and then interpolates it into the panel corners, the coupled scheme calculates the total velocity directly at the four corners on each wake panel, as shown in Figure 2. This improves the numerical accuracy and stability. The convergence study on the wake panel numbers showed that the predicted propeller performance was irrelevant to the panel number after 100 panels are used. Detailed description of the unsteady wake alignment model coupled with the FWA can be found in [12].

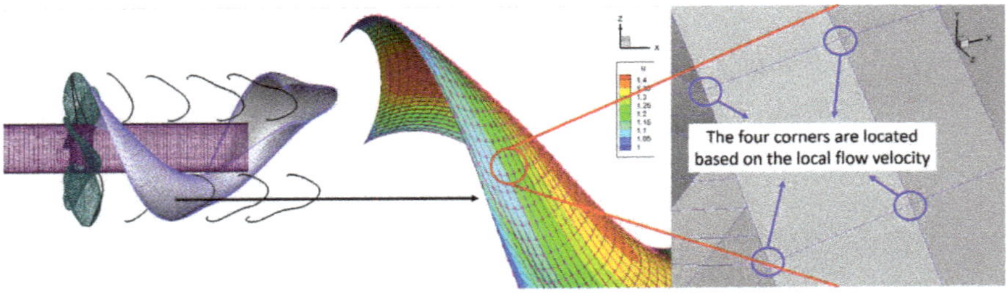
The four corners are located
based on the local flow velocity

Figure 2. Key wake with the local velocity vectors (red arrows in the middle figure) plotted on each nodal point of the wake, taken from Kim [12].

2.3. *Boundary Element/Reynolds-Averaged Navier-Stokes Interactive Scheme*

In Section 2.1, the BEM solver is used to evaluate the perturbation potential when the inflow $\vec{U}_{IN}$ is given. However, this inflow information is not always available. Some researchers monitor the flow (nominal wake) on the propeller disk plane without the existence of the propeller and then perform an adjustment to the nominal wake to predict the effective wake $\vec{U}_{EFF}$. The effective wake considers the propeller's influence on the flow and can be defined by Equation (5).

$$
\vec{U}_{EFF} = \vec{U}_T - \vec{U}_P \tag{5}
$$

To predict the effective wake, BEM solver is coupled with a RANS solver, as shown in Figure 3. In the RANS solver, the propeller is represented by a local mass source term which is added to the continuity equation and a local body force term which is added to the momentum equation. The strength of the mass source term and the body force term is determined by the BEM solver. The total flow $\vec{U}_T$ is evaluated out of the RANS solver and can be used to calculate the effective wake by Equation (5). The details of this scheme can also be found in the work of Su & Kinnas [10]. Unlike some other similar implementations, the mass source term is included in the RANS model. It is found that a consistent representation of the blockage effect in both the total flow and the propeller-induced flow is important for the accurate prediction of the effective wake field [10].

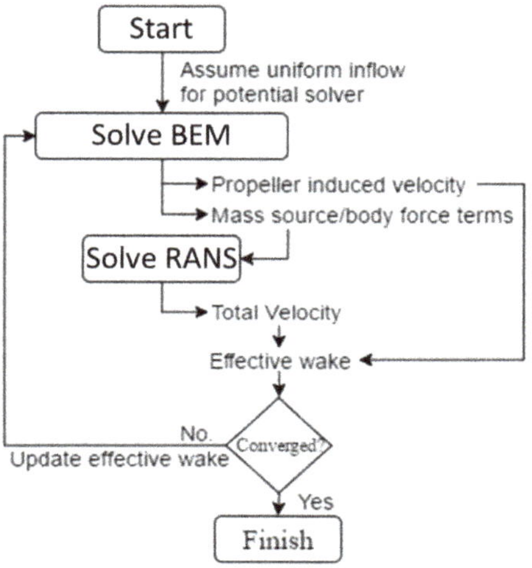

Figure 3. Numerical algorithm of the BEM/RANS[2] interactive scheme.

2.4. Boundary Element—Solver for the Oscillating Hull Pressure

According to the Bernoulli equation, the small amplitude pressure oscillation $P^{(t)}$ can be represented by the steady velocity potential $\Phi^{(t)}$, as shown in Equation (6). In this equation, ω is the angular velocity of the propeller, Z is the number of blades, and Φ_n the velocity potential magnitude at a certain frequency.

$$P^{(t)} = -\rho\frac{\partial\Phi^{(t)}}{\partial t} = -\rho \sum_{n=1,2,\ldots} nZ\omega\Phi_n i e^{inZ\omega t} \tag{6}$$

Instead of solving the oscillating pressure field, the velocity potential field Φ_n is solved at different frequencies. The potential field Φ_n is governed by the Helmholtz equation. For the near-field small-amplitude pressure fluctuation caused by marine propellers, an infinite sound speed can be assumed so that the Helmholtz equation is reduced to a Laplace equation, as shown in Equation (7).

$$\nabla^2\Phi_n = 0 \tag{7}$$

Similar to the hydrodynamic BEM, the nth total velocity potential field Φ_n can be decomposed to a radiated potential field $\Phi_n^{(R)}$ (taken equal to the potential due to the propeller flow in the absence of the

2 BEM is the abbreviation of the boundary element method; RANS is the abbreviation of the Reynolds-averaged Navier-Stokes method.

hull) and a diffraction potential field $\Phi_n^{(D)}$. The radiated potential fields $\Phi_n^{(R)}$ is the Fourier series of the unsteady perturbation potential field in the hydrodynamic BEM model which represents the propeller. The lowest frequency for the Fourier decomposition is the blade-passing frequency $Z(\omega/2\pi)$.

Based on the boundary element method, the total potential field Φ_n can be solved by Equation (8) where S_H is the hull surface and S_I is the image of the hull.

$$\frac{1}{2}\Phi_n = \Phi_n^{(R)} - \frac{1}{4\pi}\iint_{S_H+S_I} \Phi_n \frac{\partial G}{\partial n} dS \tag{8}$$

Different from the hydrodynamic BEM model, the total potential, instead of the diffraction potential, is solved. This eliminates the source-induced potentials term in the boundary integral equation which saves computational time. However, the method cannot be used for the hydrodynamic BEM model because the total velocity field can be vortical and not governed by the Laplace equation. Equation (8) can be solved the same way as in Section 2.1. Constant strength dipole panels are placed on the ship hull and the rudder. The effect from the top tunnel wall is included by an image model. Finally, the unsteady pressure field on the ship hull can be calculated by Equation (8).

3. Numerical and Experimental Models

3.1. Experimental Model

Numerical schemes are validated by comparing them with the model test results [2,6]. In the experiment, a 1:20 model hull is mounted on the ceiling of a cavitation tunnel, as shown in Figure 4. Eight pressure transducers are placed on the model hull above the propeller. The arrangement and the numbers of each transducer are also shown in Figure 4.

(a)

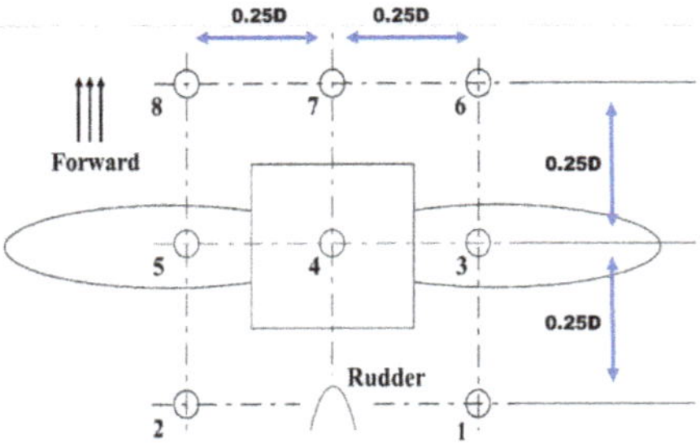

(b)

Figure 4. (**a**) Photo of the model test facility and (**b**) pressure transducer arrangement and numbering.

3.2. Hydrodynamic Model

In the hydrodynamic model, BEM is coupled with RANS to predict the effective wake as well as the propeller performance. Figure 5a shows the hydrodynamic BEM model with 80 by 20 panels on a single blade. 4-bladed P2772 propeller [2] is adopted, and the counterclockwise direction about the center of hub is defined to be the negative x direction. Both the steady and the unsteady wake alignment models are tested in fully-wetted flow. Figure 6 shows the RANS model as well as the boundary conditions. The domain size of the RANS model matches the dimensions of the experiment tank. As a result, the effect of the tank is included which can be helpful for reducing the different between the experiment measurement and the numerical prediction. Around 2 million cells are used in this RANS model and the mass/body force zone is shown in Figure 5b.

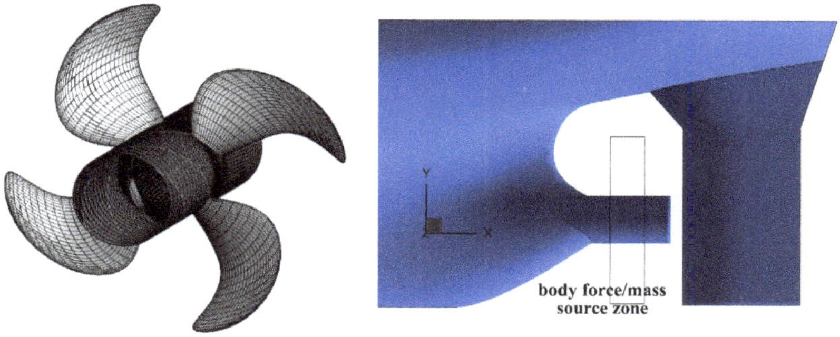

(a) (b)

Figure 5. Propeller geometry generated via (a) BEM and (b) Mass/Body force zone in RANS.

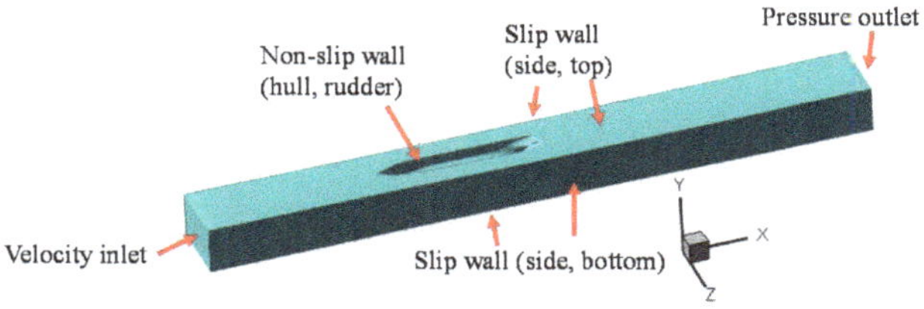

Figure 6. The RANS model in the BEM/RANS scheme.

3.3. Pressure Boundary Element Model

In the pressure-BEM model, an external flow problem is solved, which means the BEM panels need to be placed on the surface of a closed body. In reality, since the fore part of the hull has very little influence over the propeller-induced pressure, only the aft part of the hull needs to be included, as shown in Figure 7. An image hull is considered to represent the free surface effect. Therefore, the pressure on the free surface needs not to be solved.

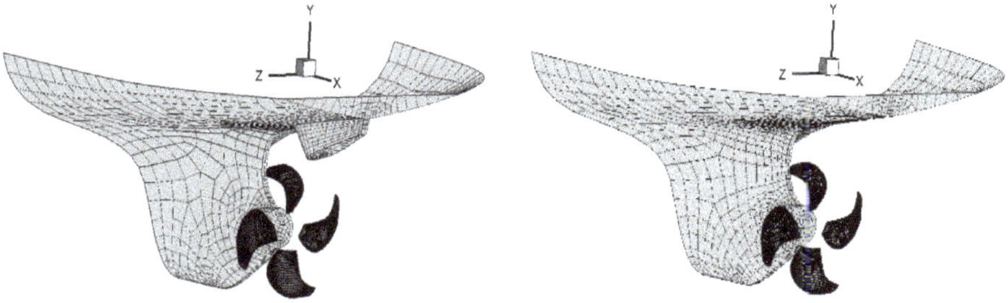

(a) (b)

Figure 7. Pressure-BEM model (a) with and (b) without the upper part of the rudder (the propeller is not a part of this model).

In the current scheme, the wake alignment model does not consider the influence from the rudder. Therefore, it is almost certain that the blade trailing wake penetrates the rudder. Even though the numerical simulation of the trailing wake/rudder interactions was already proposed by He & Kinnas [14], this modeling is not included in this paper. It is because the interaction between the blade trailing wake and the rudder may cause singular radiated pressures on the rudder surface where it touches the edges of the wake panels. To reduce this singular behavior, only the upper part of the rudder is included in the pressure-BEM model. To study the effect of the rudder, a test is also made, in which the rudder is totally neglected in the pressure-BEM model, as shown in Figure 7. Similarly, the length of the hub is reduced so that it does not extend to the propeller zone. It should be noted that the time-averaged interaction of the rudder with the propeller is already included through the presented BEM/RANS algorithm.

Although this rudder geometry is not realistic, it is still used to test and study the validity of the pressure-BEM model in this study. In the future, the hydrodynamic BEM model can be improved so that the wake alignment scheme includes the effect of the rudder. As a result, the full rudder geometry can be used in the pressure-BEM model.

The mesh convergence of the pressure-BEM model has been extensively conducted by Hwang et al. [15]. In that convergence study, a similar mesh density is used on the ship hull. The pressure-BEM model is found to be invariant to the panel numbers used to discretize the hull surface.

3.4. Unsteady Reynolds-Averaged Navier-Stokes Model

In addition to the experimental data, unsteady RANS (URANS) simulation is used to provide a second reference. In the URANS model, the hull, the 4-blade propeller, and the rudder are modeled while sliding interfaces are used to handle the motion of the propeller. Around 6 million cells are used in this model and it takes 48 h on 240 × 2.7 GHz CPUs to solve this problem in ANSYS Fluent[3]. Eight pressure monitors are placed at the same locations as the pressure transducers in the real experiment.

4. Results and Comparison with Experiment

In this paper, all the dimensionless numbers are defined based on the propeller diameter D, ship speed V_s, and the propeller rps n. The advance ratio J_S, the thrust coefficient K_T, the torque coefficient K_Q, the cavitation number σ, and the pressure coefficient C_P are defined by the following equations.

$$J_S = \frac{V_s}{nD} \tag{9}$$

$$K_T = \frac{T}{\rho n^2 D^4} \tag{10}$$

$$K_Q = \frac{Q}{\rho n^2 D^5} \tag{11}$$

$$\sigma = \frac{P - P_V}{1/2\rho n^2 D^2} \tag{12}$$

$$C_P = \frac{P - P_\infty}{1/2\rho n^2 D^2} \tag{13}$$

First, the BEM/RANS interactive scheme is applied to different load conditions by changing advance ratios. The predicted propeller forces are compared to the experimental data, as shown in Figure 8. In the experiment and the BEM/RANS model, the propeller is working behind the ship hull at different advance ratios. A good agreement is obtained between the two. Figure 9 shows the predicted effective wake field for different advance ratios. In Figure 9, the axial effective wake velocity is represented by the gray scale and the in-plane effective wake velocity is plotted by vectors. Since the effective wake is defined at the center of the blade panels, only the mid-chord disk of the 3-dimensional wake field is plotted.

---

3 The simulation is performed in the Texas Advanced Computing Center (TACC) at The University of Texas at Austin (Austin, TX 78703, USA). URL: http://www.tacc.utexas.edu.

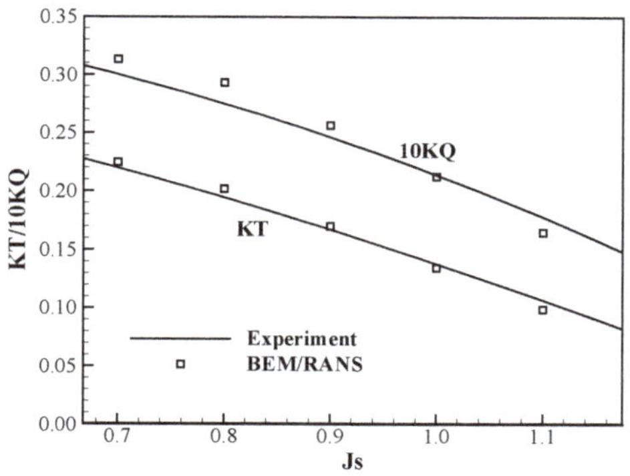

Figure 8. Comparison between the propeller force predicted by the BEM/RANS scheme and that from the experimental data[4].

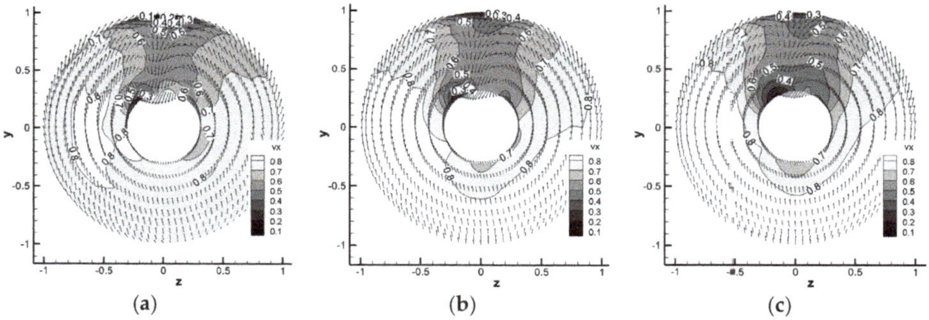

Figure 9. The predicted effective wake field at the advance ratios of (**a**) Js = 0.7; (**b**) Js = 0.9; and (**c**) Js = 1.1. Only the effective wake distribution on the mid-chord disk is shown. The actual effective wake field may vary in the axial direction.

Then, the same scheme is applied to another load condition in which the advance ratio Js is 0.808 and the cavitation number σ is 7.34. The solution of the BEM/RANS scheme is imported to the pressure-BEM model to evaluate the hull pressure fluctuation on the eight different locations where the pressure transducers are placed. For this load condition, several comparisons are made.

The first comparison study is on the different wake alignment models. Both the steady wake alignment model and the unsteady wake alignment model are tested. In the steady wake alignment scheme, all the non-axisymmetric components of the effective wake are neglected so that the wake geometry, as shown in Figure 10a, does not change with time. More importantly, the positive velocity in the vertical direction of the effective wake, as shown in Figure 9, is neglected. Therefore, the steady wake goes along the axial direction without any inclination. On the other hand, the unsteady wake alignment scheme considers the non-axisymmetric part of the effective wake so that the wake geometry is a function of time (or blade angle), as shown in Figure 10b,c. All the geometries in Figure 10 are plotted in a rotating coordinate which is fixed to the propeller shaft. When the BEM solver is used to predict the propeller mean performance, the difference between the steady and unsteady wake

[4] KT is the abbreviation of thrust coefficient and is defined by Equation (10). KQ is the abbreviation of torque coefficient and is defined by Equation (11).

alignment model are not significant. However, if the results of the BEM solver are used to evaluate the hull pressure fluctuations, the accurate prediction of the wake geometry becomes more vital.

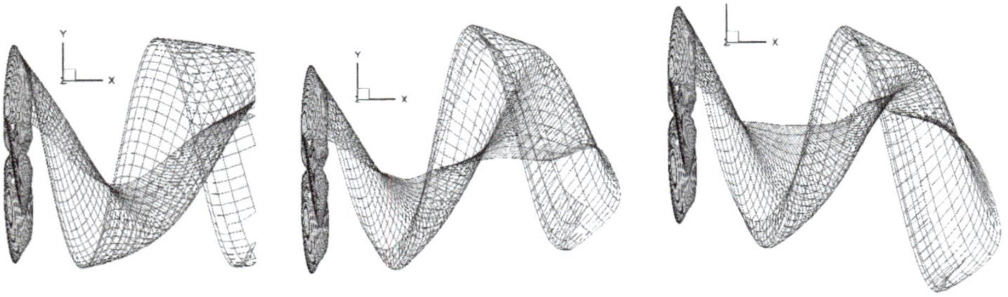

Figure 10. Comparison of steady wake geometry and unsteady blade wake geometries at different blade angles. The blade-angle ranges between 0 and 2π and is defined by the angle the blade has passed starting from the "upright position". The "upright position" means the mid-camber point of the root section is located on the vertical axis above the hub ($+y$ axis). (**a**) steady wake; (**b**) unsteady wake 0 degrees; (**c**) unsteady wake 180 degrees. Please note the wake is shown relative to the propeller fixed system. So the hull is physicallly located above the propeller in (**b**), but "below" the propeller in (**c**). Thus, in both cases, (**b**) and (**c**), the propeller wake appears to be aligned in a direction parallel to the hull surface.

To study the influence of different wake alignment models, the pressure history monitored by the pressure transducers and by the URANS model are compared with the pressure fluctuation predicted by our scheme with either a steady wake or an unsteady wake. Results are plotted in Figure 11. The unsteady RANS results have a good agreement with the experimental data on transducer T3, T5, T6, and T8. It also correctly predicted the peak pressure pulse amplitude while the location of the peak pressure is shifted towards upstream (see pressure on location T4 and T7). URANS severely under-predicted the pressure pulse at downstream locations (T1 and T2). This can be explained by the numerical diffusion caused by the unstructured mesh. Because the pressure pulses at T1 and T2 highly rely on the behavior of the blade trailing wakes, too much downstream diffusion can smooth out the vortices and diminish the hull pressure pulse.

In terms of the pressure pulses predicted by the pressure-BEM method, the influence of different wake alignment models is not significant for upstream transducer locations (T3–T8), due to their relative long distance to the trailing wake. For downstream points (T1 and T2), however, the difference becomes bigger and unsteady wake alignment improves the results. This is because the non-axisymmetric component of the effective wake pushes the wake towards the hull so that the wake has a stronger influence on the hull pressure.

On transducers T1, T3, T6, T7, and T8, a good agreement is obtained between the BEM-predicted pressure pulses and the experimental data. On transducer T2, the fluctuation amplitude is correctly predicted but the phase angle of the pressure peak is not accurate. This might be attributable to the ignored rudder effect in the unsteady wake alignment model. In other words, although the tip vortex of a blade is close enough to the hull to induce a correct pressure fluctuation amplitude, the time when the tip vortex reaches the closest point to the hull might not be well predicted due to the ignored rudder effect. The complete description of the developed tip vortex cavity in BEM is provided in Lee & Kinnas [16]. On transducer T4 and T5, current numerical scheme under-predicts the pressure fluctuation amplitudes. Unlike the pressure pulses from the URANS model and from the experimental data, the pressure fluctuation amplitude predicted by the pressure-BEM method is not symmetric about the ship center; the scheme tends to under-predict the pressure pulses on the port side. This phenomenon can be traced back to the non-symmetric pattern of the effective wake, as can be seen in Figure 9. As a result, the loading on the blade tip is relatively smaller on the port side and larger on the starboard side.

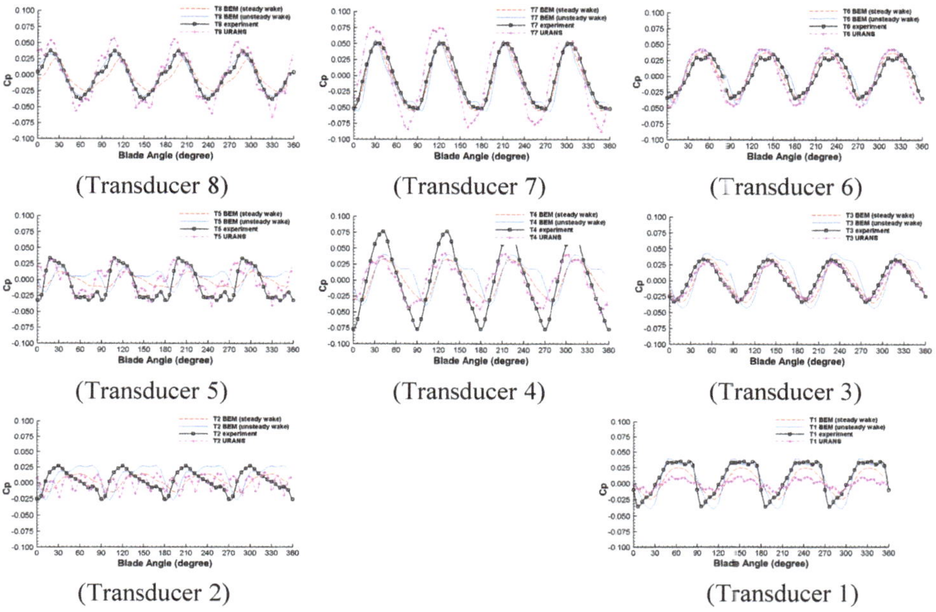

Figure 11. Comparison among the experimental pressure data, Unsteady Reynolds-averaged Navier-Stokes (URANS) method pressure history, and the pressure history predicted by pressure-BEM solver with either a steady or an unsteady wake.

To understand how the hull pressure is affected by the different ways of evaluating the effective wake, another study was made. Two cases were tested in this study and the results are compared with URANS/experiments, as shown in Figure 12. In the first case, the effective wake is evaluated at the center of every BEM panel on the blade. In the second case, the axial-varying effective wake field from the first case is extrapolated to a disk near the leading edge. The disk conforms to the leading edge and has a constant distance (2% of max radius) to the leading edge. The details of both schemes can be found in Tian et al. [9].

Based on Figure 12, the change of locations for evaluating the effective wake can significantly change the predicted hull pressures, especially for downstream monitors. For monitors T1, T2, and T8, defining the effective wake at the center of every blade panel shows a clear advantage over the other method which defines the effective wake at a leading-edge disk. At other monitors (T3 to T7), there is no clear advantage for either of methods. In summary, the hull pressure pulses are shown to be very sensitive to the effective wake field. Although our current axial-varying effective wake shows some advantages, it still needs to be improved due to the inconsistency in the pressure of several monitors between the BEM results and the experimental data.

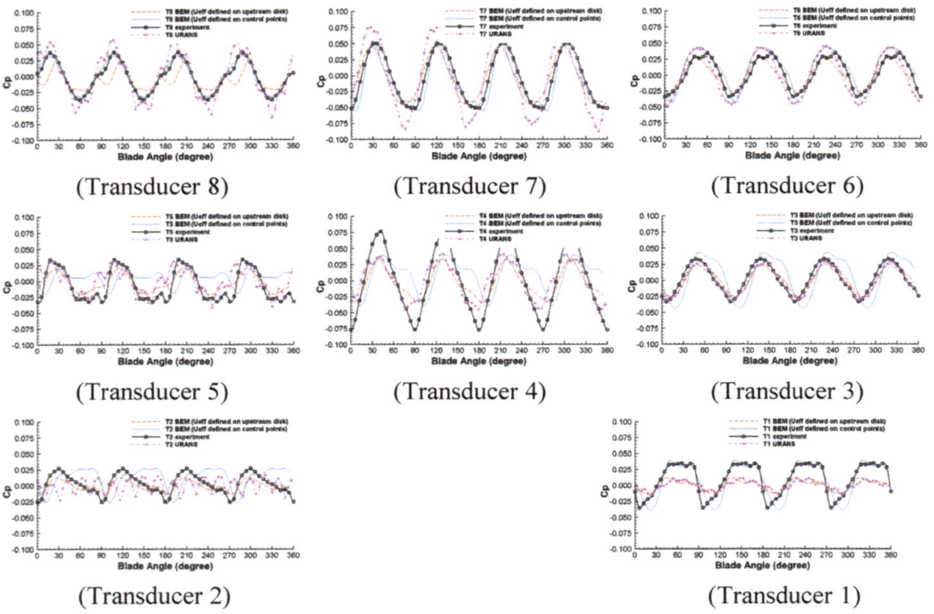

Figure 12. Comparison of hull pressures predicted by the pressure-BEM solver with different treatment of the effective wake field: (a) Effective wake field is evaluated at the center of every blade BEM panel (blue solid line); (b) The effective wake field is extrapolated to the upstream curved surface close to the blade leading edge (red dash line).

The last comparison study focuses on the effect of the rudder. According to Section 3.3, in order to maintain numerical stability, only the upper part of the rudder is used in the pressure-BEM model, as can be seen in Figure 7a. The underlying assumption is that the lower part of the rudder is far from the pressure transducer locations and therefore, has a negligible influence towards the hull pressure fluctuation. To study whether this is a good approximation, we tried two different rudder geometries in the pressure-BEM model: one is the original geometry which is shown in Figure 7a; the other geometry totally neglects the rudder part, as can be seen in Figure 7b. The predicted hull pressures from both cases are compared in Figure 13. According to the figure, the difference is negligible on T4 and T7 while the pressure on T1 is affective by the rudder geometry. This is because T1 is at a downstream location which is close to the leading edge of the rudder. We can also further claim that by including the upper part of the rudder, only the local pressure field is noticeably affected. This supports our original assumption that the lower part of the rudder would have an even smaller influence on the predicted hull pressures, given its relatively longer distance from the hull.

It is worth noting that a more complete way for solving the numerical stability issue is to include the wake-rudder interaction in both the unsteady wake alignment model and the pressure-BEM model. This can be a future topic of this research.

Another neglected effect is from the blade boundary layer. The boundary layer effect, similar to the blade thickness, can be represented by source panels in the BEM model. The BEM solver can be used together with a boundary integral solver to determine the displacement thickness [17,18]. Then, the thickness can be converted into source strength and applied to the BEM solver. The displacement thickness can be as large as 15% of the blade thickness. Also, the source induced pressure decays slower ($\sim R^{-1}$) with distance compared to the dipole induced pressure ($\sim R^{-2}$). Therefore, it is possible that the boundary layer effect contributes to the downstream hull pressure fluctuation. The effect of boundary layer is another future topic of this research.

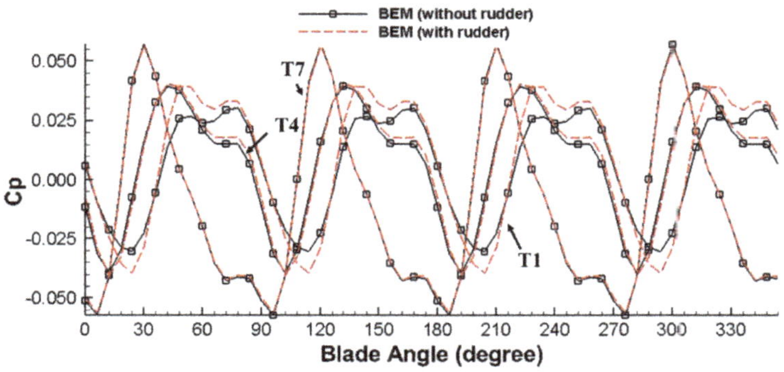

Figure 13. Study of the level of simplification on the rudder geometry in the pressure-BEM model. The blade-angle ranges between 0 and 2π and is defined by the angle the blade has passed starting from the "upright position". The "upright position" means the mid-camber point of the root section is located on the vertical axis above the hub ($+y$ axis).

5. Conclusions and Future Work

In this paper, several schemes are introduced to predict the propeller-induced hull pressure fluctuations. A BEM/RANS solver is first used to determine the effective wake and the propeller performance under the effective wake. Then, a pressure-BEM model is used to calculate the pressure on the ship hull. In the pressure-BEM model, only the upper part of the rudder is used to avoid singular radiated pressure on the rudder. A comparison study was made which indicated that excluding the lower part of the rudder is a good approximation. The effect of wake alignment model is also studied. The unsteady wake alignment scheme behaves better in terms of predicting the downstream hull pressures. Including the non-axisymmetric effective wake and using the unsteady wake alignment scheme improved the accuracy of the location of the wake and subsequently the accuracy of predicted the hull pressures.

Future work includes studying the blade boundary layer effect on the hull pressure, incorporating the unsteady wake–rudder interaction in the numerical scheme, and studying the effect of a finite speed of sound. The scheme of predicting the effective wake can also be improved in terms of accuracy.

Author Contributions: Y.S. participated in the development and improvement of the BEM pressure prediction model. S.K. participated in the development of different types of wake alignment models. S.A.K. provided guidance throughout the research.

Acknowledgments: Support for this research was provided by the U.S. Office of Naval Research (Grant Nos. N00014-14-1-0303 and N00014-18-1-2276; Ki-Han Kim) partly by Phases VII and VIII of the "Consortium on Cavitation Performance of High Speed Propulsors".

Conflicts of Interest: The authors declare no conflicts of interest.

References

1. Van Wijngaarden, E. Recent developments in predicting propeller-induced hull pressure pulses. In Proceedings of the 1st International Ship Noise and Vibration Conference, London, UK, 20–21 June 2005.
2. Tani, G.; Viviani, M.; Hallander, J.; Johansson, T.; Rizzuto, E. Propeller underwater radiated noise: A comparison between model scale measurements in two different facilities and full scale measurement. *Appl. Ocean Res.* **2016**, *56*, 48–66. [CrossRef]

3. Salvatore, F.; Testa, C.; Greco, L. Coupled hydrodynamics-hydroacoustics BEM modeling of marine propellers operating in a wakefield. In Proceedings of the First International Symposium on Marine Propulsors (SMP '09), Trondheim, Norway, 22–24 June 2009; pp. 537–547.
4. Kellett, P.; Turan, O.; Incecik, A. A study of numerical ship underwater noise prediction. *Ocean Eng.* **2013**, *66*, 113–120. [CrossRef]
5. Seol, H.; Jung, B.; Suh, J.-C.; Lee, S. Prediction of non-cavitating underwater propeller noise. *J. Sound Vib.* **2002**, *257*, 131–156. [CrossRef]
6. Li, D.-Q.; Hallander, J.; Johansson, T.; Karlsson, R. Cavitation dynamics and underwater radiated noise signature of a ship with cavitating propeller. In Proceedings of the VI International Conference on Computational Methods in Marine Engineering (MARINE 2015), Rome, Italy, 15–17 June 2015.
7. Lee, K.; Lee, J.; Kim, D.; Kim, K.; Seong, W. Propeller sheet cavitation noise source modeling and inversion. *J. Sound Vib.* **2014**, *333*, 1356–1368. [CrossRef]
8. Seol, H.; Suh, J.-C.; Lee, S. Development of hybrid method for the prediction of underwater propeller noise. *J. Sound Vib.* **2005**, *288*, 345–360. [CrossRef]
9. Tian, Y.; Jeon, C.H.; Kinnas, S.A. Effective Wake Calculation/Application to Ducted Propellers. *J. Ship Res.* **2014**, *58*, 1–13. [CrossRef]
10. Su, Y.; Kinnas, S.A. A Generalized Potential/RANS Interactive Method for the Prediction of Propulsor Performance. *J. Ship Res.* **2017**, *61*, 214–229. [CrossRef]
11. Tian, Y.; Kinnas, S.A. A Wake Model for the Prediction of Propeller Performance at Low Advance Ratios. *Int. J. Rotating Mach.* **2012**, *2012*, 372364. [CrossRef]
12. Kim, S. An Improved Full Wake Alignment Scheme for the Prediction of Open/Ducted Propeller Performance in Steady and Unsteady Flow. Master's Thesis, Ocean Engineering Group, CAEE, The University of Texas at Austin, Austin, TX, USA, August 2017.
13. Lee, H. Modeling of Unsteady Wake Alignment and Developed Tip Vortex Cavitation. Ph.D. Thesis, Ocean Engineering Group, CAEE, The University of Texas at Austin, Austin, TX, USA, August 2002.
14. He, L.; Kinnas, S.A. Numerical simulation of unsteady propeller/rudder interaction. *Int. J. Nav. Archit. Ocean Eng.* **2017**, *9*, 677–692. [CrossRef]
15. Hwang, Y.; Sun, H.; Kinnas, S.A. Prediction of Hull Pressure Fluctuations Induced by Single and Twin Propellers. In Proceedings of the Society of Naval Architects and Marine Engineers Propellers/Shafting 2006 Symposium, Williamsburg, VA, USA, 12–13 September 2006.
16. Lee, H.; Kinnas, S.A. Application of BEM in the Prediction of Unsteady Blade Sheet and Developed Tip Vortex Cavitation on Marine Propellers. *J. Ship Res.* **2004**, *48*, 15–30.
17. Drela, M. XFOIL: An analysis and design system for low Reynolds number airfoils. In *Low Reynolds Number Aerodynamics*; Springer: Berlin/Heidelberg, Germany, 1989; Volume 54.
18. Kinnas, S.A.; Yu, X.; Tian, Y. Prediction of Propeller Performance under High Loading Conditions with Viscous/Inviscid Interaction and a New Wake Alignment Model. In Proceedings of the 29th Symposium on Naval Hydrodynamics, Gothenburg, Sweden, 26–31 August 2012.

Journal of
Marine Science and Engineering

MDPI

Article

Influence of Propulsion Type on the Stratified Near Wake of an Axisymmetric Self-Propelled Body

Matthew C. Jones [1,*] and Eric G. Paterson [2]

[1] Department of Aerospace and Ocean Engineering, Virginia Polytechnic Institute and State University, Blacksburg, VA, 24061 USA

[2] Department Head and Rolls-Royce Commonwealth Professor of Marine Propulsion, Department of Aerospace and Ocean Engineering, Virginia Polytechnic Institute and State University, Blacksburg, VA 24061, USA; egp@vt.edu

* Correspondence: mattjones@vt.edu; Tel.: +1-626-421-8803

Received: 1 March 2018; Accepted: 23 April 2018; Published: 1 May 2018

Abstract: To better understand the influence of swirl on the thermally-stratified near wake of a self-propelled axisymmetric vehicle, three propulsor schemes were considered: a single propeller, contra-rotating propellers (CRP), and a zero-swirl, uniform-velocity jet. The propellers were modeled using an Actuator-Line model in an unsteady Reynolds-Averaged Navier–Stokes simulation, where the Reynolds number is $Re_L = 3.1 \times 10^8$ using the freestream velocity and body length. The authors previously showed good comparison to experimental data with this approach. Visualization of vortical structures shows the helical paths of blade-tip vortices from the single propeller as well as the complicated vortical interaction between contra-rotating blades. Comparison of instantaneous and time-averaged fields shows that temporally stationary fields emerge by half of a body length downstream. Circumferentially-averaged axial velocity profiles show similarities between the single propeller and CRP in contrast to the jet configuration. Swirl velocity of the CRP, however, was attenuated in comparison to that of the single propeller case. Mixed-patch contour maps illustrate the unique temperature distribution of each configuration as a consequence of their respective swirl profiles. Finally, kinetic and potential energy is integrated along downstream axial planes to reveal key differences between the configurations. The CRP configuration creates less potential energy by reducing swirl that would otherwise persist in the near wake of a single-propeller wake.

Keywords: actuator line; near wake; stratified; net-zero momentum; self-propelled; mixed patch; energy budget; axisymmetric

1. Introduction

Experiments show that propeller-driven wakes evolve from a complicated near wake with discernible propeller-blade features, to a far wake, in which these features have mixed together to form a nearly-axisymmetric field [1,2]. Sirviente and Patel [3] show that the near-wake region transitions to the far wake in roughly twelve initial wake diameters, but the development of the far wake can be delayed by appendages on the body [4]. This transition is influenced by the Reynolds number, body geometry, and operation of the propulsor [5], which itself has a large impact on the ingested stern boundary layer and downstream turbulence [6,7]. The swirling propeller induces helical vortices that are shed from the roots and tips of the individual blades. In the near wake, these vortices break down, which is a topic of extensive study [8]. Although experiments show the contribution of swirl [9], its role in the evolution from near to far wake is not well-characterized.

In a stratified wake, a mixed patch is formed by swirl from the propeller, turbulent mixing, and potential effects from the upstream body [10]. This mixed patch can further modify the far wake in the event of a mixed-patch collapse when buoyancy forces are large [11–14]. Numerous experiments have explored the interaction between stratification and wake evolution with close observation to the generation of internal gravity waves and coherent structures [15,16]. Direct Numerical Simulation (DNS) provides further insight into the physics of the flow, particularly with its turbulence properties [17,18]. Background turbulence increases the turbulent kinetic energy and energy transfer in the wake, which in turn lowers the mean velocity and increases horizontal spreading [19]. Excess momentum leads to changes in increased turbulent kinetic energy and qualitative changes in the wake dynamics, particularly in downstream vortical structures [20]. High levels of stratification in the wake create a non-equilibrium region in which the mean velocity decay is reduced [21,22]. By reducing the level of potential energy in the near wake thermal-haline distribution, the effects of buoyancy in the far wake can be reduced.

Originally studied as a disc-with-center-jet [23] and later with self-propelled axisymmetric bodies [24,25], the net-zero-momentum wake functions as a theoretical model of a self-propelled marine vehicle. Beyond experiment, the study of self-propelled wakes includes several numerical methods. Ordered by increasing fidelity and computational expense, these methods include [26]: panel/lattice methods, actuator models, and fully resolved rotating geometry. Generalized actuator models include the Actuator Disk (AD), Actuator Line (AL), and Actuator Surface (AS) models. Each of these models imposes a body force over a volume in a Computational Fluid Dynamics (CFD) simulation to simulate the effects of a propeller on the surrounding fluid. Although a fully resolved propeller may offer more fidelity, its computational requirements are often large, so an actuator model provides a cost-effective alternative [27].

In a self-propelled near wake, the mixed-patch structure and overall potential energy depend largely on the propulsor. A single propeller will mix fluid unopposed within the swirling region of the wake. Contra-rotating propellers of equivalent thrust will modify the initial swirl profile thereby changing the shape of the downstream mixed patch and reducing its potential energy. Contra-rotating propeller blades add additional complexity to the interaction between root and tip vortices and reduce the swirling kinetic energy of the wake. These influences on the near wake may be compared to the simplified case of a zero-swirl, jet-propelled configuration with uniform-velocity, which results in the smallest generation of the potential energy in the wake.

The present study is an extension of Jones and Paterson [28]. The unsteady Reynolds-Averaged Navier–Stokes (URANS) equations are solved to examine the near-wake evolution of the stratified, turbulent, net-zero-momentum propeller wake of the axisymmetric Iowa Body using three different propulsion schemes: single propeller, dual contra-rotating propellers (CRP), and a zero-swirl, uniform-velocity jet. The propellers are simulated using the AL model. The Iowa Body hull geometry is chosen for comparison to the non-stratified experiment by Hyun and Patel [2], which is the only known experiment to have phase-averaged propeller data for a self-propelled axisymmetric body. The authors have previously shown good agreement to this experiment for towed and self-propelled configurations [27]. Flow visualization reveals the interaction between propeller-root and tip vortices and the additional complexity introduced by CRP. Comparison between instantaneous and time-averaged cross-plane profiles demonstrates the transition from near- to far-wake regions. Circumferentially-averaged profiles of velocity reveal the evolution of momentum, with observations drawn in comparison to the theoretical disc-with-center-jet that is often used in far-wake simulations [11]. Mixed-patch velocity and temperature-deviation cross-plane profiles show the structure of kinetic and potential energy in the developed wake. Finally, the relative growth, decay, or persistence of integrated kinetic and potential energy of each propulsion scheme is considered. Compared to the single-propeller configuration, the CRP configuration is more effective at reducing potential and swirling kinetic energy in the wake, with potential energy reductions similar to that of the zero-swirl jet.

2. Approach

2.1. Governing Equations

This fluid-flow problem is defined by the unsteady Reynolds-averaged Navier–Stokes (RANS) equations in Boussinesq form with an additional body force term f_p to account for the propeller model.

$$\frac{\partial U_j}{\partial x_j} = 0 \tag{1}$$

$$\frac{\partial U_i}{\partial t} + \frac{\partial (U_i U_j)}{\partial x_j} = -\frac{1}{\rho_0}\frac{\partial \hat{p}}{\partial x_i} + \nu \frac{\partial^2 U_i}{\partial x_j \partial x_j} + \frac{\partial}{\partial x_j}\overline{u_i' u_j'} + \frac{\Delta\rho}{\rho_0} g_j \delta_{ij} + \frac{1}{\rho_0} f_p \tag{2}$$

The equations are written in terms of the non-inertial velocity U_i. In the equations, t is time, ν is the kinematic viscosity, and ρ is density. The density is expressed as $\rho = \rho_0 + \Delta\rho$, where ρ_0 is a reference value and $\Delta\rho$ is the deviation from that value. The gravitational vector g_j points downward in the negative z direction, where z is the upward-positive, vertical position. This formulation includes the piezometric pressure, $\hat{p} = p - \rho_0 g z$ where g is the magnitude of the gravitational vector.

The governing equations are solved using a custom solver written with the CFD framework OpenFOAM. This custom solver takes into account salinity and temperature transport and the corresponding turbulent fluctuations. The transport of temperature T and salinity S in the stratified environment are determined through the following equations with the diffusion coefficients κ_T and κ_S.

$$\frac{\partial T}{\partial t} + \frac{\partial (U_j T)}{\partial x_j} = \kappa_T \frac{\partial^2 T}{\partial x_j \partial x_j} + \frac{\partial}{\partial x_j}\overline{u_j' t'} \tag{3}$$

$$\frac{\partial S}{\partial t} + \frac{\partial (U_j S)}{\partial x_j} = \kappa_S \frac{\partial^2 S}{\partial x_j \partial x_j} + \frac{\partial}{\partial x_j}\overline{u_j' s'} \tag{4}$$

The Reynolds stresses $\overline{u_i' u_j'}$ and turbulent fluxes $\overline{u_j' t'}$ and $\overline{u_j' s'}$ are determined using a linear eddy-viscosity closure model.

$$-\overline{u_i' u_j'} = 2\nu_t S_{ij} - \frac{2}{3} k \delta_{ij} \tag{5}$$

$$-\overline{u_j' t'} = \frac{\nu_t}{\sigma_T}\frac{\partial T}{\partial x_j} \tag{6}$$

$$-\overline{u_j' s'} = \frac{\nu_t}{\sigma_S}\frac{\partial S}{\partial x_j} \tag{7}$$

In these equations, ν_t is the eddy viscosity, S_{ij} is the mean rate of strain, and k is the turbulent kinetic energy. For this study, the $k - \omega$ SST turbulence model is chosen to compute ν_t due to its ease of implementation and relative advantage in computing the attached flow over a body [29]. Production terms in the $k - \omega$ equations are modified to include buoyancy effects, but in the near wake they are small in comparison to the production due to shear. Wall functions are used in the computation of k and specific turbulence dissipation ω at wall boundaries to relax mesh requirements near the hull in the high Reynolds-number flow.

Density is computed by solving the UNESCO seawater equation of state [30]. For the given problem, it is appropriate to approximate the secant bulk modulus as constant at sea-level conditions, even though it is a function of salinity, temperature, and pressure. Thus, the secant bulk modulus is $K(S, T, p) = K(0, 20, p_{atm})$ where p_{atm} is atmospheric pressure. Additionally, substituting the hydrostatic pressure for the total pressure, the equation of state becomes,

$$\rho(S, T, p) = \frac{\rho(S, T, 0)}{1 - p/K(S, T, p)} \tag{8}$$

$$
\begin{aligned}
\rho(S,T,0) = & \left(a_0 + a_1 T + a_2 T^2 + a_3 T^3 + a_4 T^4 + a_5 T^5\right) \\
& + \left(b_0 + b_1 T + b_2 T^2 + b_3 T^3 + b_4 T^4\right) S \\
& + \left(c_0 + c_1 T + c_2 T^2\right) S^{3/2} \\
& + d_0 S^2
\end{aligned}
\tag{9}
$$

where a_n, b_n, c_n and d_0 terms are empirical coefficients given in Table 1. Because the environment in the present study is isohaline, only changes in temperature from the thermally-stratified background affect changes in density.

Table 1. Coefficients in UNESCO equation of state for seawater.

Coefficient	Value	Coefficient	Value	Coefficient	Value
a_0	9.998425×10^{2}	b_0	8.2449×10^{-1}	c_0	-5.7247×10^{-3}
a_1	6.793952×10^{-2}	b_1	-4.0899×10^{-3}	c_1	1.0227×10^{-4}
a_2	-9.095290×10^{-3}	b_2	7.6438×10^{-5}	c_2	-1.6546×10^{-6}
a_3	1.001685×10^{-4}	b_3	-8.2467×10^{-7}	d_0	4.8314×10^{-4}
a_4	-1.120083×10^{-6}	b_4	5.3875×10^{-9}		
a_5	6.536332×10^{-9}				

2.2. Kinetic and Potential Energy

The evolution and transfer of energy in the wake is examined in the form of kinetic and potential energy defined as,

$$ke = \frac{1}{2}\rho U^2, \quad pe = -\frac{1}{2}\frac{g}{\partial \rho_0/\partial z}(\rho - \rho_0)^2 \tag{10}$$

$$KE = \iint_A ke\, dA, \quad PE = \iint_A pe\, dA, \tag{11}$$

The per-unit-volume energy ke, and pe may be integrated over an axial slice of area A in the wake to find energy per-unit-length, KE and PE, as functions of downstream distance. Kinetic energy is computed for the magnitude of velocity and also individually for each component of velocity in cylindrical coordinates. The potential energy per-unit-volume pe follows the formulation of Holliday and McIntyre [31].

2.3. Actuator-Line Model

The unsteady propeller for each non-BOR hull form is simulated using an AL model from the Simulator fOr Wind Farm Applications (SOWFA) library [32]. The AL model projects a distributed line of force f_p in the place of each propeller blade,

$$f_p(r) = \frac{F_p}{\varepsilon^3 \pi^{3/2}} \exp\left[-\left(\frac{r}{\varepsilon}\right)^2\right] \tag{12}$$

where F_p is the actuator element force composed of contributions from lift F_L and drag F_D. The distance between CFD cell center and actuator point is r, and ε controls the Gaussian width. This function decays to 1% of its maximum value when $\varepsilon = 2.15r$. If ε is too small, numerical oscillations arise, and if ε is too large, the applied body forces will be smoothed considerably. Troldborg [33] recommends $\varepsilon \equiv 2\Delta x$ where Δx is the grid spacing at the actuator position. Martınez et al. [34] developed best practices for AL modeling and suggested $\varepsilon > 2\Delta x$. For the present study, $\varepsilon \equiv 4\Delta x$ was selected because it eliminated the numerical instabilities that arose when $\varepsilon \equiv 2\Delta x$ was assigned. The Cartesian-mesh region of the propeller was refined to a resolution such that, $\Delta x/\Delta b = 0.74$, where Δb is the width of each hydrofoil section. Martınez et al. [34] suggests a value smaller than 0.75. Lift and drag at each section are computed from a lookup table of lift and drag coefficients C_ℓ and C_d as functions of α,

$$F_L = \frac{1}{2} C_\ell(\alpha) \rho U_{rel}^2 c\, w, \quad F_D = \frac{1}{2} C_d(\alpha) \rho U_{rel}^2 c\, w \tag{13}$$

where ρ is the density, U_{rel} is the local flow speed, c is the chord and w is the width of the actuator section. The relationship between C_ℓ and C_d with α must be predetermined from experiment, simulation, or theory for each hydrofoil section. Figure 1 shows the magnitude of the projected propeller body force $|f_p|/(\rho_0 R_p \text{ rps}^2)$ on the AL propeller plane of the single-propeller case, where R_p is the propeller radius and rps is the propeller rotations per second.

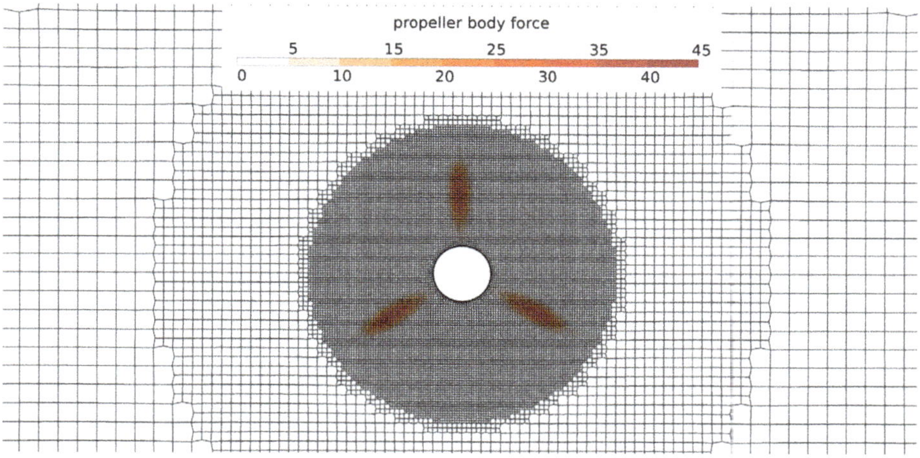

Figure 1. Non-dimensional body force on mesh slice at propeller plane $|f_p|/(\rho_0 R_p \text{ rps}^2)$ for the single-propeller case.

2.4. Iowa Body

The axisymmetric Iowa Body, described in the experiment by Hyun and Patel [2], is shown in Figure 2 for the standard, single-propeller case. This geometry is representative of a typical marine vehicle without appendages. Features of this geometry are listed in Table 2 where L is the body length, D is the body diameter, D_p is the propeller diameter, and D_h is the hub diameter.

Minor modifications are made to the Iowa Body hull for the CRP and jet configurations. For the CRP configuration, the hub is extended by the length of the rotating portion so that a second propeller may be placed directly downstream of the first. For the jet configuration, the hub is truncated at the propeller plane to function as an exhaust port. In effect, the zero-swirl, uniform-velocity jet exhausts with an initial diameter of D_h.

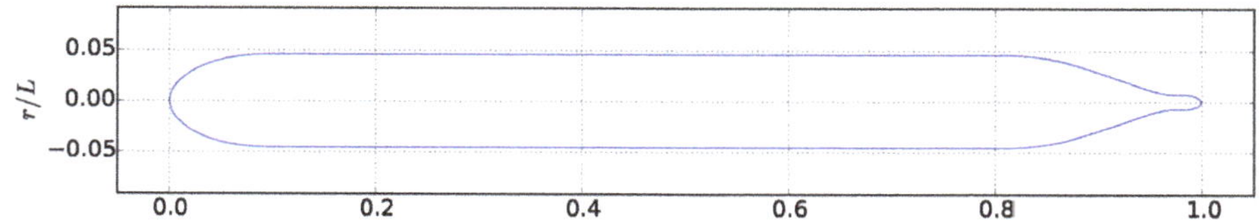

Figure 2. Standard Iowa Body profile.

Table 2. Iowa Body geometry.

Feature	Value
L/D	10.90
D/D_p	1.369
D_p/D_h	6.266
Hub location	$0.9688 < x/L < 0.9832$
Propeller location	$x/L = 0.9755$
Number of blades	3
Propeller hydrofoil	NACA 66-Modified

2.5. *Iowa Body Propeller*

The Iowa Body propeller is defined by 36 discrete sections to account for variations in radial propeller-blade geometry. Sectional lift C_ℓ is computed using the analytic expression of Brockett [35] for the NACA 66-Modified foil,

$$C_\ell = 2\pi(1 - 0.83\tau)(\alpha + 2.05f) \tag{14}$$

where α is the local flow angle of attack, τ is the maximum thickness ratio, and f is the maximum camber ratio. Sectional drag C_d is imposed by combining viscous and induced drag at each section,

$$C_d = C_{d0} + \frac{C_\ell^2}{\pi e \mathrm{AR}} \tag{15}$$

where C_{d0} is the viscous drag, e is the efficiency factor, and AR is the aspect ratio.

For the present unsteady simulations, α at each section of the propeller blades remains below 3° at every instant in time. Because α remains small, these analytic expressions do not require additional conditions for stall. Pitch, chord, thickness, and camber distributions for the Iowa Body propeller blade are tabulated in Hyun [1]. The Iowa Body propeller has zero rake and zero skew.

2.6. *Computational Mesh*

The three computational meshes were generated using the software *cfMesh* [36]. Cells are focused near the body, the propulsor region, and in the wake. The hull is located at a depth of one body length. The inlet, outlet, and far-field boundaries are located two body lengths away from the hull. Comparison to simulations from spatially-larger meshes showed that the boundaries of the computational domain did not affect the solution. Mesh design and quality features are listed in Table 3. Because wall functions are used in the computation of turbulence variables, the dimensionless wall distance requirement of $y^+ < 100$ can capture the boundary layer effects and the viscous drag of the hull even for boundary cells where $y^+ \approx 100$. A grid-refinement study of the propeller- and wake-region cells showed that 100 cells/D_p adequately resolved the AL model and downstream wake cross-plane profiles. These meshes are also visualized in Figure 3. Cutting planes reveal the distribution of cells surrounding the hull and in the wake region. Views of the propulsor region show how the mesh is modified for each configuration. A single AL-modeled propeller is implemented within the highlighted region for the single-propeller case. For the CRP configuration, the hub is extended with one AL-modeled propeller placed behind the first. There is no AL model for the jet configuration since there is no propeller. Instead, fluid is exhausted from the truncated hub.

Table 3. Mesh design and quality features.

Mesh Feature	Value
Boundary layer cells	> 20
Near-wall mesh spacing	$y^+ < 100$
Propulsor and wake cells/D_p	100
Wake region extends to	$x/L = 1.6$
Total number of cells	2×10^7
Maximum aspect ratio	AR < 170
Maximum non-orthogonality	$< 45°$
Maximum skewness	< 0.8

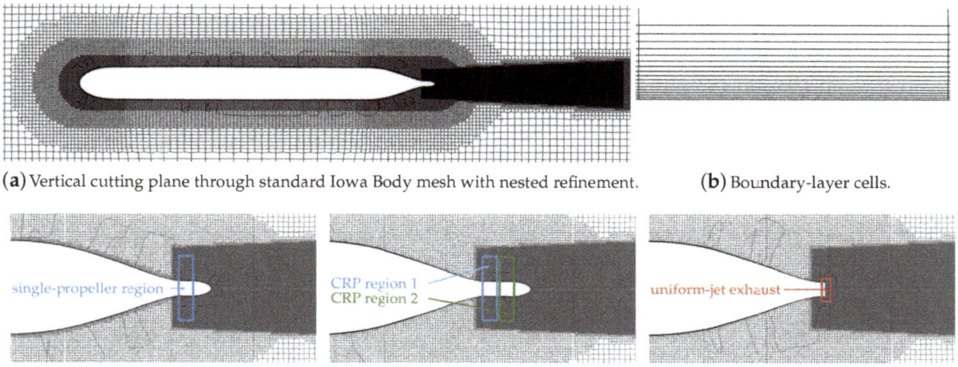

(a) Vertical cutting plane through standard Iowa Body mesh with nested refinement. **(b)** Boundary-layer cells. **(c)** Propulsor region for standard mesh. **(d)** Propulsor region for CRP mesh. **(e)** Propulsor region for jet mesh.

Figure 3. Computational meshes generated for each configuration.

2.7. Numerical Methods

The Navier–Stokes unsteady mass and momentum equations are solved using the Pressure-Implicit with Splitting of Operators (PISO) method [37]. This segregated approach decouples operations on pressure and velocity variables. At each time step, the following procedure is followed in the customized OpenFOAM solver. First, the momentum equations are solved to provide velocity by using pressure from the previous time step. Next, the pressure-Poisson equation is solved iteratively with corrections to velocity to conserve mass. Three inner iterations are are used in the present study, each with an additional mesh non-orthogonality correction step. After completion of these inner iterations, turbulence quantities are solved for, followed by salinity and temperature. The time step is then advanced.

Implicit, second-order, backward differencing is used in temporal discretization, while the cell-centered finite volume method is used in spatial discretization. A second-order, linear-upwind scheme is applied to the advective term of the momentum equations. A first-order, upwind scheme is applied to turbulence quantities, and a second-order, linear scheme is applied to all other divergence terms. Laplacian terms are discretized using a second-order, linear scheme that is partially-limited to correct for mesh non-orthogonality.

Two iterative methods are employed to solve the resulting systems of algebraic equations. The pressure equation is solved using the Preconditioned Conjugate Gradient (PCG) method with a residual tolerance of 10^{-6}. The momentum, scalar transport, and turbulence equations are solved using the Pre-Conditioned Bi-Conjugate Gradient (PBiCG) scheme with a residual tolerance of 10^{-8}.

2.8. Initial and Boundary Conditions

Several boundary conditions are employed. Velocity at the inlet is set to the freestream velocity U_0 through a Dirichlet boundary condition. The no-slip condition is set on the hull boundary, and the slip condition is set in the far field. Zero-gradient conditions are specified for velocity and pressure in the outlet. Background turbulence values of k and ω are computed assuming a turbulence intensity of 1% and eddy viscosity ratio ν_t/ν of 100. Turbulence variables on the hull boundary are computed with wall functions. Other variables satisfy the zero-gradient Neumann boundary condition.

Initial conditions for pressure and velocity are computed by solving the potential flow equations. The PISO algorithm is then used in the transient simulation. Distributed body forces from the propeller-blades rotate at each time step at the propeller's rotation rate. The simulation is then run until initial-transient flow features advect far downstream and a periodic wake flow field is found.

2.9. Flow Field Analysis

This study examines primary flow variables including: deviation of temperature from the background ΔT and the axial, radial, and azimuthal velocities U_x, U_r, and U_θ, respectively. The second invariant of the velocity gradient tensor Q is computed to visualize vortical structures. The cross-plane-integrated kinetic and potential energies are examined, where the kinetic energy is considered exclusively for each of the three components of velocity KE_x, KE_r, and KE_θ. Data are extracted in axial planes within $0.9755 \leq x/L \leq 1.5$ where x is the downstream distance from the bow of the body and L is the body length.

2.10. Flow Coefficients and Case Studies

Several of the important flow coefficients for this propeller-driven flow are the Reynolds number Re_L, advance ratio J, thrust coefficient C_T, and torque coefficient C_Q. An alternate expression for the thrust coefficient C_{T^*} is computed for comparison to the jet configuration.

$$Re_L = \frac{U_0 L}{\nu}, \quad J = \frac{U_0}{n D_p}, \quad C_T = \frac{F_T}{\rho_0 n^2 D_P^4}, \quad C_{T^*} = \frac{F_T}{\frac{1}{2}\rho_0 U_0^2 \pi R^2}, \quad C_Q = \frac{F_Q}{\rho_0 n^2 D_P^5} \tag{16}$$

For these expressions, U_0 is the freestream velocity, ν is kinematic viscosity, D_p is the diameter of the propeller, R is the radius of the Iowa Body, n is the propeller speed in revolutions per second, F_T is the thrust, and F_Q is the torque. The thrust-to-drag ratio is F_T/F_D. The Reynolds number for this study is $Re_L = 3.1 \times 10^8$, a typical operating condition in the ocean. Other coefficients are listed in Table 4. The fore and aft propellers are listed individually for the CRP case, and total thrust is equivalent for all cases.

Table 4. Flow coefficients.

Configuration	J	C_T	C_T^*	C_Q	F_T/F_D
Single	0.86	0.047	0.084	0.011	0.99
CRP (fore)	0.90	0.024	0.041	0.0071	0.50
CRP (aft)	0.86	0.023	0.041	0.0072	0.50
Jet	-	-	0.082	0	1.07

The Froude number Fr provides a measure of the density stratification, where an infinite Fr means zero stratification, and a small Fr means high levels of stratification. The present study considers a linearly varying temperature stratification, typical of an ocean environment, with a Froude number of $Fr = 350$, where,

$$Fr = \frac{1}{N}\frac{U_0}{D}, \text{ and } \quad N = \frac{1}{2\pi}\sqrt{\frac{-g}{\rho_0}\frac{\partial \rho}{\partial z}}. \tag{17}$$

In these expressions N is the Brunt Väisälä frequency, g is acceleration due to gravity, and z is the vertical coordinate. The influence of buoyancy on the near-wake fluid dynamics is often small and can be quantified by the Richardson number Ri, which is the ratio of buoyancy to flow gradient terms [15].

$$Ri = \frac{g}{T_0} \frac{d\overline{T}/dz}{(d\overline{U}/dz)^2} \tag{18}$$

where T_0 is a reference temperature, $\overline{T}$ is the mean temperature, and $\overline{U}$ is the mean velocity. For the single-propeller case, $Ri \approx 2.54 \times 10^{-3}$ which indicates that the near-wake inertial forces of the propeller dominate the buoyancy forces. As the local velocity U_x decays, Ri increases and buoyancy forces become important further downstream in the far wake, beyond the geometric bounds of these simulations.

3. Results

3.1. Near-Wake Transition

Individual vortices are visualized using the second invariant of the velocity gradient tensor Q. Figure 4 shows contour surfaces of the non-dimensionalized $(L/U_0)^2 Q = 16.9$ with a vertical cutting plane colored by axial velocity defect $U_x/U_0 - 1$ that extends to half of a body length downstream where $x/L = 1.5$. For the single-propeller case, root and tip vortices induced by the propeller are apparent. These vortices follow a helical path and disappear by $x/L \approx 1.25$. For the CRP case, additional complexity is introduced by the interaction between the two propellers. Complicated vortical structures are visible until $x/L \approx 1.35$. Negligible vortical structures are found in the case of the zero-swirl jet. This figure illustrates the complexity introduced by contra-rotating propellers in comparison to the other two cases.

To better understand the transition from near to far wake regimes in the propeller-driven cases, axial planes behind in the propulsor are examined. Figure 5 compares the instantaneous and time-averaged axial velocity defect field $U_x/U_0 - 1$ for the single propeller case. Instantaneous fields are taken at an arbitrary time long after initial-transient features have disappeared from the simulation and the flow field has become periodic. Time-averaging occurs over temporal interval of two periods of the propeller. Near the propellers, at $x/L = 1.01$ and $x/L = 1.10$, individual propeller-blade wakes can be seen in the instantaneous field. These blade wakes follow the azimuthal motion of the propellers directly upstream. The time-averaged field, by contrast, is axisymmetric. Further downstream at $x/L = 1.3$, only a small variation is seen between the instantaneous and time-averaged fields, and by $x/L = 1.45$, the two contour maps are nearly identical showing that the wake is steady and axisymmetric. By half of a body length downstream, the flow is stationary in time and space when viewed from a body-fixed reference frame.

For the CRP case shown in Figure 6, similar observations are drawn. Unsteadiness is apparent for $x/L \leq 1.10$, however by $x/L = 1.45$ the cross-plane profile is temporally stationary. In this case, a unique hexagonal shape is formed due to the interaction between the two opposing three-bladed propellers. This shape is still present at half of a body length downstream of the propulsor.

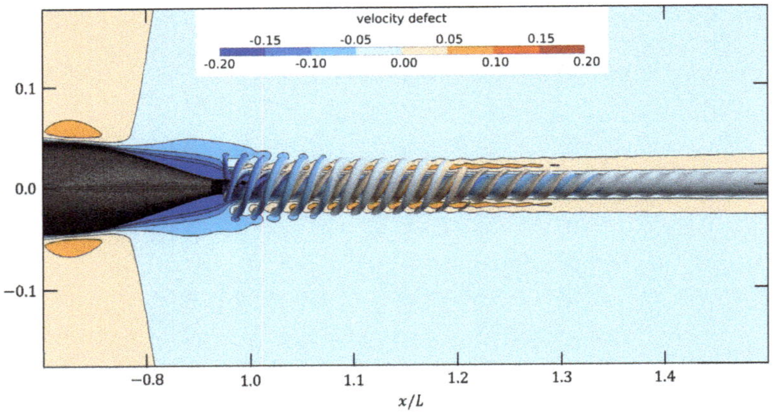

(**a**) Single propeller.

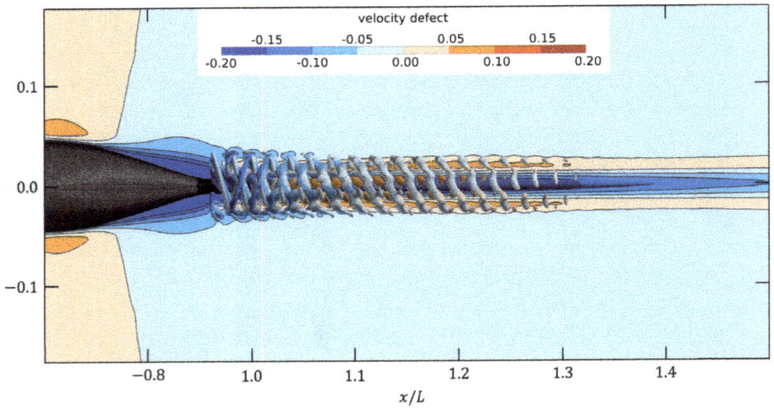

(**b**) Contra-rotating propellers.

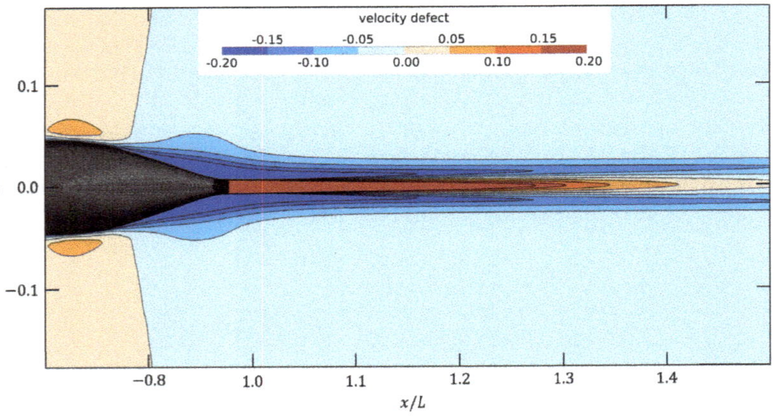

(**c**) Jet.

Figure 4. Flow visualization for $x/L \leq 1.5$ using Q-criterion visualization non-dimensionalized as $(L/U_0)^2 Q = 16.9$ colored by $U_x/U_0 - 1$ with vertical cutting plane through mesh.

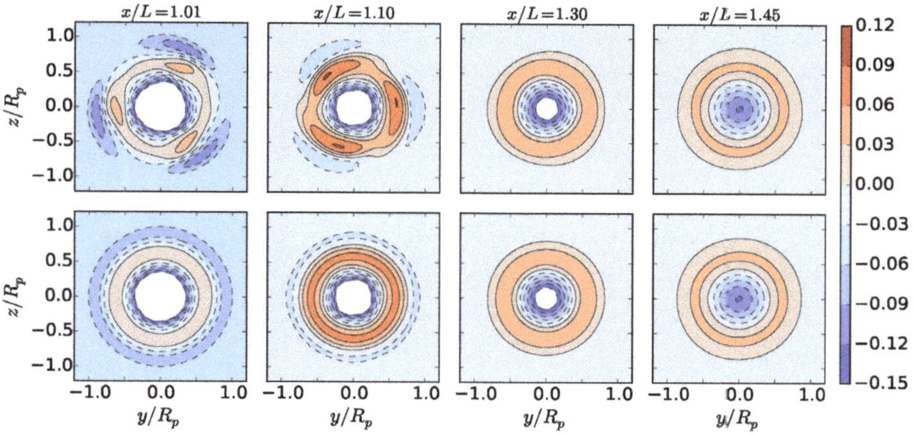

Figure 5. Instantaneous (**top**); and time-averaged (**bottom**) velocity defect $U_x/U_0 - 1$ for the single propeller case.

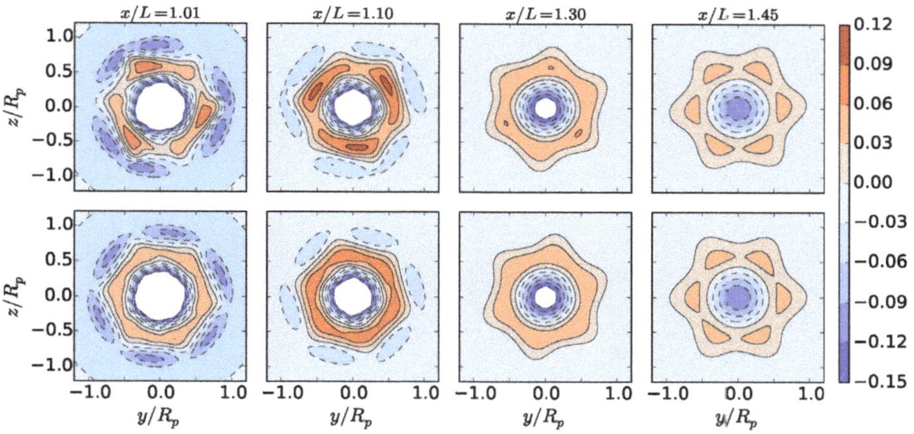

Figure 6. Instantaneous (**top**); and time-averaged (**bottom**) velocity defect $U_x/U_0 - 1$ for the CRP case.

3.2. Velocity Profiles

The evolution of the near wake may also be described by circumferentially-averaged velocity profiles. Figure 7 shows the circumferentially-averaged velocity defect profiles U_x/U_0 for the three self-propelled configurations at the downstream positions $x/L = 1.01$, $x/L = 1.3$, and $x/L = 1.5$. Just behind the propulsor at $x/L = 1.01$, the jet is shown to have uniform, positive velocity leaving the exhaust port, while negative velocity due to drag appears for $r/R_p > 0.2$. By $x/L = 1.3$ and further at $x/L = 1.5$, the jet profile appears as a classical net-zero-momentum wake and may be described using the analytical formulation of the disc-with-center-jet.

Circumferentially averaged profiles of the swirl component of velocity U_θ are shown in Figure 8. The jet configuration profile exhibits zero swirl because there are no sources of swirl for this case. The jet exhaust contains uniform axial velocity, the body is axisymmetric, and buoyancy forces are relatively small. The single-propeller configuration at $x/L = 1.01$, shows swirl imparted by the propeller and rotating hub. By $x/L = 1.3$ and $x/L = 1.5$, most of the momentum due to swirl exists in a region centered at $r/R_p \approx 0.4$. In the case of the CRP, regions of positive and negative swirl develop due to interference between the opposing, contra-rotating blades. Throughout the near wake, the CRP swirl magnitude is attenuated, remaining less than half of that of the single propeller, which is explained by the interactions of opposing azimuthal forces of the two propellers.

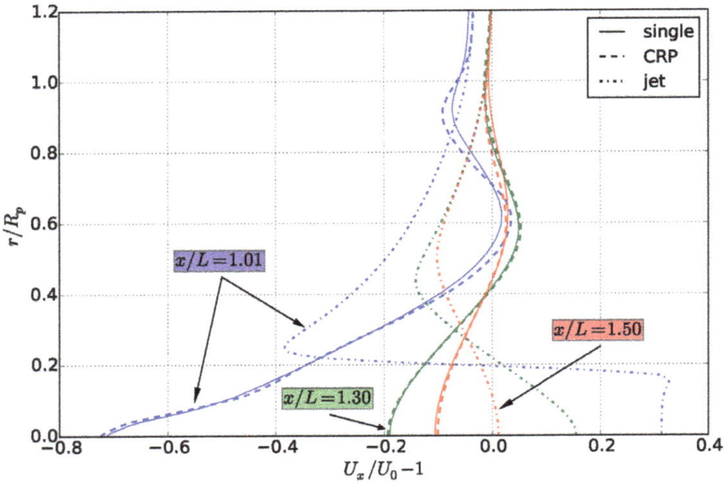

Figure 7. Circumferentially-averaged velocity defect $U_x/U_0 - 1$ profiles for each configuration at various distances downstream.

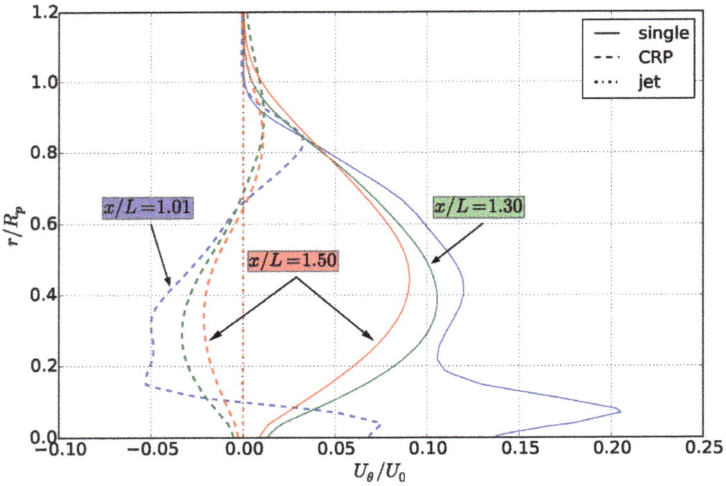

Figure 8. Circumferentially-averaged swirl velocity U_θ/U_0 profiles for each configuration at various distances downstream.

The propeller-driven cases show their own unique profiles. Positive momentum from the propellers exists in a region near $r/R_p \approx 0.6$, while negative momentum due to drag from the body exists near the center and further outward. These two circumferentially-averaged profiles are nearly equivalent and decay at similar rates, which is explained by the similar distributions of axial momentum. They also may be defined analytically using a process described in Jones and Paterson [27]. For all configurations, the positive momentum decays more quickly than the negative momentum, which is a feature of idealized wakes described by Tennekes and Lumley [38]. The theoretical, axisymmetric drag wake decays with the power of $-2/3$, while the theoretical, axisymmetric jet decays with the power of -1.

3.3. *Velocity and Temperature Fields in the Mixed-Patch*

The wake is mixed by half of a body length downstream of the stern. Unsteadiness from the propulsor has disappeared and axial gradients are small in comparison to the transverse. This location is significant because a cross-plane profile may be considered as the initial data plane (IDP) for further far-wake simulations, which is beyond the scope of this paper. Unique features of the IDP cross-plane profiles of velocity and temperature deviation are presented.

3.3.1. Velocity Field

Cross-plane contour maps of the axial velocity defect $U_x/U_0 - 1$ at $x/L = 1.5$ are shown for the single propeller and CRP cases in Figure 9. For the single propeller, axial velocity is axisymmetric and has previously been fit to an analytical curve as a function of radial distance [27]. The CRP, however, is not axisymmetric, and a steady, hexagonal profile is formed. The geometric shape is attributed to the two three-bladed propellers interacting with one another. Unlike the single-propeller profile, the CRP profile is a function of both radial and azimuthal positions. To create an analytical expression as a function of radial position alone, the profile must first be circumferentially averaged.

The swirl component of velocity U_θ/U_0 is shown in Figure 10. Again, the single-propeller velocity is axisymmetric, while the CRP velocity has discernible geometry. The magnitude of swirl velocity is much higher for the single propeller case, since the CRP propulsor imparts opposing azimuthal forces. Swirl velocity from the CRP is less than half of that of the single propeller and varies in azimuthal sign.

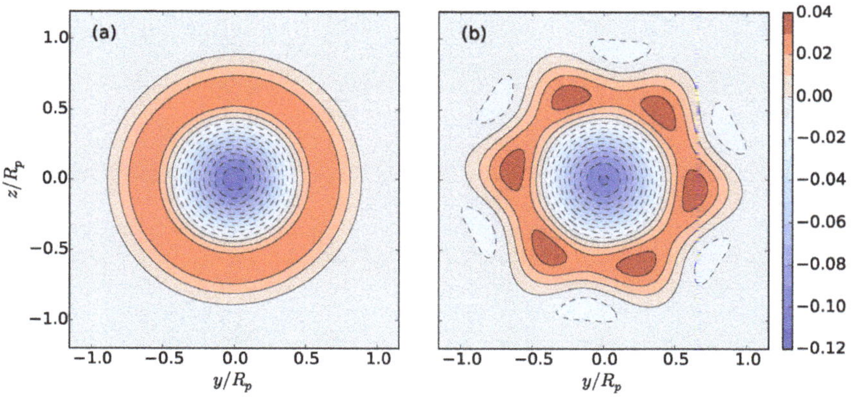

Figure 9. Velocity defect $U_x/U_0 - 1$ profiles at $x/L = 1.5$ for single propeller (**a**) and CRP (**b**) cases.

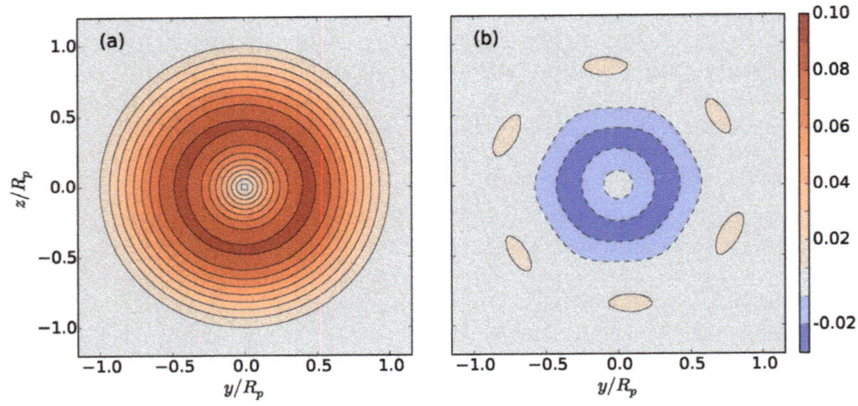

Figure 10. Swirl velocity U_θ/U_0 profiles at $x/L = 1.5$ for single propeller (**a**) and CRP (**b**) cases.

3.3.2. Temperature Field

Because these simulations take place in a thermally-stratified environment, any vertical redistribution of the fluid will generate potential energy. Mixing in the wake plays an important role in the redistribution of temperature T. Given that the background temperature field T_b is initially linearly stratified, mixing from the wake develops a temperature deviation $\Delta T = T - T_b$. This field is non-dimensionalized by the linear change in temperature over the depth of one propeller-blade length ΔT_{R_p}.

Figure 11 shows $\Delta T/\Delta T_{R_p}$ for the single-propeller case. Additional radial profiles help to visualize how the field varies in polar coordinates. A unique cross-plane profile shape is formed that is steady in time. Colder fluid has been driven to the top, while warmer fluid has been driven to the bottom of the wake. A maximum is found near the center of the warm region and a minimum is found near the center of the colder region. Two "tails" are shown trailing off of the warm and cold regions as a result of the counter-clockwise swirling motion of the fluid due to the propeller.

The mixed-patch $\Delta T/\Delta T_{R_p}$ cross-plane profile of the CRP case is shown in Figure 12. Compared to the single-propeller case, the magnitude of $\Delta T/\Delta T_{R_p}$ is less than half. The profile is split between an inner region where clockwise-swirling fluid dominates and an exterior region where counter-clockwise-swirling fluid dominates as shown previously in Figure 8. The complexity and lower $\Delta T/\Delta T_{R_p}$ magnitude in the profile arise directly from the initial interactions of the opposing azimuthal forces of the contra-rotating blades that drive the swirling fluid. While a single propeller can transport the temperature field across the swirling wake region unimpeded, the addition of an opposing propeller directly counters this effect. The net-swirl in the wake is reduced and regions of both positive and negative swirl exist. The inner, negative-swirl region forms a $\Delta T/\Delta T_{R_p}$ profile that mirrors the single-propeller case because of the sign difference in swirl. The outer, positive swirl region shares the same sign of $\Delta T/\Delta T_{R_p}$ as the single-propeller case. In effect, the net loss in swirl reduces the overall potential energy.

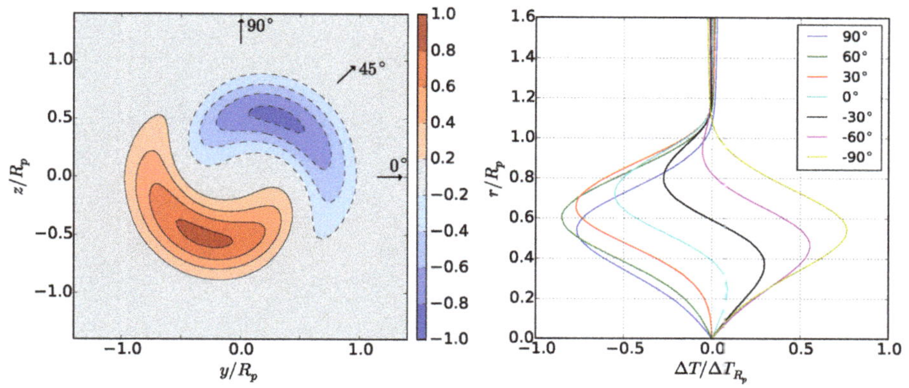

(**a**) Cross-plane contours.

(**b**) Profile in radial directions.

Figure 11. Single propeller $\Delta T/\Delta T_{R_p}$ at $x/L = 1.5$.

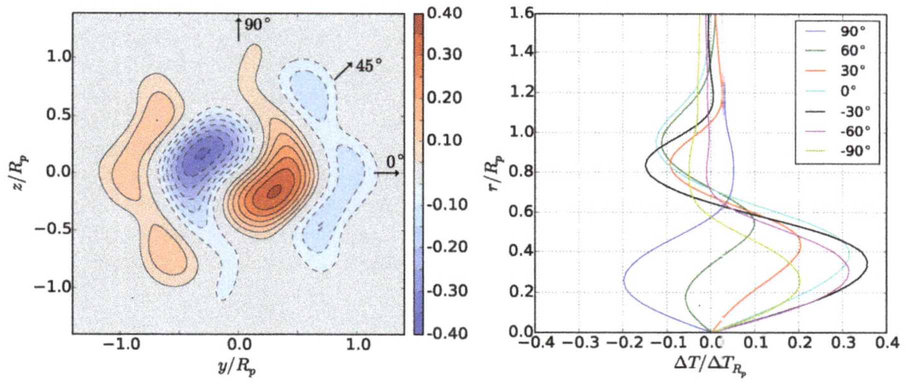

(**a**) Cross-plane contours.

(**b**) Profile in radial directions.

Figure 12. CRP $\Delta T/\Delta T_{R_p}$ at $x/L = 1.5$.

Finally, the mixed-patch $\Delta T/\Delta T_{R_p}$ cross-plane profile of the jet case is shown in Figure 13. For this configuration, the magnitude of $\Delta T/\Delta T_{R_p}$ is the smallest due to the absence of swirl. Instead, the shearing of axial momentum and potential effects from the body control the shape of the profile. The positive-momentum "jet" core entrains fluid from the negative-momentum "drag" periphery of the wake. Potential effects from the body further influence the temperature distribution. The resulting transport of $\Delta T/\Delta T_{R_p}$ suspends warmer fluid above colder fluid in the center, and the reverse in sign in the periphery. Without swirl, the distribution of $\Delta T/\Delta T_{R_p}$ in the central core of the jet wake is opposite in sign to that of the single-propeller case.

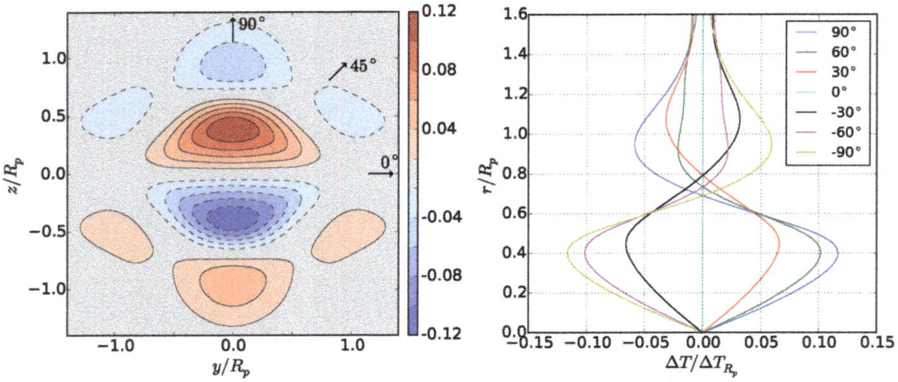

(**a**) Cross-plane contours. (**b**) Profile in radial directions.

Figure 13. Jet configuration $\Delta T/\Delta T_{R_p}$ at $x/L = 1.5$.

3.3.3. Comparison to a Perfectly-Mixed Temperature Field

The simulated cross-plane profiles may be compared to the idealized, perfectly-mixed profile shown in Figure 14. This conceptual profile assumes perfect mixing such that the temperature distribution T is uniform up until the boundary of the wake disc, beyond which $T = T_b$. The temperature deviation $\Delta T = T - T_b$, however, varies because of changes from the background stratification within the disc. Given the linear background stratification, ΔT also varies linearly in the vertical direction. The single propeller case relates most closely to this idealized profile, but, because it does not perfectly-mix T, differences can be observed. The upper and lower regions of colder and warmer fluid are shifted from the centerline, and the maximum temperature deviations are not on the wake boundary but instead closer inward. Results from the single propeller case show that T is not mixed uniformly within the disc of the swirling wake.

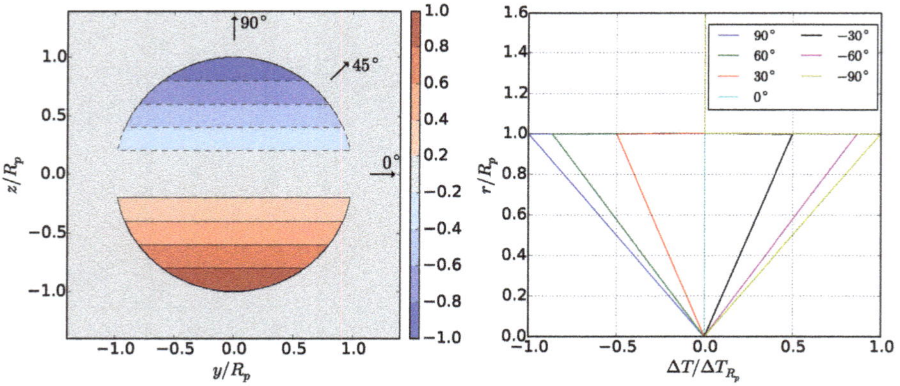

(**a**) Cross-plane contours. (**b**) Profile in radial directions.

Figure 14. Idealized mixed-patch profile of $\Delta T/\Delta T_{R_p}$ for a wake region of constant T.

3.4. Potential and Kinetic Energy Evolution

Kinetic and potential energy is integrated along axial planes downstream for the three configurations, as shown in Figure 15. Kinetic energy KE is computed individually for the three components of velocity, namely radial, swirl, and axial as KE_r, KE_θ, and KE_x, respectively, as well as for the velocity magnitude, KE. Downstream distance is described both by x/L measured from the bow and by x'/D_p measured from the stern.

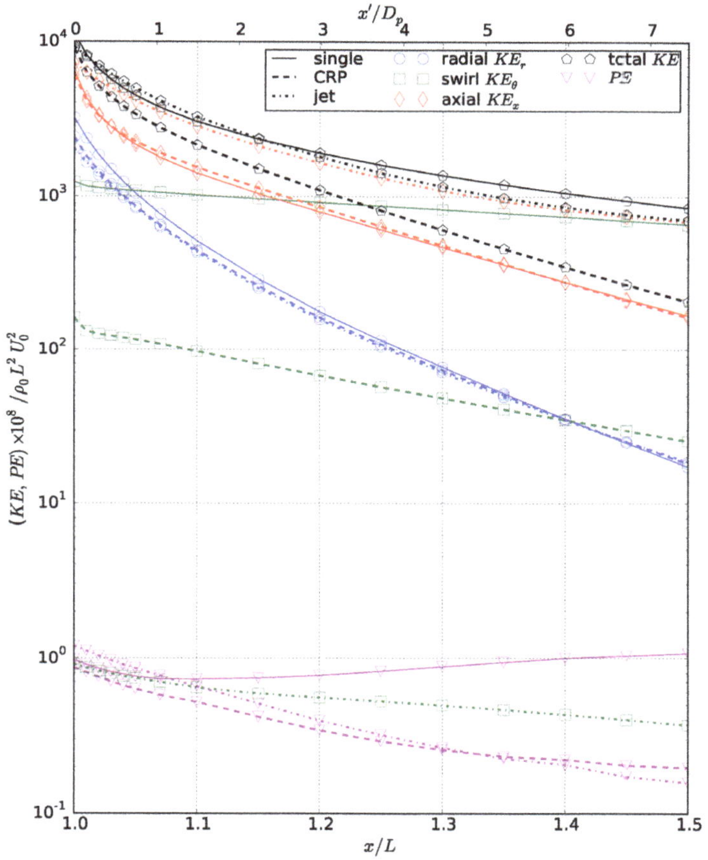

Figure 15. Integrated energy evolution downstream of the vehicle for each configuration, with x measured from the bow of the hull and x' measured from the stern.

For all three cases, KE_r decays more rapidly than KE_x and KE_θ. The swirl component notably exhibits the slowest decay in the near wake, an observation consistent with Sirviente and Patel [9]. For the single-propeller case, this relative persistence leads to a rise in PE due to expansion of the wake and entrainment of the surrounding passive scalar T, indicating a change in density. For the CRP, the PE does not grow due to the opposing regions of positive and negative swirl velocity. Instead, the PE decays with a rate similar to the zero-swirl jet. This result shows that the contra-rotating blades can effectively reduce PE, which will reduce the strength of buoyancy effects in the far wake.

Comparing the swirl component for the three cases, KE_θ of the CRP is an order of magnitude lower than that of the single propeller. Counteracting azimuthal momentum leads to a reduction in the swirl kinetic energy. The reduced KE_θ of the CRP is consistent with its reduced PE. The jet KE_θ is small because of the initial absence of swirl.

Additionally, comparison of the total KE shows that the CRP is the most-effective configuration for the total reduction of energy by $x/L = 1.5$. This increased decay rate suggests that the CRP far wake will decay more quickly than the single propeller and jet cases. The single propeller is less effective than the jet due to persisting KE_θ from its unidirectional swirling velocity.

4. Conclusions

The influence of swirl on the evolution of self-propelled, stratified near wakes and the development of the mixed patch has not previously been well-characterized. In this study, the linearly stratified near wake of the Iowa body was investigated with three separate propulsor configurations: single propeller, contra-rotating propellers, and a zero-swirl, uniform-velocity jet. Unsteady, rotating propeller blades were simulated using an AL model in a URANS computation. Comparison between the configurations revealed unique differences in the evolution of the near wake.

While clear root and tip vortices were visible in the single-propeller case, the CRP disrupted these structures, introducing additional complexity in the wake evolution. Nevertheless, by half of a body length downstream, the wake flow fields were steady in time. The single-propeller and CRP cases shared similar circumferentially-averaged axial velocity defect profiles due to similar spanwise loading in the propulsor. Swirl velocity, however, varied between the two propeller-driven cases, with the CRP introducing both positive and negative swirl regions exhibiting half of the magnitude of the single-propeller case. Furthermore, by half of a body length downstream, the magnitude of the temperature deviation $\Delta T/\Delta T_{R_p}$ for the CRP was less than half of that of the single propeller. The jet $\Delta T/\Delta T_{R_p}$ magnitude was the smallest, due to the absence of swirl. Contour maps of velocity revealed that the single propeller has an axisymmetric profile, whereas the CRP exhibits a unique hexagonal structure as a result of its two three-bladed propellers. The evolution of kinetic and potential energy varied as a direct result of the swirl imparted by each propulsor. Because of the interaction of positive and negative swirl, the CRP configuration showed an order of magnitude lower swirling kinetic energy compared to the single propeller configuration. Additionally, its potential energy was similar in decay and magnitude to that of the swirl-free jet, and the total kinetic energy decayed most rapidly out of the three propulsion schemes. These results indicate that the CRP can effectively reduce potential energy that would otherwise develop from a single-propeller configuration. By removing potential energy, buoyancy effects in the far wake will be weakened.

Author Contributions: Matthew C. Jones implemented the SOWFA actuator-line method into in-house code, conducted and analyzed the simulations, and wrote the paper. Eric G. Paterson is Principal Investigator of the research and dissertation adviser to the first author .

Acknowledgments: The first author has been supported by a Rolls-Royce Commonwealth Graduate Fellowship at Virginia Tech. Virginia Tech Advanced Research Computing is also acknowledged for providing high-performance computing resources. Finally, Matt Churchfield is acknowledged for advising in the implementation of the actuator-line model code from the SOWFA library.

Conflicts of Interest: The authors declare no conflict of interest.

Abbreviations

The following abbreviations are used in this manuscript:

AL	Actuator line
DNS	Direct Numerical Simulation
CRP	Contra-rotating propeller
SOWFA	Simulator fOr Wind Farm Applications
URANS	unsteady Reynolds-Averaged Navier–Stokes

References

1. Hyun, B.S. Measurements in the Flow Around a Marine Propeller at the Stern of an Axisymmetric Body. Ph.D. Thesis, University of Iowa, Iowa City, IA, USA 1990.
2. Hyun, B.S.; Patel, V.C. Measurements in the flow around a marine propeller at the stern of an axisymmetric body. *Exp. Fluids* **1991**, *11*, 33–44. [CrossRef]
3. Sirviente, A.; Patel, V.C. Wake of a self-propelled body, part 1: Momentumless wake. *AIAA J.* **2000**, *38*, 613–619. [CrossRef]
4. Posa, A.; Balaras, E. A numerical investigation of the wake of an axisymmetric body with appendages. *J. Fluid Mech.* **2016**, *792*, 470–498. [CrossRef]
5. Felli, M.; Camussi, R.; Di Felice, F. Mechanisms of evolution of the propeller wake in the transition and far fields. *J. Fluid Mech.* **2011**, *682*, 5–53. [CrossRef]
6. Posa, A.; Balaras, E. Large-eddy simulations of a propelled submarine model. In Proceedings of the APS Division of Fluid Dynamics Meeting Abstracts, Boston, MA, USA, 22–24 November 2015.
7. Posa, A.; Balaras, E. Large–eddy simulations of a notional submarine in towed and self–propelled configurations. *Comput. Fluids* **2018**, *165*, 116–126. [CrossRef]
8. Lucca-Negro, O.; O'Doherty, T. Vortex breakdown: A review. *Progress Energy Combust. Sci.* **2001**, *27*, 431–481. [CrossRef]
9. Sirviente, A.; Patel, V.C. Wake of a self–propelled body, part 2: Momentumless wake with swirl. *AIAA J.* **2000**, *38*, 620–627. [CrossRef]
10. Esmaeilpour, M.; Martin, J.E.; Carrica, P.M. Near-field flow of submarines and ships advancing in a stable stratified fluid. *Ocean Eng.* **2016**, *123*, 75–95. [CrossRef]
11. Hassid, S. Collapse of turbulent wakes in stably stratified media. *J. Hydronaut.* **1980**, *14*, 25–32. [CrossRef]
12. Chernykh, G.; Voropayeva, O. Numerical modeling of momentumless turbulent wake dynamics in a linearly stratified medium. *Comput. Fluids* **1999**, *28*, 281–306. [CrossRef]
13. Meunier, P.; Spedding, G.R. Stratified propelled wakes. *J. Fluid Mech.* **2006**, *552*, 229–256. [CrossRef]
14. Brucker, K.A.; Sarkar, S. A comparative study of self–propelled and towed wakes in a stratified fluid. *J. Fluid Mech.* **2010**, *652*, 373–404. [CrossRef]
15. Lin, J.; Pao, Y.H. Wakes in stratified fluids. *Ann. Rev. Fluid Mech.* **1979**, *11*, 317–338. [CrossRef]
16. Spedding, G. Wake signature detection. *Ann. Rev. Fluid Mech.* **2013**, *46*, 273–302. [CrossRef]
17. Pal, A.; de Stadler, M.B.; Sarkar, S. The spatial evolution of fluctuations in a self–propelled wake compared to a patch of turbulence. *Phys. Fluids* **2013**, *25*, 095106. [CrossRef]
18. Pal, A. *Dynamics of Stratified Flow Past a Sphere: Simulations Using Temporal, Spatial and Body Inclusive Numerical Models*; University of California: San Diego, CA, USA, 2016.
19. Pal, A.; Sarkar, S. Effect of external turbulence on the evolution of a wake in stratified and unstratified environments. *J. Fluid Mech.* **2015**, *772*, 361–385. [CrossRef]
20. De Stadler, M.B.; Sarkar, S. Simulation of a propelled wake with moderate excess momentum in a stratified fluid. *J. Fluid Mech.* **2012**, *692*, 28–52. [CrossRef]
21. Spedding, G. The evolution of initially turbulent bluff–body wakes at high internal Froude number. *J. Fluid Mech.* **1997**, *337*, 283–301. [CrossRef]
22. Pal, A.; Sarkar, S.; Posa, A.; Balaras, E. Direct numerical simulation of stratified flow past a sphere at a subcritical reynolds number of 3700 and moderate froude number. *J. Fluid Mech.* **2017**, *826*, 5–31. [CrossRef]
23. Naudascher, E. Flow in the wake of self–propelled bodies and related sources of turbulence. *J. Fluid Mech.* **1965**, *22*, 625–656. [CrossRef]
24. Lin, J.; Pao, Y. *Velocity and Density Measurements in the Turbulent Wake of a Propeller-Driven Slender Body in a Stratified Fluid*; Flow Research Inc.: Middlesex, MA, USA, 1974.
25. Schetz, J.; Jakubowski, A. Experimental studies of the turbulent wake behind self–propelled slender bodies. *AIAA J.* **1975**, *13*, 1568–1575. [CrossRef]
26. Sanderse, B.; Pijl, S.; Koren, B. Review of computational fluid dynamics for wind turbine wake aerodynamics. *Wind Energy* **2011**, *14*, 799–819. [CrossRef]
27. Jones, M.C.; Paterson, E.G. Evolution of the propeller near-wake and potential energy in a thermally-stratified environment. In Proceedings of the OCEANS 2016 MTS/IEEE Monterey, Monterey, CA, USA, 19–23 September 2016; pp. 1–10.

28. Jones, M.C.; Paterson, E.G. Influence of propulsion type on the near-wake evolution of kinetic and potential energy. In Proceedings of the Fifth International Symposium on Marine Propulsors, Helsinki, Finland, 12–15 June 2017, pp. 1–8.
29. Menter, F.R. Two-equation eddy-viscosity turbulence models for engineering applications. *AIAA J.* **1994**, *32*, 1598–1605. [CrossRef]
30. Fofonoff, N.P.; Millard, R.C. Algorithms for computation of fundamental properties of seawater. *UNESCO Tech. Pap. Mar. Sci.* **1983**, *44*, 1–53.
31. Holliday, D.; McIntyre, M.E. On potential energy density in an incompressible, stratified fluid. *J. Fluid Mech.* **1981**, *107*, 221–225. [CrossRef]
32. NWTC Information Portal (SOWFA). Available online: https://nwtc.nrel.gov/SOWFA (accessed on 21 April 2016).
33. Troldborg, N. Actuator Line Modeling of Wind Turbine Wakes. Ph.D. Thesis, Technical University of Denmark, Lyngby, Denmark, 2008.
34. Martınez, L.A.; Leonardi, S.; Churchfield, M.J.; Moriarty, P.J. A comparison of actuator disk and actuator line wind turbine models and best practices for their use. In Proceedings of the 50th AIAA Aerospace Sciences Meeting including the New Horizons Forum and Aerospace Exposition, Nashville, TN, USA, 9–12 January 2012.
35. Brockett, T. *Minimum Pressure Envelopes for Modified Naca-66 Sections with Naca a = 0.8 Camber and Buships Type 1 and Type 2 Sections*; Technical Report; DTIC Document: Fort Belvoir, VA, USA, 1966.
36. cfMesh 1.1.1. Available online: http://www.cfmesh.com/ (accessed on 16 February 2016).
37. Issa, R.I. Solution of the implicitly discretised fluid flow equations by operator-splitting. *J. Comput. Phys.* **1986**, *62*, 40–65. [CrossRef]
38. Tennekes, H.; Lumley, J.L. *A First Course in Turbulence*; MIT Press: Cambridge, MA, USA, 1972.

Journal of
Marine Science and Engineering

MDPI

Article

Modelling a Propeller Using Force and Mass Rate Density Fields

David Hally

Defence Research and Development Canada, P.O. Box 1012, Dartmouth, NS B2Y 3Z7, Canada;
david.hally@drdc-rddc.gc.ca; Tel.: +1-902-407-0348

Received: 23 February 2018; Accepted: 3 April 2018; Published: 12 April 2018

Abstract: A method to replace a propeller by force and mass rate density fields has been developed. The force of the propeller on the flow is calculated using a boundary element method (BEM) program and used to generate the force and mass rate fields in a Reynolds-averaged Navier–Stokes (RANS) solver. The procedures to calculate the fields and to allocate them to the cells of a RANS grid are described in detail. The method has been implemented using the BEM program PROCAL and the RANS solver OpenFOAM and tested using the propeller DTMB P4384 operating in open water. Close to the design advance coefficient, the time-average flow fields generated by PROCAL and by OpenFOAM with the force and mass rate fields match to within 1.5% of the inflow speed over almost all of the flow field, including the swept volume of the blades. At two-thirds of the design advance coefficient, the match is about 4% of the inflow speed. The sensitivity of the method to several of its free parameters is investigated.

Keywords: propeller; force field; mass rate field; RANS; BEM

1. Introduction

Although it is now feasible to use Reynolds-averaged Navier–Stokes (RANS) solvers to calculate the flow around a propeller operating in the vicinity of its supporting structure, such calculations are still complex and long. A simpler alternative is to replace the propeller in the RANS calculation by a time-averaged force field which accelerates the fluid in the same manner. The force field can be determined by an independent propeller analysis program typically using the boundary element method (BEM). To determine the mutual interaction between the propeller and its support, the flow calculation proceeds in an iterative manner with successive calculations of the flow using the RANS solver and the BEM propeller program. The RANS solver determines the inflow for the BEM program; the BEM program determines the force field to be used in the RANS solver.

The complete calculation is simpler than a full RANS calculation because the details of the propeller need not be reflected in the RANS grid, nor is there a need to have a separate rotating domain within the RANS grid. In addition, since the force field is averaged in time, the RANS calculation is usually steady resulting in much decreased computation times. Even if the RANS simulation is unsteady, when the time scales associated with the propeller are much shorter than those of the RANS flow (as may be the case, for example, with a manoeuvering ship), significant computational savings may be obtained by using time-averaged forced fields to model the propeller.

However, it should be noted that time-averaging is not necessary. For example, Greve et al. [1] have modelled the propeller of a ship in a seaway using a time-varying force field. Nevertheless, the time-saving advantages of the coupling scheme are not nearly so significant when the RANS solution is unsteady. In the remainder of this paper it will be assumed that the force fields are averaged in time.

The RANS-BEM coupling scheme can be implemented whenever the interaction between the propeller and its immediate surroundings is important. The problem that has motivated this research is that of a propeller operating on a ship and its interaction with the hull and its appendages. The inflow into the BEM propeller program, including the interaction effects with the hull, is called the effective wake.

The use of a force field to model the propeller has a long history. Actuator disk theory dates back to Rankine [2] and Froude [3]. In 1964, Hough and Ordway [4] determined an analytic solution for an actuator disk with constant radial load. Their expressions were generalized further by Conway [5,6]. The purpose of an actuator disk is to provide a rudimentary model of the action of the propeller so its effect on the flow around the ship can be determined. Typically, the mutual interaction of the disk and the ship is ignored although a linear theory of the interaction of an actuator disk and an axisymmetric body was developed by Sparenberg [7].

The converse problem of determining the effect of the ship on the propeller, i.e., calculating the effective wake, was addressed by Huang and Groves [8] with their V-shaped Segments (VSM) method. It applied to axisymmetric hulls but was later generalized to arbitrary hulls by Bujnicki [9]. The Force Field Method (FFM) was developed at Maritime Institute of the Netherlands (MARIN) for the Cooperative Research Ships organization [10]. Neither VSM nor FFM account for the effect of the propeller on the flow around the ship.

The mutual interaction of hull and propeller was modelled by Choi and Kinnas [11] using an Euler solver initialized one propeller diameter upstream of the propeller disk and a vortex lattice propeller code. Coupling of a RANS code for the flow past the ship and a vortex lattice code for the flow past the propeller has been reported by Stern et al. [12]. Three methods were tried for adding the effect of the propeller in the RANS solution: a force field method, a method in which extra boundary conditions were prescribed at the surfaces of an internal volume containing the propeller, and a method in which the propeller-induced flow field was prescribed. They found that the force-field approach gave the best results.

In the RANS-BEM coupling scheme, since both programs calculate the propeller induction, it must be subtracted from the flow field calculated by the RANS program before it is used as the inflow to the BEM program. To avoid errors in the BEM inflow, it is important that the inductions calculated by RANS and BEM match well. Hally and Laurens [13] showed that a good match cannot be achieved unless the blockage of the flow by the propeller is taken into account; because the propeller blades displace some of the fluid volume, they tend to retard the flow upstream, an effect that is not captured by the force field since it displaces no fluid. They accounted for this effect by calculating the flow inside the propeller blades in the BEM calculation to produce a force field that included the retarding effect. However, the flow inside the blades has high velocities making the method sensitive to the panelling of the propeller blades.

Starke and Bosschers [14] accounted for the blockage effects in a different way. Using the Morino formulation [15] of the BEM, they argued that the blockage is caused by the source terms distributed on the panels while the propeller induction is caused by the dipole terms. Therefore, when calculating the propeller induction, they simply set the sources to zero and include the effect of the dipoles alone. The match between RANS and BEM inductions is then good, but the upstream flow is not correct since any interaction with the hull caused by the retardation of the flow will not be present.

Hally [16,17] described a third method for accounting for the propeller blockage by adding a source term to the continuity equation in the RANS solver; an additional term must also be added to the momentum equations to account for the change in momentum induced by the injection of mass. This method has subsequently been used by Su et al. [18]. It is comparatively insensitive to the blade panelling and includes the retarding effects in the propeller–hull interaction.

The current paper expands on the description by Hally [16,17] of the modelling of the propeller by force and mass rate density fields. The expressions for the sources in the momentum and continuity equations are derived more completely and an error in the momentum source induced by the injection of mass is corrected. The method for allocating the force and mass rate densities to RANS cells is described in detail and the sensitivities to various parameters are investigated. An example of the use of the method in a self-propulsion calculation can be found in Ref. [17].

2. Theory

2.1. The Boundary Element Method

Boundary element methods for propellers represent the flow as a known background velocity, $\mathbf{V}$, plus the gradient of a potential, ϕ, which satisfies Laplace's Equation:

$$\mathbf{v} = \mathbf{V} + \nabla\phi; \qquad \nabla^2\phi = 0. \tag{1}$$

Using Green's Second Theorem, the equation for ϕ can be recast as a Fredholm integral equation of the second kind:

$$-4\pi\phi = \int_S \left(\Delta\phi\frac{\partial G}{\partial n} - G\frac{\partial\Delta\phi}{\partial n}\right) da, \tag{2}$$

where $\mathcal{S}$ is the surface of the body in the flow, G is a Green function for the Laplacian, $\partial/\partial n$ denotes a directional derivative along an outward pointing normal to the surface and $\Delta\phi = \phi_o - \phi_i$ is the difference between the values of the potential on the inner and outer surfaces of the body. Equation (2) is often rewritten as

$$\phi = \int_S \left(\sigma G - \mu\frac{\partial G}{\partial n}\right) da; \qquad \sigma = \frac{1}{4\pi}\frac{\partial\Delta\phi}{\partial n}; \qquad \mu = \frac{\Delta\phi}{4\pi}. \tag{3}$$

with σ known as the source strength and μ the dipole strength. The condition that there be no flux of fluid through the surface of the body requires that

$$\frac{\partial\phi_o}{\partial n} = -\hat{n}\cdot\mathbf{V} \tag{4}$$

on the outer surface of the body, where $\hat{n}$ is an outward pointing normal to the surface.

Equation (2) is valid for the flow both inside and outside the body. It is possible to specify the value of either σ or μ arbitrarily; provided that the no-flux boundary condition is satisfied, the flow outside the body remains the same; only the flow inside is changed. In the Morino formulation [15], the source strength is chosen to be

$$\sigma = -\frac{\hat{n}\cdot\mathbf{V}}{4\pi}. \tag{5}$$

On the inside surface, $\partial\phi_i/\partial n = 0$ implying that ϕ is constant, and consequently $\mathbf{v} = \mathbf{V}$, everywhere inside the body.

BEM codes approximate the integral equation for ϕ by splitting the surface of the body into panels on which the value of μ is unknown. The values of μ are determined so that the no-flux boundary condition is satisfied. If the surface generates lift, extra rows of panels are shed into the wake of the body. On each row, μ is the same for each panel; its value is determined so that the Kutta condition is satisfied where the panels detach from the body. The source strengths on the wake panels are zero.

The wake panels are convected by the flow and so there should be no flux of fluid through them. Since their source and dipole strengths are constrained, this can only be achieved by adjusting their locations iteratively until the no flux condition is satisfied. In practice, empirical formulae are often used to impose the locations of the panels. The importance of different methods for placing the wake panels is discussed further in Section 5.2.

2.2. Representing the Propeller as Mass Rate and Force Density Fields

If the propeller is modelled in the RANS solution using only a force field, the effect of the physical bulk of the blades—the blade blockage—is ignored. One way to account for the blade blockage is to include a source term in the continuity equation as well as a source term in the momentum equations. In effect, the RANS solution is determined so that it mimics the BEM flow both inside and outside the blades.

Consider a small volume, δV, in the RANS flow (hence in a ship-fixed non-rotating coordinate system) through which the surface, $\mathcal{S}$, of a blade passes: see Figure 1. Let the area of $\mathcal{S}$ within δV be δs and let ω be the angular speed of the blade (radians per unit time), r the distance from the propeller axis and $\hat{\theta}$ a unit vector in the direction of travel of δs. The normal velocity relative to the blade surface is $\hat{n} \cdot (\mathbf{v} - \omega r\hat{\theta}) = 0$ on the outside of the blade and $\hat{n} \cdot (\mathbf{V} - \omega r\hat{\theta})$ on the inside. Therefore, the flux of mass out of δV due to the convection of mass across δs is

$$\delta M = \rho\big(\hat{n} \cdot (\mathbf{V} - \omega r\hat{\theta}) - \hat{n} \cdot (\mathbf{v} - \omega r\hat{\theta})\big)\delta s = \rho\hat{n} \cdot (\mathbf{V} - \omega r\hat{\theta})\delta s. \tag{6}$$

Similarly, there is a flux of momentum equal to

$$\delta \mathbf{F} = \rho\Big(\mathbf{V}\big(\hat{n} \cdot (\mathbf{V} - \omega r\hat{\theta})\big) - \mathbf{v}\big(\hat{n} \cdot (\mathbf{v} - \omega r\hat{\theta})\big)\Big)\delta s = \rho\mathbf{V}\big(\hat{n} \cdot (\mathbf{V} - \omega r\hat{\theta})\big)\delta s = \delta M\mathbf{V}. \tag{7}$$

Consequently, if a source term, M, is included in the continuity equation to simulate an injection of mass, a source term $M\mathbf{V}$ must be included in the momentum equations to account for the corresponding injection of momentum. Because this momentum source is associated solely with the flow inside the blades, it is the velocity inside the blades, i.e., $\mathbf{V}$, that appears in this term. The RANS equations to be solved then become the following:

$$\rho\nabla \cdot \mathbf{v} = M, \tag{8}$$

$$\rho\nabla \cdot (\mathbf{v}\mathbf{v}) = -\nabla p + \nabla \cdot (\mu_t \nabla \mathbf{v}) + M\mathbf{V} + \mathbf{F}, \tag{9}$$

where M is the rate of increase of mass per unit volume and $\mathbf{F}$ is the force density associated with the thrust of the propeller (i.e., caused by the pressure distribution over $\mathcal{S}$). These equations have been misquoted by Hally [16,17] and Su, Kinnas and Jukola [18]. In each case, the momentum source associated with the mass source was given, either explicitly or implicitly (because the momentum equations were cast in weak conservation form), as $M\mathbf{v}$: i.e., the total velocity was used, not the background flow from the BEM solution.

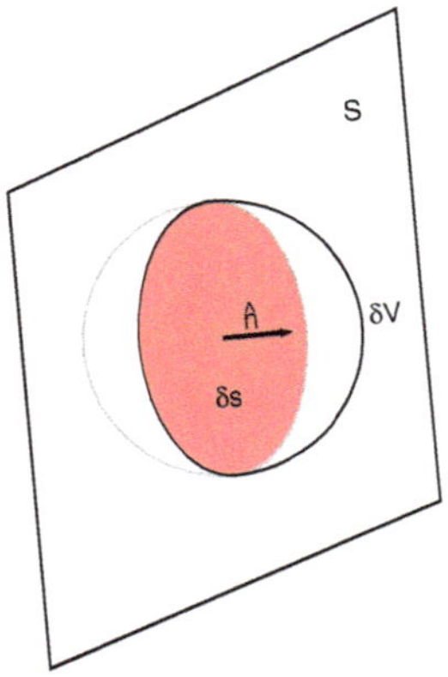

Figure 1. The intersection of the blade surface with a small volume of fluid.

The mass rate is independent of the propeller rotation rate and therefore independent of the advance coefficient. Therefore, as the advance coefficient decreases and the propeller becomes more heavily loaded, the contribution of the mass rate to the velocity field becomes less significant relative to the contribution from the force field. The mass rate term has the most effect at the inner radii where the blade sections are thickest.

Although frictional forces are not included in the BEM solution, BEM solvers often include simple predictions of the skin friction on the blades using boundary layer theory. It would be straightforward to include the force due to the skin friction in **F**, but this should usually not be done when coupling RANS and BEM solvers. Although the force and torque might be corrected for the skin friction in the BEM solver, the velocity usually is not. Adding the effect of the skin friction in the RANS solver would then cause an extra source of error when comparing the RANS and BEM induced velocities.

3. Calculating the Mass and Force Densities from BEM Output

3.1. The Force Density and Rate of Change of Mass

Consider a small area δs on the surface of a blade. The force of the blade on the fluid caused by this area is $p\hat{n}\delta s$ where p is the pressure and $\hat{n}$ is an outward pointing unit normal to the surface. In a small time δt, the rotation of the blade causes the surface to sweep out a small volume $\omega r|\hat{n}\cdot\hat{\theta}|\delta s\,\delta t$ where ω is the angular speed of the blade (radians per unit time), r is the distance from the propeller axis and $\hat{\theta}$ is a unit vector in the direction of travel of δs. The change in momentum of the fluid during that interval is $p\hat{n}\,\delta s\,\delta t$. Now consider this volume of fluid to be fixed in space. During one full revolution of the propeller, the volume will not experience any additional change in momentum due to that blade, but it will experience a similar change from each of the other blades as they pass through it, so its change in momentum due to the action of the propeller over one full revolution is $Zp\hat{n}\,\delta s\,\delta t$, where Z is the number of blades. The change of momentum per unit volume in one propeller revolution is therefore $Zp\hat{n}/\omega r|\hat{n}\cdot\hat{\theta}|$. Since the period of the propeller revolution is $T = 2\pi/\omega$, the force density (equivalent to the average rate of change of momentum per unit volume) is $Zp\hat{n}/2\pi r|\hat{n}\cdot\hat{\theta}|$. At any point in the flow, the force density is then obtained by determining when a blade surface will pass through the point (if ever) and at which time. In general, there will be two surface points per blade: one on the face and one on the back. The force density **F** at any point in the flow is then

$$\mathbf{F} = \sum_{f,b} \frac{Zp\hat{n}}{2\pi r|\hat{n}\cdot\hat{\theta}|}, \tag{10}$$

where the sum is over points on the face and back of a single blade and it is understood that p is evaluated at the moment that the surface passes through the point in the flow.

Similarly, during the time δt, the discontinuity in velocity across the blade surface causes a mass of $-\rho(\mathbf{V}-\omega r\hat{\theta})\cdot\hat{n}\,\delta s\,\delta t$ to be introduced into the volume. The average rate of change of mass is then

$$M = \sum_{f,b} -\frac{Z\rho(\mathbf{V}-\omega r\hat{\theta})\cdot\hat{n}}{2\pi r|\hat{n}\cdot\hat{\theta}|} = \sum_{f,b} -\frac{Z\rho\mathbf{V}\cdot\hat{n}}{2\pi r|\hat{n}\cdot\hat{\theta}|} + \sum_{f,b} \frac{Z\rho\omega\,\mathrm{sgn}(\hat{\theta}\cdot\hat{n})}{2\pi} = \sum_{f,b} -\frac{Z\rho\mathbf{V}\cdot\hat{n}}{2\pi r|\hat{n}\cdot\hat{\theta}|}, \tag{11}$$

the term in ω vanishing because $\hat{\theta}\cdot\hat{n}$ will have opposite signs on the face and the back.

Note that both **F** and M will be infinite at locations near the leading edge where $\hat{n}\cdot\hat{\theta}$ is zero. As was pointed out by Starke and Bosschers [14], this can cause problems if the force densities are simply interpolated either to the nodes of the RANS grid or to the centroids of the RANS grid cells. Their solution was to smooth the force density while keeping the total thrust constant. This has the difficulties that

- since the force density is sampled before it is smoothed, an infinite density may still be encountered by an unlucky coincidence of a RANS sampling point where $\hat{n}\cdot\hat{\theta} = 0$,
- the amount of smoothing required is not known a priori and may be different for different grids and different propellers, and
- the flow is altered to some extent by the redistribution of force.

Similar difficulties occur if the same approach is taken for the distribution of the rate of change of mass. A different approach is considered here; it is described in the next section.

3.2. Allocation of the Force Density and Rate of Change of Mass to the RANS Equations

Propeller analysis programs that use BEM divide the blade surfaces into panels, typically but not necessarily quadrilaterals. In the analysis that follows, it will be assumed that the panels are quadrilaterals. Triangular panels can be treated by making two panel vertices coincident, while panels with more sides can be treated by splitting them into several quadrilateral or triangular sub-panels.

When the flow is unsteady because $\mathbf{V}$ is not uniform, the rotation of the propeller is broken into N_t equal time steps per revolution. It will also be assumed that the BEM solution supplies the pressure at each panel vertex at each time step. In the common case where the pressure is considered constant over the panel, the values at each vertex will be the same.

Although the method has been presented as following from a BEM solution of the propeller flow, it can also be used to convert a RANS solution of the propeller flow to force and mass rate fields. The surface grid on the blades would then be used in lieu of the BEM panels. In this case, it is more common that the pressures and velocities have distinct values at the corner points of each surface element.

During each time interval $\Delta t = T/N_t = 2\pi/\omega N_t$, a panel sweeps out a small hexahedral prism in a cylindrical coordinate system aligned with the propeller axis. If it is assumed that the force density is trilinear over this volume (by trilinear interpolation from the values at the corners), then the force imparted to the flow during this interval can be obtained exactly. Similarly, integrating M over the volume of the prism yields the total rate of change of mass caused by the panel.

To implement the source terms in the RANS equations, the cells of the RANS grid that intersect each prism are determined. The force and mass rate associated with the prism is then split among the RANS cells in a conservative way. A similar allocation scheme is described by Kinnas et al. [19] but has the drawback that the RANS grid must be axisymmetric over the swept volume of the propeller blades. The method described in Section 3.2.2 can be used on arbitrary RANS grids.

Section 3.2.1 gives expressions for the integrals of the mass rate and force density over a BEM prisms. These expressions are derived in detail in the appendix. Section 3.2.2 provides details of the method for allocating the force and mass rate to the RANS grid.

3.2.1. Integrating the Force Density and Mass Rate over the Swept Volume of a BEM Panel

Let the corner points in cylindrical coordinates of the hexahedron swept out by the panel in one time step be $\mathbf{x}_{ijk} = (x_{ijk}, r_{ijk}, \theta_{ijk})$ for i, j and k equal to 0 or 1. The k subscript represents the $\hat{\theta}$ direction. The i and j indices are not associated with any coordinate direction, but they are chosen so that for each k, $\mathbf{x}_{00k}$, $\mathbf{x}_{10k}$, $\mathbf{x}_{11k}$ and $\mathbf{x}_{01k}$ are ordered cyclically around one end of the hexahedron.

In Appendix A, it is shown that the mass rate obtained by integrating M over the hexahedron is

$$M_\square \equiv \int_\square M\, r\, dr\, d\theta\, dx = \sum_{i=0}^{1}\sum_{j=0}^{1}\sum_{k=0}^{1}\left(B_{xij}V_{xijk} + B_{rij}V_{rijk} + B_{\theta ij}V_{\theta ijk}\right) = \sum_{i=0}^{1}\sum_{j=0}^{1}\sum_{k=0}^{1}\mathbf{B}_{ij}\cdot\mathbf{V}_{ijk}, \tag{12}$$

where the symbol $\square$ indicates that the integration is done over the volume of the hexahedron, $\mathbf{V}_{ijk}$ are the velocity vectors in cylindrical coordinates at the corners of the hexahedron, and $\mathbf{B}_{ij} = B_{xij}\hat{x} + B_{rij}\hat{r} + B_{\theta ij}\hat{\theta}$ are geometrical coefficients depending only on the panel corners and the time step: see Equations (A23), (A25) and (A27).

Appendix A also shows that the force obtained by integrating $\mathbf{F}$ over the hexahedron is

$$\mathbf{F}_\square \equiv \int_\square \mathbf{F}\, r\, dr\, d\theta\, dx = \sum_{i,j,k}\mathbf{F}_{ijk}p_{ijk}, \tag{13}$$

where p_{ijk} are the values of the pressure at the corners of the hexahedron and $\mathbf{F}_{ijk}$ are again geometrical coefficients depending only on the panel corners and the time step: see Equations (A41) and (A42).

3.2.2. Allocation of Force Density and Mass Rate to RANS Cells

Like the method of force allocation described by Hally and Laurens [13], each hexahedron is split into smaller hexahedra until the force and mass rate from each can be assigned to a cell in the RANS grid. Each time the hexahedron is split, each of its edges in one direction are simply bisected. If all the corner points of a hexahedron are contained in a single RANS cell, its entire mass rate and force can be assigned to that cell. While more and more hexahedra will be wholly contained by a single RANS cell as the number of splits increases, there will always be some hexahedra that span more than one RANS cell. Therefore, one must impose a maximum number of splits for each cell and decide how to allocate the force for hexahedra that span more than one cell. In the algorithm described here, the maximum number of splits can be specified by the user. When a hexahedron still spans more than one RANS cell, one eighth of the force and mass rate of the hexahedron is allocated to each cell containing a corner point of the hexahedron. This simple scheme can be made as accurate as necessary by increasing the maximum number of splits.

The expressions in Equations (12) and (13) for the mass rate and force for a hexahedron are linear in the values of p and $\mathbf{V}$ at its corners. When a hexahedron is split, the values of p and $\mathbf{V}$ at the corners of the new hexahedra are linear combinations of the values at the corners of the original hexahedron. Therefore, there is a linear relation between the values of p and $\mathbf{V}$ at the panel corner points at each time step, and the force and mass rate in each RANS cell. If p_m is an array of values of pressure at the panel corner points and $\mathbf{F}_n$ is an array of force densities in the RANS cells, then

$$\mathbf{F}_n = \frac{1}{\mathcal{V}_n} \sum_m \mathbf{P}_{nm} p_m, \tag{14}$$

where $\mathbf{P}_{nm}$ is a sparse matrix mapping the pressures at the panel corner points to the forces in the RANS cells and $\mathcal{V}_n$ is the volume of RANS cell n. Similarly, one can define a sparse matrix $\mathbf{Q}_{nm}$ mapping values of $\mathbf{V}$ at the panel corner points to the mass rate densities of each RANS cell:

$$M_n = \frac{1}{\mathcal{V}_n} \sum_m \mathbf{Q}_{nm} \cdot \mathbf{V}_m. \tag{15}$$

The matrices $\mathbf{P}_{nm}$ and $\mathbf{Q}_{nm}$ depend only on the panelling of the propeller blades, the RANS grid, and the time step used for the propeller rotation. If they are saved, they can be reused when the flow environment is changed. For a RANS-BEM coupling with a large RANS grid, this can lead to considerable savings in time.

Therefore, the allocation procedure is split into two parts: first, the $\mathbf{P}_{nm}$ and $\mathbf{Q}_{nm}$ are calculated and saved, then they are used to determine the force and mass rate in each RANS cell. The latter step can be repeated many times provided that the panelling, RANS grid and time step remain the same.

The values at the corner points of a split hexahedron are linear combinations of the values at the original hexahedron from which it was split. Therefore, for each hexahedron, we store a tensor, L_{ijkqrs}, which represents the mapping from the original values at corner (q, r, s) to the values at corner (i, j, k) of the split hexahedron. For the original hexahedron, the mapping is the identity: $L_{ijkqrs} = \delta_{iq}\delta_{jr}\delta_{ks}$. When a hexahedron is split, say in the i direction, then two new hexahedra are created. The lower hexahedron, from $i = 0$ to $\frac{1}{2}$, has mapping, L^{ℓ}_{ijkqrs}, given by

$$L^{\ell}_{0jkqrs} = L_{0jkqrs}; \qquad L^{\ell}_{1jkqrs} = \tfrac{1}{2}\left(L_{0jkqrs} + L_{1jkqrs}\right); \qquad \text{for all } j, k, q, r \text{ and } s. \tag{16}$$

Similarly, the mapping of the upper hexahedron, from $i = \frac{1}{2}$ to 1, has mapping, $L^u_{ijk0qrs}$, given by

$$L^u_{0jkqrs} = \tfrac{1}{2}\left(L_{0jkqrs} + L_{1jkqrs}\right); \qquad L^u_{1jkqrs} = L_{1jkqrs}; \qquad \text{for all } j,\, k,\, q,\, r \text{ and } s. \tag{17}$$

Splitting in the j and k directions is similar.

Let m_{qrs} be the index of the panel corner point at time t that corresponds with corner (q, r, s) of the unsplit hexahedron. Then, the pressures at the corners of the unsplit hexahedra are $p_{m_{qrs}}$ and the velocities are $\mathbf{V}_{m_{qrs}}$. The mass rate and force associated with any of the split hexahedra is then

$$M = \sum_{i,j,k,q,r,s} \mathbf{B}_{ij} \cdot L_{ijkqrs}\mathbf{V}_{m_{qrs}}; \qquad \mathbf{F} = \sum_{i,j,k,q,r,s} \mathbf{F}_{ijk} L_{ijkqrs} p_{m_{qrs}}. \tag{18}$$

When the splitting has finished, one eighth of the values of M and $\mathbf{F}$ are assigned to the RANS cell containing each corner point of the hexahedron. If corner (i, j, k) of the split hexahedron is contained in RANS cell n, then the corner (i, j, k) contributes

$$\Delta\mathbf{P}_{nm_{qrs}} = \tfrac{1}{8}\mathbf{F}_{ijk} L_{ijkqrs} \tag{19}$$

to $\mathbf{P}_{nm_{qrs}}$ and

$$\Delta\mathbf{Q}_{nm_{qrs}} = \tfrac{1}{8}\mathbf{B}_{ij} L_{ijkqrs} \tag{20}$$

to $\mathbf{Q}_{nm_{qrs}}$.

The basic algorithm to calculate $\mathbf{P}_{nm}$ and $\mathbf{Q}_{nm}$ is shown in Algorithm 1. The loop to populate the list `contiguous_hex_list` ensures that the force and mass rate fields cover a contiguous volume of cells in the RANS grid and prevents 'lumpiness' in their distribution due to some cells being missed because they intersect the centre of a hexahedron and not its corners.

In theory, the sum of mass rate over all the RANS cells should be zero. While not exactly zero, the sum is typically very small. For example, for the calculations described in Section 5, the total mass rate was 3×10^{-7} times the flux of mass carried through the propeller disk by the inflow velocity. In the code implementing Algorithm 1, provision has been made to adjust the mass rate density so its sum is zero to within machine accuracy, but it normally is not necessary.

In general, the thrust obtained when summing the x-component of the force density from all the RANS cells will not exactly equal the thrust predicted by the BEM program. This is because BEM programs usually integrate the pressure on the panels in the Cartesian coordinate system while integration here is done in the cylindrical coordinate system. In the cylindrical coordinate system, the panel edges are slightly curved leading to small differences in the results. The difference in thrust will depend on the sizes of the panels but is typically a fraction of 1%. If desired, it is straightforward to correct for the mismatch by scaling the values of the force density in the RANS cells.

Algorithm 1 Basic algorithm to calculate $\mathbf{P}_{nm}$ and $\mathbf{Q}_{nm}$.

1: Make a list of RANS cells, `blade_cells`, intersecting the swept area of the blades.
2: Make sparse matrices to represent $\mathbf{P}_{mn}$ and $\mathbf{Q}_{mn}$, each initialized to zero.
3: Make a list of all the hexahedra swept out by the blades during the N_t time steps required for one revolution.
4: **for all** `hex` in the list of hexahedra **do**
5:     Find the cells in `blade_cells` that intersect `hex` and store them in `hex_cells`.
6:     Make `hex_list`, a list of hexahedra containing only `hex`.
7:     Make an empty list of hexahedra, `contiguous_hex_list`.
8:     **while** `hex_list` is not empty **do**
9:         Remove the first element of `hex_list`, `hex1`.
10:         **if** the two points at the end of any edge of `hex1` do not lie in the same or neighbouring
11:                 RANS cells **then**
12:             Split `hex1` along the edge to create two new cells.
13:             Add the two new cells to `hex_list`.
14:         **else**
15:             Set the split level of `hex1` to zero.
16:             Add `hex1` to `contiguous_hex_list`.
17:         **end if**
18:     **end while**
19:     **while** `contiguous_hex_list` is not empty **do**
20:         Remove the first element of `contiguous_hex_list`, `hex1`.
21:         **if** all corner points of `hex1` lie in the same RANS cell
22:         **or** if its split level is the maximum allowed **then**
23:             For each corner point of the cell, update $\mathbf{P}_{nm}$ and $\mathbf{Q}_{nm}$ using Equations (19) and (20).
24:         **else**
25:             Split `hex1` along the longest edge whose end points do not lie in the same cell.
26:             Set the split level of the two new cells to one higher than the split level of `hex1`.
27:             Add the two new cells to `contiguous_hex_list`.
28:         **end if**
29:     **end while**
30: **end for**
31: Save $\mathbf{P}_{mn}$ and $\mathbf{Q}_{mn}$.

3.2.3. Finding the RANS Cell Containing a Point

For Algorithm 1 to be efficient, it is necessary to have a fast method for determining which RANS cell contains an arbitrary point in space. The reason for restricting the lists of RANS cells in steps 1 and 5 is to speed up the search. It can be made even faster by constructing a hierarchical tree of bounding boxes containing the RANS cells in the list to be searched. Each level of the tree contains a sub-tree for half of its RANS cells and a second sub-tree for the remaining RANS cells. It also contains a bounding box, which is the union of the bounding boxes for each of the sub-trees. To find the cell containing a point, each tree is traversed by traversing each of its sub-trees, but only if the bounding box of the sub-tree contains the point. When the lowest level of the tree is reached containing a single RANS cell, that cell is tested to see whether it contains the point. If there are N RANS cells to be searched, this scheme is roughly of order $\log N$ rather than order N for a simple cell by cell search.

3.3. *Run Times*

Although the CPU time taken to calculate $\mathbf{P}_{mn}$ and $\mathbf{Q}_{mn}$ depends on the BEM panelling, the RANS grid, and the maximum split level, it is typically shorter than the time taken to run the RANS solver, even when the allocation algorithm is run in series and the RANS solver is run in parallel. For open water calculations such as the one described in Section 5, a typical run time would be a few minutes. For more realistic cases in which the RANS cells resolve the boundary layer down to the wall, a typical run time is about an hour. Moreover, Algorithm 1 can easily be parallelized over the outer loop starting at step 4. Care only needs to be taken when accumulating the values of $\mathbf{P}_{mn}$ and $\mathbf{Q}_{mn}$ since more than one hexahedron can contribute to these values. A parallelized version of the algorithm can reduce the run time back to a few minutes.

4. An Implementation Using OpenFOAM and PROCAL

The algorithms described in the previous sections have been implemented using OpenFOAM [20] as the RANS solver and PROCAL, a BEM program developed by the Cooperative Research Ships organization [21]. For the steady flow problems considered here, OpenFOAM uses the SIMPLE algorithm. It has been modified to include the source terms to both the continuity equation and the momentum equation. One also must be careful not to use the 'bounded' forms of the difference schemes used for the convective term of the momentum equation; these schemes assume that the divergence of the velocity is zero, which is not the case when the mass rate field is present.

PROCAL is an implementation of the Morino formulation [15] of the BEM described in Section 2. Its output files give the pressures and velocities at the corner points of the panels at each time step. The radial contraction and pitch of the wake panels can be determined from empirical formulae or by an iteration, which approximates the convection of the panels by the flow field.

5. Comparison of RANS and BEM Flow Fields in Open Water

If the propeller operates in open water, the time-averaged flow calculated using the BEM code and the flow calculated using the RANS code with force and mass rate fields should be the same. Therefore, open water flow provides a test of the ability of the force and mass rate fields to mimic the action of the propeller. Here, the flow past propeller DTMB P4384 is presented as an example. This propeller is part of the NSRDC skewed propeller series described by Boswell [22]. It is the most skewed of the series having 108° of unbalanced skew: see Figure 2. There is no generator rake for this propeller, but the skew induced rake is $0.257D$, making the angle of skew induced rake 33° (measured from the generator line at the hub through the tip).

The points at which the RANS and BEM flow fields are evaluated and compared will normally depend on the details of the RANS-BEM coupling procedure. Often, the comparison is made at one or more planes upstream of the propeller but Hally [17], Kinnas et al. [19] and Su et al. [18] have advocated making the comparison at points within the swept volume of the blades. Here, the RANS and BEM flow fields will be compared throughout the flow field: upstream, in the swept volume of the blades, and in the slipstream downstream.

The BEM panelling, also shown in Figure 2, used 40 panels between the leading edge and trailing edge on each side of the blades, and 30 between the root and tip: 2400 panels in total per blade. The panel heights were uniform, but their widths decreased near the leading edge and trailing edge to resolve the blade geometry better in those locations. The hub was a cylinder extending one propeller radius upstream and two propeller radii downstream. There were 150 panels along the length of the hub and 15 panels between each pair of blades.

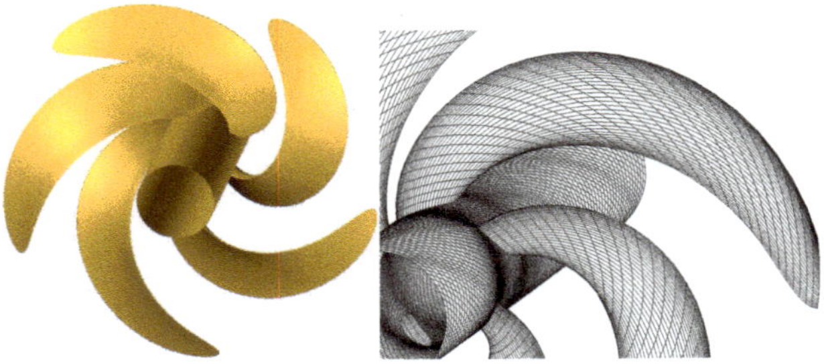

Figure 2. Propeller DTMB P4384. The panelling used in PROCAL is shown at the right.

PROCAL has several different methods for determining the locations of the wake panels in the slipstream of the propeller. For this comparison, the wake contraction was determined from an empirical equation derived from a fit to experimental data reported by Min [23]; an iterative method was used determine the pitch of the panels. Details of these models are given by Bosschers [24]. The effect of the wake panels on the solution will be discussed further in Section 5.2.

The BEM solution provides values of the source and dipole strengths on each panel at each time step. The velocity at any point in the flow field can then be obtained by summing the influence of each panel at the point. Analytic expressions for the influence of a panel at an arbitrary point are given by Morino et al. [15].

To determine the time-averaged BEM velocities at points between the blades, the method described by Hally [17] was used: induced velocities at each time step are calculated at a sequence of N_p points joining the centroids of panels on neighbouring blades, then interpolated and averaged in time. The portion of time that the evaluation point is inside one of the blades is included in the average but contributes nothing to it as the induced velocity is zero inside the blades. A similar method is used to determine the velocities in the wake. For these calculations, N_p was set to 51.

The RANS grid covered an annular region between the cylindrical hub at $r = 0.2R$ and an outer cylinder at $10R$. It extended $10R$ upstream and downstream and consisted of a single axisymmetric structured block having 152 nodes axially, 89 nodes radially and 90 nodes around the circumference: 1,217,520 points and 1,195,920 hexahedral cells in all. Over the swept volume of the blades, the size of the cells in the axial and radial directions was $0.025R$ and in the circumferential direction increased from $0.014R$ at the hub to $0.07R$ at the tip. The node separation increased upstream, downstream and radially beyond the blade tips.

When allocating the forces and mass rates to the RANS cells, a maximum split level of 9 was used. Figure 3 shows the distribution of $M/\rho V$ and $\mathbf{F}/\rho n^2 D$, where V is the speed of inflow, n is the rotation rate in revolutions per unit time and D is the propeller diameter. The non-dimensionalized mass rate is independent of the advance coefficient; the non-dimensional force is shown for $J = 0.6$. The distributions are shown in a single xr plane (the distributions are axisymmetric) with the flow coming from left to right. The mass rate is positive at the leading edge as $\mathbf{V}$ is directed into the blades increasing the mass there, negative near the trailing edge where $\mathbf{V}$ is directed out of the blades. The axial force is positive when directed downstream. The circumferential force is positive in the direction of rotation of the blades.

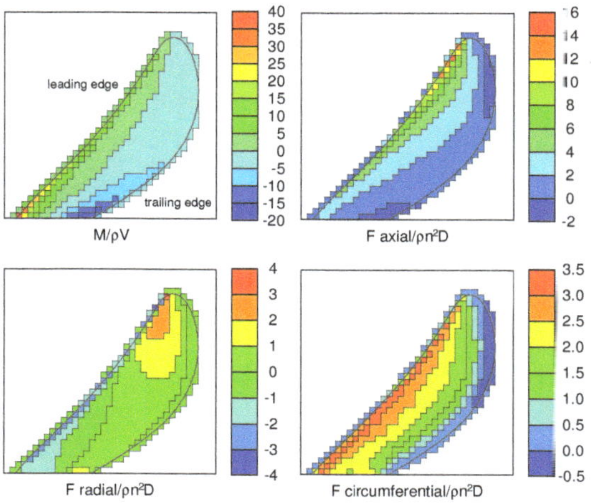

Figure 3. The distribution of $M/\rho V$ and $\mathbf{F}/\rho n^2 D$ for DTMB P4384 at $J = 0.6$. In the white regions, M and $\mathbf{F}$ are exactly zero. The grey line is the outline of the swept volume of the blades.

Figure 3 also provides an indication of the size of the RANS cells. Because the match between the RANS cells and BEM panels is imperfect, the fields can extend up to one cell width outside the swept volume of the blades. There are 43,290 cells in which the mass rate and force density are non-zero.

On the inflow plane, the velocity was set to match the inflow in the BEM computation; the normal pressure gradient was set to zero. The outflow plane was divided into two annuli at $r = 1.5R$. On both annuli, the normal gradient of the velocity was set to zero. The pressure was set to zero on the outer annulus, but its normal gradient was set to zero on the inner annulus. Both the hub and the outer cylinder were treated as free slip walls.

At $J = 0.9$, the thrust obtained from integration of the RANS force field was 0.2% higher than that obtained from summing the pressure-induced force over the BEM panels; at $J = 0.6$ it was 0.1% higher. The torques obtained in a similar way were 0.3% higher for $J = 0.9$ and 0.2% higher at $J = 0.6$. In the velocity field comparisons shown below, no correction was made for this small mismatch.

Figures 4–6 show the components of the velocity field, normalized relative to the inflow speed V, generated by PROCAL and OpenFOAM at advance coefficients $J = 0.6$ and 0.9. The latter is very close to the design advance coefficient of $J = 0.889$. The difference between the RANS and BEM velocities is also shown (the BEM velocity field is subtracted from the RANS velocity field). The flow is displayed in a single xr plane with the outline of the swept volume of the blades shown as a grey line. In these figures, the velocity components are consistent with the components of force in Figure 3: i.e., the axial velocity is positive if it is directed downstream and the circumferential velocity is positive in the direction of rotation of the blades.

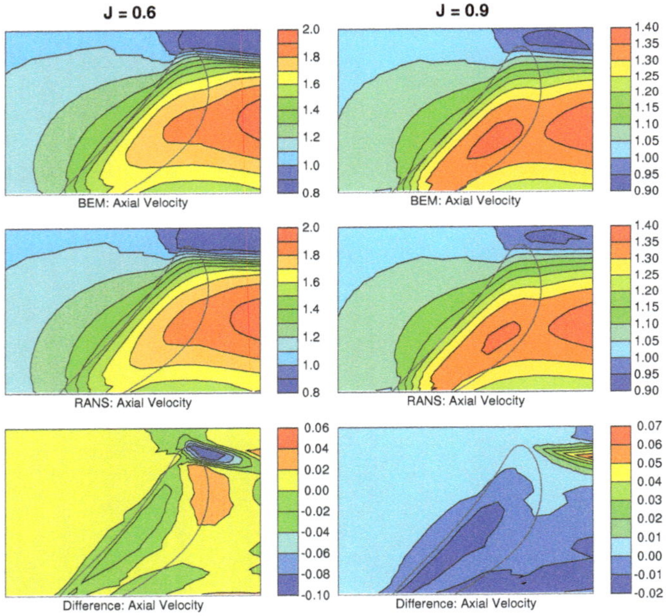

Figure 4. RANS and BEM axial velocity fields for DTMB P4384 at $J = 0.6$ and 0.9. The velocities are normalized using the inflow speed V.

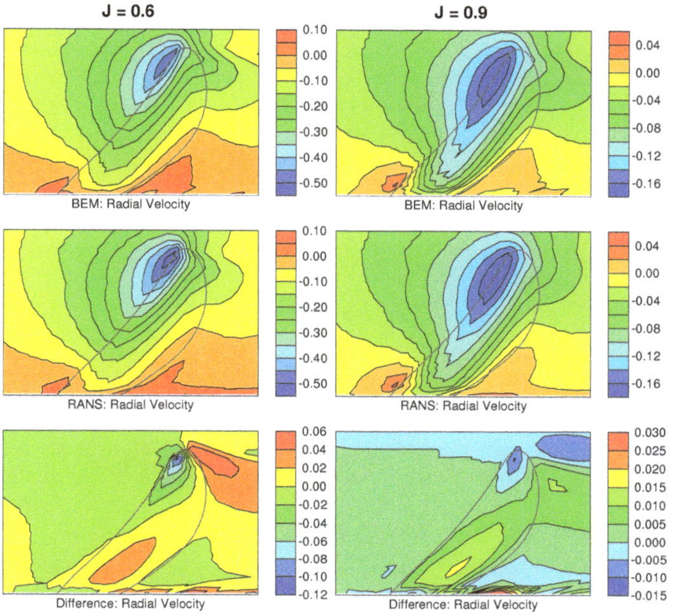

Figure 5. RANS and BEM radial velocity fields for DTMB P4384 at $J = 0.6$ and 0.9. The velocities are normalized using the inflow speed V.

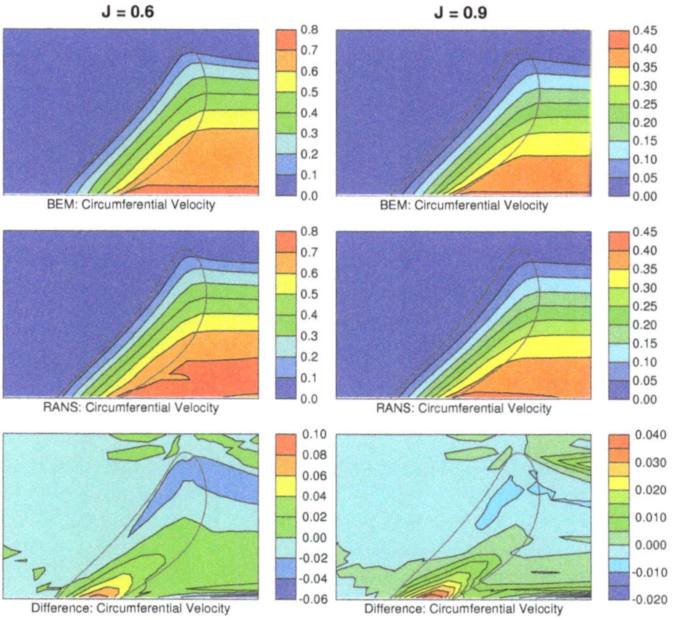

Figure 6. RANS and BEM circumferential velocity fields for DTMB P4384 at $J = 0.6$ and 0.9. The velocities are normalized using the inflow speed V.

At both advance coefficients, the largest differences between the RANS and BEM flow fields occur in the wake at the radial extremes of the wake panels: i.e., close to the hub and close to the location of the tip vortex. Discrepancies in these locations are not important for the RANS-BEM coupling procedure as the RANS and BEM flow fields will only be compared either upstream or within the swept volume of the blades. There is also a relatively large mismatch in the circumferential velocity where the blade meets the hub.

For $J = 0.9$, the differences are less than $0.01V$ everywhere upstream of the propeller and less than $0.015V$ over almost the entire swept volume of the blades. For the lower advance coefficient, the differences are less than $0.01V$ everywhere upstream of the propeller and less than $0.04V$ over almost the entire swept volume of the blades. The degradation in accuracy at higher loading has been found to be consistent for a wide variety of propellers.

5.1. Sensitivity to the Maximum Split Level

The flow field predicted by the RANS solver will depend to some extent on the maximum split level used when allocating the cells. The higher the maximum split level, the more accurately the force and mass rate are distributed among the RANS cells. A choice which is too small will cause the flow field between the blades to fluctuate in an unphysical manner. However, the most appropriate choice of maximum split level will also depend on the mean size of the cells in the RANS grid close to the propeller. The more refined the grid, the smaller the maximum split level can be.

The sensitivity to the split level was tested by generating force and mass rate fields with maximum split levels of 0, 3, 6 and 9. Figure 7 shows the axial velocity predicted by OpenFOAM at three locations: about one tenth of the chord length upstream of the leading edge, along the generator line, and about one half of the chord length downstream of the trailing edge.

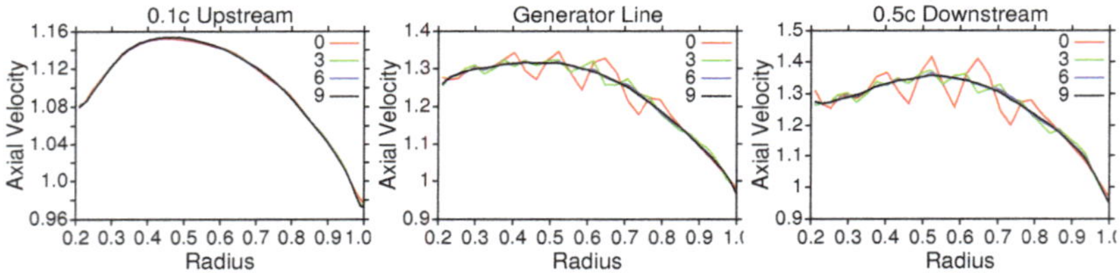

Figure 7. The variation of the axial velocity with split level.

Upstream of the propeller, the flow field is insensitive to the maximum split level; in the swept volume of the blades and downstream, the flow field is quite sensitive. If the RANS and BEM wakes are to be compared upstream, then a maximum split level of 3 should suffice. For comparisons between the blades, the maximum split level should be at least 6.

5.2. Sensitivity to Wake Panels

The sensitivity to the location of the wake panels was tested by choosing among several of the different wake model options provided by PROCAL. There are two empirical models for the wake contraction: the Hoshino model [25] and a fit to experimental data reported by Min [23]. Details of these models are given by Bosschers [24]. In addition, there is an iterative method that adjusts the contraction of the tip vortex so it is parallel to the flow. The iterative model is not very robust and could not be made to converge, so it was not considered in this study. The calculations shown in Figures 3–6 used the Min method.

For this propeller, the Min method was much better than the Hoshino method. At $J = 0.9$, the mismatch in RANS and BEM velocities at the outer edge of the slipstream was reduced by about 50%. At $J = 0.6$, the reduction was about 70%. There were similar reductions upstream of the blades. In the swept volume of the blades, the difference between the two methods was significantly smaller than the difference between the RANS and BEM velocities: about $0.002V$ at $J = 0.9$ and $0.007V$ at $J = 0.6$.

The Min method is a fit to experimental data for four propellers, all of which are closely related to DTMB P4384; they are the other three propellers from the NSRDC skewed propeller series, P4381, P4382 and P4383, as well as a fourth propeller equivalent to P4383 but with extra rake. Therefore, it is perhaps not surprising that the Min method performs well for P4384. Whether it does so consistently for other propellers is not yet clear.

The pitch of the wake panels can be set using three different empirical formulae depending on the pitch at each section, the advance coefficient and the skew angle at the tip (details given by Bosschers [24]). There is also an iterative method that adjusts the pitch of the panels to be parallel to the flow.

At $J = 0.9$, the choice of method for the pitch of the wake panels resulted in very small differences to the flow: typically less than $0.003V$ and significantly smaller than the differences between the RANS and BEM flow fields. Near the design advance coefficient the choice of pitch model is not very important.

At $J = 0.6$, the iterative wake model performed significantly better than the empirical models in the slipstream, reducing the RANS-BEM discrepancy in axial velocity by about 80%, though the differences in the tangential velocity components were not significant. Upstream of the blades there was also a significant reduction in the discrepancy of both the axial and radial velocity components. In the swept volume of the blades, the RANS-BEM difference in axial velocity was consistently about $0.02V$ lower than for the empirical methods. This difference is a significant portion of the total RANS-BEM mismatch but meant that the empirical methods performed better near the leading edge while the iterative method performed better near the trailing edge. Overall, the iterative method gave the best results.

5.3. Sensitivity to RANS Cell Size

The sensitivity to the cell size in the RANS grid was checked by making a second grid with mean cell size of $0.0125R$ radially and axially and with 180 cells around the circumference. For $J = 0.9$, the difference in velocity was less than $0.003V$ over almost the entire flow field. For $J = 0.6$, the change in all velocity components was less than $0.005V$ almost everywhere.

For both advance coefficients, the highest differences occurred in the slipstream near the hub and at its outer edge. At the outer edge of the slipstream where the velocity gradient is very high, the differences were caused by increased diffusion on the coarser grid; the differences in this region were considerably larger (about $0.05V$) at $J = 0.6$ than at $J = 0.9$ (about $0.01V$). Near the hub, the radial and circumferential velocities showed small oscillations when the coarser grid was used causing differences of about $0.03V$; these were much improved by the finer grid.

For both advance coefficients, there was also a small region where the leading edge of the blade meets the hub for which the differences were about $0.01V$.

5.4. Sensitivity to Other Parameters

The following sensitivity tests were also performed.

- The sensitivity to N_p, the number of points used to average the flow field between the blades, was checked by repeating the calculations with 11, 21 and 51 points.
- To check that the dimensions of the RANS flow domain were sufficiently large, the RANS flow field for the $J = 0.6$ case was recalculated on a grid whose outer dimensions were increased from $10R$ to $20R$.
- Two additional PROCAL blade panellings were used:
  - 20 panels root to tip and 30 panels from leading to trailing edge;
  - 40 panels root to tip and 60 panels from leading to trailing edge.

In all cases, the changes in the RANS-BEM mismatch were found to be insignificant.

6. Conclusions

The BEM solution of the time-averaged flow past a propeller can be simulated to high accuracy in a RANS solution by replacing the propeller with force and mass rate fields. When the methods described in Section 3 are used for open water flow, the difference in the flow fields upstream of the propeller and in the swept volume of the blades is typically about 1–2% of the inflow speed when the advance coefficient is close to design. Downstream, the accuracy is similar except at the extremes of the slipstream. As the advance coefficient is decreased, the accuracy of the method begins to degrade. It is likely that the accuracy is somewhat less for the more complex flow when the propeller operates behind a ship but it is not possible to measure that directly.

At higher loading, the choice of the wake model can affect the accuracy of the RANS-BEM match. This is especially true both upstream of the blades and in the slipstream. When the BEM solver provides different treatments for the wake panels, it is recommended that RANS-BEM procedures should start with an open water test to determine the most appropriate wake model for the propeller being evaluated.

Acknowledgments: This work and its publication were entirely funded by Defence Research and Development Canada.

Author Contributions: This paper was written solely by the author. All calculations described in it were performed by the author.

Conflicts of Interest: The author declares no conflict of interest.

Abbreviations

The following abbreviations are used in this manuscript:

BEM	Boundary Element Method
DTMB	David Taylor Model Basin
FFM	Force Field Method
NSRDC	Naval Ship Research and Development Center
RANS	Reynolds-Averaged Navier–Stokes
VSM	V-shaped Segments Method
Symbols	
θ	Angular coordinate around the propeller axis
μ	Dipole strength
ν_t	Turbulent kinematic viscosity
ρ	Fluid density
σ	Source strength
ϕ	Velocity potential
ω	Propeller rotation rate (rads/sec)
$\mathbf{B}_{ij}$	Coefficients mapping velocity components to the mass rate density for a hexahedron
D	Propeller diameter
$\mathbf{F}$	Force density
$\mathbf{F}_{\square}$	Force due to a hexahedron
$\mathbf{F}_{ijk}$	Coefficients mapping pressure to the force density density for a hexahedron
J	Advance coefficient
L_{ijkqrs}	Coefficients mapping values at corner points of a split hexahedron to the values on the original hexahedron
M	Rate of increase of mass per unit volume
$M_{\square}$	Rate of increase of mass due to a hexahedron
N_t	Number of time steps in one propeller rotation

n	Propeller rotation rate (revs/sec)
$\mathbf{n}$	A normal to a blade panel
$\mathbf{P}_{mn}$	Coefficients mapping pressures at panel corner points to force densities for RANS cells
p	Pressure
$\mathbf{Q}_{mn}$	Coefficients mapping velocities at panel corner points to mass rate densities for RANS cells
R	Propeller radius
r	Distance to the propeller axis
T	Period of propeller rotation
$\mathbf{t}_u$, $\mathbf{t}_v$	Tangents to a blade panel
u, v, w	Parameters used to interpolate over a hexahedron
$\mathbf{v}$	Fluid velocity
$\mathbf{V}$	Background fluid velocity in BEM solution
V	Inflow speed for a propeller in open water
$\mathcal{V}_n$	The volume of RANS cell n
x	Distance along the propeller axis
Z	Number of propeller blades

Appendix A. Integrating the Force Density and Mass Rate over the Swept Volume of a BEM Panel

Let the corner points in cylindrical coordinates of the hexahedron swept out by the panel in one time step be $\mathbf{x}_{ijk} = (x_{ijk}, r_{ijk}, \theta_{ijk})$ for i, j and k equal to 0 or 1. The k subscript represents the $\hat{\theta}$ direction. The i and j indices are not associated with any coordinate direction, but they are chosen so that $\mathbf{x}_{00k}$, $\mathbf{x}_{10k}$, $\mathbf{x}_{11k}$ and $\mathbf{x}_{01k}$ are order cyclically around one end of the hexahedron.

For any quantity, f, whose values are known at the corners of the hexahedron, we can form a trilinear interpolant in terms of three parameters u, v and w as follows:

$$f(u,v,w) = (1-w)f_0(u,v) + wf_1(u,v), \tag{A1}$$

$$\begin{aligned} f_k(u,v) &= \tfrac{1}{4}\left[(1-u)(1-v)f_{00k} + (1+u)(1-v)f_{10k} + (1-u)(1+v)f_{01k} + (1+u)(1+v)f_{11k}\right] \\ &= \overline{f}_k + u\overline{f}_{ku} + v\overline{f}_{kv} + uv\overline{f}_{kuv}, \end{aligned} \tag{A2}$$

with $u \in [-1,1]$, $v \in [-1,1]$, $w \in [0,1]$ and where

$$\overline{f}_k \equiv \tfrac{1}{4}(\ f_{00k} + f_{10k} + f_{01k} + f_{11k}) \equiv \sum_{i,j} A_{ij} f_{ijk}, \tag{A3}$$

$$\overline{f}_{ku} \equiv \tfrac{1}{4}(-f_{00k} + f_{10k} - f_{01k} + f_{11k}) \equiv \sum_{i,j} A_{uij} f_{ijk}, \tag{A4}$$

$$\overline{f}_{kv} \equiv \tfrac{1}{4}(-f_{00k} - f_{10k} + f_{01k} + f_{11k}) \equiv \sum_{i,j} A_{vij} f_{ijk}, \tag{A5}$$

$$\overline{f}_{kuv} \equiv \tfrac{1}{4}(\ f_{00k} - f_{10k} - f_{01k} + f_{11k}) \equiv \sum_{i,j} A_{uvij} f_{ijk} \tag{A6}$$

We will also use

$$\tilde{f}(w) \equiv (1-w)\overline{f}_0 + w\overline{f}_1, \tag{A7}$$

$$\tilde{f}_u(w) \equiv (1-w)\overline{f}_{0u} + w\overline{f}_{1u}, \tag{A8}$$

$$\tilde{f}_v(w) \equiv (1-w)\overline{f}_{0v} + w\overline{f}_{1v}, \tag{A9}$$

$$\tilde{f}_{uv}(w) \equiv (1-w)\overline{f}_{0uv} + w\overline{f}_{1uv}, \tag{A10}$$

so that

$$f(u,v,w) = \tilde{f}(w) + \tilde{f}_u(w)u + \tilde{f}_v(w)v + \tilde{f}_{uv}(w)uv. \tag{A11}$$

Since the propeller is rotating around the x-axis, the x coordinates of the panel corners do not change as it rotates. Therefore, when f is the coordinate x, we have $\overline{x}_0 = \overline{x}_1$ so that $\tilde{x}(w)$, $\tilde{x}_u(w)$, $\tilde{x}_v(w)$ and $\tilde{x}_{uv}(w)$ are actually independent of w. Similarly, $\tilde{r}(w)$, $\tilde{r}_u(w)$, $\tilde{r}_v(w)$ and $\tilde{r}_{uv}(w)$ are independent of w.

The coordinate θ satisfies $\theta_{ij1} - \theta_{ij0} = \Delta\theta$ so that

$$\overline{\theta}_1 = \overline{\theta}_0 + \Delta\theta; \qquad \overline{\theta}_{1u} = \overline{\theta}_{0u}; \qquad \overline{\theta}_{1v} = \overline{\theta}_{0v}; \qquad \overline{\theta}_{1uv} = \overline{\theta}_{0uv} \tag{A12}$$

and $\tilde{\theta}_u(w)$, $\tilde{\theta}_v(w)$ and $\tilde{\theta}_{uv}(w)$ are independent of w, but $\tilde{\theta}(w)$ is not. Therefore, when represented in cylindrical coordinates, $\tilde{\mathbf{x}}_u(w)$, $\tilde{\mathbf{x}}_v(w)$ and $\tilde{\mathbf{x}}_{uv}(w)$ are independent of w. Henceforth, they will be written without the argument as $\tilde{\mathbf{x}}_u$, $\tilde{\mathbf{x}}_u$ and $\tilde{\mathbf{x}}_{uv}$.

Two tangents to the panel at (u, v, w) are

$$\mathbf{t}_u(u,v) = \hat{x}\frac{\partial x}{\partial u} + \hat{r}\frac{\partial r}{\partial u} + r\hat{\theta}\frac{\partial \theta}{\partial u} = \hat{x}(\tilde{x}_u + v\tilde{x}_{uv}) + \hat{r}(\tilde{r}_u + v\tilde{r}_{uv}) + \hat{\theta}(\tilde{\theta}_u + v\tilde{\theta}_{uv})r(u,v), \tag{A13}$$

$$\mathbf{t}_v(u,v) = \hat{x}\frac{\partial x}{\partial v} + \hat{r}\frac{\partial r}{\partial v} + r\hat{\theta}\frac{\partial \theta}{\partial v} = \hat{x}(\tilde{x}_v + u\tilde{x}_{uv}) + \hat{r}(\tilde{r}_v + u\tilde{r}_{uv}) + \hat{\theta}(\tilde{\theta}_v + u\tilde{\theta}_{uv})r(u,v), \tag{A14}$$

where $\hat{x}$, $\hat{r}$ and $\hat{\theta}$ are unit vectors in the direction of increasing x, r and θ, respectively, and

$$r(u,v) = \tilde{r} + \tilde{r}_u u + \tilde{r}_v v + \tilde{r}_{uv} uv. \tag{A15}$$

A normal to the panel at (u, v, w) is then

$$\begin{aligned} \mathbf{n}(u,v) = \mathbf{t}_u(u,v) \times \mathbf{t}_v(u,v) &= \hat{x}((\tilde{r}_u\tilde{\theta}_v - \tilde{r}_v\tilde{\theta}_u) + u(\tilde{r}_u\tilde{\theta}_{uv} - \tilde{r}_{uv}\tilde{\theta}_u) + v(\tilde{r}_{uv}\tilde{\theta}_v - \tilde{r}_v\tilde{\theta}_{uv}))r(u,v) \\ &+ \hat{r}((\tilde{\theta}_u\tilde{x}_v - \tilde{\theta}_v\tilde{x}_u) + u(\tilde{\theta}_u\tilde{x}_{uv} - \tilde{\theta}_{uv}\tilde{x}_u) + v(\tilde{\theta}_{uv}\tilde{x}_v - \tilde{\theta}_v\tilde{x}_{uv}))r(u,v) \\ &+ \hat{\theta}((\tilde{x}_u\tilde{r}_v - \tilde{x}_v\tilde{r}_u) + u(\tilde{x}_u\tilde{r}_{uv} - \tilde{x}_{uv}\tilde{r}_u) + v(\tilde{x}_{uv}\tilde{r}_v - \tilde{x}_v\tilde{r}_{uv})). \end{aligned} \tag{A16}$$

Therefore, the components of the normal in cylindrical coordinates are

$$n_x(u,v) = n_x^*(u,v)\, r(u,v); \qquad n_r(u,v) = n_y^*(u,v)\, r(u,v); \qquad n_\theta(u,v) = n_\theta^*(u,v), \tag{A17}$$

with

$$\mathbf{n}^*(u,v) = \tilde{\mathbf{x}}_u \times \tilde{\mathbf{x}}_v + u\tilde{\mathbf{x}}_u \times \tilde{\mathbf{x}}_{uv} + v\tilde{\mathbf{x}}_{uv} \times \tilde{\mathbf{x}}_v. \tag{A18}$$

The Jacobian of the transformation from (x, r, θ) to (u, v, w) is

$$\frac{\partial(x,r,\theta)}{\partial(u,v,w)} = \frac{\partial(x,r)}{\partial(u,v)}\frac{\partial\theta}{\partial w} = \begin{vmatrix} \tilde{x}_u + v\tilde{x}_{uv} & \tilde{x}_v + u\tilde{x}_{uv} \\ \tilde{r}_u + v\tilde{r}_{uv} & \tilde{r}_v + u\tilde{r}_{uv} \end{vmatrix} \frac{1}{\Delta\theta} = \frac{\hat{\theta}\cdot\mathbf{n}}{\Delta\theta}, \tag{A19}$$

the last step following from Equation (A16). The integral of the mass density over the hexahedron is then

$$\begin{aligned} M_\square \equiv \int_\square M\, r\, dr\, d\theta\, dx &= \int_\square \frac{Z\rho\mathbf{V}\cdot\hat{n}}{2\pi r|\hat{\theta}\cdot\hat{n}|} r\, dr\, d\theta\, dx \\ &= \frac{Z\rho}{2\pi}\int_0^1\int_{-1}^1\int_{-1}^1 \frac{\mathbf{V}\cdot\hat{n}}{|\hat{\theta}\cdot\hat{n}|}\left|\frac{\partial(x,r,\theta)}{\partial(u,v,w)}\right| du\, dv\, dw \\ &= \frac{Z\rho\Delta\theta}{2\pi}\int_0^1\int_{-1}^1\int_{-1}^1 \mathbf{V}\cdot\mathbf{n}\, du\, dv\, dw \\ &= \frac{Z\rho\Delta\theta}{4\pi}\sum_{k=0}^1\int_{-1}^1\int_{-1}^1 \mathbf{V}_k(u,v)\cdot\mathbf{n}(u,v)\, du\, dv, \end{aligned} \tag{A20}$$

the last step being possible since the components of $\mathbf{n}$ in cylindrical coordinates are independent of w. Notice that the terms in $|\hat{\theta}\cdot\hat{n}|$ have cancelled out so that $M_\square$ is finite even when $\hat{\theta}\cdot\hat{n} = 0$. Therefore,

$$M_\square = \frac{Z\rho\Delta\theta}{4\pi}\sum_{k=0}^{1}\int_{-1}^1\int_{-1}^1\Big[\Big(V_{xk}(u,v)n_x^*(u,v) + V_{rk}(u,v)n_r^*(u,v)\Big)r(u,v) + V_{\theta k}(u,v)n_\theta^*(u,v)\Big]\, du\, dv. \tag{A21}$$

We have

$$\begin{aligned}
&\tfrac{Z\rho\Delta\theta}{4\pi}\int_{-1}^{1}\int_{-1}^{1}V_{xk}(u,v)n_x^*(u,v)r(u,v)\,du\,dv \\
&\quad= \tfrac{Z\rho\Delta\theta}{4\pi}\int_{-1}^{1}\int_{-1}^{1}(\overline{V}_{xk}+\overline{V}_{xku}u+\overline{V}_{xkv}v+\overline{V}_{xkuv}uv)(\overline{r}+\overline{r}_u u+\overline{r}_v v+\overline{r}_{uv}uv)(\overline{n}_x^*+\overline{n}_{xu}^*u+\overline{n}_{xv}^*v)\,du\,dv \\
&\quad= \tfrac{Z\rho\Delta\theta}{\pi}\Big[\overline{V}_{xk}\left(\overline{r}\,\overline{n}_x^*+\tfrac{1}{3}\overline{r}_u\overline{n}_{xu}^*+\tfrac{1}{3}\overline{r}_v\overline{n}_{xv}^*\right)+\overline{V}_{xku}\left(\tfrac{1}{3}\overline{r}\,\overline{n}_{xu}^*+\tfrac{1}{3}\overline{r}_u\overline{n}_x^*+\tfrac{1}{9}\overline{r}_{uv}\overline{n}_{xv}^*\right) \\
&\qquad+\overline{V}_{xkv}\left(\tfrac{1}{3}\overline{r}\,\overline{n}_{xv}^*+\tfrac{1}{3}\overline{r}_v\overline{n}_x^*+\tfrac{1}{9}\overline{r}_{uv}\overline{n}_{xu}^*\right)+\tfrac{1}{9}\overline{V}_{xkuv}\left(\overline{r}_u\overline{n}_{xv}^*+\overline{r}_v\overline{n}_{xu}^*+\overline{r}_{uv}\overline{n}_x^*\right)\Big] \\
&\quad= \textstyle\sum_{i,j}B_{xij}V_{xijk},
\end{aligned} \tag{A22}$$

with

$$\begin{aligned}
B_{xij} &= \tfrac{Z\rho\Delta\theta}{\pi}\Big[A_{ij}\left(\overline{r}\,\overline{n}_x^*+\tfrac{1}{3}\overline{r}_u\overline{n}_{xu}^*+\tfrac{1}{3}\overline{r}_v\overline{n}_{xv}^*\right)+A_{uij}\left(\tfrac{1}{3}\overline{r}\,\overline{n}_{xu}^*+\tfrac{1}{3}\overline{r}_u\overline{n}_x^*+\tfrac{1}{9}\overline{r}_{uv}\overline{n}_{xv}^*\right) \\
&\quad+A_{vij}\left(\tfrac{1}{3}\overline{r}\,\overline{n}_{xv}^*+\tfrac{1}{3}\overline{r}_v\overline{n}_x^*+\tfrac{1}{9}\overline{r}_{uv}\overline{n}_{xu}^*\right)+\tfrac{1}{9}A_{uvij}\left(\overline{r}_u\overline{n}_{xv}^*+\overline{r}_v\overline{n}_{xu}^*+\overline{r}_{uv}\overline{n}_x^*\right)\Big],
\end{aligned} \tag{A23}$$

where A_{ij}, A_{uij} and A_{vij} are defined by Equations (A3)–(A5). Similarly,

$$\frac{Z\rho\Delta\theta}{4\pi}\int_{-1}^{1}\int_{-1}^{1}V_{rk}(u,v)n_r^*(u,v)r(u,v)\,du\,dv=\sum_{i,j}B_{rij}V_{rijk}, \tag{A24}$$

with

$$\begin{aligned}
B_{rij} &= \tfrac{Z\rho\Delta\theta}{\pi}\Big[A_{ij}\left(\overline{r}\,\overline{n}_r^*+\tfrac{1}{3}\overline{r}_u\overline{n}_{ru}^*+\tfrac{1}{3}\overline{r}_v\overline{n}_{rv}^*\right)+A_{uij}\left(\tfrac{1}{3}\overline{r}\,\overline{n}_{ru}^*+\tfrac{1}{3}\overline{r}_u\overline{n}_r^*+\tfrac{1}{9}\overline{r}_{uv}\overline{n}_{rv}^*\right) \\
&\quad+A_{vij}\left(\tfrac{1}{3}\overline{r}\,\overline{n}_{rv}^*+\tfrac{1}{3}\overline{r}_v\overline{n}_r^*+\tfrac{1}{9}\overline{r}_{uv}\overline{n}_{ru}^*\right)+\tfrac{1}{9}A_{uvij}\left(\overline{r}_u\overline{n}_{rv}^*+\overline{r}_v\overline{n}_{ru}^*+\overline{r}_{uv}\overline{n}_r^*\right)\Big]
\end{aligned} \tag{A25}$$

and

$$\begin{aligned}
&\tfrac{Z\rho\Delta\theta}{4\pi}\int_{-1}^{1}\int_{-1}^{1}V_{\theta k}(u,v)n_\theta^*(u,v)\,du\,dv \\
&\quad= \tfrac{Z\rho\Delta\theta}{4\pi}\int_{-1}^{1}\int_{-1}^{1}(\overline{V}_{\theta k}+\overline{V}_{\theta ku}u+\overline{V}_{\theta kv}v+\overline{V}_{\theta kuv}uv)(\overline{n}_\theta^*+\overline{n}_{\theta u}^*u+\overline{n}_{\theta v}^*v)\,du\,dv \\
&\quad= \tfrac{Z\rho\Delta\theta}{\pi}\left[\overline{V}_{\theta k}\overline{n}_\theta^*+\tfrac{1}{3}\overline{V}_{\theta ku}\overline{n}_{\theta u}^*+\tfrac{1}{3}\overline{V}_{\theta kv}\overline{n}_{\theta v}^*\right] \\
&\quad= \textstyle\sum_{i,j}B_{\theta ij}V_{\theta ijk},
\end{aligned} \tag{A26}$$

with

$$B_{\theta ij} = \frac{Z\rho\Delta\theta}{\pi}\left[A_{ij}\overline{n}_\theta^*+\tfrac{1}{3}A_{uij}\overline{n}_{\theta u}^*+\tfrac{1}{3}A_{vij}\overline{n}_{\theta v}^*\right]. \tag{A27}$$

Therefore,

$$M_\square=\sum_{i=0}^{1}\sum_{j=0}^{1}\sum_{k=0}^{1}(B_{xij}V_{xijk}+B_{rij}V_{rijk}+B_{\theta ij}V_{\theta ijk})=\sum_{i=0}^{1}\sum_{j=0}^{1}\sum_{k=0}^{1}\mathbf{B}_{ij}\cdot\mathbf{V}_{ijk}, \tag{A28}$$

where $\mathbf{B}_{ij}=B_{xij}\hat{x}+B_{rij}\hat{r}+B_{\theta ij}\hat{\theta}$.

The force induced by the hexahedron can be determined in a similar fashion:

$$\begin{aligned}
\mathbf{F}_\square \equiv \int_\square \mathbf{F}\,r dr\,d\theta\,dx &= \int_\square \tfrac{Z\rho p\hat{n}}{2\pi r|\hat{n}\cdot\hat{\theta}|}\,r dr\,d\theta\,dx=\tfrac{Z\rho}{2\pi}\int_0^1\int_{-1}^{1}\int_{-1}^{1}\tfrac{p\hat{n}}{|\hat{n}\cdot\hat{\theta}|}\left|\tfrac{\partial(x,r,\theta)}{\partial(u,v,w)}\right|du\,dv\,dw \\
&= \tfrac{Z\rho\Delta\theta}{2\pi}\int_0^1\int_{-1}^{1}\int_{-1}^{1}p\mathbf{n}\,du\,dv\,dw.
\end{aligned} \tag{A29}$$

Note, however, that here the normal does not appear in a dot product and, since its direction changes with θ, it is not independent of w: i.e., while n_x, n_r, n_θ and $\hat{x}$ are independent of w, $\hat{r}$ and $\hat{\theta}$ are not. Since, for any cylindrical coordinate system,

$$\frac{d\hat{r}}{d\theta} = \hat{\theta}; \qquad \frac{d\hat{\theta}}{d\theta} = -\hat{r}, \tag{A30}$$

and, since $\theta = \theta_0 + w\Delta\theta$, we have

$$\frac{d\hat{r}}{dw} = \frac{d\hat{r}}{d\theta}\frac{d\theta}{dw} = \hat{\theta}\Delta\theta; \qquad \frac{d\hat{\theta}}{dw} = \frac{d\hat{\theta}}{d\theta}\frac{d\theta}{dw} = -\hat{r}\Delta\theta. \tag{A31}$$

Therefore,

$$\int_0^1 \hat{r}\,dw = -\frac{1}{\Delta\theta}\int_0^1 \frac{d\hat{\theta}}{dw}\,dw = -\frac{(\hat{\theta}_1 - \hat{\theta}_0)}{\Delta\theta}, \tag{A32}$$

$$\int_0^1 \hat{\theta}\,dw = \frac{1}{\Delta\theta}\int_0^1 \frac{d\hat{r}}{dw}\,dw = \frac{\hat{r}_1 - \hat{r}_0}{\Delta\theta}, \tag{A33}$$

$$\int_0^1 w\hat{r}\,dw = -\frac{1}{\Delta\theta}\int_0^1 w\frac{d\hat{\theta}}{dw}\,dw = -\frac{\hat{\theta}_1}{\Delta\theta} + \frac{\hat{r}_1 - \hat{r}_0}{\Delta\theta^2}, \tag{A34}$$

$$\int_0^1 w\hat{\theta}\,dw = \frac{1}{\Delta\theta}\int_0^1 w\frac{d\hat{r}}{dw}\,dw = \frac{\hat{r}_1}{\Delta\theta} + \frac{\hat{\theta}_1 - \hat{\theta}_0}{\Delta\theta^2}, \tag{A35}$$

and

$$\begin{aligned}
\textstyle\int_0^1 p\mathbf{n}\,dw &= \textstyle\int_0^1 (p_0 + w(p_1 - p_0))(n_x\hat{x} + n_r\hat{r} + n_\theta\hat{\theta})\,dw \\
&= p_0\left[n_x\hat{x} - \tfrac{n_r(\hat{\theta}_1 - \hat{\theta}_0)}{\Delta\theta} + \tfrac{n_\theta(\hat{r}_1 - \hat{r}_0)}{\Delta\theta}\right] \\
&\quad + (p_1 - p_0)\left[\tfrac{1}{2}n_x\hat{x} + \tfrac{n_r}{\Delta\theta}\left(-\hat{\theta}_1 + \tfrac{(\hat{r}_1 - \hat{r}_0)}{\Delta\theta}\right) + \tfrac{n_\theta}{\Delta\theta}\left(\hat{r}_1 + \tfrac{(\hat{\theta}_1 - \hat{\theta}_0)}{\Delta\theta}\right)\right] \\
&= \tfrac{1}{2}(p_0 + p_1)n_x\hat{x} - \tfrac{\hat{x}\times(p_1\mathbf{n}_1 - p_0\mathbf{n}_0)}{\Delta\theta} + \tfrac{(p_1 - p_0)(\mathbf{n}_1 - \mathbf{n}_0)}{\Delta\theta^2}.
\end{aligned} \tag{A36}$$

Equation (A29) now becomes

$$\mathbf{F}_\square = \frac{Z\Delta\theta}{2\pi}\int_{-1}^1\int_{-1}^1 \left[\tfrac{1}{2}(p_0 + p_1)n_x\hat{x} - \frac{\hat{x}\times(p_1\mathbf{n}_1 - p_0\mathbf{n}_0)}{\Delta\theta} + \frac{(p_1 - p_0)(\mathbf{n}_1 - \mathbf{n}_0)}{\Delta\theta^2}\right] du\,dv. \tag{A37}$$

To perform the integration, the normals $\mathbf{n}_0$ and $\mathbf{n}_1$ are converted to Cartesian coordinates (denoted with a superscript c) and interpolated over the panel from the values at the corner points. Note that, in general, $\mathbf{n}^c_{uv}$ is not zero, unlike $\mathbf{n}_{uv}$. For any trilinear interpolants f and g, let

$$\begin{aligned}
I(f,g) &\equiv \textstyle\int_{-1}^1\int_{-1}^1 (\overline{f} + u\overline{f}_u + v\overline{f}_v + uv\overline{f}_{uv})(\overline{g} + u\overline{g}_u + v\overline{g}_v + uvg_{uv})\,du\,dv \\
&= 4\overline{f}\overline{g} + \tfrac{4}{3}\overline{f}_u\overline{g}_u + \tfrac{4}{3}\overline{f}_v\overline{g}_v + \tfrac{4}{9}\overline{f}_{uv}\overline{g}_{uv}.
\end{aligned} \tag{A38}$$

Then,

$$\begin{aligned}
\mathbf{F}_\square &= \tfrac{Z\Delta\theta}{2\pi}\left[\tfrac{1}{2}\hat{x}I(p_0 + p_1, n^c_x) - \tfrac{\hat{x}\times\left(I(p_1,\mathbf{n}^c_1) - I(p_0,\mathbf{n}^c_0)\right)}{\Delta\theta} + \tfrac{I(p_1 - p_0,\mathbf{n}^c_1 - \mathbf{n}^c_0)}{\Delta\theta^2}\right] du\,dv \\
&= \tfrac{Z}{\pi}\Big[\Delta\theta\hat{x}\left((p_0 + p_1)n^c_x + \tfrac{1}{3}(p_{0u} + p_{1u})n^c_{xu} + \tfrac{1}{3}(p_{0v} + p_{1v})n^c_{xv} + \tfrac{1}{9}(p_{0uv} + p_{1uv})n^c_{xuv}\right) \\
&\quad - 2\hat{x}\times\left((p_1\mathbf{n}^c_1 - p_0\mathbf{n}^c_0) + \tfrac{1}{3}(p_{1u}\mathbf{n}^c_{1u} - p_{0u}\mathbf{n}^c_{0u}) + \tfrac{1}{3}(p_{1v}\mathbf{n}^c_{1v} - p_{0v}\mathbf{n}^c_{0v}) + \tfrac{1}{9}(p_{1uv}\mathbf{n}^c_{1uv} - p_{0uv}\mathbf{n}^c_{0uv})\right) \\
&\quad + \tfrac{2}{\Delta\theta}\Big((p_1 - p_0)(\mathbf{n}^c_1 - \mathbf{n}^c_0) + \tfrac{1}{3}(p_{1u} - p_{0u})(\mathbf{n}^c_{1u} - \mathbf{n}^c_{0u}) + \tfrac{1}{3}(p_{1v} - p_{0v})(\mathbf{n}^c_{1v} - \mathbf{n}^c_{0v}) \\
&\quad + \tfrac{1}{9}(p_{1uv} - p_{0uv})(\mathbf{n}^c_{1uv} - \mathbf{n}^c_{0uv})\Big)\Big],
\end{aligned} \tag{A39}$$

where 0 and 1 subscripts have been left off n^c_x since $n^c_{x0} = n^c_{x1}$.

Equation (A28) shows that $M_\square$ is linear in the components of $\mathbf{V}$ at the corner points of the hexahedron. Similarly, $\mathbf{F}_\square$ is linear in the values of p at the corner points of the hexahedron:

$$\mathbf{F}_\square = \sum_{i,j,k} \mathbf{F}_{ijk} p_{ijk}, \tag{A40}$$

with

$$\begin{aligned}\mathbf{F}_{ij0} &= \frac{Z}{\pi}\Big[\Delta\theta\hat{x}\left(A_{ij}n_x^c + \tfrac{1}{3}A_{uij}n_{xu}^c + \tfrac{1}{3}A_{vij}n_{xv}^c + \tfrac{1}{9}A_{uvij}n_{xuv}^c\right)\\ &+ 2\hat{x}\times\left(A_{ij}\mathbf{n}_0^c + \tfrac{1}{3}A_{uij}\mathbf{n}_{0u}^c + \tfrac{1}{3}A_{vij}\mathbf{n}_{0v}^c + \tfrac{1}{9}A_{uvij}\mathbf{n}_{0uv}^c\right)\\ &- \tfrac{2}{\Delta\theta}\left(A_{ij}(\mathbf{n}_1^c - \mathbf{n}_0^c) + \tfrac{1}{3}A_{uij}(\mathbf{n}_{1u}^c - \mathbf{n}_{0u}^c) + \tfrac{1}{3}A_{vij}(\mathbf{n}_{1v}^c - \mathbf{n}_{0v}^c) + \tfrac{1}{9}A_{uvij}(\mathbf{n}_{1uv}^c - \mathbf{n}_{0uv}^c)\right)\Big],\end{aligned} \tag{A41}$$

$$\begin{aligned}\mathbf{F}_{ij1} &= \frac{Z}{\pi}\Big[\Delta\theta\hat{x}\left(A_{ij}n_x^c + \tfrac{1}{3}A_{uij}n_{xu}^c + \tfrac{1}{3}A_{vij}n_{xv}^c + \tfrac{1}{9}A_{uvij}n_{xuv}^c\right)\\ &- 2\hat{x}\times\left(A_{ij}\mathbf{n}_1^c + \tfrac{1}{3}A_{uij}\mathbf{n}_{1u}^c + \tfrac{1}{3}A_{vij}\mathbf{n}_{1v}^c + \tfrac{1}{9}A_{uvij}\mathbf{n}_{1uv}^c\right)\\ &+ \tfrac{2}{\Delta\theta}\left(A_{ij}(\mathbf{n}_1^c - \mathbf{n}_0^c) + \tfrac{1}{3}A_{uij}(\mathbf{n}_{1u}^c - \mathbf{n}_{0u}^c) + \tfrac{1}{3}A_{vij}(\mathbf{n}_{1v}^c - \mathbf{n}_{0v}^c) + \tfrac{1}{9}A_{uvij}(\mathbf{n}_{1uv}^c - \mathbf{n}_{0uv}^c)\right)\Big].\end{aligned} \tag{A42}$$

References

1. Greve, M.; Wöckner-Kluwe, K.; Abdel-Maksoud, M.; Rung, T. Viscous-Inviscid Coupling Methods for Advanced Marine Propeller Applications. *Int. J. Rotat. Mach.* **2012**, *2012*, 743060. [CrossRef]
2. Rankine, W.J.M. On the Mechanical Principle of the Action of Propellers. *Trans. Inst. Nav. Arch.* **1865**, *6*, 13.
3. Froude, R.E. On the Part Played in Propulsion by Differences of Fluid Pressure. *Trans. Inst. Nav. Arch.* **1889**, *30*, 390.
4. Hough, G.R.; Ordway, D.E. (Eds.) *Developments in Theoretical and Applied Mechanics*; Elsevier: Pergamon, Turkey, 1965; pp. 317–336.
5. Conway, J.T. Analytical solutions for the actuator disk with variable radial distribution of load. *J. Fluid Mech.* **1995**, *297*, 327–355. [CrossRef]
6. Conway, J.T. Exact actuator disk solutions for non-uniform heavy loading and slipstream contraction. *J. Fluid Mech.* **1998**, *365*, 235–267. [CrossRef]
7. Sparenberg, J.A. On the Potential theory of the Interaction of an Actuator Disk and a Body. *J. Ship Res.* **1972**, *16*, 271–277.
8. Huang, T.T.; Groves, N.C. Effective Wake: Theory and Experiment. In Proceedings of the 13th Symposium on Naval Hydrodynamics: The Shipbuilding Research Association of Japan, Tokyo, Japan, 6–10 October 1980; pp. 651–673.
9. Bujnicki, A. *V-SHAPE: Effective Wake Calculation; Documentation of Computer Program and Test Cases*; Report No. 83-0293; Det Norske Veritas: Oslo, Norway, 1983.
10. Van Gent, W. A model for propeller-ship wake interaction. In Proceedings of the International Symposium on Propeller Cavitation, Wuxi, China, 8–12 April 1986.
11. Choi, J.K.; Kinnas, S.A. Prediction of Non-Axisymmetric Effective Wake by a Three-Dimensional Euler Solver. *J. Ship Res.* **2001**, *45*, 13–33.
12. Stern, F.; Kim, H.T.; Patel, V.C.; Chen, H.C. A Viscous-Flow Approach to the Computation of Propeller-Hull Interaction. *J. Ship Res.* **1988**, *32*, 263–284.
13. Hally, D.; Laurens, J.M. Numerical Simulation of Hull-Propeller Interaction Using Force Fields within Navier–Stokes Computations. *Ship Technol. Res.* **1998**, *45*, 28–36.
14. Starke, A.R.; Bosschers, J. Analysis of scale effects in ship powering performance using a hybrid RANS/BEM approach. In Proceedings of the 29th Symposium on Naval Hydrodynamics, Gothenburg, Sweden, 26–31 August 2012.
15. Morino, L.; Chen, L.T.; Suciu, E.O. Steady and oscillatory subsonic and supersonic aerodynamics around complex configurations. *AIAA J.* **1975**, *13*, 368–374.
16. Hally, D. Propeller analysis using RANS/BEM coupling accounting for blade blockage. In Proceedings of the 4th International Symposium on Marine Propulsors, Austin, TX, USA, 1–4 June 2015; pp. 296–303.

17. Hally, D. A RANS-BEM coupling procedure for calculating the effective wakes of ships and submarines. In Proceedings of the 5th International Symposium on Marine Propulsors, Espoo, Finland, 12–15 June 2017.
18. Su, Y.; Kinnas, S.A.; Jukola, H. Application of a BEM/RANS Interactive Method to Contra-Rotating Propellers. In Proceedings of the 5th International Symposium on Marine Propulsors, Espoo, Finland, 12–15 June 2017.
19. Kinnas, S.A.; Chang, S.H.; Tian, Y.; Jeon, C.H. Steady And Unsteady Cavitating Performance Prediction of Ducted Propulsors. In Proceedings of the 22nd International Offshore and Polar Engineering Conference, Rhodes, Greece, 17–22 June 2012.
20. The OpenFOAM Foundation. Available online: http://openfoam.org (accessed on 31 January 2018).
21. Cooperative Research Ships. Available online: http://www.crships.org (accessed on 31 January 2018).
22. Boswell, R.J. *Design, Cavitation Performance, and Open-Water Performance of a Series of Research Skewed Propellers*; Report 3339; Naval Ship Research and Development Centre: Washington, DC, USA, 1971.
23. Min, K.S. *Numerical and Experimental Methods for the Prediction of Field Point Velocities Around Propeller Blades*; Report 78-12; Massachusetts Institute of Technology: Cambridge, MA, USA, 1978.
24. Bosschers, J. *PROCAL v2.0 Theory Manual*; MARIN Report No. 20834-7-RD; MARIN: Wageningen, The Netherlands, 2009.
25. Hoshino, T. A surface panel method with deformed wake model to analyze hydrodynamic characteristics of propellers in steady flow. In *Mitsubishi Technical Bulletin*; Mitsubishi Heavy Industries: Tokyo, Japan, 1991.

Journal of
Marine Science and Engineering

MDPI

Article

Numerical Analysis of Azimuth Propulsor Performance in Seaways: Influence of Oblique Inflow and Free Surface

Nabila Berchiche, Vladimir I. Krasilnikov * and Kourosh Koushan

SINTEF Ocean, P.O. Box 4762 Torgard, N-7465 Trondheim, Norway; nabila.berchiche@gmail.com (N.B.); Kourosh.Koushan@sintef.no (K.K.)

* Correspondence: Vladimir.Krasilnikov@sintef.no; Tel.: +47-92-09-08-84

Received: 28 February 2018; Accepted: 27 March 2018; Published: 5 April 2018

Abstract: In the present work, a generic ducted azimuth propulsor, which are frequently installed on a wide range of vessels, is subject to numerical investigation with the primary focus on performance deterioration and dynamic loads arising from the influence of oblique inflow and the presence of free surface. An unsteady Reynolds-Averaged Navier-Stokes (RANS) method with the interface Sliding Mesh technique is employed to resolve interaction between the propulsor components. The VOF formulation is used to resolve the presence of free surface. Numerical simulations are performed, separately, in single-phase fluid to address the influence of oblique inflow on the characteristics of a propulsor operating in free-sailing, trawling and bollard conditions, and in multi-phase flow to address the influence of propulsor submergence. Detailed comparisons with experimental data are presented for the case of a propulsor in oblique flow conditions, including integral propulsor characteristics, loads on propulsor components and single blade loads. The results of the study illustrate the differences in propulsor performance at positive and negative heading angles, reveal the frequencies of dynamic load peaks, and provide quantification of thrust losses due to the effect of a free surface without waves. The mechanisms of ventilation inception found at different propulsor loading conditions are discussed.

Keywords: CFD; RANS; azimuth propulsor; unsteady loads; ventilation

---

1. Introduction

Pushing type ducted azimuth propulsors have become a very popular solution for propulsion system on such vessels as tugboats, anchor handlers, offshore supply vessels, research vessels and others. They are found to combine the advantages of ducted propellers under heavy loading with the superior maneuvering characteristics of pod propulsors at low speed operation, while offering reasonable efficiency in free sailing. At the same time, azimuth propulsors may experience considerable deterioration in their performance characteristics and come at risk of failure due to high dynamic loads arising from various factors. Heavy propeller loading in combination with large azimuth angles is the frequent scenario at low speed operation that results in unsteady loads of high amplitude on the whole propulsor as well as on the individual propeller blades. These loads put extra demands on the structural strength of propulsor, and reliability of blade bearings and transmission mechanisms. The operation conditions are further complicated by vessel drift, presence of sea current, orbital velocities induced by free surface waves, and wave induced ship motions. Massive flow separation develops on propulsor components such as pod housing and ducts, which makes the inflow on a pushing propeller highly inhomogeneous. As a consequence, propeller blades may experience sheet and vortex cavitation of an unsteady nature, causing increased levels of noise and vibration, and posing the danger of blade erosion. Operation in the vicinity of free surface is associated with the risk of ventilation. The aforementioned aspects of propulsor operation are the focus of the ongoing R&D Project "INTER-THRUST" supported by the MARTEC II ERA-NET program (Maritime Technologies), where the authors participate on behalf of SINTEF Ocean (formerly known as MARINTEK).

The complexity of hydrodynamic interactions taking place between the components of a ducted azimuth propulsor has been noted in a number of studies [1,2]. Advanced numerical methods such as those of Computational Fluid Dynamics (CFD) need to be employed when targeting improvements in propulsor design and its operational reliability. For the successful implementation of CFD results in design practice, validation is however crucial. While there is a considerable amount of validation material on pod propulsors with open propellers [3–6] and shaft ducted propellers [7,8], for ducted pod propulsors, the available data are quite limited. The majority of CFD models currently employed for the numerical simulation of pod propulsors are based on the unsteady Reynolds-Averaged Navier-Stokes (RANS) method with one or another isotropic turbulence closure model [9–11]. The SST k-ω turbulence model is by far the most popular choice in this type of simulation, whether one considers an open propeller case [3] or a ducted propeller case [8,12]. In model scale analyses, the GaReTheta transition model based on the SST k-ω model formulation is more and more frequently employed to address the influence of laminar-turbulent flow transition at low Reynolds numbers [6,13,14]. More advanced turbulence modelling approaches such as. for example, Detached Eddy Simulation (DES) gradually find their way into the engineering analyses, in particular, to address the scenarios where complex unsteady interaction and massive separation are the salient flow features, as in the case of a podded thruster turning with constant azimuth speed [15]. As regards computational mesh, many authors favour unstructured meshing approaches that offer ease of mesh generation and greater flexibility compared to structured meshes, when dealing with complex geometries. Adaptive mesh techniques are shown to have certain advantages in modelling pod propulsors operating at very large heading angles [16]. In the present work, the authors attempt a detailed validation study with the generic model of a ducted azimuth thruster tested earlier at the SINTEF Ocean laboratory. The comparisons between the numerical simulations and experiments cover a range of operational conditions from bollard to free sailing, and a range of propulsor heading angles, at fixed propeller RPM and zero azimuth speed, in unbounded flow. Both the total forces and moments acting on the unit and its components such as the propeller, duct and gear housing, and loads on single blade are compared. Separately, CFD calculations are performed with the same thruster operating in the presence of an initially undisturbed free surface, at different magnitudes of submergence and different propulsor loading conditions, in order to quantify the impact of free surface presence and ventilation phenomenon on propulsor thrust losses and individual blade loads. The findings regarding the trends in propulsor characteristics are supported by the analysis of underlying flow mechanisms.

2. Propulsor Model and Simulated Conditions

The main characteristics of the generic model of the azimuth ducted pushing thruster tested in the towing tank at SINTEF Ocean and used in the CFD simulations are given in Table 1. A picture showing the general view of the test setup is presented in Figure 1. A more detailed model specification, including propeller blade drawings, is found in [17]. Regarding the propeller model, it has to be noted that the blade design represents a compromise solution between an open propeller and a ducted propeller, since the same propeller model and the same pod unit have been tested earlier in both the ducted and open configurations [3,18].

The model tests in oblique flow are conducted at constant propeller shaft immersion of 345 mm (1.38 Dp, where Dp is the propeller diameter). A plate is mounted on top of the thruster to avoid wave generation and aeration. Nevertheless, at high heading angles propeller rate of revolution had to be reduced, in order to avoid aeration and mitigate excessive free surface disturbance. The propeller was driven by an electric motor installed on top of the thruster via a right-angle transmission gear. A six-component balance was positioned on top of the thruster unit to measure total loads on the thruster, including propeller, but excluding the duct. A cylindrical rod of 35 mm diameter connected the duct to the two force transducers, which separately measured the axial and transverse forces acting on the duct. A propeller balance measured propeller thrust and torque, the measured signals being wirelessly transmitted to the acquisition system. During the blade load measurements, one of the propeller blades was mounted on a blade dynamometer placed inside the hub. A pulse meter on the shaft provided propeller rate revolutions, while another pulse meter inside the thruster registered the angular position of the reference blade. An angular meter was installed on top of the rig to measure the heading angle of the unit.

Table 1. Main particulars of the tested propulsor model.

Thruster Particular	Value
Vertical distance from propeller centre to upper end of strut (mm)	342
Gondola length (mm)	181
Strut chord length (mm)	86
Propeller P-1374 direction of rotation	Right-handed
Propeller diameter, Dp (mm)	250
Hub diameter (mm)	60
Design pitch ratio, $P_{0.7}/Dp$	1.1
Blade skew (degrees)	25
Expanded blade area ratio	0.6
Number of blades	4
Duct type	19A
Duct length (mm)	125
Duct inner diameter (mm)	252.78
Blade tip clearance, tc/Dp	0.0056
Duct max. outer diameter (mm)	303.96
Duct leading edge radius (mm)	2.78
Duct trailing edge radius (mm)	1.39

Figure 1. General view of model test setup.

The tests are performed at different heading angles ranging from −90° to +90°, at both negative and positive advance coefficients. The measured forces and moments include propeller thrust and torque, duct thrust and side force, total thrust and total side force, steering moment, blade thrust and torque, and blade spindle and bending moments. The conditions selected for the numerical simulations and comparisons with experimental data in oblique unbounded flow are summarized in Table 2. The simulation conditions in the presence of free surface are reduced in Table 3.

Table 2. Simulation conditions in oblique unbounded flow.

Advance Coefficient, J [1]	Heading Angle (°)
0.6 (free sailing)	±60, ±35, ±15, 0
0.3 (trawling)	±35, ±15, 0
0 (bollard)	0

[1] J = V/(n × Dp).

Table 3. Simulation conditions in presence of free surface.

Advance Coefficient, J	Heading Angle (°)	Submergence, H/Dp [1]
0.6 (free sailing)	0	2, 1.7, 1.5, 1.3, 1.0, 0.9, 0.8, 0.6, 0.5
0.2 (trawling)	0	1.5, 1.0, 0.9, 0.8, 0.7
0 (bollard)	0	1.0, 0.9, 0.8, 0.7

[1] H is the vertical distance from propeller center to water surface level; Dp is the propeller diameter.

The thrust force, T, is defined along the propeller shaft axis, and is positive forward. The side force, S, is the horizontal force perpendicular to propeller shaft axis, and is positive to portside. The steering moment, MY, is positive in an anti-clockwise direction when viewing from above. The blade loads are defined in the local Cartesian blade coordinate system with the origin at the blade root, and Yb-axis along the Blade Reference Line. The blade thrust, Tb, is positive in the direction of the propulsor thrust force. The blade torque, MXb, the blade spindle moment, MYb, and the blade bending moment, MZb, are positive in an anti-clockwise direction when viewing from the end of the respective axes of the blade coordinate system. Positive heading angles correspond to crossflow from portside. Propulsor submergence is defined as the vertical distance from propeller center to water surface level. These definitions and sign conventions are illustrated in Figure 2.

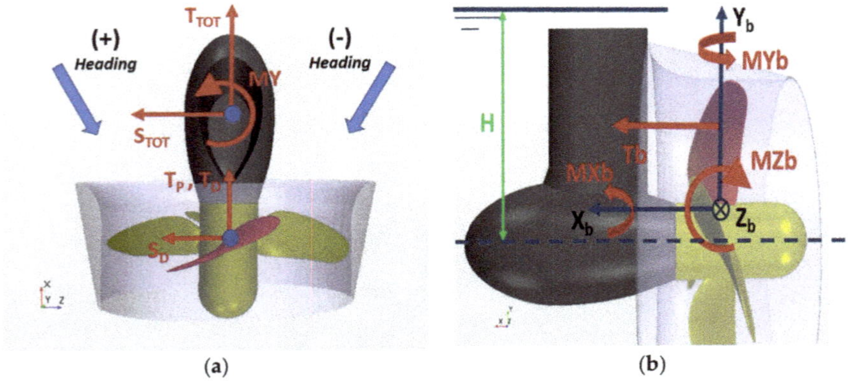

Figure 2. Definitions of forces and moments and sign conventions: (**a**) propulsor loads; (**b**) blade loads.

The standard definition of advance coefficient is used as follows: J = V/(n × Dp), V being the speed of inflow, and n being the propeller rate of revolution. In all simulations, the blade pitch setting is P(0.7)/Dp = 1.10, which corresponds to the design pitch. At this pitch setting, the operation point J = 0.6 selected for free sailing conditions is near the maximum of propulsor efficiency. All numerical simulations are performed in model scale conditions, and propeller rate of revolution is fixed to 11 (rps), according to model tests. The speed of inflow V corresponding to the speed of towing carriage in model tests equals to V = 1.65 (m/s) at J = 0.6, and it equals 0.55 (m/s) at J = 0.2.

3. Numerical Method

The unsteady RANS equations are solved in the commercial solver STAR-CCM+ (Version 11.02.010), using a finite volume method, to simulate the flow around the studied propulsor. For the resolution of interaction between the rotating and stationary components of propulsor, the interface sliding mesh technique is employed, which allows the mesh region containing the propeller to rotate at the same rotational speed as that of the propeller with respect to the outer region of stationary fluid. The implicit unsteady method is used with a first-order temporal discretization scheme. The flow is considered fully turbulent, and the Shear-Stress-Transport (SST) k-ω model is used for turbulence closure in most of the simulations, except additional test calculations in straight flow condition, where the GaReTheta transition model based on the same SST k-ω model formulation has been used. The turbulence models are used in combination with All-Y+ wall treatment approach as implemented in STAR-CCM+. In the problem, involving the presence of free surface, an Eulerian Multiphase Mixture formulation is employed, with the Volume of Fluid (VOF) method and the Flat VOF Wave model to set up the free surface conditions. The High-Resolution Interface Capturing (HRIC) scheme is used for tracking the water-air interface. No cavitation model is included in the analyses, since the intention was to separate, in the present study, the pure effects of free surface and ventilation.

The computational domain consists of two sub-domains (mesh regions): (1) a rotating sub-domain that contains the propeller, part of the hub and part of the duct above the propeller; and (2) a stationary sub-domain that includes the pod gondola and strut, and parts of the duct and hub, which are not in the rotating propeller region. The rotating sub-domain is a cylinder surrounding the propeller. Both its upstream and downstream boundaries are located at 0.15 Dp, as shown in Figure 3. In the present approach, the circumferential boundary of the rotating propeller region coincides with the duct inner surface. Therefore, only two sliding interfaces—upstream and downstream—are created between the stationary and rotating mesh regions.

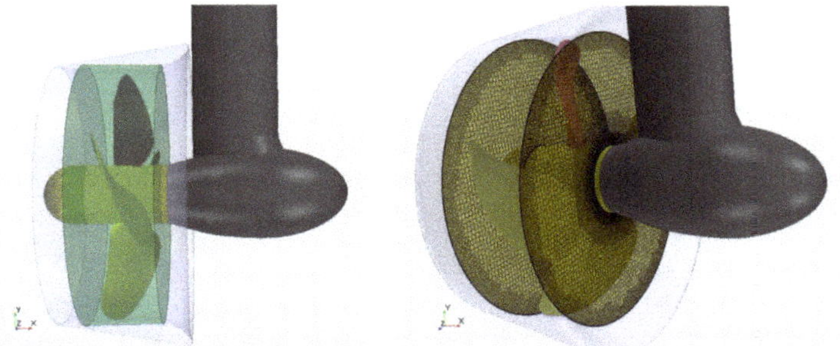

(a) (b)

Figure 3. Rotating propeller region and sliding mesh interfaces: (**a**) propeller region; (**b**) sliding interfaces.

The boundary condition on the duct inner surface inside the rotating region is satisfied by setting the wall rotation velocity to zero. In earlier studies, this approach has been tested against a more traditional approach that implies the use of a complete cylindrical region around the propeller, with additional circumferential sliding interface in the gap between the blade tip and duct inner surface. Both the aforementioned approaches bring very close results in terms of integral (forces, moments) and distributed (pressure, velocities) characteristics. However, the former approach with only two sliding interfaces shows somewhat lower levels of solution residuals and lower oscillations in the overall pattern of force convergence. It is also better suited for meshing the cases with more complex configurations of the duct interior surface.

The computational grid is generated in STAR-CCM+ using polyhedral cells in the rotating mesh block and hexahedral trimmed cells in the stationary mesh block. The boundary layer mesh consists of 10 prismatic layers applied on all solid surfaces, except the extended part of the strut, which is protruded to the upper boundary of computation domain. The values of the stretching factor and total thickness of the boundary layer have been set in order to achieve the target range of wall Y+ from 0 to 5 in model scale calculations, and at the same time, to ensure sufficiently smooth transition from the boundary layer mesh to the core mesh. The inlet and outlet boundaries of the computation domain are placed at 22 propeller diameters from the propeller center by applying prismatic extrusion upstream and downstream from the core mesh, while the upper, bottom and side boundaries are placed at 10 Dp. The mesh refinement is regulated by means of curve, surface and volume controls. The most important areas for mesh refinement include the domain around the propulsor, leading and trailing edges of the propeller blades, duct and strut, junction between the strut and gondola of the pod housing, gap between the pod gondola and propeller hub, blade tip clearance, and wake downstream of the pod housing. The gap between the rotating propeller hub and stationary gondola of the pod housing is modelled explicitly, with about 25 cells placed across the gap. A similar number of cells is accommodated in the clearance between the blade tip and duct inner surface. In the simulations involving free surface, additional refinement is applied in a vertical direction at the level of free surface everywhere in the domain, and in all three directions above the propulsor, where deformations of free surface and, eventually, ventilation are expected. Some details of surface mesh on propeller and duct and volume mesh around the propulsor and free surface can be seen in Figure 4. The total cell count in the computation mesh is about 21.5 million cells, of which about 4 million cells are in the propeller region. The applied meshing approach is based on best practices elaborated by the authors in their earlier studies regarding the CFD simulations of ducted propellers [7,19] and pod propulsors [20,21].

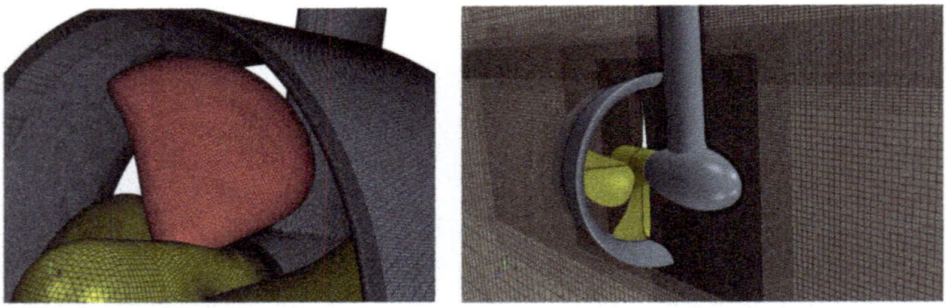

(a) (b)

Figure 4. Details of computation mesh: (**a**) surface mesh on propeller and duct; (**b**) volume mesh around propulsor and free surface.

4. Propulsor Operation in Oblique Flow Conditions

All the simulations discussed below have been run in two stages. At the first stage, a converged steady-state solution was attained using the Moving Reference Frame (MRF) method applied in the propeller region. At the second stage, the implicit unsteady solution has been activated from the solution of the first stage, and the case has been run for 40 complete propeller revolutions, with the time-step corresponding to 1 degree of propeller rotation. For comparison with the measurements, the computed forces and moments are expressed as non-dimensional coefficients by dividing them by ($\varrho \times n^2 \times D^4$) and ($\varrho \times n^2 \times D^5$), respectively. When post-processing the simulation results, the force and moment coefficients are time-averaged over the last 20 propeller revolutions.

4.1. Integral Forces and Moments on the Propulsor

The computed and measured propeller thrust, KTP, propeller torque, KQP, duct thrust, KTD, and total unit thrust force, KTTOT, are compared in Figures 5 and 6. The numerical simulations reproduce fairly well the variation of propeller thrust and torque with heading angle, for all loading conditions. The propeller torque is predicted somewhat better than thrust in most cases. At heavier loading (J = 0.3), KQP is in a very good agreement with the measurements for the whole range of heading angles. At J = 0.6, the computed KQP values are also close to the measured data except the largest heading angles of ±60°, where KQP is over-predicted. The latter scenarios are associated with massively separated flow over the pod and duct, and therefore they represent considerable challenges for numerical simulations. The numerical predictions of duct thrust, KTD, and total unit thrust, KTTOT, are found to be in good agreement with the measurements in the whole range of studied conditions.

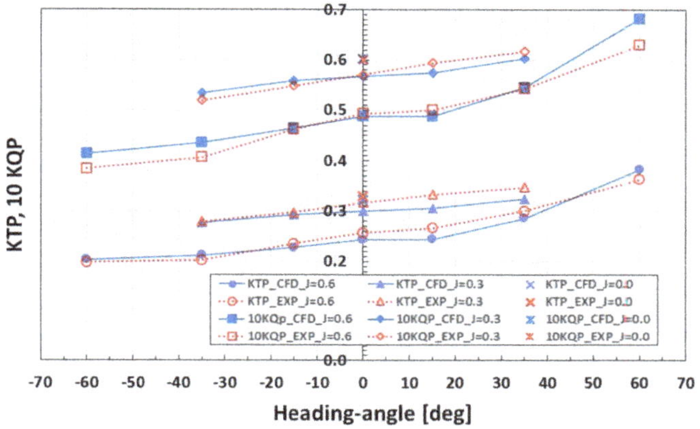

Figure 5. Measured (red) and computed (blue) propeller thrust, KTP, and propeller torque, KQP, at different heading angles and loading conditions.

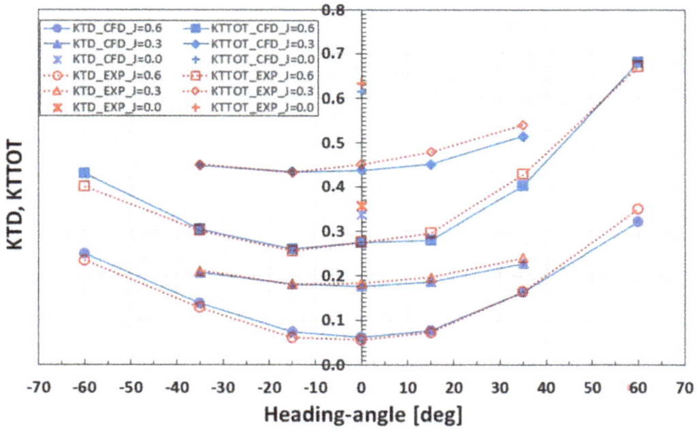

Figure 6. Measured (red) and computed (blue) duct thrust, KTD, and total unit thrust, KTTOT, at different heading angles and loading conditions.

Regarding the prediction of KTP, at least two important phenomena are believed to play a role: (1) effect of the gap between the rotating hub and stationary pod; and (2) pattern of the flow over the duct. In order to address the first effect, additional CFD calculations have been performed in straight flow for different loadings and at two propeller pitch ratios, P/D. The simulations predict lower pod resistance and lower KTP, compared to the measurements, while the computed total thrust, KTTOT, and propeller torque, KQP, are predicted close to the measured data, as it can be concluded from Figures 7 and 8, similar to the observations from Figures 5 and 6.

While, as mentioned above, in the CFD model the gap width is included in exact accordance with the test setup, the details of shafting, bearings and seals are not modelled. Furthermore, there are also uncertainties associated with the pressure levels in the gap during model tests and their effect on the measured propeller thrust KTP. At the same time, one needs to point out that without including the hub/pod gap in the numerical model, one cannot achieve a realistic prediction of pod resistance and hub contribution to propeller thrust, due to the uncertainties associated with pressure integration over the sides of the pod and hub facing the gap. In general, the influence of gap flow on the results of CFD/EFD comparisons is relatively higher for podded propellers than for conventional shaft propellers, either ducted or open. The "gap effect" is more pronounced for conical hub configurations compared to cylindrical hub configurations, and it is also found dependent on the location of the gap with respect to the cone, as well as on propeller operation mode with respect to the pod housing, or dynamometer shaft, i.e., pulling more or pushing mode. For example, when using the numerical model similar to the one employed in the present study, the predictions of propeller thrust for shaft propellers featuring cylindrical hubs was found to be in close agreement with the measurements done on ducted propellers [7] and on open propellers [21]. It is worth noting that the same propeller P1374 as used in the present study was also used in the referred papers.

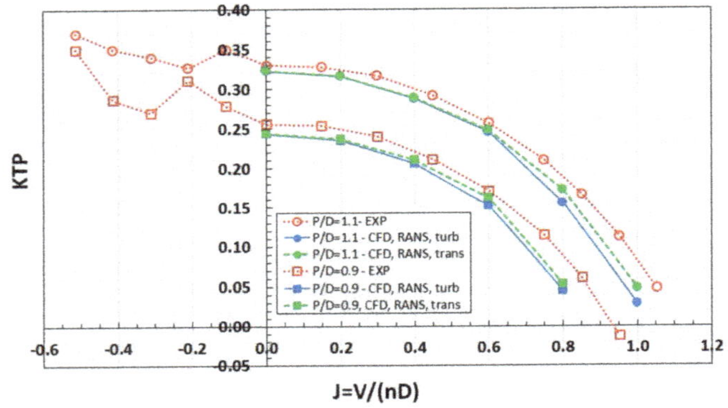

Figure 7. Measured (red) and computed (blue—fully turbulent model, green—transition model) propeller thrust, KTP, at the two propeller pitch settings, P/D, and different loading conditions.

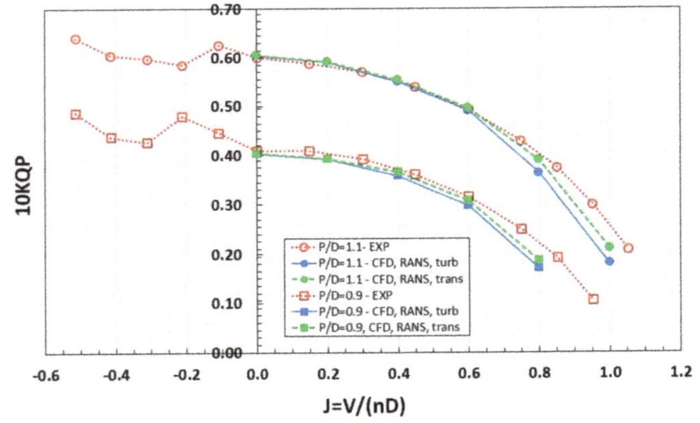

Figure 8. Measured (red) and computed (blue—fully turbulent model, green—transition model) propeller torque, KQP, at the two propeller pitch settings, P/D, and different loading conditions.

The second phenomenon is attributed to the flow over the duct at low Re numbers in model scale, and it is closely related to flow regime transition on the outer side of the duct. Unlike the phenomenon of gap effect, it is mostly evident in light loading conditions. An estimation of characteristic Reynolds number Re_D for the duct flow at J = 0.6, where the aforementioned flow transition effect becomes appreciable, results in the following figures: $Re_D = 1.8 \times 10^5$ for the outer duct surface, and $Re_D = 3.0 \times 10^5$ for the outer duct surface. Flow transition affects the extent of flow separation on the duct, and also the flow pattern over the duct trailing edge. It is mainly the trailing edge flow that influences the flow velocity through the duct and results in changes in propeller thrust and torque. In most cases, for conventional shaft ducted propellers, it has been observed that transition leads to a somewhat larger expansion rates of propeller slipstream, and hence lower velocity through the propeller compared to fully turbulent flow conditions [7]. It leads to an increase of propeller thrust and torque as observed in Figures 7 and 8, where the results of calculation using the transition model show higher values of KTP and KQP at high advance coefficients. At lower advance coefficients (J < 0.6), in straight flow, propeller torque is very well predicted, while at higher advance coefficients (J > 0.6), KQP is under-predicted, as it can be seen from Figure 8. Since propeller torque is not influenced by the gap flow effect, one can conclude that in the range of loading conditions $J \leq 0.6$, the influence of flow transition is minor, and the main reason for the differences in KTP should probably be attributed to uncertainties associated with the gap flow in both the physical and numerical models. In oblique flow, the flow around the gap is no longer axisymmetric, and consequently, the contributions of gap walls to pod resistance and propeller thrust change. In such cases, the gap flow is also strongly influenced by the configurations of hub and pod in the vicinity of the gap, as well as by the vortices shed from the pod housing. Therefore, in the case of oblique flow, even small differences in flow pattern between the simulations and tests may result in large differences in contributions to thrust from the gap walls.

Figure 5 shows that propeller thrust and torque increase with the increase in positive heading angle, while they decrease with the increase in negative heading angle, in comparison with their values at zero heading. This asymmetry of forces with respect to zero heading is a characteristic feature of pushing podded units, and it is mainly associated with the interaction between the rotating propeller and cross-flow in the separated wake behind the strut. This phenomenon has been discussed in detail in [9]. For ducted pushing propulsors, the force asymmetry is smaller than for open pushing propulsors, due to the flow-straightening effect of the duct. Furthermore, as it can be observed from Figure 5, the asymmetric influence of cross-flow altered by the strut wake on KTP is larger for lighter loadings.

As mentioned above, the numerical predictions of duct thrust, KTD, and total unit thrust, KTTOT, compare well with the measurements in the whole studied range. As one can see from Figure 6, KTD exhibits nearly symmetric behavior with respect to zero heading angle, for the heading angles within $\pm 35^\circ$, since the duct experiences very similar conditions in terms of inflow. At larger heading angles, some asymmetry is introduced in the distribution of KTD by the differences in propeller loading at positive and negative heading angles, as discussed above.

The comparisons of the measured and computed side forces acting on the duct, KSD, and on the whole unit, KSTOT, are presented in Figures 9 and 10, for the loading conditions of J = 0.6 and J = 0.3, respectively. Figure 10 also includes the results obtained for bollard condition, J = 0.0.

The major contribution to the total side force, KSTOT, is given by the duct at all heading angles, and as the heading angle increases, the contributions by the pod and propeller are added. It has to be noted that due to the flow straightening effect of the duct on propeller flow, the contribution by the propeller to the total side force is much smaller than it is in the case of open pushing thrusters. The duct straightens both the oblique inflow onto the propeller and the slipstream downstream of the propeller. These two effects result in a reduction of propeller side force.

The numerical simulations appear to under-predict the total side force for all tested conditions. This is obviously due to the under-prediction of the duct side force, KSD. It has been found from earlier studies in the cases where flow separation develops on the duct in straight flow, that a fully turbulent calculation, especially with the SST k-ω model in use, results in a greater separation extent than obtained from the calculations using the transition model [7,19]. Under the conditions of oblique flow, flow separation on the duct leeward side becomes massive, which puts even heavier demands on the turbulence closure. The use of a transition model does not remedy the situation in these cases.

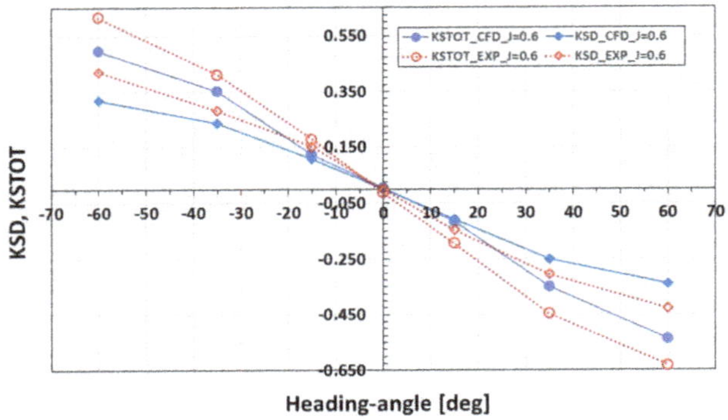

Figure 9. Measured (red) and computed (blue) duct side force, KSD, and total unit side force, KSTOT, at different heading angles and loading condition of J = 0.6.

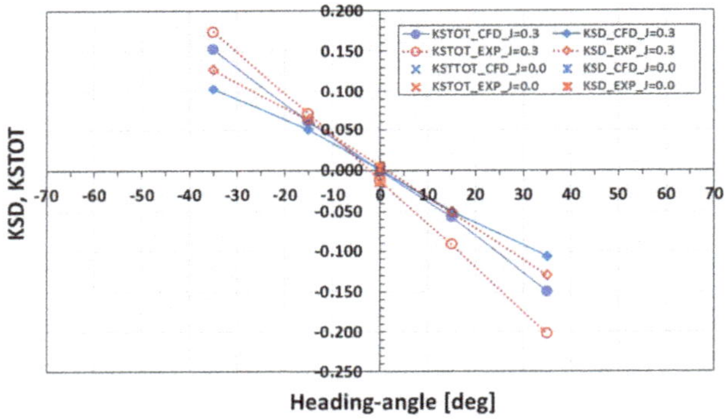

Figure 10. Measured (red) and computed (blue) duct side force, KSD, and total unit side force, KSTOT, at different heading angles and loading conditions of J = 0.3 and J = 0.0.

The comparison between the measurements and calculations in terms of the steering moment, KMY, have been done for the moment quantity, excluding the duct contribution, since only this quantity is measured in the tests. This comparison is presented in Figure 11, while the computed distributions of the steering moment acting on the duct, KMYD, are shown in Figure 12. As one can see from Figure 11, quite large deviations are observed between the measurements and calculations, particularly at a high advance coefficient of J = 0.6, and at large heading angles. However, the overall variation of the steering moment with the loading and heading angle appears qualitatively well reproduced by the CFD simulations, in particular at lower J = 0.3. The quantitative agreement with the measurements is more difficult to achieve, since the values of KMY are small, because KMY is the result of opposing contributions from the propeller, KMYP, and from the pod, KMYG. At higher speeds, the experimental data also appear affected by considerable scattering. At zero heading angle, KMY is entirely determined by the contribution from the propeller. Minor difference in the location of resulting force centers on the pod and propeller may lead to considerable difference in the contributions from respective forces to KMY due to the force arm.

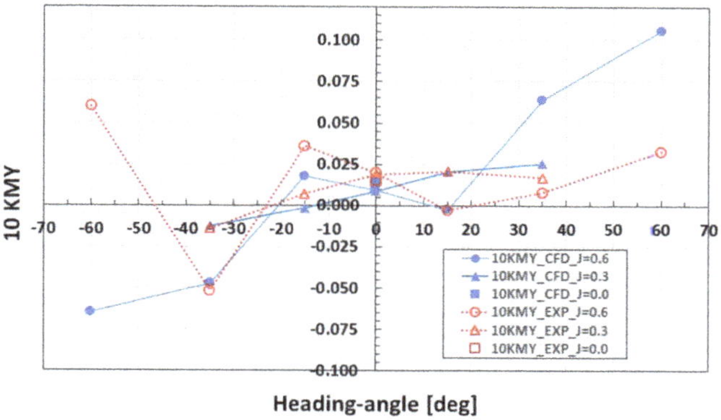

Figure 11. Measured (red) and computed (blue) steering moment, KMY, excluding duct contribution, at different heading angles and loading conditions.

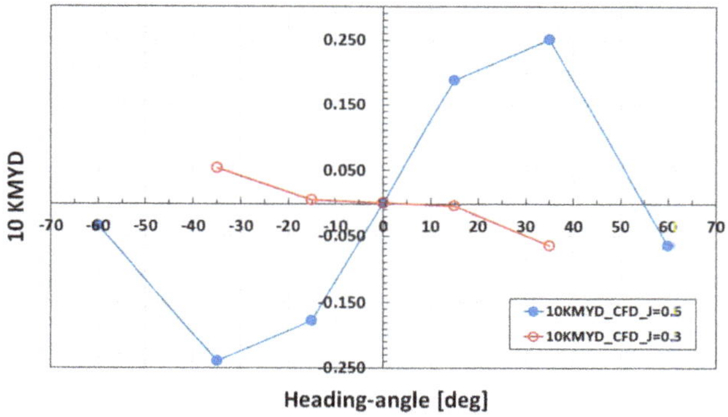

Figure 12. Computed steering moment on the duct, KMYD, at different heading angles and loading conditions.

Both the side force on the propeller, KSP, and side force on the pod, KSG, act in the direction of cross-flow caused by the heading angle. At zero heading, the propeller develops a small side force acting in the direction of portside (for the right-handed propeller studied herewith) due to its interaction with the wake of the pod housing. This small side force causes a deviation of the total reaction force, resulting in a non-zero positive steering moment acting on the unit. The moment KMY is larger at lower J values, where propeller loading is heavier. This is an important phenomenon that has direct impact on course-keeping characteristics of vessels equipped with pushing and dual-end pod propulsors.

At heading angles, the major contribution to the total steering moment is given by the duct. It can be observed from Figure 12 that, for the same heading angle, the steering moment on the duct (KMYD) may have different signs, depending on loading condition. More specifically, for the positive heading angle of +35 degrees, at $J = 0.6$ the moment KMYD is positive, i.e., it tends to turn the rear part of the unit in the direction of cross-flow (opposite clockwise), whereas at $J = 0.3$ this moment acts in the opposite direction. This result is explained by the differences in the contributions to KMYD originating from the duct thrust and duct side force at different loading conditions. The side force developed on the duct tends to turn the rear part of the unit in the direction of cross-flow, similar to the side force on the propeller. The force center of the duct reaction in oblique flow is however not located at the propeller shaft but is shifted towards the windward side of the duct (side facing the cross-flow), and thus the duct thrust tends to turn the rear part of the unit in the direction opposite to cross-flow. At lighter loading conditions (J = 0.6), the duct side force gives a greater contribution to KMYD. At heavier loading conditions (J = 0.3), it is on the contrary the duct thrust that gives a greater contribution to KMYD, which explains the differences in sign of this quantity observed in the results. As the heading angle increases, a massive flow separation develops on the duct at higher speeds (J = 0.6). It slows down the growth of duct side force with heading angle, while duct thrust continues to increase with approximately the same gradient as at lower heading angles.

Thus, the ratio KSD/KTD is reduced, and due to the redistribution of pressure on the duct, the arm of the KSD contribution is reduced as well. These changes result in a rapid decrease in the steering moment produced by the duct as seen at the heading angles of ±60 degrees. Further, at these heading angles, flow separation on the pod housing develops to such extent that it affects the inflow on the duct.

Separately, a frequency analysis has been performed on the computed time histories of propulsor forces and moments, using the Fast Fourier Transform (FFT). This analysis captured, in addition to the propeller blade frequency of 44 Hz, a lower frequency peak in the range of 3.9 to 4.2 Hz, as shown in Figure 13 for duct thrust. This low frequency peak is found to be related to the vortex shedding process that accompanies flow separation on the leeward side of the duct. Because of the interaction, the low frequency peaks in the same range are also observed for the loads on other propulsor components. The examples of flow patterns around the propulsor at large heading angles are shown in Figure 14. This figure shows flow streamlines coloured by the magnitude of dimensionless axial flow velocity in the horizontal plane located approximately at 0.4 Rp, for the top vertical position of the propeller blade (Rp being the propeller radius).

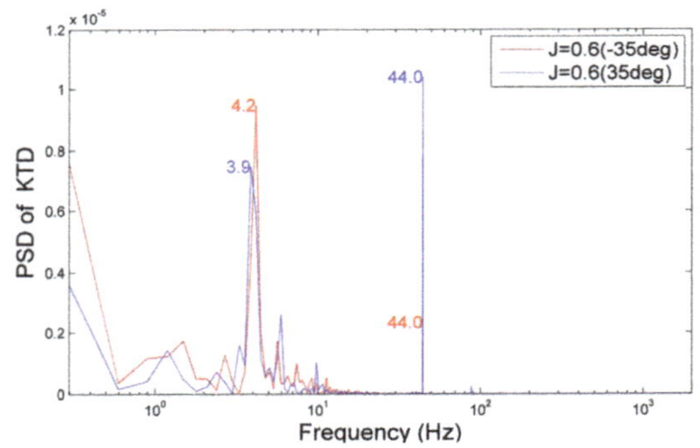

Figure 13. Power spectral density of duct thrust, KTD, at J = 0.6, for the positive and negative heading angles of 35 degrees.

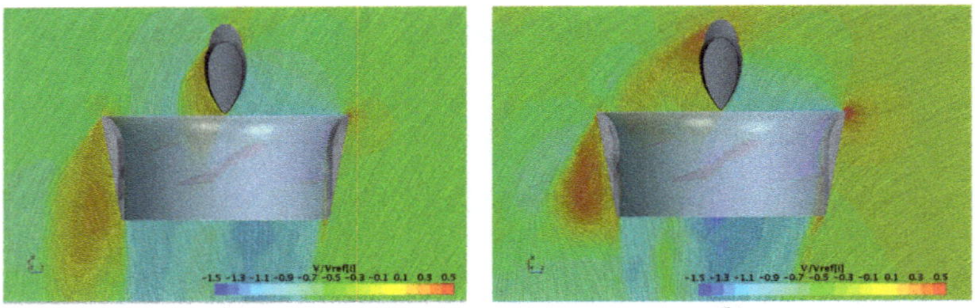

(a) **(b)**

Figure 14. Examples of flow pattern around propulsor at large heading angles, J = 0.6: (**a**) −35 degrees; (**b**) −60 degrees. View from top.

4.2. Single Blade Loads

The computed and measured loads on a single blade defined in the local blade coordinate system with the origin at the blade root are compared in Tables 4 and 5 in terms of the time-averaged values and their standard deviations, for the advance coefficients J = 0.6 and J = 0.3, respectively. The standard deviations, σ, are expressed as a percentage of the time-averaged values. Similar to the integral propulsor loads, the blade loads are given as dimensionless force and moment coefficients, for blade thrust, KTb, blade torque, KQb, blade spindle moment, KMYb, and blade bending moment, KMZb. The numerical predictions compare quite well with the experimental data, especially in terms of blade thrust, KTb, and blade bending moment, KMZb. The values of spindle moment, KMYb, are very small and, therefore, sensitive to small variations. The computed fluctuations of blade loads are close to the measured values and, as expected, they increase as the magnitude of the heading angle increases due to the propeller being subject to increasingly more intensive flow separation from the pod housing. The examples of computed time histories of blade thrust and bending moment from 5 complete propeller revolutions at J = 0.6 are shown in Figures 15 and 16 for the two heading angles of +35° and −35°, respectively. The measured values are also plotted for comparison.

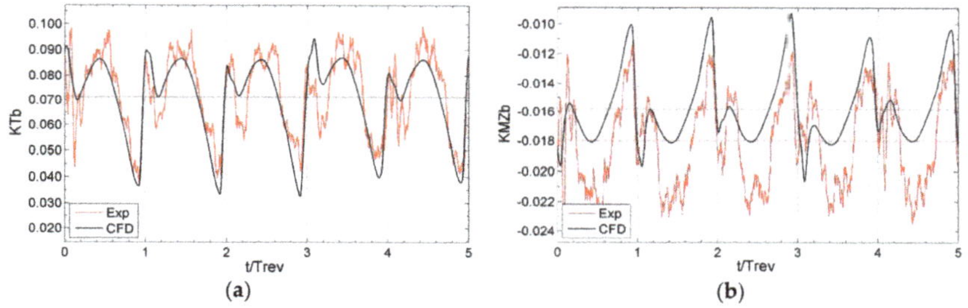

Figure 15. Measured and computed time histories of blade loads at J = 0.6 for the heading angle +35°: (**a**) blade thrust, KTb; (**b**) blade bending moment, KMZb.

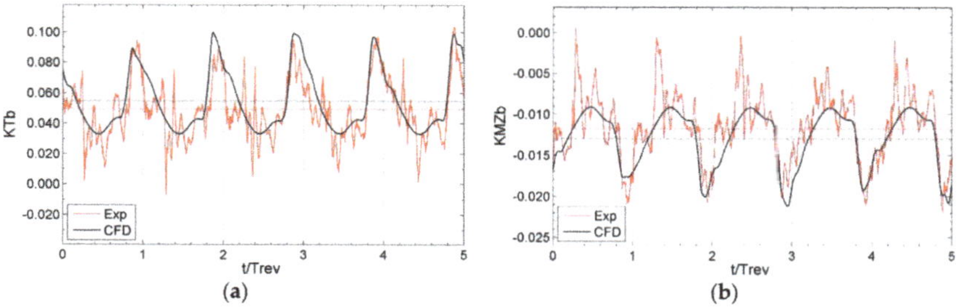

Figure 16. Measured and computed time histories of blade loads at J = 0.6 for the heading angle −35°: (**a**) blade thrust, KTb; (**b**) blade bending moment, KMZb.

The horizontal lines represent the time-averaged values and the vertical lines represent the 12 o'clock positions of the reference blade. A good agreement is observed between the calculations and measurements, particularly for KTb, in terms of the amplitudes of maximum blade loads and corresponding angular positions of the blade. Similar observations are made for all the other tested conditions. The experimental patterns exhibit higher frequency oscillations which are not evident in the CFD results. These oscillations are presumably related to the inherent measurement "noise", and also to the presence in the flow of intermittent vortex structures, which originate from the flow separation on propulsor components, and whose scale is beyond the resolution capabilities of the

RANS method. Figure 17 illustrates the computed contours of the axial velocity together with the tangential velocity vectors on a control plane upstream of the propeller for the positive and negative heading angles of 35°.

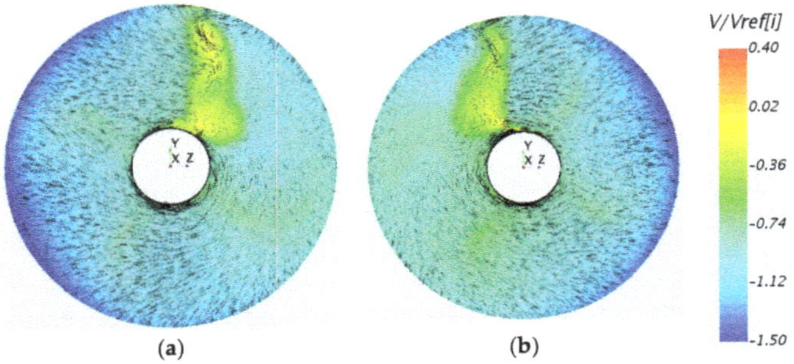

Figure 17. Computed velocity fields on the control plane upstream of the propeller at J = 0.6 for positive and negative heading angles: (**a**) +35 degrees; (**b**) −35 degrees. View from downstream.

Table 4. Measured and computed single blade loads at J = 0.6 for different heading angles.

	−35°		−15°		0°		15°		35°	
Quantity	**CFD**	**EXP**	**CFD**	**EXP**	**CFD**	**EXP**	**CFD**	**EXP**	**CFD**	**EXP**
KTb	0.055	0.049	0.058	0.056	0.062	0.059	0.061	0.063	0.071	0.071
σ (KTb) %	36.4	38.9	16.8	17.5	5.8	6.3	11.6	13.1	22.1	22.0
10 KQb	0.071	0.086	0.075	0.089	0.079	0.098	0.078	0.095	0.087	0.087
σ (KQb) %	24.8	23.5	12.3	12.1	4.2	4.8	8.0	8.5	15.1	19.0
10 KMYb	0.021	0.016	0.021	0.015	0.019	0.008	0.020	0.011	0.016	0.012
σ (KMYb) %	126.3	194.4	51.2	95.7	20.6	64.9	53.6	100.1	157.6	247.0
10 KMZb	−0.130	−0.117	−0.136	−0.152	−0.143	−0.171	−0.140	−0.170	−0.158	−0.180
σ (KMZb) %	26.5	35.4	13.4	15.6	4.6	4.6	8.5	10.0	15.7	17.5

Table 5. Measured and computed single blade loads at J = 0.3 for different heading angles.

	−35°		−15°		0°		15°		35°	
Quantity	**CFD**	**EXP**	**CFD**	**EXP**	**CFD**	**EXP**	**CFD**	**EXP**	**CFD**	**EXP**
KTb	0.071	0.070	0.074	0.075	0.076	0.071	0.077	0.078	0.081	0.079
σ (KTb) %	11.9	12.0	7.1	8.0	4.0	4.7	5.3	5.6	8.1	10.8
10 KQb	0.087	0.099	0.091	0.096	0.092	0.112	0.093	0.108	0.097	0.101
σ (KQb) %	8.3	9.8	5.4	6.9	3.0	3.6	3.7	4.9	5.3	9.3
10 KMYb	0.014	0.008	0.012	0.001	0.012	−0.006	0.011	−0.004	0.010	0.000
σ (KMYb) %	80.8	165.2	52.1	689.6	29.2	48.4	52.4	106.2	105.9	3776.1
10 KMZb	−0.167	−0.176	−0.173	−0.198	−0.175	−0.205	−0.177	−0.206	−0.185	−0.222
σ (KMZb) %	8.6	9.5	5.8	6.4	3.4	3.5	4.1	5.3	5.8	8.4

At the positive heading angle, the blade of a right-handed pushing propeller would experience the heaviest loads at the 6 o'clock position, where it encounters the cross-flow induced by the inflow. Around the 12 o'clock position, the same cross-flow would result in a decrease in blade loads, but the separated wake from the strut produces velocities opposite to the cross-flow, which increases the blade loading. Accordingly, the two peaks are observed in the time histories of blade loads at positive heading, as seen from Figure 15.

At the negative heading angle, the heaviest blade loads due to cross-flow are observed around the 12 o'clock position, where they appear reduced by wake induced velocities whose direction is opposite to the cross-flow. One peak is observed in the time histories of blade loads, around this location. In addition to the above, there is also the influence of flow retardation in the pod wake, which results in the local increase of blade loading when it enters the wake area. However, the effect of flow retardation is comparable at positive and negative headings. Thus, it can be concluded that the major contribution to the unsteadiness of blade loads is related to the interaction of the propeller with the separated wake from the pod housing. The degree of unsteadiness of blade loads increases with the increase of heading angles, as the separation becomes more intensive. As one can see from Figure 14, at the heading angle of 60°, separation on the pod also affects the inflow on the duct.

5. Propulsor Operation in the Presence of Free Surface

Under certain conditions, such as, for example, those of vessels in ballast, propulsor may operate close to the free surface. The said conditions pose an increased risk of ventilation and, consequently, loss of thrust. The main objective of the present study is to determine critical propulsor submergence under different loading conditions, where ventilation occurs, and to quantify associated thrust losses. Further analyses are carried out to assess the impact of ventilation on the magnitude of single blade loads oscillations and to gain additional insight into the ventilation inception mechanisms. Similar to the studies in oblique flow, the operation conditions corresponding to free sailing (J = 0.6), trawling (J = 0.2) and bollard (J = 0.0) are selected for investigation. Only the numerical results are presented in this section. As mentioned above, the numerical simulations are performed without inclusion of the cavitation model to separate the pure effects of free surface and ventilation.

5.1. Implications of Numerical Simulation in Multiphase Flow

When conducting numerical simulations of the propulsor in the presence of free surface according to the VOF method, one needs, apart from the obvious considerations regarding the mesh refinement, to pay attention to the selection of time-step and simulation time. The time step is decided by the propeller rate of revolution, and it is usually sufficiently small to resolve the required flow features. However, it has been noticed that, at early runtime instances, the air is sucked from the air-water interface toward the propulsor even for deep submergences, thus resulting in non-physical ventilation events. Only after a sufficiently large number of propeller revolutions does the propeller become free of the air fraction, and a converged solution for free surface is attained. Figure 18 shows the contours of volume fraction around the propeller at different solution time instances for the submergence of H/Dp = 1, J = 0.6. Only after about 20 revolutions, no air fraction and no significant changes in the flow pattern around the free surface and propulsor are observed. Therefore, the first 20 propeller revolutions should not be included in the analysis of simulation results.

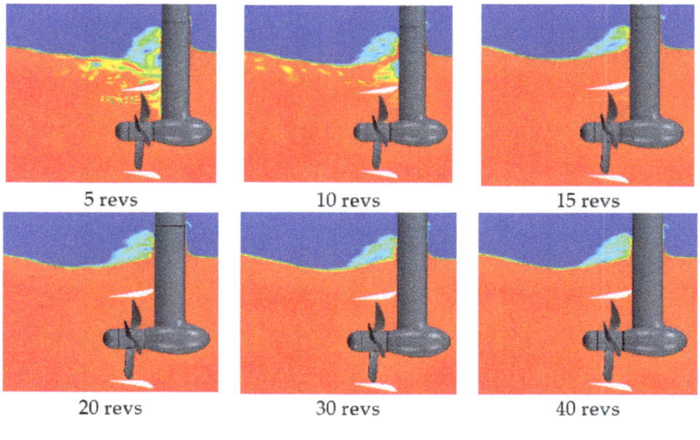

Figure 18. Contours of volume fraction at different solution time instances for propulsor submergence H/Dp = 1.0, at J = 0.6.

5.2. Propulsor Ventilation at Free Sailing Conditions

The variations in propeller thrust, propeller torque and duct thrust with the magnitude of submergence are illustrated in Figures 19 and 20, respectively. The time-averaged values of these quantities are related to KTP_0, KQP_0 and KTD_0, which refer to the corresponding values obtained from the solution without the influence of free surface at J = 0.6. Both the propeller thrust and torque remain almost unchanged until H/Dp = 1.3, and then they decrease gradually as the propulsor gets closer to the water surface. The reduction in loading becomes appreciable at H/Dp = 0.8, where both the KTP and KQP are reduced to about 3% of their values in unbounded flow. Under the above conditions, and until H/Dp = 0.7, the propulsor is still free of ventilation.

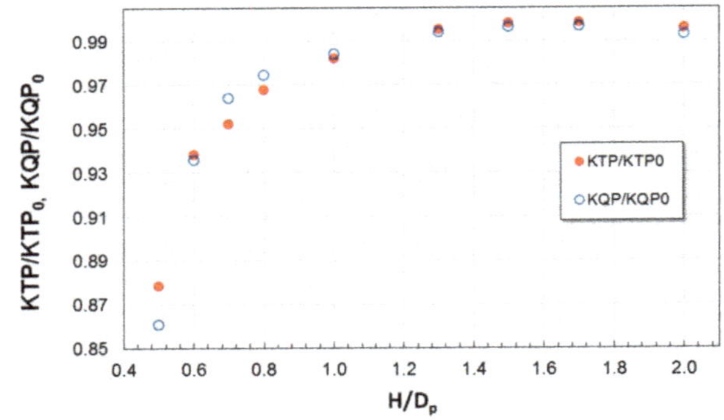

Figure 19. Variations in propeller thrust, KTP, and propeller torque, KQP, with propulsor submergence, H/Dp, at J = 0.6. (KTP_0, KQP_0 refer to the corresponding values in unbounded flow).

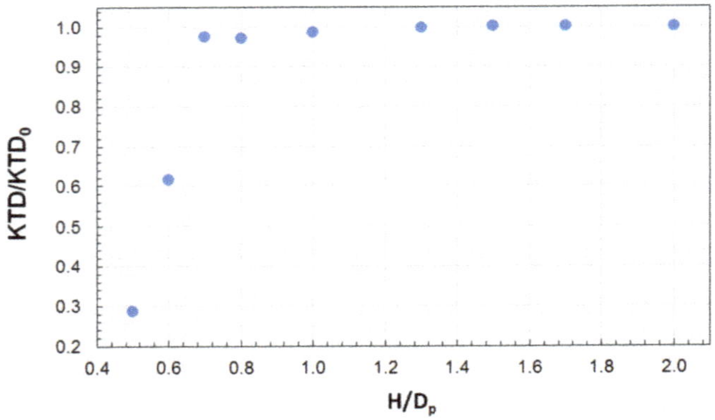

Figure 20. Variations in duct thrust, KTD, with propulsor submergence, H/Dp, at J = 0.6. (KTD_0 refers to the corresponding value in unbounded flow).

At lower submergences, a more rapid decrease in KTP and KQP is observed, and after H/Dp = 0.6 these quantities drop for 6%, while the duct thrust drops almost instantly by 36%, from the respective values in unbounded flow. This magnitude of submergence corresponds to a rapid onset of a fully ventilated condition, in the form of a pocket-like deformation of the free surface above the propulsor. At the lowest studied submergence (H/Dp = 0.5), the propeller thrust appears reduced by 12%, while the duct thrust is reduced by 70%. The process of ventilation inception with the decrease in propulsor submergence at the loading condition of J = 0.6 is illustrated in Figure 21.

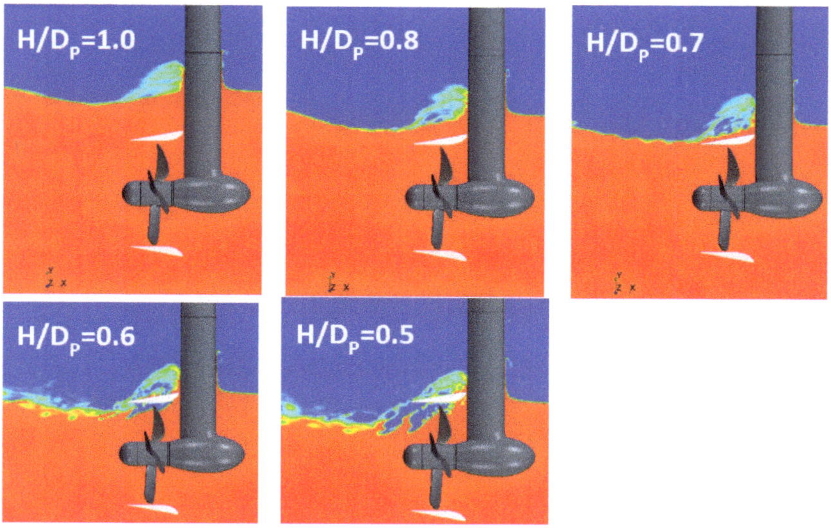

Figure 21. Ventilation inception with decrease in propulsor submergence, H/Dp, at J = 0.6.

Under the conditions of developed ventilation (H/Dp $\leq$ 0.6), the upper part of the duct is exposed to air, which causes a very large loss of duct thrust. However, under light propeller loading, there is no significant air entrainment by propeller blades on the duct interior side, so that the air fraction remains limited to the upper part of the propeller disk area. This phenomenon can be seen more clearly from the images presented in Figure 22. Thus, propeller forces experience more gradual and less significant decrease than the duct thrust.

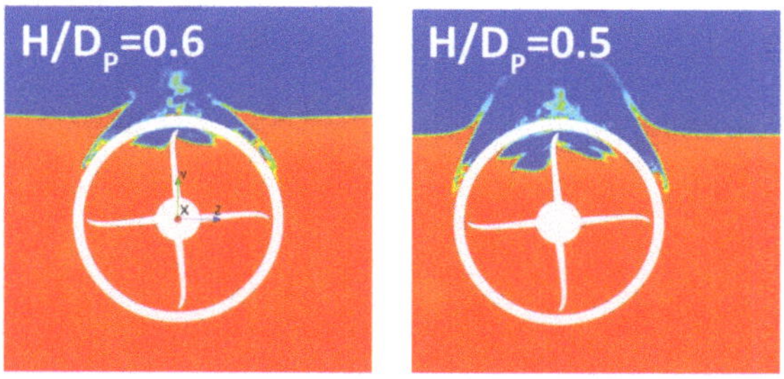

Figure 22. Contours of volume fraction under fully ventilated conditions at the submergence values of H/Dp = 0.6 and 0.5, J = 0.6.

The reduction of KTP, KQP and KTD observed with the decrease of submergence at non-ventilated conditions is attributed to the changes in propeller inflow and flow over the duct. Flow analyses indicate that the inflow on the propeller does not change until the submergence of H/Dp = 1.0, where the changes firstly occur at the inner blade sections, close to the propeller hub. They become more pronounced as submergence decreases. In addition, the vertical extent of flow retardation domain in the wake behind the strut is reduced, resulting in higher velocities on the propeller. These changes are the result of interaction between the propulsor and free surface, which causes redistribution of the pressure field, primarily on the upper side of the duct and in the area of the strut wake.

The influence of propulsor submergence on single blade loads is presented in Table 6. In non-ventilated conditions, the averaged blade loads and their standard deviations from mean values do not show considerable variation with submergence. Around the submergence magnitudes H/Dp = 0.8 ÷ 0.7, where the propulsor is close to free surface, but still free of ventilation, the load amplitudes are reduced slightly due to the damping effect of free surface. The onset of ventilation is associated with a rapid increase in load oscillations, which is noticeable in Table 6 by the large values of standard deviation. The discussed trends are also illustrated by the time histories of single blade thrust, KTb, presented in Figure 23. The maxima of blade loads occur at the time instances just before the blade enters the air domain in the upper part of the propeller disk. Once the blade enters the said domain, the blade load drops instantly, causing a large load gradient.

Table 6. Computed single blade loads at J=0.6 for different magnitudes of propulsor submergence.

Quantity	w/o FS	H/Dp = 2	H/Dp = 1.5	H/Dp = 1	H/Dp = 0.8	H/Dp = 0.7	H/Dp = 0.6	H/Dp = 0.5
KTb	0.062	0.062	0.062	0.061	0.060	0.059	0.058	0.056
σ (KTb) %	5.8	5.5	5.7	5.4	4.5	3.6	15.0	35.4
10 KQb	0.079	0.078	0.079	0.078	0.077	0.076	0.073	0.067
σ (KQb) %	4.1	4.2	4.6	4.3	3.7	3.1	14.0	34.7
10 KMYb	0.020	0.020	0.020	0.021	0.021	0.022	0.011	0.000
σ (KMYb) %	19.8	20.2	19.9	19.6	17.5	14.5	85.5	3481.4
10 KMZb	−0.143	−0.142	−0.143	−0.141	−0.139	−0.137	−0.130	−0.120
σ (KMZb) %	4.5	4.0	4.5	4.0	3.0	2.4	17.0	40.1

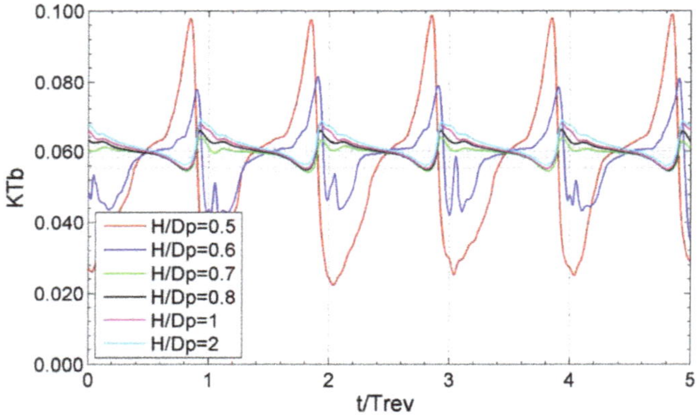

Figure 23. Computed time histories of single blade thrust, KTb, for different magnitudes of submergence, H/Dp, at J = 0.6.

5.3. Propulsor Ventilation at Trawling and Bollard Conditions

Propulsor operation near the free surface at low speed, where propeller loading is heavy, shows some distinct differences from the case of free sailing. The ventilation event occurs at deeper submergence, and the ventilation inception mechanism is different from that in free sailing. As one can see from Figure 24, under the trawling operation condition (J = 0.2), the presence of an air fraction inside the duct is observed already at the submergence of H/Dp = 0.9, while at H/Dp = 0.8 a fully ventilated condition is established. At the bollard condition (J = 0.0), ventilation inception is delayed to slightly lower magnitudes of submergence, around H/Dp = 0.8, where it takes the form very similar to the one observed for the trawling condition.

The ventilation inception mechanism at heavy loading conditions is associated with the formation of a vortex, originating on the free surface and extending to the suction side of the propeller blade, or at early instances, to the inner side of the duct, near the blade tip. Such a vortex, serving as air transport way, may firstly develop on one side of the propulsor strut, as observed in the case of J = 0.2, H/Dp = 0.9. In this case, the said vortex shows unstable behaviour, and it does not appear at all propeller revolutions. At lower magnitudes of submergence, and at heavier loading of the propeller, such as at bollard, the ventilation vortex develops on both sides of the strut, and it exists all the time as a stable structure attached to the suction side of the propeller blade. Three-dimensional images of ventilation vortices derived from the CFD simulations are presented in Figure 25, for the conditions discussed above.

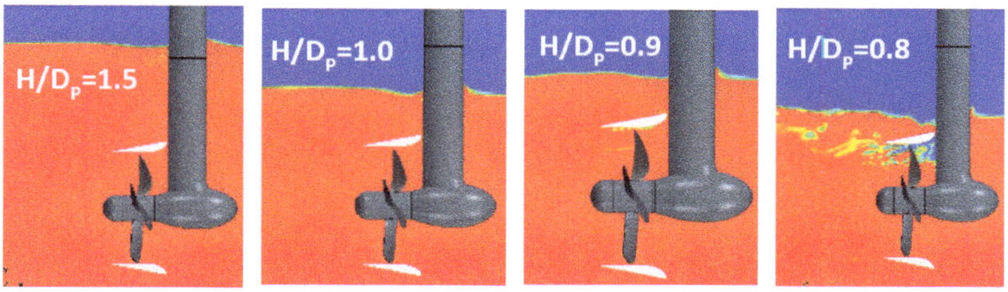

Figure 24. Ventilation inception with decrease in propulsor submergence, H/Dp, at J = 0.2.

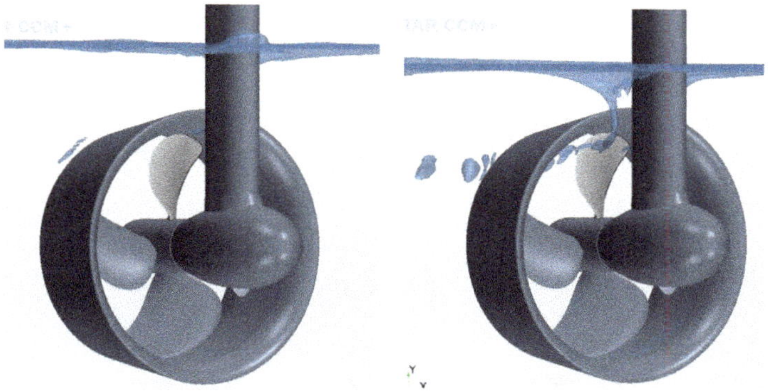

(**a**) (**b**)

Figure 25. Ventilation inception vortex at the conditions of heavy propulsor loading: (**a**) H/Dp = 0.9, at J = 0.2 (trawling); (**b**) H/Dp = 0.8, at J = 0.0 (bollard).

The described mechanism of ventilation inception in the form of an air-transport vortex has earlier been discussed in the literature addressing ventilation on heavily loaded propellers as well as on tunnel thrusters. In particular, reference [18] presents images of ventilation on the same ducted thruster as studied in the present work, showing a very similar picture of two vortices existing on both sides of the strut, as shown in Figure 25b, at an instance during a heave cycle at bollard operation (J = 0.0).

The influence of free surface and ventilation on the propeller thrust and duct thrust is illustrated in Figures 26 and 27, respectively. As above, the presented thrust values are related to KTP_0 and KTD_0, which refer to the corresponding values obtained from the solution in unbounded flow.

In non-ventilated conditions and at early stages of ventilation inception, the effect of submergence on propeller thrust and duct thrust is more pronounced at J = 0.2 than at J = 0.0, which one can see for example, from a greater reduction of KTP and KTD at H/Dp = 0.8. However, once ventilation has developed, the loss of thrust takes place more rapidly at the bollard condition, J = 0.0. More precisely, at J = 0.2, H/Dp = 0.8, KTP drops by about 4%, while KTD drops by about 34% of the respective values in unbounded flow. At J = 0.0, and the same submergence, KTP drops by about 2%, while KTD drops by about 3% of the respective values in unbounded flow. At the submergence H/Dp = 0.7 (fully ventilated condition at both loadings), KTP is reduced by17% at J = 0.2, and by20% at J = 0.0, and KTD is reduced by 55% at both advance coefficients. The trends in propeller torque are similar to those of propeller thrust.

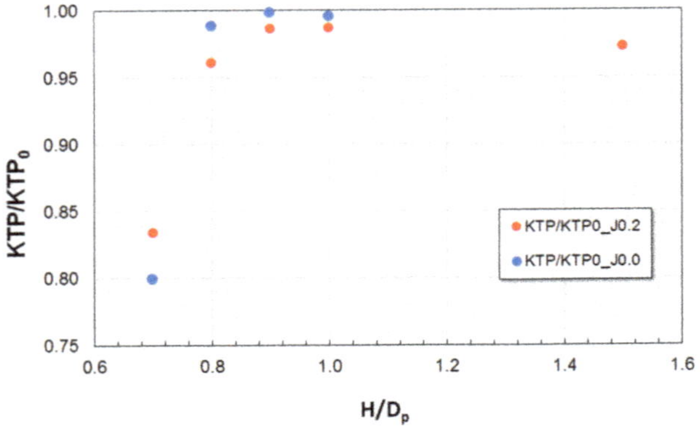

Figure 26. Variations in propeller thrust, KTP, with propulsor submergence, H/Dp, at J = 0.2 and J = 0.0. (KTP_0 refers to the corresponding value in unbounded flow).

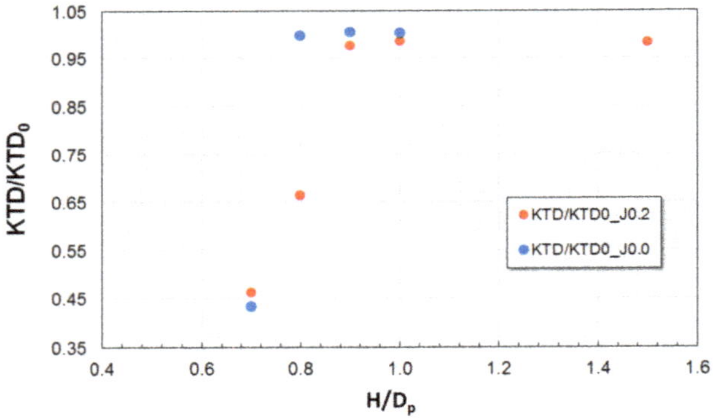

Figure 27. Variations in propeller thrust, KTD, with propulsor submergence, H/Dp, at J = 0.2 and J = 0.0. (KTD_0 refers to the corresponding value in unbounded flow).

Unlike the case of free sailing, at low speed operation during the ventilation event there observed a significant air entrainment by propeller blades along the inner side of the duct, toward the bottom part of propeller disk, as illustrated in Figure 28. A greater percentage reduction of propeller forces registered at low speed operation condition is, to a large degree, related to the aforementioned air entrainment phenomenon.

The influence of propulsor submergence on single blade loads is presented in Table 7, for the trawling condition, J = 0.2. Similar to the observations made in the case of free sailing, in non-ventilated conditions the influence of free surface on blade loads is very small. From H/Dp = 0.8, where ventilation inception begins in the form of a vortex, one can notice an increase in load amplitudes (higher values of standard deviation), and a reduction in the averaged values. In fully ventilated conditions (H/Dp = 0.7), the load amplitudes increase further, while the averaged loads continue decreasing, except for the blade spindle moment, KMYb. The noticed trend in KMYb, which is different from the trends of other blade loads, and also from the trends in KMYb in free sailing, is explained by the effect of air entrainment by propeller blades, which results in redistribution of pressure on the blade. Spindle moment is the quantity that is very sensitive to changes in pressure distribution, especially in the vicinity of blade edges.

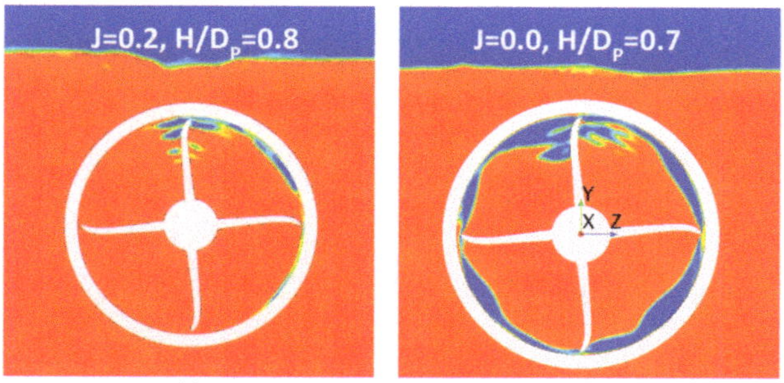

(a) (b)

Figure 28. Contours of volume fraction under fully ventilated conditions at low speed operation conditions: (**a**) J = 0.2, H/Dp = 0.8; (**b**) J = 0.0, H/Dp = 0.7.

Table 7. Computed single blade loads at J = 0.2 for different magnitudes of propulsor submergence.

Quantity	w/o FS	H/Dp = 1.5	H/Dp = 1	H/Dp = 0.9	H/Dp = 0.8	H/Dp = 0.7
KTb	0.079	0.077	0.078	0.078	0.077	0.069
σ (KTb) %	3.6	3.3	3.6	3.7	11.3	26.7
10KQb	0.096	0.093	0.095	0.094	0.090	0.077
σ (KQb) %	2.8	2.7	3.0	3.1	11.2	28.3
10KMYb	0.011	0.012	0.012	0.012	-0.005	0.025
σ (KMYb) %	25.7	24.9	26.8	30.2	203.2	52.2
10KMZb	−0.183	−0.179	−0.181	−0.181	−0.170	−0.143
σ (KMZb) %	3.3	2.6	3.0	3.0	13.9	33.9

The computed time histories of single blade thrust, KTb, are presented for different magnitudes of submergence at the trawling condition (J = 0.2) in Figure 29. The occurrence of high-frequency oscillations of KTb observed at H/Dp = 0.8 is associated with ventilation inception in the form of an unstable free surface vortex. At H/Dp = 0.7, where ventilation is fully developed, the oscillations of KTb increase significantly, and the periodic pattern of time history is destroyed.

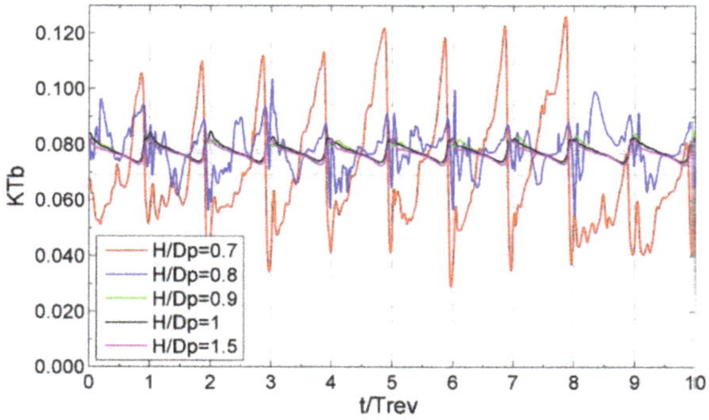

Figure 29. Computed time histories of single blade thrust, KTb, for different magnitudes of submergence, H/Dp, at J = 0.2.

6. Conclusions

A detailed verification and validation study is prerequisite for numerical simulation of complex flow phenomena associated with dynamic loads on azimuth propulsors. The results of the present work show that a good agreement can be achieved between the CFD calculations by an unsteady RANS method and experimental data for a ducted pushing thruster operating in straight flow and oblique flow, in the range of relevant loading conditions in the first quadrant. The above conclusion stands for both the integral forces and moments on the unit and individual blade loads, including their amplitudes and time histories, with the exception of steering moment, where larger differences are found. The steering moment, as a quantity of smaller order of magnitude compared to other loads on the thruster, is very sensitive to small differences in flow pattern, resulting in changes in the locations of the force centers, and hence force arms.

The forces and moments on the ducted azimuth thruster reveal asymmetry with respect to positive and negative heading angles, which is typical for pushing pod units. The said asymmetry is associated with the interaction between the rotating propeller and cross-flow in separated wake behind the strut. Due to the flow straightening effect of the duct, the said asymmetry of loads acting on the ducted unit is found to be smaller compared to the case of an open pushing thruster. For a unit with a right-handed propeller, the loads are smaller at negative heading angles, where the oblique flow on the propulsor is from starboard. At the same time, both the calculation results and experimental measurements show that, at negative headings, the amplitudes of single blade loads are larger than at positive headings, except the blade spindle moment, which shows an opposite trend.

The mentioned flow straightening effect of the duct results in relatively small side forces produced by the propeller, especially at small heading angles, where the main contributions to the side force and steering moment are given by the duct.

At zero heading angle, the propeller develops a small side force acting in the direction of portside (for a right-handed propeller) due to its interaction with the wake of the pod housing. This small side force causes a deviation in the total reaction force and results in a non-zero steering moment acting on the unit. The steering moment is larger at lower J values, where propeller loading is heavier. This phenomenon has a direct impact on the course-keeping characteristics of vessels equipped with pushing pod propulsors.

The frequency analysis of forces and moments acting on the propulsor at large heading angles reveals, in addition to the blade frequency, the presence of a lower frequency peak, which is related to the vortex shedding process that accompanies flow separation on the leeward side of the duct.

The presence of free surface causes important changes in the performance characteristics of the ducted azimuth thruster. In free sailing, the influence of free surface becomes appreciable at the relative submergence, H/Dp, around 0.8, where both the propeller thrust and duct thrust appear reduced by about 3% from their respective values in unbounded flow, without the occurrence of ventilation. The onset of fully ventilated conditions at $H/Dp = 0.7$ is associated with a 6% loss of propeller thrust and 36% loss of duct thrust. Thrust losses continue increasing as submergence decreases. Under the conditions of light propeller loading in free sailing, propulsor ventilation develops as a sudden pocket-like deformation of the free surface above the propulsor. There is no air entrainment by propeller blades, so that the air fraction remains limited to the upper part of the propeller disk.

The ventilation event begins at deeper submergence under the conditions of heavy loadings, such as those of trawling and bollard operation. In the present study, ventilation inception is registered at $H/Dp = 0.9$ for the trawling condition, and at $H/Dp = 0.8$ for the bollard condition. In fully ventilated conditions, the propeller thrust loss amounts to 20%, and the duct thrust loss amounts to 55% from their respective values in unbounded flow. In the case of heavy propulsor loading, the mechanism of ventilation inception is associated with the formation of a vortex extending from the free surface to the suction side of the propeller blade, or to the inner side of the duct near the blade tip. The occurrence of the ventilation event is accompanied by air entrainment by the propeller blades on the duct interior surface and a very large increase in unsteady blade loads.

Acknowledgments: The present work has been carried out within the frameworks of the trans-national R&D Project "INTER-THRUST" supported by the MARTEC II ERA-NET program (Maritime Technologies). The "INTER-THRUST" consortium members are: SINTEF Ocean (Norway), Hamburg University of Technology (Germany), Voith Turbo Schneider Propulsion GmbH & Co. KG (Germany), Jastram GmbH und Co. KG (Germany), and Havyard Group AS (Norway). The authors acknowledge the use of the commercial software STAR-CCM+ by Siemens in all CFD simulations performed in this study.

Author Contributions: Vladimir Krasilnikov conceived and developed the numerical setup used in the CFD simulations with the propulsor in open water conditions. Nabila Berchiche extended the aforementioned setup to the case of propulsor operating in presence of free surface and conducted all numerical simulations and post-processing. Kourosh Koushan contributed with the experimental data and description of the test setup. Vladimir Krasilnikov and Nabila Berchiche analyzed the data and performed comparisons. All the authors contributed to the discussion about the results. Vladimir Krasilnikov wrote the paper.

Conflicts of Interest: The authors declare no conflict of interest.

References

1. Funeno, I. Hydrodynamic Optimal Design of Ducted Azimuth Thrusters. In Proceedings of the First International Symposium on Marine Propulsors SMP'09, Trondheim, Norway, 22–24 June 2009.
2. Palm, M.; Jürgens, D.; Bendl, D. Numerical and Experimental Study on Ventilation for Azimuth Thrusters and Cycloidal Propellers. In Proceedings of the Second International Symposium on Marine Propulsors SMP'11, Hamburg, Germany, 17–18 June 2011.
3. Koushan, K.; Krasilnikov, V.I. Experimental and Numerical Investigation of Open Thrusters in Oblique Flow Conditions. In Proceedings of the 27th Symposium on Naval Hydrodynamics, Seoul, Korea, 5–10 October 2008.
4. Achkinadze, A.S.; Berg, A.; Krasilnikov, V.I.; Stepanov, I.E. Numerical Analysis of Podded and Steering Systems Using a Velocity Based Source Boundary Element Method with Modified Trailing Edge. In Proceedings of the Propellers/Shafting'2003 Symposium, Virginia Beach, VA, USA, 17–18 September 2003.
5. Heinke, H.-J. Investigations about the Forces and Moments at Podded Drives. In Proceedings of the T-POD04—1st International Conference on Technological Advances in Podded Propulsion, Newcastle upon Tyne, UK, 14–16 April 2004.
6. Grygorovicz, M.; Szantyr, J.A. Open Water Experiments with Two Pod Propulsor Models. In Proceedings of the T-POD04—1st International Conference on Technological Advances in Podded Propulsion, Newcastle upon Tyne, UK, 14–16 April 2004.
7. Bhattacharyya, A.; Krasilnikov, V.I.; Steen, S. Scale Effect on Open Water Characteristics of a Controllable Pitch Propeller Working within Different Duct Designs. *Ocean Eng.* **2016**, *112*, 226–242. [CrossRef]
8. Krasilnikov, V.I.; Sun, J.; Zhang, Z.; Hong, F. Mesh Generation Technique for the Analysis of Ducted Propellers Using a Commercial RANSE Solver and Its Application to Scale Effect Study. In Proceedings of the 10th Numerical Towing Tank Symposium NuTTS'07, Hamburg, Germany, 23–25 October 2007.
9. Krasilnikov, V.I.; Zhang, Z.; Hong, F. Analysis of Unsteady Propeller Blade Forces by RANS. In Proceedings of the First International Symposium on Marine Propulsors SMP'09, Trondheim, Norway, 22–24 June 2009.
10. Sanchez-Caja, A.; Ory, E. Simulation of Incompressible Viscous Flow around a Tractor Thruster in Model and Full Scale. In Proceedings of the 8th International Conference on Numerical Ship Hydrodynamics, Busan, Korea, 5–8 August 2003.
11. Junglewitz, A.; El Moctar, O.M. Numerical Analysis of the Steering Capabilities of a Podded Drive. *Ship Technol. Res.* **2004**, *51*, 134–145.
12. Abdel-Maksoud, M.; Heinke, H.-J. Scale Effects on Ducted Propellers. In Proceedings of the 24th Symposium on Naval Hydrodynamics, Fukuoka, Japan, 8–13 July 2002.
13. Bhattacharyya, A.; Neitzel, J.C.; Steen, S.; Abdel-Maksoud, M.; Krasilnikov, V.I. Influence of Flow Transition on Open and Ducted Propeller Characteristics. In Proceedings of the Forth International Symposium on Marine Propulsors SMP'15, Austin, TX, USA, 31 May–4 June 2015.
14. Baltazar, J.M.; Rijpkema, D.; Falcão de Campos, J. On the Use of the γ-Reθ Transition Model for the Prediction of the Propeller Performance at Model-Scale. In Proceedings of the Fifth International Symposium on Marine Propulsors SMP'17, Espoo, Finland, 12–15 June 2017.
15. Schiller, P.; Wang, K.; Abdel-Maksoud, M.; Palm, M. Unsteady Loads on an Azimuth Thruster in Off-Design Conditions. In Proceedings of the Fifth International Symposium on Marine Propulsors SMP'17, Espoo, Finland, 12–15 June 2017.

16. Islam, M. Modeling Techniques of Puller Propulsor in Extreme Conditions. *J. Ship Res.* **2017**, *61*, 230–255. [CrossRef]
17. Bhattacharyya, A.; Krasilnikov, V.I.; Steen, S. Scale Effect on a 4-Bladed Propeller Operating in Ducts of Different Design in Open Water. In Proceedings of the Forth International Symposium on Marine Propulsors SMP'15, Austin, TX, USA, 12–15 June 2015.
18. Koushan, K. Dynamics of propeller blade and duct loadings on ventilated ducted thrusters operating at zero speed. In Proceedings of the T-POD06—2nd International Conference on Technological Advances in Podded Propulsion, Brest, France, 3–5 October 2006.
19. Koushan, K. Dynamics of ventilated propeller blade loading on thrusters due to forced sinusoidal heave motion. In Proceedings of the 26th Symposium on Naval Hydrodynamics, Rome, Italy, 17–22 September 2006.
20. Krasilnikov, V.I.; Sileo, L.; Steinsvik, K. Numerical Investigation into Scale Effect on the Performance Characteristics of Twin-Screw Offshore Vessels. In Proceedings of the Fifth International Symposium on Marine Propulsors SMP'17, Espoo, Finland, 12–15 June 2017.
21. Krasilnikov, V.I.; Sileo, L.; Joung, T.-H. Investigation into the Influence of Reynolds Number on Open Water Chracteristics of Pod Propulsors. In Proceedings of the Forth International Symposium on Marine Propulsors SMP'15, Austin, TX, USA, 31 May–4 June 2015.

Journal of
Marine Science and Engineering

MDPI

Article

Experimental and Numerical Investigation of Propeller Loads in Off-Design Conditions

Fabrizio Ortolani [1,*], Giulio Dubbioso [1], Roberto Muscari [1], Salvatore Mauro [1] and Andrea Di Mascio [2]

1 CNR-INSEAN, National Research Council—Marine Technology Research Institute, Via di Vallerano 139, 00128 Rome, Italy; giulioantonino.dubbioso@cnr.it (G.D.); roberto.muscari@cnr.it (R.M.); salvatore.mauro@cnr.it (S.M.)

2 CNR-IAC—Istituto per le Applicazioni del Calcolo, Via dei Taurini 19, 00185 Rome, Italy; andrea.dimascio@cnr.it

* Correspondence: fabrizio.ortolani@cnr.it; Tel.: +39-6-5029-9215

Received: 27 February 2018; Accepted: 4 April 2018; Published: 24 April 2018

Abstract: The understanding of the performance of a propeller in realistic operative conditions is nowadays a key issue for improving design techniques, guaranteeing safety and continuity of operation at sea, and reducing maintenance costs. In this paper, a summary of the recent research carried out at CNR-INSEAN devoted to the analysis of propeller loads in realistic operative scenarios, with particular emphasis on the in-plane loads, is presented. In particular, the experimental results carried out on a free running maneuvering model equipped with a novel force transducer are discussed and supported by *CFD* (Computational Fluid Dynamics) analysis and the use of a simplified propeller model, based on Blade Element Momentum Theory, with the aim of achieving a deeper understanding of the mechanisms that govern the functioning of the propeller in off-design. Moreover, the analysis includes the scaling factors that can be used to obtain a prediction from model measurements, the propeller radial force being the primary cause of failures of the shaft bearings. In particular, the analysis highlighted that cavitation at full scale can cause the increment of in-plane loads by about 20% with respect to a non-cavitating case, that that in-plane loads could be more sensitive to cavitation than thrust and torque, and that Reynolds number effect is negligible. For the analysis of cavitation, an alternative version of the *BEMT* solver, improved with cavitation linear theory, was developed.

Keywords: off-design; propeller radial force; free running experiments; CFD maneuvering simulations; propeller models; scale effects; cavitation

1. Introduction

The performance of marine propellers during realistic and off-design conditions nowadays represents a topic worth of investigation for the perspective of both scientific research and marine technology development. Traditionally, the propeller performance was analyzed considering the wake of the ship advancing at the desired speed in calm water. However, a seagoing ship is often subjected to environmental perturbations or desired control maneuvering actions that depart it from the ideal straight ahead motion. In these circumstances, the inflow to the propeller is altered with respect to the one adopted for the blade shape optimization and, consequently, the developed loads may change considerably. During motion in waves or maneuvering, the cross flow induced by the motion and the consequent modification of the viscous wake of the hull alter the blade hydrodynamics and, consequently, the resultant propeller loads. In particular, in addition to the modification of thrust and torque, undesired in-plane forces and moments that further stress the propulsion system (shaftline and bearings) and hull structure are generated.

An extensive investigation carried out by different Classification Societies (i.e., DNV, ABS, Germanisher Lloyd) stressed that the failure of the tail shaft bearing resulted one of the most frequent accident of the propulsion system for on–going vessels. In fact, due to the the higher power installed to achieve higher speed in modern designs, very large radial loads are developed by the propeller during maneuvers [1,2]. Moreover, this aspect is central for the safe mechanical sizing of orientable propulsive devices; these novel propulsive devices replace the rudder to control and maneuver the ship and, hence, might operate at relatively high drift angles. Moreover, free running CFD maneuvering simulations highlighted that the in-plane forces acting on the propeller are not negligible during large amplitude motions and, therefore, their quantification is fundamental to correctly predict the response of the vessel during maneuvering [3–5] or in motion in waves.

Recently, propeller design procedures oriented to account for realistic ship operative scenario were explored by the NATO STO Task Group AVT-204, "Assess the Ability to Optimize Hull Forms of Sea Vehicles for the Best Performance in a Sea Environment" [6]. A marine propeller design optimization procedure was proposed accounting for hull wakes calculated by CFD at different relative position of the hull with respect to the wave profile. A systematic analysis on the effect of waves on propeller performance, cavitation and pressure pulses was addressed in [7–9]. In these works, it became apparent that changes of the wake distribution on the propeller disk play a crucial role in the performance of the propulsion device and the related problems of fluid structure interactions with the hull.

Although an increased level of attention is paid to the understanding and quantification of the propeller performance in off-design condition, studies (both numerical and experimental) are limited and mainly focused on idealized conditions for the isolated propeller [10–18]. In fact typical size of ship models makes it challenging the installation of dedicated devices to measure all propeller loads in behind hull conditions; from the computational perspective, although CFD techniques were successfully applied to many topics of naval architecture, including off-design scenario such as ship maneuvering, the direct simulation of rotating propeller in behind hull is beyond the computational availability of day-by-day design applications.

Numerical simulations employing the discretization of the rotating propeller for self–propulsion or ship maneuvering studies are limited in the open literature [19–23]. Moreover, even if at disposal, the detailed description of the propeller behavior is not provided and the results on hub and blade loads are not validated with experiments. In turn, the availability of experimental data or validated, high accurate data can be useful to improve low order propeller models usually coupled with CFD solvers for ship motion predictions, this being an established approach to speed-up the simulations.

A series of free running model tests were carried out at CNR–INSEAN to investigate systematically these issues. In particular, twin screw configuration was the subject of the investigation, because the inflow conditions for the propellers cause their off-design operation more critical with respect to single screw ships. In fact, in twin screw propulsive configurations, the propellers are located at larger distance from wall of the hull and, during maneuvering, the "shadow" effect of hull is weaker; for this reason, the external propeller experiences an inflow inclined by an equivalent angle given by the kinematic of the model (during tight maneuver this angle might be of the order of 30°). On the contrary, the internal propeller has the higher probability to be impinged by coherent structures detached from the fore portion of the hull or appendages that are convected toward the leeward side. Moreover, the quantification of the asymmetric behavior of the propellers is of utmost importance in case of cross-connect configurations, because the reduction gear can experience very large fatigue loads. In particular, in [24] the analysis was entirely focused to monitor the overloading and unbalancing phenomena of propeller thrust and torque during maneuvering conditions for a modern twin screw ship. The same experimental set-up and the model was also considered for a campaign dedicated to the quantification of the in-plane loads developed by the propeller during steady and unsteady maneuvers [25,26]; for this purpose, a novel 2–component transducer was developed. The relation between the motion and propeller loads were preliminary assessed by simple momentum theory. The analysis was then broadened by CFD computation and the use of simplified propeller modeling in order to clarify the effects of the wake evolution on propeller loads [27–29]. In the present work, a synthesis of the experimental and numerical activity of these works is proposed to recap the key aspects that drive the propeller performance in off-design functioning. In particular, the asymmetric behavior of the propellers of the twin screw configuration is related to the wake morphology obtained by CFD computations analyzed by blade element propeller solver. Since the problem of damages of the bearing loads is a critical aspect for the reduction of maintenance costs of a vessel, the problem to full scale extrapolation of model test data was tackled. For this purpose, the blade element theory was modified in order to analyze the effects of Reynolds number and cavitation number on blade loads, with particular emphasis on the in-plane forces and moments. Finally, the future research activities inspired by the present work on the off–design propeller performance are briefly introduced.

2. Test Case

In this section, the experimental and numerical set-up are described.

2.1. Experimental Set-Up

The experimental activities were carried out at the outdoor maneuvering basin of the CNR–INSEAN located at the Nemi lake. This location is ideal due to frequent long-term dead-calm water conditions and environmentally protected area. The water surface is large enough to allow the execution on any kind of manoeuvring test regardless the model size and speed.

The ship selected for present analysis is a fast twin screw/twin rudder ship, similar to those analyzed in previous studies [24]. The model was equipped with bilge keels, propeller shafts with brackets, a centerline skeg and two all–movable rudders; the propellers are inward rotating from the top. In Table 1, principal geometric characteristics of the model and the propeller are reported and an overview of the hull geometry and stern appendages configuration is shown in Figure 1a,b.

Table 1. Geometric details of the model.

HULL	
L/B	7.5
B/T	3.25
C_B	0.5
PROPELLER	
N. blades, Z	5
P/D, Pitch to diam. ratio	1.35
Expanded area ratio	0.79
Hub ratio	0.25

The layout of a typical set-up and the measurements devices employed in free manevering model tests is sketched in Figure 2a. The model is equipped with IMU for the reconstruction of the *6DoF* motion, DGPS (differential GPS) and real time data transmission devices. Each propeller shaft is driven by a dedicated electric brush–less motor and is equipped with a load cell for the measurement of thrust and torque. The overall energy demand of the on–board instruments was provided by a diesel electric generator.

The bearing radial force components were measured on the starboard shaft by an in-house developed, pass-through hole 2–component transducer, positioned between the propeller and the tail of the bossing of the astern V-brackets (see Figure 2b). The installation of the transducer at the tail shaft and the shaftline configuration are visualized in Figure 3a. The transducer was made waterproof, because the layout of the propulsive shaft and the limited useful space inside the shaft bossing or stern tube, constrain the transducer to be submerged in water (see Figure 3b).

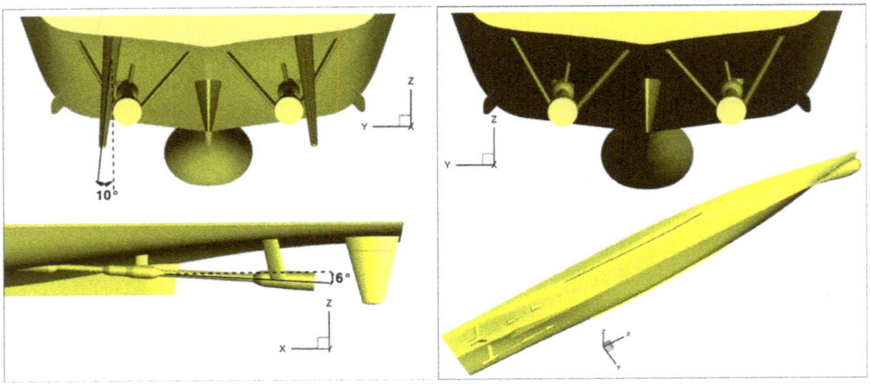

(**a**) (**b**)

Figure 1. (**a**) hull and appendage geometry (**b**) geometry considered in the computation.

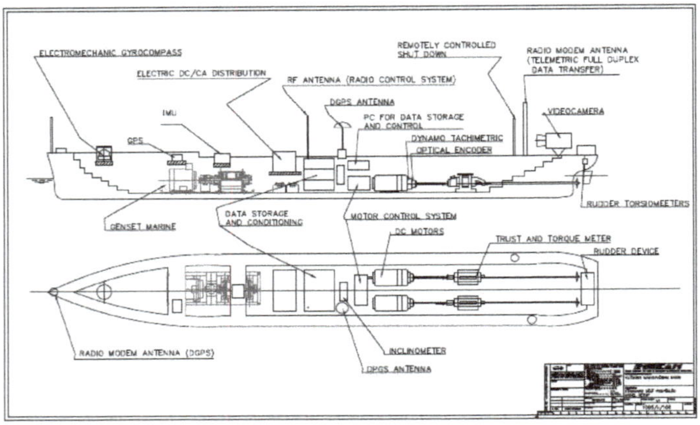

(**a**)

(**b**)

Figure 2. (**a**) sketch of the instrumented model. (**b**) radial force transducer.

The transducer is composed by two systems of opposite cantilever beams that are the sensing element of the device. Each couples of beams are connected each other by a pierced rigid beam through which the propeller shaft passes. With this configuration the deflection of the shaft caused by the forces generated by the propulsor is transmitted to the sensing beams that are equipped with strain gauge sensors. The rigid beams are properly designed in order to obtain a decoupled measure of the orthogonal components of the radial force. This configuration allows to directly obtain the radial load with respect to a fixed frame and its set-up does not require any temporal synchronization of the measure with the angular position of the shaft, as it is needed in case of transducer rotating with the shaft [10].

(**a**) (**b**)

Figure 3. (**a**) installation at the tail shaft (**b**) final configuration.

In Table 2 the test matrix of the experimental activity is listed; various kind of maneuvers were carried out in order to investigate as broader as possible realistic operative scenario. In addition to self–propulsion tests, the maneuvers considered both quasi–steady (turning circles) and unsteady motions (zig–zag trajectories) at different rudder angles. All tests were carried out at constant propeller rate of revolution and in a calm environment, in order to avoid external disturbances. The present work does not consider the unsteady cases, the interested reader is referred to [26].

Table 2. Test matrix of the experiments.

Free Running Tests		
Test	**Speed**	**Rudder Angle [deg]**
self–propulsion	$0.05 < F_N < 0.45$	$\delta = 0°$
turning circle (with pull–out)	$F_N = 0.218, 0.31$	$\delta = \pm 15°, \pm 25°, \pm 35°$
turning circle (with pull–out)	$F_N = 0.35$	$\delta = \pm 35°$
zig–zag	$F_N = 0.218, 0.31, 0.35$	$\delta = \pm 10°, \pm 20°, \pm 35°$

2.2. Numerical Set-Up

The numerical computations were performed on the geometry with rudder and propeller removed (see Figure 1b), since the task was the evaluation of the nominal wake. In particular, the computations focused both on quasi steady and transient periods of a turning maneuver; the core of the present review is mainly focused on the steady turning maneuvers, because the phenomenology is driven by similar propeller wake interactions also during unsteady motions [28].

The domain has been discretized by 186 body–fitted patched and overlapped blocks, for a total of about 10M cells. Grid distribution is such that the thickness of the first cell on the wall is always below one in wall unit, and at least 30 cells are within the boundary layer thickness ($y^+ = O(1)$, with $\Delta / L_{pp} = O(20/Re)$, Δ being the thickness of the cell adjacent to the wall). A four level multi–grid technique is exploited in order to speed–up convergence, each level being obtained from the next finer one by removing every other point along each spatial direction. Details of the cells distribution are listed in Table 3.

Table 3. Details of grid cells distribution.

Domain	Cells	Percent
HULL	2.81 M	27.52%
BILGE KEELS	1.42 M	13.90%
PROP. SHAFT	1.50 M	14.69%
FORWARD BRACKETS	0.33 M	3.20%
ASTERN BRACKETS	1.35 M	13.20%
SKEG	0.80 M	16.70%
STERN REFINEMENT	1.54 M	7.80%
BACKGROUND BUFFER	0.36 M	3.50%
BACKGROUND	0.11 M	0.10%
TOTAL	10.21 M	

Mesh topology and particular of the overlapping grids in the stern region of the model is sketched in Figure 4a,b, respectively.

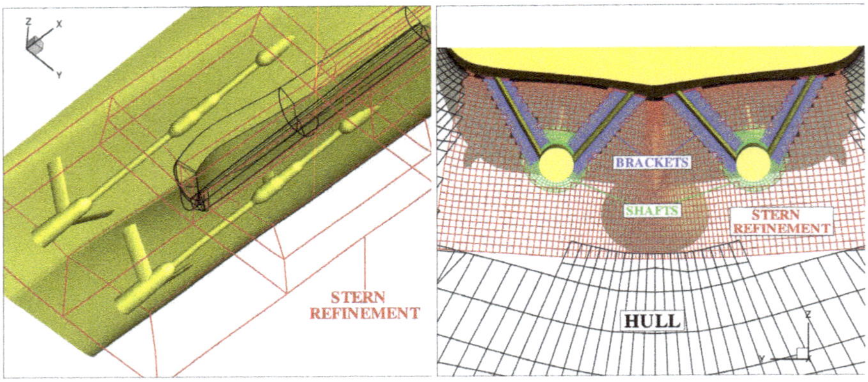

(**a**) (**b**)

Figure 4. (**a**) Topology. (**b**) Mesh.

The numerical simulations corresponded to straight ahead and steady turning maneuvers with rudder deflection angles $\delta = 15^\circ, 25^\circ$ and 35° at the lowest velocity considered in the experiments, i.e., $Fr = 0.26$, for which $Re = 1.60 \cdot 10^7$; without loss of generality of the phenomenological analysis, the numerical simulations were carried out at the lowest speed. This had the advantage to reduce the computational burden of the simulations, neglecting the modeling of the free surface. In fact, the analysis of the experimental data highlighted that the variation of the in-plane loads was not altered with the speed; moreover, at this speed the wave pattern of the model was weak and the transom stern was completely wet and, therefore, ventilation phenomena that could have affected the propeller loads were avoided.

The maneuvering simulations were carried out prescribing the angular velocity, drift angle and turning radius monitored during the experiments. These parameters are listed in Table 4 for each rudder angle considered.

Table 4. Turning circle parameters.

Free Running Tests		
Rudder	**Yaw Rate** $\cdot \frac{L_{PP}}{V_{ref}}$	**Drift Angle [deg]**
$\delta = 15°$	0.236	7.5
$\delta = 25°$	0.333	12.5
$\delta = 35°$	0.363	13.5

3. Models and Methodologies

The CFD code solves the Navier-Stokes Equations for unsteady high Reynolds number (turbulent) free surface–flows around complex geometries. The interested reader is referred to [17,19,30–34] for detailed description of the features of the solver.

The basic concepts of the propeller model are briefly described, because are useful to follow the discussion of the results and the correlation of the propeller loads with the wake features. The propeller solver is based on the blade element method. In the present work, the basic solver described in [27,35] was modified to account for scale effects; in this paragraph the basic algorithm is briefly reviewed (the interested reader can refer to [27,35]) and the modified one is discussed in Section 6.

The blade element method synthesizes the blade performance by the contribution provided by its cross sections, each one formally treated independent from the others by the $2D$ airfoil representation. Formally, the propeller loads are obtained by summation of the contribution provided by each blade section once the "polar" characteristics, i.e., the coefficients of lift (dL) and drag (dD), and the inflow (velocity magnitude and incidence angle α) are known. Referring to a representative section of the propeller blade (see Figure 5a), the inflow and the elemental forces are described in Figure 5b. According to the $2D$ representation, we neglect the radial components. Without loss of generality, the representation neglects the effects of the self–induced velocity. In particular:

$$\alpha = \Theta - \gamma \qquad \gamma = tan^{-1}\left[\frac{u}{\omega r + v_\theta}\right] \tag{1}$$

where Θ is the geometric pitch of the blade and γ is the inflow angle of the section, given by the velocity system in the plane of the section (v_θ is the tangential velocity given by the combination of the components of the wake in the plane of the propeller in the azimuthal direction, see Figure 5a). In addition, therefore, the loads are given by:

$$dL, dD \propto C_{L,D}(\alpha)V^2 \tag{2}$$

where $C_{L,D}$ is the lift and drag coefficients. Once lift and drag forces generated by the foil are determined, the elemental contribution to thrust dF_x and tangential force dF_t can be derived:

$$dF_x = dLcos\gamma - dDsin\gamma \qquad dF_t = dLsin\gamma + dDcos\gamma \tag{3}$$

The step from a pure 2D representation to a three dimensional one that tries to model the key physic of the blade (self–induced velocity due to trailing vortices and blade–to–blade interaction) can be achieved by solving the non linear Prandtl–Betz relation for every section:

$$\frac{Zc}{16r}\frac{\partial C_L}{\partial \alpha}(\Theta - \phi - \alpha_i) = acos\left[exp\left(-\frac{k - kr}{2sin\alpha_i}\right)\right] tan\alpha_i sin(\phi + \alpha_i) \tag{4}$$

where Z is the number of blades, r and c are the spanwise position and chord of the foil, respectively, Θ is the picth angle of the section and ϕ is inflow angle of the nominal flow (see Figure 5b); the equation is non-linear and it is solved iteratively in terms of the self induction angle α_i. The flow chart reported on the left half of Figure 6, explains the pseudo-code of the non-cavitating propeller solver: given the nominal incidence and the geometric details of the section, the Prandtl-Betz equation is solved to determine the self-induced velocity field. The strict coupling between the lift (circulation) generated by the section and the induced velocity field give rise to the non linear nature of the Prandtl-Betz equation.

It has to be stressed that the *BEMT* model is inherently quasi steady, because the Prandtl–Betz condition implicitly assumes the balance of the load developed by the blade strip and the vorticity (constant) carried by trailing vortex. However, as long as the inflow is time–varying, this equilibrium condition is not suddenly met and its lag depends on the time rate of circulation detached at the trailing edge of the airfoil, because it is responsible of an additional perturbation to the blade section [36].

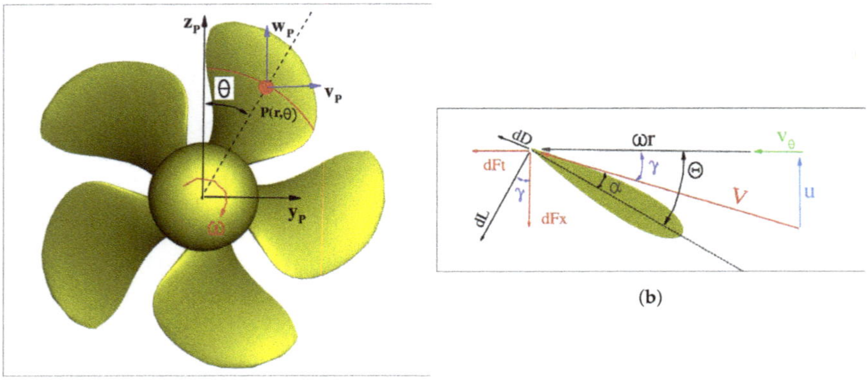

(a) (b)

Figure 5. (**a**) hub and blade reference system. (**b**) Inflow and forces generated on a generic blade section.

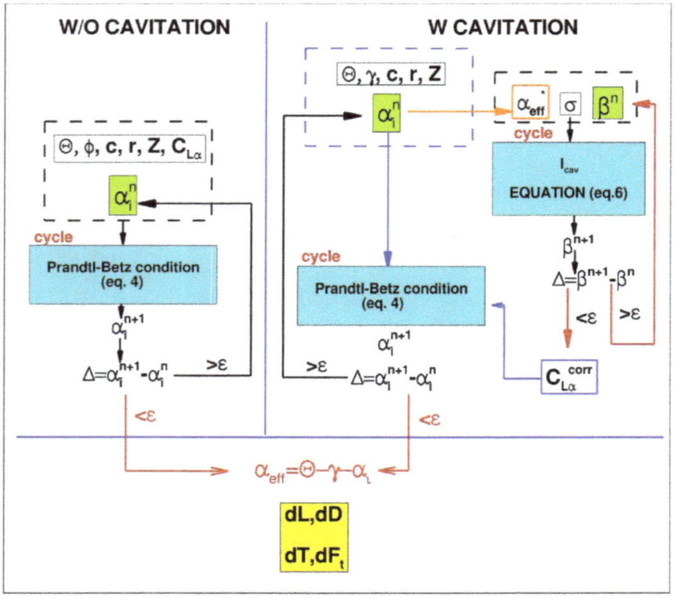

Figure 6. Solution algorithm for the blade element solvers.

4. Results

In this section, the experimental results are synthesized with the complementary CFD results and wake analysis by the BEMT solver. The discussion is carried out qualitatively to focus on the key mechanisms that govern the off-design performance of the propeller and a comparison of the numerical and experimental work is not reported. BEMT is reliable to capture the trends and the order of magnitude of the loads, although absolute errors were rather high. These discrepancies were ascribed, aside from the approximation of the propeller solver, to the off-line analysis of the nominal wake. By this approach, the modification of the wake due to the self–induction effect of the propeller is obviously neglected and, consequently, the prediction of the loads might be affected. In particular, on the internal side, this interaction can modify the strength and distribution of the swirling flow and, on the external shaft, the flow around the appendages and the shaft. For this reasons, weakly-coupled $uRANSE - BEMT$ simulations in transverse plane should be carried out to achieve a more accurate prediction of these loads; moreover, the real attitude in the transverse plane (roll angle) should be considered in the simulation, because the in-plane loads are sensitive to the wake and load distribution over the propeller disk as described in the following analysis. The interested reader is referred to [27,28] for a deeper discussion on these aspects.

The velocity profiles are referred to the hub reference system, the propeller disk is viewed from the stern, and the circumferential coordinate θ is positive signed when oriented in the same sense of propeller rotation (that is, increases clockwise and anticlockwise for the port and starboard propellers, respectively). In the discussion, the propeller disk is divided into four sectors, the first one being defined by $0^\circ < \theta < 90^\circ$, and so on. Finally, consistently with [25,26], results for propeller loads are summarized in unified graphs, where positive and negative rudder angles correspond respectively to internal and external propellers.

In the following, all the quantities are made non dimensional using as reference quantities the length between perpendiculars ($L_{ref} = L_{pp}$), the approach speed at model scale ($U_{ref} = U_\infty$), and the density of water ($\rho_{ref} = 1000$ kg/m^3). The propeller loads are expressed in terms of coefficients defined, as usual, by dividing the thrust and torque by the factors $\rho N^2 D^4$ and $\rho N^2 D^5$ respectively, where D is the propeller diameter and N is the propeller rate of revolution (RPS). Without loss of generality, kinematic values, forces and moments are always presented in terms of ratio with respect to values in the approach phase (identified with the subscript "0").

Experimental Results

The experimental results are reported in Figure 7 in terms of thrust, torque and radial force components ratio for the three different F_N. The results refer to the starboard propeller.

In general, a marked asymmetry of the loads generated by the external and internal propeller is experienced. Moreover, the effect of speed is negligible: this fact is indicative of the limited effect of the F_N on the maneuvering response, as was also reported in terms of kinematic response and macroscopic parameter of the maneuver in [25]. In other words, the propeller inflow (and therefore, the propeller performance) would be morphologically similar for maneuvers carried out at the same rudder angle.

The thrust and torque show a similar behavior; on the external propeller ($\delta < 0^\circ$) a remarkable increase of almost 80% and 60% of the value in the approach phase, can be evidenced for thrust and torque, respectively. The increase is almost linear with δ. On the internal side, the overloading is considerably weaker (less than 10% for both thrust and torque) and the trend non-linear, the slope being negative for $\delta = 15^\circ$ and then positive.

The side force, Figure 7c, is always oriented to stabilize the vessel, i.e., it is oriented toward the centre of curvature of the trajectory. During straight ahead motion, this force is oriented toward the plane of simmetry of the model. On the internal side, the force shows a non-linear trend, it slightly increases up to $\delta = 25^\circ$ and then diminishes in case of the tighter maneuver. The average value over the range of rudder angles corresponds to about about 15% of K_{T0}. On the external side, this force increases almost linearly up to $\delta = -25^\circ$ (where it reaches a value doubled with respect to the internal shaft) and at $\delta = -35°$ slightly diminishes. The external propeller develops a side force that is more than doubled with respect to internal propeller.

The vertical component, reported in Figure 7d, resembles the same character of K_{Ty}. At $\delta = 0°$ the force is upwards oriented due to the inclination of the shafts and the mean direction of the flow in the stern region; on the internal side, the trend is markedly non linear, K_{Tz} experiencing first a maximum at $\delta = 15°$ and thereby a linear decrease til $\delta = 35°$ (50% lower than the value observed during the approach phase). On the external shaft, this component gradually weakens and achieves a negligible value with respect to the side force at the maximum rudder angle. In general, it is interesting to observe that, although the absolute magnitude of the radial force (given by the combination of K_{Ty} and K_{Tz}) is similar for the internal and external shaft, it is developed at completely different maneuvering conditions that lead to completely different inflow conditions to the propeller, as described in the next section.

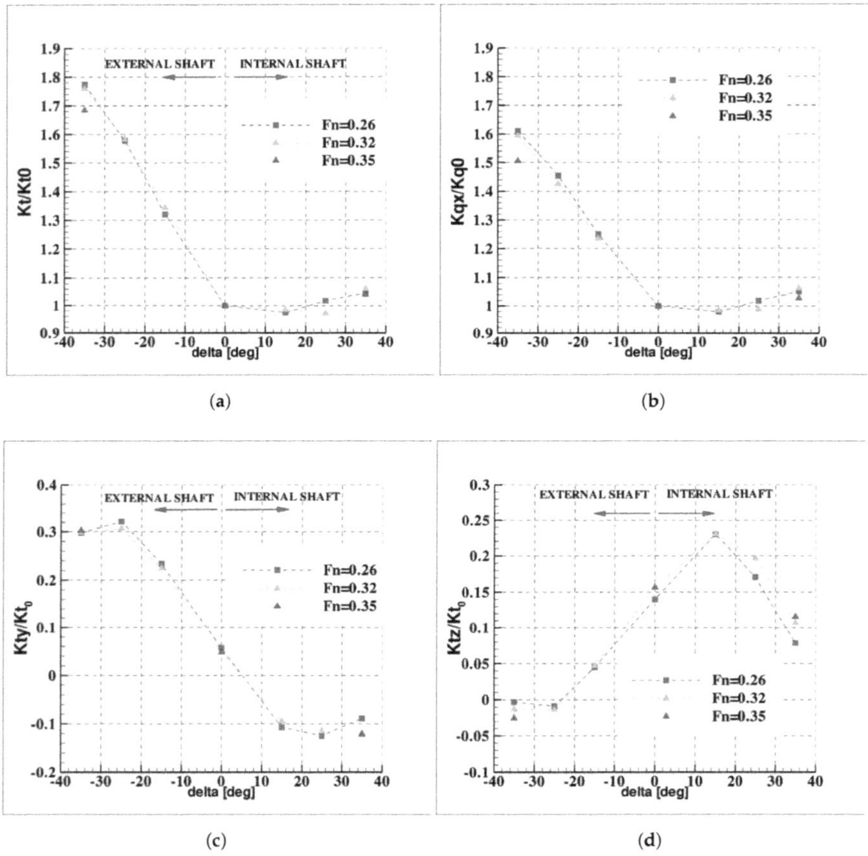

Figure 7. (**a**) Thrust. (**b**) Torque. (**c**) Side force. (**d**) Vertical force.

5. Discussion

The different behavior of the propeller loads during the maneuvers, synthesized by a marked asymmetry between the internal and external propeller and a linear variation of the loads on the external side by contrast to a non-linear trend on the internal side, has to be ascribed to the different evolution of the inflow to the propellers. The nominal wake experienced by the propellers in straight ahead motion and during maneuvering conditions at small and high turning rates ($\delta = 15°$ and $35°$) is described in terms of axial velocity contours and tangential velocity and field in Figures 8 and 9a, 10a. In order to relate the flow field with the system of vortices detached from the hull and appendages, Figures 9b and 10b visualize the coherent structures obtained by the vortex identification method based on the λ_2–criterion [37].

In rectilinear motion (see Figure 8) the inflow to the propeller is perturbed only on the upper half of the disk, the velocity defect being associated to the wake past the brackets and the boundary layer of the hull; moreover, the tangential flow on the propeller disk (represented with vectors) is directed upwards due to the inclination of the shaft and the mean orientation of the flow along the stern cut-off of the hull [25,27].

During maneuvering conditions, the wake is gradually modified and the asymmetry between the internal and external propeller are evident also for the weaker maneuver at $\delta = 15°$. In fact, on the external shaft, the flow always resembles a pure oblique flow, the axial flow being almost homogeneous and the tangential flow oriented horizontally; it can be noticed that at $\delta = 15°$, the propeller is moderately affected by the disturbance of a vortex detached from the bilge keel in the 1st quadrant, whereas at $\delta = 35°$ the disk is weakly affected by smaller structures detached from the propeller shaft (see Figures 9b and 10b). On the contrary, the inflow to the internal propeller develops in relation to the evolution and interaction of the vortical structures detached from the hull and appendages. In fact, at $\delta = 15°$, the strong recirculation region in the 1st quadrant of the port propeller (see Figure 9a) is associated to the passage of the vortex detached at the skeg and deflected toward the propeller disk (see Figure 9b). At $\delta = 35°$, the propeller is completely immersed in a large vortical structure formed by the combination of skeg and bilge keel vortices; this large structure interacts also with the shaft and brackets and causes the separation of smaller structures, whose trace is for example evidenced by the recirculation region in the upper half of the disk (see Figure 10a).

The relation of the wake with the propeller loads can be effectively established focusing on the single blade performance. Moreover, in the framework of blade element description (see Section 3), a very effective and simple method can be introduced to explain the onset of the in-plane loads and moments and to better synthesize the blade loads during the period of revolution [27]. For example, the in-plane force, i.e., the lateral force, is due to the imbalance of the force developed by the blade in the upper and lower half of the disk. Similarly, considering the thrust force, this asymmetric variation of the blade loads generates the pitching moment. Similar conclusion can be easily derived for the components relative to the vertical axis.

The thrust, side and vertical forces developed by the single blade and their distribution over the disk are reported in Figures 11–13; the integral on single halves that support the simplified description of the generation of the in-plane loads are reported in Figures 14 and 15.

At $\delta = 0°$, the blade experiences higher local incidence angle in the 1st and 2nd quadrants (downstroke cycle), $0° < \theta < 180°$ (red line in Figure 11a,b) due to the velocity defect and the upwards tangential velocity that acts to increase the local angle of incidence of the blade sections (according to a blade element representation). Consistently, the blade develops higher thrust in this sector, as clearly described by its distribution over the disk at the top half of Figure 11, causing the generation of in-plane moments (see Figure 14); considering the starboard propeller (and the hub frame of reference in Figure 5a), these consist of a positive pitching moment and a negative yaw moment. Moreover, the vertical force is dominant with respect to the side force, because the asymmetry of the load distribution is stronger between the left–right halves of the disk. This aspect is quantitatively supported by the imbalance of K_{Ty} and K_{Tz} in the bottom/upper and left/right halves of the disk in Figure 15a,b and by the trend of the single blade loads at the correspondent sectors (red line in Figures 12b and 13b).

In maneuvering conditions, the performance of the external propeller is markedly affected by the cross flow, its behavior being similar to the functioning of a propeller in pure oblique flow. In general, the load distribution over the disk is very different with respect to the rectilinear motion and shows an imbalance between the lower half (2nd and 3rd quadrants) and upper half (1nd and 2rd quadrants) of the disk. The blade develops higher load in the lower half of the disk with respect to the opposite half. In fact, the cross flow induced by the motion, being opposite to the blade rotation, acts to increase the tangential flow experienced by the blade and consequently the developed loads; differently, this mechanism is inverted in the upper half. This behavior is similar for the maneuvers considered and it gradually amplifies with the increase of rudder angle δ, as it can be clearly evidenced by the distribution of the loads over the disk and the trend of the single blade loads (thrust and in-plane forces) visualized in the figures. Specifically, the imbalance of the thrust causes a negative pitching moment that is prevalent with respect to the yaw moment (see Figure 14); the side force, oriented in the positive direction, is originated in the lower half of the disk, its partial contribution increasing faster with respect to the upper half (where it is almost constant) (see Figure 15a). Obviously, the vertical force is negligible, the counteracting contributions from the left and right halves of the disk being almost coincident. It has to be stressed that the effects of the cross flow is the primary element that determines the behavior of the external propeller synthesized in Figure 7a–d [27]. In fact, the self-similar character of the blade hydrodynamics determines the linear increase of thrust, torque and side force and, conversely, the smooth drop of the vertical force; moreover, the marked overload with respect to the internal propeller has to be also ascribed to the increase of the sectional incidence angle by the cross flow, since the mean velocity defect is stronger on the internal propeller (see Figures 9a and 10a).

On the contrary, on the internal side the blade hydrodynamics is more complicated, because the disturbance due to the wake of the hull and appendages is superposed to the cross flow. As a matter of fact, the blade hydrodynamics is affected by the wake and interaction with vortical structures in the upper half of the disk, while by cross-flow in the lower half. At $\delta = 15^\circ$, interaction of the blade with the skeg vortex (it acts to increase the tangential speed relative to the section, and consequently, the loads) causes the localized increase of thrust at about $\theta = 60^\circ$, and amplifies its fluctuation during the cycle (see Figure 11a). This interaction also provokes the increase of the magnitude of in-plane moments and the relevant peak of the vertical force reported in Figure 7d. In particular, the increase of the vertical force has to be ascribed to this localized phenomenon, because the trend of this load developed by the single blade is similar to the one experienced in rectilinear motion with the exception of the passage about $\theta = 60^\circ$. Moreover, because of the reduction of blade load in the lower half of the disk due to cross flow, the side force increases (see Figure 15a). At the maximum rudder angle, $\delta = 35^\circ$, the localized effect of the skeg vortex is weakened, since the vortex moved further to the lee side. The further buildup of thrust is originated in the left half of the disk, as it is proved by the increase of the single blade load for $180^\circ < \theta < 360^\circ$ (see Figure 7a); this is associated to the combination of velocity defect and speed drop (experienced by the model during the maneuver), since the tangential flow, upwards oriented (see Figure 10a), acts to reduce the incidence angle of the blade sections. The increase of the pitch moment and the yaw moment is weakened, as demonstrated by the inversion of the trend of the load in the lower and left halves of the disk (see Figure 14). Consistently, the in-plane loads are also reduced: the drop of K_{Tz} has to be ascribed to the weakening of left/right imbalance of the load associated to the increase of the magnitude of vertical force in the left half of the disk (see Figures 7d and 15b); similarly, the increase of the side force in the lower half of the disk counteracts the force on the upper half and, hence, the asymmetry with respect to the horizontal axis is reduced.

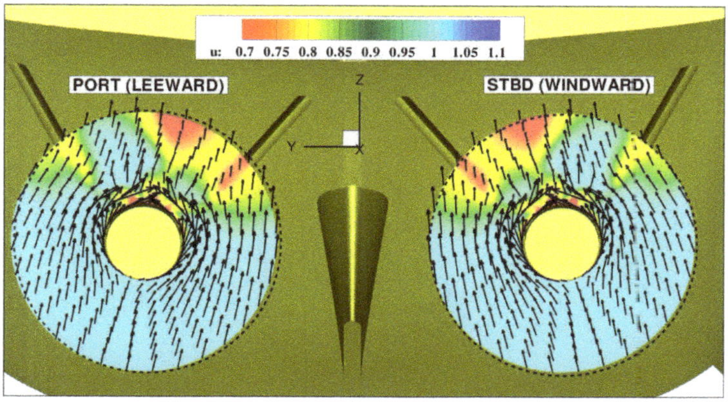

Figure 8. Nominal wake during approach (rectilinear motion).

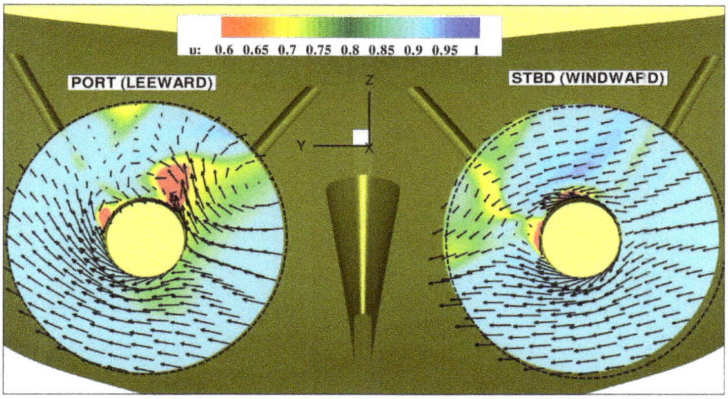

(**a**)

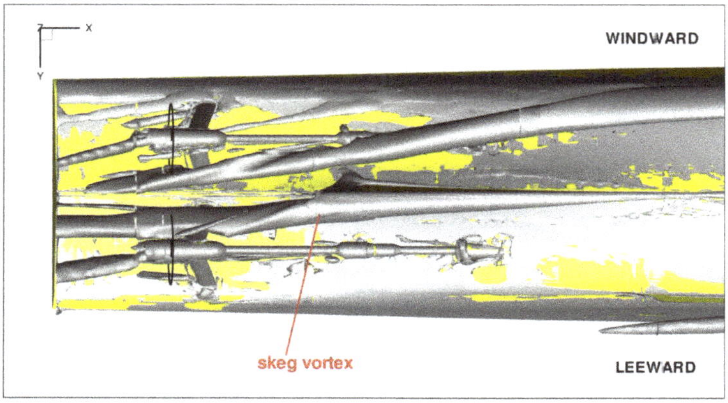

(**b**)

Figure 9. (**a**) Nominal wake during turn at $\delta = 15°$. (**b**) Vortex structures.

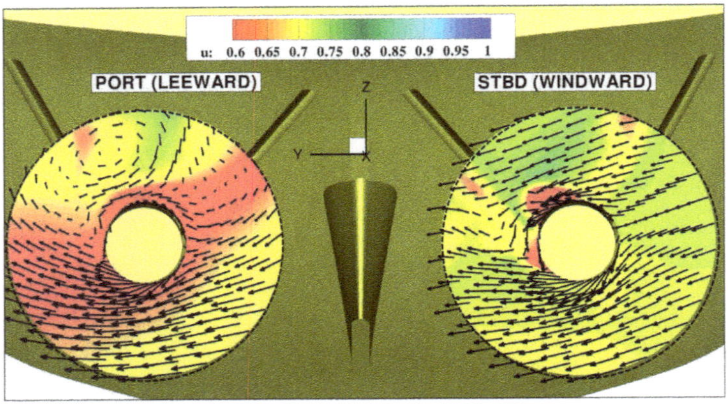

(**a**)

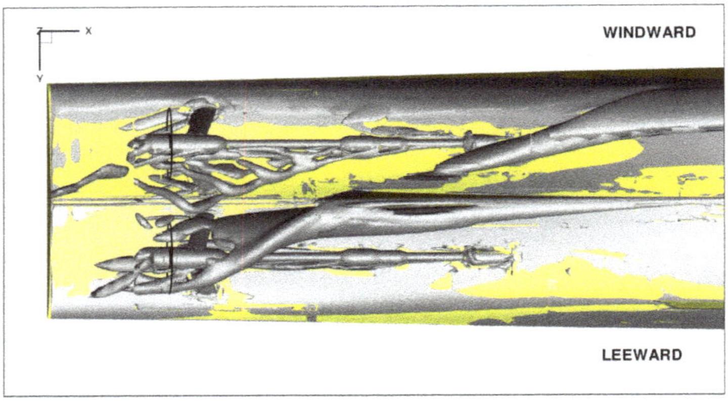

(**b**)

Figure 10. (**a**) Nominal wake during turn at $\delta = 35°$. (**b**) Vortex structures.

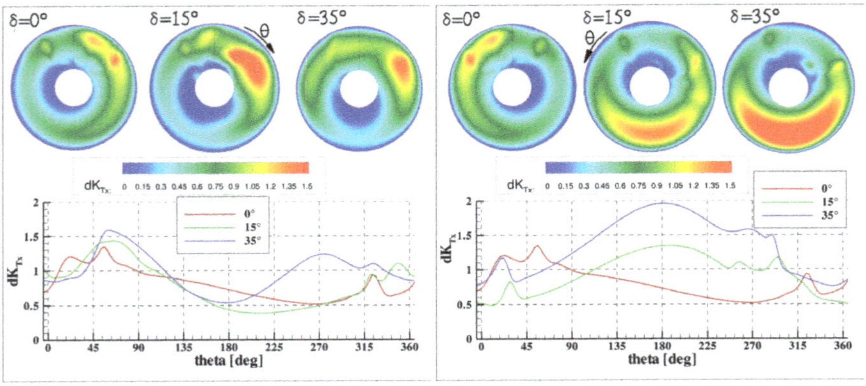

(**a**) (**b**)

Figure 11. Thrust (**a**) Internal propeller. (**b**) External propeller.

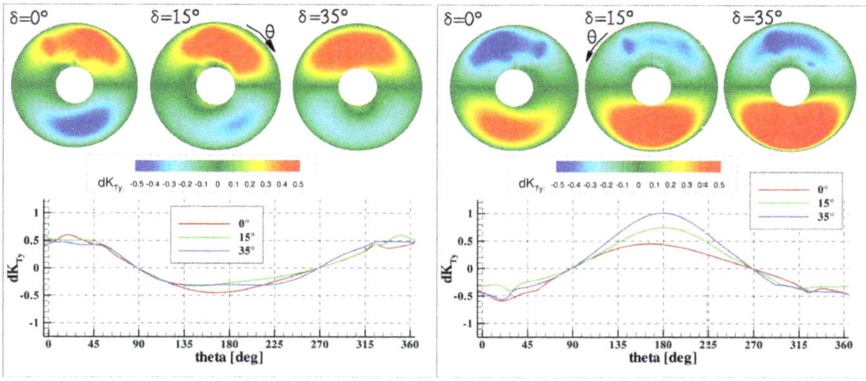

(**a**) (**b**)

Figure 12. Side force (**a**) Internal propeller. (**b**) External propeller.

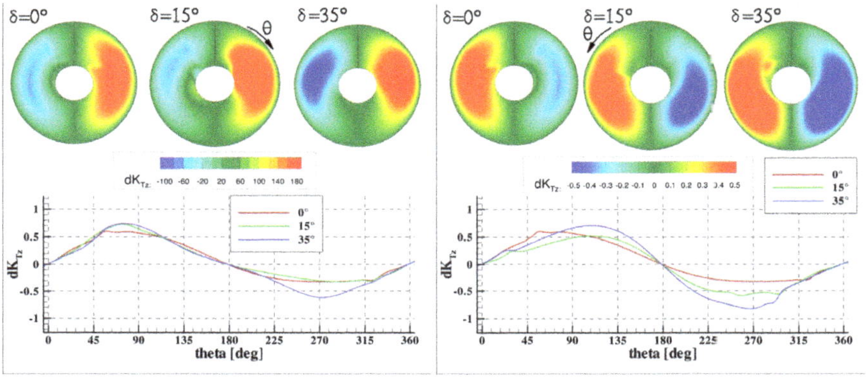

(**a**) (**b**)

Figure 13. Vertical force (**a**) Internal propeller. (**b**) External propeller.

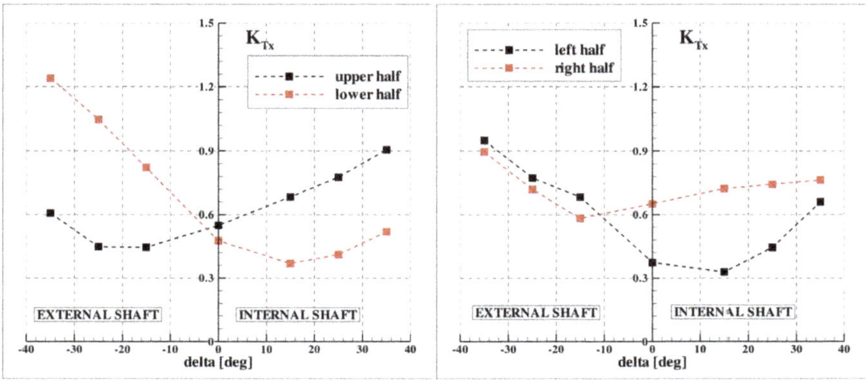

(**a**) (**b**)

Figure 14. (**a**) imbalance of thrust on the upper/lower half of the disk. (**b**) imbalance of thrust on the left/right half of the disk.

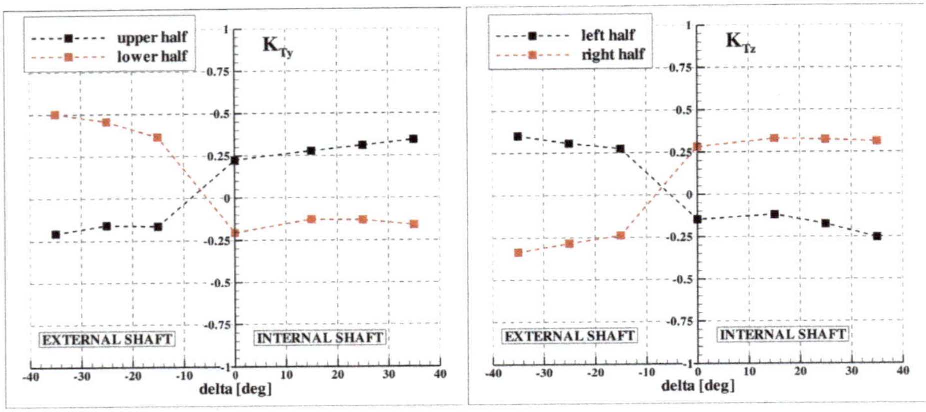

(a) (b)

Figure 15. (**a**) imbalance of side force on the upper/lower half of the disk. (**b**) imbalance of vertical force on the left/right half of the disk.

6. Semi Empirical Method for Full Scale Prediction of in-plane loads

Scaling of results from model to full scale represents the most long been debated topic in marine hydrodynamics since ever. The scaling to full scale during maneuvering is further complicated (with respect to rectilinear motion) by the fact that the attitude of the model is also affected by scale effects, influencing the distribution of the wake and the inflow over the propeller disk [25]. In this investigation, only the effects of Reynolds number and cavitation on propeller loads are analyzed by a simplified approach based on a modified versions of the blade element solver described in Section 3. However, the same approach could be equally used together with a more advanced solver (i.e., Boundary Element Method) and with an increasing level of accuracy by considering also the wake distribution at full scale.

6.1. Effect of Reynolds Number

The scale effect due to different Reynolds number was accounted for by the ITTC procedure. The method (the formulation is not reported for the sake of conciseness) provides the corrections for thrust and torque of a propeller operating in open water conditions referring to the section at $r/R = 0.75$. In our study, since the propeller sections meet a variable flow during a period of revolution, the method was implemented in the $BEMT$ solver in two different ways. In particular, the corrections were applied both for each blade section and only for the one at $r/R = 0.75$; in both cases, the total correction was the result of the integral performed along the span and the complete revolution, and the complete revolution only, respectively. In the propeller solver, the torque correction was properly transformed to a modification of the tangential force developed by element in order to compute the scale effects also for the in-plane loads. The outcome of the study was that the viscous correction provided a negligible contribution to the propeller loads, therefore the results were not reported. The difference between the two implemented approaches was small; this was due to the fact that the size of the propeller model is sufficiently big to guarantee that the local Reynolds number over the largest portion of the blade is grater than the suggested one (i.e., 2×10^5) [38]. It has to be pointed out that the ITTC scaling does not account for the effects of propeller loading and this aspect can be important during maneuvering because the propeller loading increases.

6.2. Effect of Cavitation Number

The $BEMT$ model is improved to account for the effects of cavitation extension on blade loads by means of the 2D partially cavitating foil linearized theory [39]. In fact, the lift coefficient of a partially cavitating foil depends on the cavitation length l_{CAV} by:

$$C_L = \pi\alpha\left(1 + \frac{1}{sin\beta}\right) \tag{5}$$

where β is parameter that defines the cavitation length ($l_{CAV} = cos^2\beta$); l_{CAV} is made non dimensional with respect to the chord of the foil. To obtain β, the following equation must be solved:

$$\frac{\alpha}{\sigma} = \frac{1}{2}tan\beta\frac{1 - sin\beta}{1 + sin\beta} \tag{6}$$

where σ is the cavitation number ($\sigma = p - p_{vap}/0.5\rho V^2$), and α is the (geometric) angle of attack of the foil. Moreover, the volume of the attached bubble V_{CAV} can be also determined:

$$V_{CAV} = \frac{\pi\alpha}{16}cot\beta\left(1 + 4sin\beta - sin^2\beta - 4sin^3\beta\right) \tag{7}$$

The algorithm of the modified propeller solver is explained in Figure 6. In particular, the solution for the cavitation length introduces an additional iterative cycle (Equation (6) is non linear) that is nested in the iterative solution of the Prandtl–Betz condition: in–fact, the self–induced velocity field, being strictly coupled with the circulation (lift) generated by the section, is obviously related to the cavitation extension (i.e., β, see Equation (5)). It can be observed that as long as the induced angle of incidence α_i changes to fulfill the Prandtl–Betz relation, the lift coefficient is consistently adjusted to obtain a physically consistent model. Obviously, this method presents the following drawbacks:

- the cavitation lenght l_{CAV} is overestimated because the linear theory does not account for thickness
- the maximum value of the cavity length is $l_{CAV} = 0.75$; in fact, Equation (6) admits two solutions, a long and a short bubble. The long cavity is unstable and physically unacceptable. For this reason, the iteration loop for the solution of Equation (6) was stopped when the bubble length achieved this limit
- during off-design conditions, the higher angle of incidences experienced by the blade sections may give rise to types of cavitation other than the attached sheet cavity assumed in linear theory (cloud cavitation, bubble cavitation)

In the computations, σ is calculated by considering the incident flow magnitude scaled at full scale and the nominal wake is kept the same of the model scale. The use of semi–empirical wake scaling technique was out of the aim of this approximate analysis; it has to be noted that these scaling techniques were not validated for maneuvering conditions and the most immediate way to assess or improve their capabilities to ship maneuvering problems could be the use of CFD.

The propeller forces and moments obtained with the new model are reported in Figure 16 in terms of ratio with respect to the values with cavitation switched off. The increase of the loads is qualitatively similar on the external and internal propeller. Figure 16 highlights that the in-plane forces and moments are more sensitive to cavitation phenomena with respect to the thrust and torque. This is due to the fact that cavitation amplifies the imbalance of these loads between the upper/lower and left/right halves of the disk. This aspect is supported by the distribution of l_{CAV} over the propeller disk for the internal propeller at $\delta = 15°$ and the external one $\delta = 35°$ (see Figure 17). The cavitation pattern resembles the same features of load distribution discussed previously and, in particular, the asymmetry of their distribution over the disk; therefore, the consequent amplification of the loads can be qualitatively expected by relation (5). This aspect is further supported in Figure 18a,b, that show a comparison of the propeller thrust and in-plane loads in cavitating and non cavitating conditions with the cavitation extent (Equation (7)) for the same representative maneuvers considered in the previous analysis. In the figures, the vertical and the lateral force are reported with the thrust for the internal and external propeller at $\delta = 15°$ and $35°$, respectively. In particular, it can be observed that the discrepancy between the loads during the cavitating (solid line) and the non cavitating (dashed line) conditions are experienced in correspondence of the blade/wake interaction on the leeward side and in the lower half of the disk on the windward side; moreover, in the cavitating conditions, the amplitude of the fluctuation of the blade load is higher (with consequent increase of the in-plane loads).

It is interesting to observe that on the internal propeller the increase of the loads is faster than the external propeller up to $\delta = 25°$ for all the loads with the exception of the side force (see Figure 16a). This can be ascribed to the large gradients of the inflow caused by the evolution of the wake and coherent structures detached from the appendages. The effect of the non–homogeneity of the wake on the onset and development of cavitation was also remarked in [7,8]. It has to be pointed out that the abrupt increase of the vertical force on the external shaft and the peak of the side force at $\delta = 0°$ are not relevant converted to dimensional quantities. It can be further noticed that on the external propeller K_{Tx}, K_{Qx}, K_{Ty} and K_{Qy} increase with a linear trend, the flow being unperturbed by the wake of the hull.

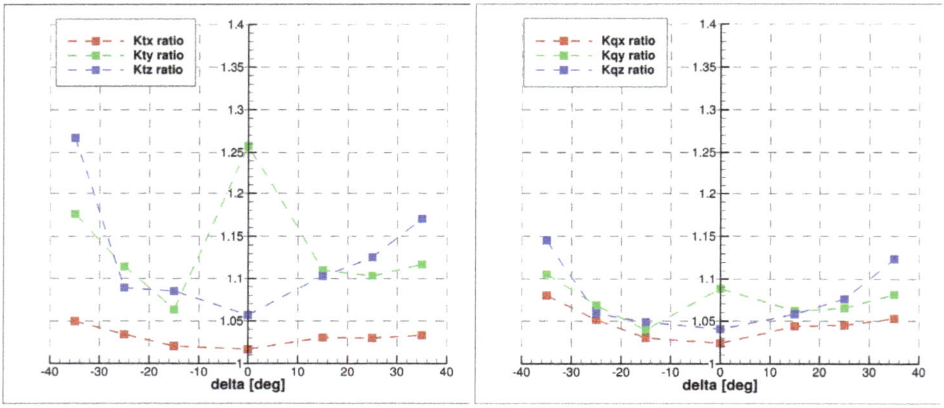

(**a**) (**b**)

Figure 16. Ratio between forces and moments in cavitating and non cavitating conditions. (**a**) forces. (**b**) moments.

In general, for the worst conditions resulted at the maximum rudder angle, $\delta = 35°$, the increase of thrust amounts to about 5%; on the other hand, the side force and the vertical force experience a 20% increment with respect to the non cavitating case, for the external and internal propeller, respectively. The discrepancy between the torque and the other moments is weaker, although the torque increase is almost halved with respect to the pitching and yawing moments. In order to obtain a full scale prediction of the experimental data, the percentage increases obtained from this numerical analysis could be imposed to the experimental data, considering also the fact that the Reynolds number effect was negligible. For the mechanical sizing of the bearings of the propulsive shaft, the maximum values of the loads exerted by the propeller should be known; extrapolation to full scale of the experimental data can be performed, according to the present analysis, considering maximum percentage increase of the load due to cavitation. Since this condition is experienced on the external propeller, characterized by a pure oblique inflow condition, a cheaper approach could be to modify experimental value on a free running model by the percentage increase of the loads for a propeller in open water set at the same incidence of the model (determined by the kinematic of the maneuver) or, alternatively, presumed at the full scale achieved numerical (by a *CFD* computation) or water tunnel experiments. In principle this approach is cheaper than a complete *CFD* analysis for the complete ship configuration.

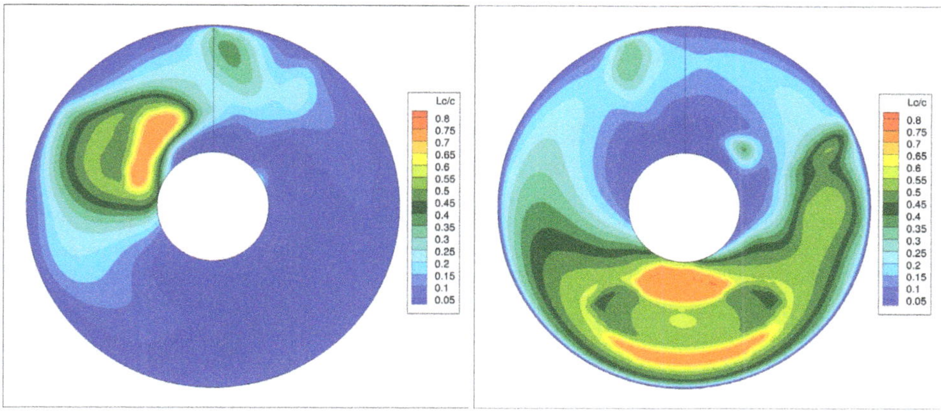

(**a**) (**b**)

Figure 17. Cavitation extent, *BEMT* solution via the linearized theory. (**a**) $\delta = 15°$, internal propeller. (**b**) $\delta = 35°$, external propeller.

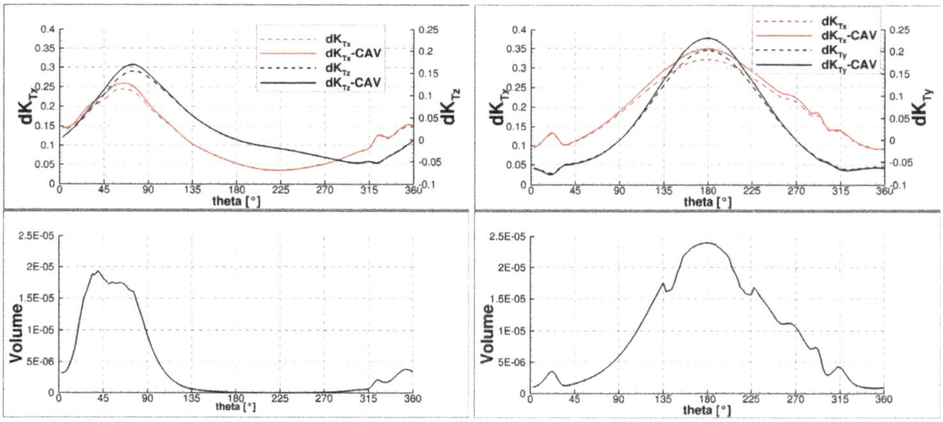

(**a**) (**b**)

Figure 18. Correlation between blade loads and cavitation volume. (**a**) $\delta = 15°$, internal propeller. (**b**) $\delta = 35°$, external propeller.

6.3. Further Remarks on the Scale Factors in Cavitating Conditions

The scaling factors were evaluated under the assumption that the maneuvering response and the inflow to the propeller at model and full scale are the same. Obviously, the knowledge of the wake at full scale is of utmost importance to obtain reliable predictions of propeller performance, in particular when cavitation analysis and related side effects (pressure pulses and noise) are of concern. In [40], it was highlighted that the nominal wake can lead to under prediction of the cavitation pattern and evolution (pressure pulses) with respect to the use of the wake at full scale. Referring to the case analyzed in this work, assuming a pure "contraction" of the nominal wake for its scaling, the gradients of the inflow experienced by the blade section of the propeller could be stronger and hence, the increase of the load be higher than the predicted one. During maneuvering, on the internal propeller, the evolution of the wake is characterized by a concurrent contribution of vortical structures and non-organized separate swirling flow, moving from smaller to higher rudder angles. Therefore, the modification of the scaling factor could be expected to be different for the various maneuvers. On the external propeller, the flow and the global propeller performance of the propeller are governed by the cross-flow induced by the motion; moreover, since these global effects are (almost) linearly amplified with the rudder, it can be expected that the eventual correction of the scaling factor changes in a similar way. For the purpose of the proposed analysis, i.e., the quantification of a scaling factor, the key point is the identification of the most critical condition characterized by the maximum amplification of the loads due to cavitation. On the internal shaft, the worst condition remains the maneuver at $\delta = 15^\circ$ in terms of absolute increment of the load; in this case, the radial force in non cavitating condition is of the same order of magnitude of the one experienced by the external propeller. The obtained scaling factors for both cases lead to similar increment of the loads. The critical phenomenology on the internal propeller is governed by a localized development of the cavitation due to the interaction between the blade and the vortex; on the external propeller, the cavitation is completely developed in the lower half of the disk and it is caused by the effect of cross-flow that cause the strong overload of the propeller in this sector. Although the contraction of the wake at full scale probably amplifies the extension of the cavitation pattern, a global effect similar to the that experienced on the external propeller seems difficult to develop (the absolute increment of loads on the external propeller are almost doubled with respect to the internal propeller), considering also the fact that the loading of the two propellers are completely different (see Figure 7a). This is particularly evident also by comparing the blade thrust in Figure 11a,b: the maximum load of the blade at $\delta = 35^\circ$ for the external propeller during its rotation in the lower half of the disk is markedly higher with respect to the load of the blade of the internal propeller. Therefore, the predicted order of magnitude for the safety factor (on the external propeller) can be considered plausible (under the hypothesis and simplification of the model).

Finally, it has to be remarked that the present conclusions are valid for this specific configuration of the propulsion system, i.e., propeller inward rotating from the top; as it was shown in [29], the inversion of the sense of revolution of the propellers completely alters the behavior of the internal and external propellers.

7. Conclusions

Present work is a contribution to the understanding of the mechanisms that rule the performance of a propeller operating in off-design conditions (maneuvering). In particular, the work consisted of a synthesis of the results of free running model tests and their phenomenological assessment with numerical computations performed with a $uRANSE$ maneuvering simulations and simplified propeller model based on Blade Element Momentum Theory. The test case for the study was a twin screw model; the propulsive configuration is particularly suitable for this study, since the regime of the propellers on the leeward and windward sides is markedly different.

In fact, the propeller on the external side operates in a pure oblique flow condition and the increase of the loads is almost linear with the rudder angle. This propeller experiences the most critical condition with respect to the internal one at maximum rudder angle. In fact, the thrust and torque increases up to 80% and 60% of the value in straight ahead sailing; moreover, since the inflow is directed almost in the horizontal direction, the propeller develops only a side force that acts to stabilize the vessel and amounts to about 30% of the thrust generated in the approach phase of the maneuver. On the other hand, on the internal side, the propeller hydrodynamics is markedly affected by the evolution of the wake detached from the hull, the interaction with vortical structures convected through the propeller disk and the cross-flow induced by the motion. The dominance of one effect with respect to the other at different maneuvering conditions is the reason of the non-linear variation of the loads with rudder angle. Although the thrust and torque did not experience a large overload

(10% with respect to the value during the approach phase), the behavior of the in-plane loads resulted sensitive to these flow conditions; indeed, the asymmetries and localized gradients of the inflow to the propeller was responsible to a marked imbalance of the periodic blade loads, with the consequence that the maximum magnitude of the radial force was comparable to the one developed on the external shaft. Moreover, the peak of the radial force was experienced during a weaker maneuver ($\delta = 15°$) and was ascribed to the interaction of propeller blades with a coherent structure developed past the appendages. These results stressed that also mild variations from the straight ahead conditions can be critical for the propulsive system, even to the same extent of heavy off-design conditions, and can possibly generate undesired noise emission, vibrations and fatigue loads on the shafting system.

Finally, a preliminary analysis on the scale effects on the propeller loads was carried out. In particular, the study considered only the scale effects for the propeller, i.e., the Reynolds number and cavitation number on propeller loads. To this aim, the *BEMT* propeller solver was modified to calculate the loads corrections by the ITTC method for the former problem: for the second issue, the propeller solver was enhanced with a linear, partially cavitating, hydrofoil theory, accounted in a consistent way with the solution for the self–induced velocity field. The study highlighted that viscous effects on the propeller loads are negligible with respect to cavitation; moreover, the in-plane loads are more sensitive to cavitation with respect to thrust and torque because it increases the asymmetry of blade loads during the cycle. The amplification of the loads due to cavitation is experienced on the external shaft at the maximum angle for the present propulsive configuration (propellers inward rotating from the top); therefore, a viable scaling procedure during the design phase could be to complement free running maneuvering tests with measurements or *CFD* computation for the isolated propeller at the critical incidence and equivalent loading condition in cavitating and non-cavitating conditions for achieving the scaling factor. A complete analysis of the propeller performance at full scale cannot be performed without considering cavitation and its related effect; *CFD* simulations with weak coupling with propeller models are feasible. More accurate analysis could be possible by a multi–phase *CFD* solver (with discretized propeller); the computational burden is prohibitive for design purposes and thus far this kind of simulation has not yet been carried out.

Reynolds number effects on the modification of the nominal wake to the propeller is a very challenging issue, since scale effects the maneuvering response and, therefore, the evolution of the wake. These effects are also important to correctly predict the cavitation volume and, in particular, it dynamics in order to predict accurately propeller hull induced vibrations and acoustic emission. In this regard, *CFD* can provide a valuable insight on the modification of the wake and, moreover, contribute to understand whether the development of scaling techniques similar to those applied in straight ahead motion are a plausible way to a safer extrapolation of the loads to full-scale. Obviously, measurements at full scale, despite difficult to set up, could provide useful hints to tune the extrapolation procedure; unfortunately, at present the availability of data is extremely poor or restricted by industry.

To better understand the complex phenomenology of the propeller behavior in off-design conditions, in particular to relate the loads to physic experienced at blade rate frequency, the set-up of the free running model has been recently improved for the characterization of the single blade loads. On the basis of these measurements, all the system of loads acting on the propeller can be determined and these measurements represent a basis for deeper investigation also on propeller side effects related to propeller vibration, pressure pulses and noise emission. In Figure 19a,b a partial overview of the ad–hoc design propeller to house the measurement system of the single blade and in the behind hull configurations are showed. The test case is a twin screw ships with a different stern geometry with respect the one considered in this study and the model size is smaller. The experiments will be carried out at the same experimental conditions on a free running, self propelled maneuvering model.

(**a**)

(**b**)

Figure 19. (**a**) Propeller with hub transducer. (**b**) Behind–hull configuration.

Author Contributions: F.O., G.D. and S.M. were involved in the experimental campaign; F.O. and G.D. analyzed and interpreted the experimental data; G.D., R.M. and A.D.M. carried out the numerical investigation. F.O., G.D., R.M. and A.D.M. synthesized the experimental and numerical data and equally contributed on writing the paper; all the authors contributed on the final drafting of the paper.

Acknowledgments: This research activity was funded by the Flagship Project RITMARE—The Italian Research for the Sea—coordinated by the Italian National Research Council, and by the Italian Ministry of Education, University and Research within the National Research Program 2011–2013. The authors would like to acknowledge Fabio Carta and Roberto Zagaglia for their valuable technical contribution and support for the development of the 2–component transducer, and adjustments and refinements to CAD models.

Conflicts of Interest: The authors declare no conflict of interest.

Abbreviations

The following abbreviations are used in this manuscript:

ITTC	International Towing Tank Conference
CFD	Computational Fluid Dynamics
BEMT	Blade Element Momentum Theory
RANSE	Reynolds Averaged Navier Stokes

References

1. Vartdal, B.; Gjestland, T.; Arvidsen, T. Lateral Propeller Forces and their effects on Propeller Shafts Bearings. In Proceedings of the Symposium on Marine Propellers and Propulsion, Trondheim, Norway, 22–24 June 2009.
2. Sverko, D.; Sestan, A. Experimental Determination of Stern Tube Journal Bearing Behaviour. *Brodogradnja/Shipbuilding* **2010**, *61*, 130–141.
3. Broglia, R.; Dubbioso, G.; Durante, D.; Di Mascio, A. Simulation of Turning Circle by CFD: Analysis of different propeller models and their effect on manoeuvering prediction. *Appl. Ocean Res.* **2013**, *39*, 1–10. [CrossRef]

4. Broglia, R.; Dubbioso, G.; Durante, D.; Di Mascio, A. Turning ability analysis of a fully appended twin screw vessel by CFD. Part I: Single rudder configuration. *Ocean Eng.* **2015**, *105*, 275–286. [CrossRef]
5. Dubbioso, G.; Broglia, R.; Durante, D.; Di Mascio, A. Turning ability analysis of a fully appended twin screw vessel by CFD. Part II: Single rudder vs Twin rudder configuration. *Ocean Eng.* **2016**, *117*, 259–271. [CrossRef]
6. Grigoropoulos, G.; Campana, E.; Diez, E.; Serani, A.; Goren, O.; Sarioz, K.; Danisman, D.; Visonneau, M.; Queutey, P.; Abdel-Maksoud, M.; et al. Misson–based hull form and propeller optimization of a transom stern destroyer for best performance in the sea environment. In Proceedings of the 7th VII International Congress on Computational Methods in Marine Engineering MARINE, Nantes, France, 15–17 May 2017.
7. Taskar, B.; Steen, S. Effect of waves on cavitation and pressure pulses. *Appl. Ocean Res.* **2016**, *60*, 61–74. [CrossRef]
8. Taskar, B.; Steen, S.; Eriksson, J. Effect of waves on cavitation and pressure pulses of a tanker with twin podded propulsion. *Appl. Ocean Res.* **2017**, *65*, 206–218. [CrossRef]
9. Taskar, B. The Effects of Waves on Marine Propellers and Propulsion. Ph.D. Thesis, Norwegian University of Science and Technology, Trondheim, Norwegian, 2017.
10. Amini, H.; Steen, S. Theoretical and experimental analysis of propeller shaft loads in oblique flow. *J. Ship Res.* **2011**, *55*, 268–288. [CrossRef]
11. Gutsche, F. *The Study of Ships' Propellers in Oblique Flow*; Technical Report; Volume 4306 DRIC Translation; Defense Research Information Centre: New Delhi, India, 1975.
12. Cassella, P.; Mandarino, M.; Scamardella, A. Systematic tests with B—Wageningen screw propellers in non-axial flow: Presentation and analysis of the experimental results. In Proceedings of the 4th International Symposium on Practical Design of Ships and Floating Structures (PRADS), Varna, Bulgaria, 23–28 October 1989.
13. Amini, H.; Steen, S. Theoretical and experimental investigation of propeller shaft loads in transient condition. *Int. Shipbuild. Progress* **2012**, *59*, 55–82.
14. Krasilinov, V.; Zhang, Z.; Hong, F. Analysis of Unsteady Propeller Blade Forces by RANS. In Proceedings of the First International Symposium on Marine Propulsors (SMP'09), Trondheim, Norway, 22–24 June 2009.
15. Liu, P.; Islam, M.; Veitch, B. Unsteady Hydrodynamics of a steering podded propeller unit. *Ocean Eng.* **2009**, *36*, 1003–1014. [CrossRef]
16. Dubbioso, G.; Muscari, R.; Di Mascio, A. Performance of a marine propeller in oblique flow. *Comput. And Fluids* **2013**, *39*, 1–10.
17. Dubbioso, G.; Muscari, R.; Di Mascio, A. Analysis of a marine propeller operating in oblique flow. Part 2: Very high incidence angles. *Comput. Fluids* **2014**, *92*, 56–81. [CrossRef]
18. Yao, J. Investigation on hydrodynamic performance of a marine propeller in oblique flow by RANSE computations. *Int. J. Naval Arch. Ocean Eng.* **2015**, *7*, 56–69. [CrossRef]
19. Muscari, R.; Felli, M.; Di Mascio, A. Analysis of the Flow Past a Fully Appended Hull with Propellers by Computational and Experimental Fluid Dynamics. *J. Fluids Eng.* **2011**, *133*, 061104. [CrossRef]
20. Castro, A.M.; Carrica, P.M.; Stern, F. Full scale self-propulsion computations using discretized propeller for the KRISO container ship KCS. *Comput. Fluids* **2011**, *51*, 35–47. [CrossRef]
21. Mofidi, A.; Carrica, P.M. Simulations of zigzag maneuvers for a container ship with direct moving rudder and propeller. *Comput. Fluids* **2014**, *96*, 191–203. [CrossRef]
22. Carrica, P.M.; Sadat-Hosseini, H.; Stern, F. CFD analysis of broaching for a model surface combatant with explicit simulation of moving rudders and rotating propellers. *Comput. Fluids* **2012**, *53*, 117–132. [CrossRef]
23. Sadat-Hosseini, H.; Carrica, P.; Stern, F.; Umeda, N.; Hashimoto, H.; Yamamura, S.; Mastuda, A. CFD, system-based and EFD study of ship dynamic instability events: Surf-riding, periodic motion, and broaching. *Ocean Eng.* **2011**, *38*, 88–110. [CrossRef]
24. Coraddu, A.; Dubbioso, G.; Mauro, S.; Viviani, M. Analysis of twin screw ships' asymmetric propeller behaviour by means of free running model tests. *Ocean Eng.* **2013**, *68*, 47–64. [CrossRef]
25. Ortolani, F.; Mauro, S.; Dubbioso, G. Investigation of the radial bearing force developed during actual ship operations. Part 1: Straight ahead sailing and turning maneuvers. *Ocean Eng.* **2015**, *94*, 67–87. [CrossRef]
26. Ortolani, F.; Mauro, S.; Dubbioso, G. Investigation of the radial bearing force developed during actual ship operations. Part 2: Unsteady maneuvers. *Ocean Eng.* **2015**, *106*, 424–445. [CrossRef]
27. Dubbioso, G.; Muscari, R.; Ortolani, F.; Di Mascio, A. Analysis of propeller bearing loads by CFD. Part I: Straight ahead and steady turning maneuvers. *Ocean Eng.* **2017**, *130*, 241–259. [CrossRef]

28. Muscari, R.; Dubbioso, G.; Ortolani, F.; Di Mascio, A. Analysis of propeller bearing loads by CFD. Part II: Transient maneuvers. *Ocean Eng.* **2017**, *146*, 217–233. [CrossRef]
29. Muscari, R.; Dubbioso, G.; Ortolani, F.; Di Mascio, A. CFD analysis of the sensitivity of propeller bearing loads to stern appendages and propulsive configurations. *Appl. Ocean Res.* **2017**, *65*, 205–219. [CrossRef]
30. Di Mascio, A.; Muscari, R.; Dubbioso, G. On the wake dynamics of a propeller wake operating in drift. *J. Fluid Mech.* **2014**, *754*, 263–307. [CrossRef]
31. Muscari, R.; Di Mascio, A.; Verzicco, R. Modelling of vortex dynamics in the wake of a marine propeller. *Comput. Fluids* **2013**, *73*, 65–79. [CrossRef]
32. Di Mascio, A.; Broglia, R.; Muscari, R. On the Application of the One-Phase Level Set Method for Naval Hydrodynamic Flows. *Comput. Fluids* **2007**, *36*, 868–886. [CrossRef]
33. Favini, B.; Broglia, R.; Di Mascio, A. Multi-grid Acceleration of Second Order ENO Schemes from Low Subsonic to High Supersonic Flows. *Int. J. Num. Meth. Fluids* **1996**, *23*, 589–606. [CrossRef]
34. Di Mascio, A.; Broglia, R.; Favini, B., A Second Order Godunov–Type Scheme for Naval Hydrodynamics. In *Godunov Methods: Theory and Applications*; Kluwer Academic/Plenum Publishers: Dordrecht, the Netherlands, 2001; pp. 253–261.
35. Phillips, W.F.; Anderson, E.A.; Kelly, Q.J. Predicting the Contribution of Running Propellers to Aircraft Stability Derivatives. *J. Aircraft* **2003**, *40*, 1107–1114. [CrossRef]
36. Amini, H.; Sileo, L.; Steen, S. Numerical calculations of propeller shaft loads on azimuth propulsors in oblique inflow. *J. Mar. Sci. Technol.* **2012**, *17*, 403–421. [CrossRef]
37. Jeong, J.; Hussain, F. On the identification of a vortex. *J. Fluid Mech.* **1995**, *285*, 69–94. [CrossRef]
38. Carlton, J. *Marine Propellers and Propulsion*, 2nd ed.; Butterworth-Heinemann: New York, NY, USA, 2006.
39. Kerwin, S.; Hadler, J. *Principles of Naval Architecture 'Series'-Propulsion*, 1st ed.; SNAME: New York, NY, USA, 2010.
40. Krasilinov, V.; Zhang, Z.; Hong, F. Nominal vs Effective Wake Fields and their influence on Propeller Performance. In Proceedings of the First International Symposium on Marine Propulsors (SMP'09), Trondheim, Norway, 22–24 June 2009.

Journal of
Marine Science and Engineering

MDPI

Article

A Semi-Empirical Prediction Method for Broadband Hull-Pressure Fluctuations and Underwater Radiated Noise by Propeller Tip Vortex Cavitation †

Johan Bosschers

Maritime Research Institute Netherlands (MARIN), Haagsteeg 2, 6708 PM Wageningen, The Netherlands; j.bosschers@marin.nl; Tel.: +31-317493425

† This article is an extended version of the paper presented at the Fifth International Symposium on Marine Propulsors (SMP'17), Espoo, Finland, 2017.

Received: 27 February 2018; Accepted: 26 April 2018; Published: 2 May 2018

Abstract: A semi-empirical method is presented that predicts broadband hull-pressure fluctuations and underwater radiated noise due to propeller tip vortex cavitation. The method uses a hump-shaped pattern for the spectrum and predicts the centre frequency and level of this hump. The principal parameter is the vortex cavity size, which is predicted by a combination of a boundary element method and a semi-empirical vortex model. It is shown that such a model is capable of representing the variation of cavity size with cavitation number well. Using a database of model- and full-scale measured hull-pressure data, an empirical formulation for the maximum level and centre frequency has been developed that is a function of, among other parameters, the cavity size. Acceptable results are obtained when comparing predicted and measured hull-pressure and radiated noise spectra for various cases. The comparison also shows differences that require adjustments of parameters that need to be further investigated.

Keywords: propeller; cavitation; tip vortex; hull pressures; underwater radiated noise

1. Introduction

Noise and vibration on board ships are important design considerations for the comfort of crew and passengers. High comfort levels are especially required by passengers on cruise vessels and owners of yachts. An important source affecting comfort is cavitation on the ship propeller. The collapse of cavitation generates pressure fluctuations that excite the ship hull above the propeller leading to on board noise and vibration. The reduction of cavitation-induced noise and vibration is usually achieved by unloading the propeller tip and decreasing sheet cavitation as much as possible, often leaving only tip vortex cavitation on the propeller. However, this type of cavitation generates low-frequency broadband hull-pressure fluctuations that may lead to ship vibration issues [1,2].

The cavitating propeller is also an important contributor to the underwater radiated noise (URN) of ships. URN is relevant for the acoustic signature of military ships, for the operation of equipment that require low self-noise such as sonar, and for the influence on marine life. The latter used to be relevant for fishery research vessels only, but, in the last decade, the impact on the marine environment of URN due to shipping in general has received considerable attention. URN by shipping is included in the environmental descriptors to achieve Good Environmental Status of European Seas in the EU Marine Strategy Framework Directive (2008/56/EC), The International Maritime Organization (IMO) has released in 2013 non-mandatory guidelines for the reduction of underwater noise from commercial shipping, and class societies DNV-GL, BV, RINA and LR have developed rules for URN of commercial ships.

The spectrum of the broadband hull-pressure fluctuations is characterized by a hump, the maximum level of which increases with ship speed, while the centre of the hump simultaneously moves to lower frequencies. This centre frequency is typically located between 30 Hz and 200 Hz. A possible explanation of the hump in the spectrum is that it is caused by a resonance frequency of the cavitating vortex [3], although experimental evidence of this is missing. The proposed criterion for resonance assumes zero group speed in the dispersion relation for cavitating vortices. Experimental evidence for this dispersion relation is presented in [4] in which further support for the criterion for resonance is also given. The relation between the broadband part and the blade rate tonals in the hull pressure spectrum is briefly discussed in [5].

It has been shown that the source level of URN due to propeller cavitation has a direct relation with the pressure levels on the hull [6,7]. Source levels can directly be estimated from measured hull pressures by correcting the hull pressures for the solid boundary factor on the hull and the distance between cavity collapse and pressure sensor. These source levels are in reasonable agreement with source levels obtained by hydrophone measurements in the far field of the ship.

The hull-pressure fluctuations and URN can be predicted in model test facilities [8–12] to evaluate propeller designs. CFD methods are rapidly developing but are still computationally expensive and not yet mature enough in this area to be used in the iterative design procedure. Rapid evaluation of propeller designs usually relies on the use of potential flow methods that can predict sheet cavitation and the resulting hull-pressure fluctuations and URN at the blade passage frequency [13–15]. The sheet cavitation extents and dynamics can also be used in semi-empirical methods to predict the related broadband URN [16,17]. Because vortices are fitted rather than captured when using potential flow methods, alternative methods are required to predict developed vortex cavitation. The results of the potential flow method can be used in computational noise prediction methods that solve a two-dimensional Rayleigh-Plesset type of equation for a vortex cavity segment [18–20] or that make use of a semi-empirical model [21,22].

The present paper proposes a semi-empirical method to predict the broadband hull pressure fluctuations and URN by propeller tip vortex cavitation that can easily be used in the propeller design process. The method is based on the semi-empirical method to predict on board noise of cavitating tip vortices by Raestad [21]. This method relates the noise to the size of the vortex cavity that is estimated from a potential flow method. In the present method, use is made of the boundary element method (BEM) PROCAL [14,23,24] to estimate the vortex strength (circulation) of the tip vortex. This strength is used in a vortex model for the radial distribution of the azimuthal velocity to predict the cavity size for a given cavitation number. The vortex model and cavity size prediction are discussed in Sections 2.1 and 2.2, respectively. The predicted cavity size has been used in an empirical relation for the centre frequency and maximum level of the broadband hump. This procedure is presented in Section 2.3. A flow chart of the method is shown in Figure 1. The spectral shape of the source level spectrum and conversion to hull pressure spectrum and URN spectrum is presented in Section 3. Comparison of predicted and measured spectra is presented in Section 4. The results of the method are discussed in Section 5.

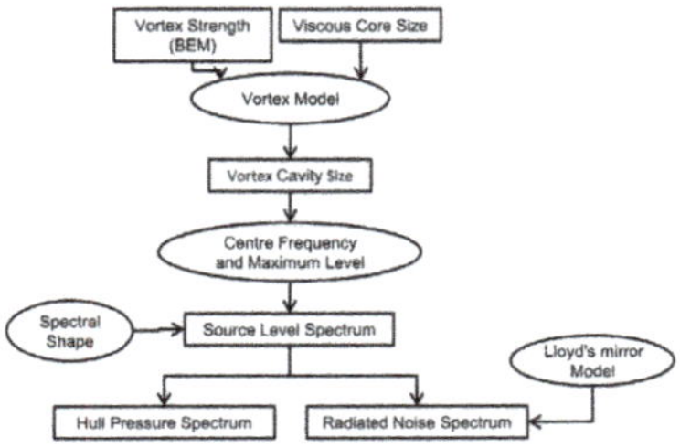

Figure 1. Flow chart of the Empirical Tip Vortex cavity (ETV) model for broadband hull pressure fluctuations and underwater radiated noise.

2. Prediction of the Maximum Source Level

2.1. Vortex Models

The relation between cavitation number and cavity size for a vortex cavity can be obtained from the radial distribution of the azimuthal velocity v_θ. It has been shown [25] that the variation of pressure p with radius r can be computed within 10 to 15% accuracy using the relation

$$p - p_\infty = -\int_r^\infty \rho \frac{v_\theta^2}{r}\, dr. \tag{1}$$

Theoretical analysis has shown that the variation of cavity size with cavitation number for an analytical solution of a columnar cavitating vortex is almost identical to that of a non-cavitating vortex if the viscous core size does not change [26]. For the non-cavitating vortex, the cavity radius was taken as the radius where pressure equals vapour pressure. Therefore, only non-cavitating vortex models will be considered in the following.

The most simple vortex model is the potential flow vortex of which the azimuthal velocity is given by

$$v_\theta(r) = \frac{\Gamma_\infty}{2\pi r}, \tag{2}$$

where Γ_∞ corresponds to the vortex strength, which is defined as the circulation Γ at large r. A disadvantage of this model, referred to in the following as inviscid vortex, is the singular behavior at the centre, which leads to an infinitely low pressure at that location. In real flow, this singular behavior does not appear due to viscous effects. An analytical solution for a columnar vortex in laminar flow is the Lamb–Oseen vortex [27] with viscous core radius r_v,

$$v_\theta(r) = \frac{\Gamma_\infty}{2\pi r}\left\{1 - \exp\left[-\varsigma(r/r_v)^2\right]\right\}. \tag{3}$$

The parameter ς is a constant that is selected such that v_θ has its maximum value at $r = r_v$, which gives $\varsigma = 1.2564$. The Lamb–Oseen vortex can be interpreted as the solution of a potential flow vortex of which the vorticity has been distributed by diffusion. For tip vortices, the distribution of vorticity is also influenced by the roll-up of the trailing vortex sheet that is generated by the spanwise circulation distribution on wing or propeller. The velocity distribution resulting from such a process has been proposed by Proctor et al. [28] and is given by

$$v_\theta(r) = \begin{cases} 1.0939\frac{\Gamma_\infty}{2\pi r}\{1 - \exp[-\beta(1.4r_v/B)^p]\}\left\{1 - \exp\left[-\varsigma(r/r_v)^2\right]\right\} & r \le 1.4r_v \\ \frac{\Gamma_\infty}{2\pi r}\{1 - \exp[-\beta(r/B)^p]\} & r > 1.4r_v \end{cases}. \tag{4}$$

In this formulation, B is the length scale related to the vorticity roll-up region, taken as the wingspan of the aircraft by Proctor. The parameters β and p are non-dimensional empirical parameters, which take the values $\beta = 10$ and $p = 0.75$. The variation of the azimuthal velocity with radius is presented in Figure 2a and the corresponding pressure distribution is presented in Figure 2b. The parameter B was taken as $B/r_v = 20$.

2.2. Prediction of Cavity Size

The vortex models described in Section 2.1 have been evaluated using experimental datasets for a wing, Pennings et al. [29], and for a propeller, Kuiper [30]. The dataset for the wing consists of velocity measurements obtained with stereo particle image velocimetry (spiv) at several stations downstream of the tip and of vortex cavity size measurements as a function of cavitation number σ_V,

$$\sigma_V = \frac{p_w - p_v}{\frac{1}{2}\rho V^2}, \tag{5}$$

with p_w the static pressure at the location of the wing, p_v the vapour pressure and V the free-stream velocity of the cavitation tunnel. The cavity size was obtained from image analysis using high speed video (hsv). The wing was of elliptical planform (half-span $B/2 = 0.15\,\mathrm{m}$) and tested at three angles of attack. The velocities were averaged in the circumferential direction and the resulting azimuthal velocity distribution for 7 deg angle of attack is presented in Figure 3a, which also shows the results of the vortex models. In these models, r_v was taken as the radius at which the measured v_θ has the maximum value. The Proctor vortex is capable of describing the measured velocity distribution very well after adjusting the coefficients for β and p. The strength for the Proctor vortex was set to 80% of the maximum value of the spanwise circulation distribution on the wing. This maximum value can be computed from the measured lift coefficient assuming that the circulation distribution is given by the analytical formulation. The resulting variation of r_c with σ_V of the Proctor vortex also shows very good agreement with experimental data, as shown in Figure 3b. The inviscid and Lamb–Oseen vortex do not describe the velocity distribution and the cavity size variation well. In the results shown, the vortex strength for these vortices has been adapted such that the average cavity size is well predicted. It is remarked that the good prediction of the variation of cavity size with cavitation number by a vortex model for non-cavitating flow is not understood from a physical point of view. Experiments for cavitating flow suggest that the viscous core size used in the vortex model changes in the presence of cavitation, which results in a different prediction of the cavity size [26].

The data set for the propeller consists of the variation of r_c with cavitation number σ_n,

$$\sigma_n = \frac{p_s - p_v}{\frac{1}{2}\rho n^2 D^2}, \tag{6}$$

with p_s the static pressure at the shaft centre, n the shaft rotation rate [rev/s] and D the propeller diameter. The propeller was four-bladed, specifically designed to show tip vortex cavitation only, and has diameter $D = 0.34\,\mathrm{m}$ and pitch ratio $P_{0.7}/D = 0.74$. The propeller was tested in open water conditions at three advance ratios by varying the free-stream velocity of the cavitation tunnel. The advance ratio J is defined as $J = V/nD$. Each advance ratio was tested at two shaft rotation rates resulting in two different Reynolds numbers. The Reynolds number was defined as $\mathrm{Re}_n = nD^2/\nu$, with ν the kinematic viscosity.

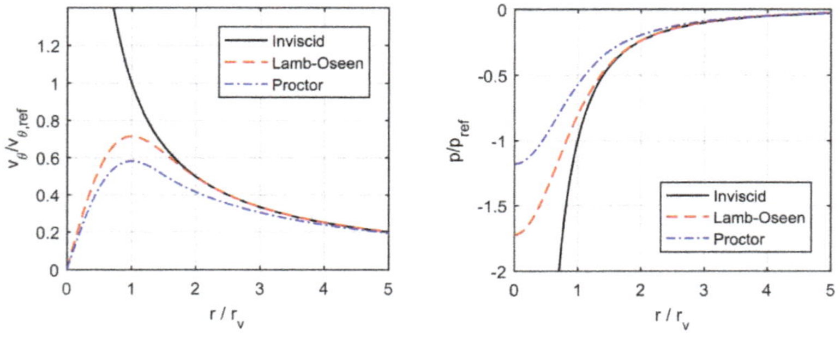

(**a**) Non-dimensional azimuthal velocity distribution (**b**) Non-dimensional pressure distribution

Figure 2. Comparison between various vortex models. The reference values are taken from the values for the inviscid vortex at $r = r_v$.

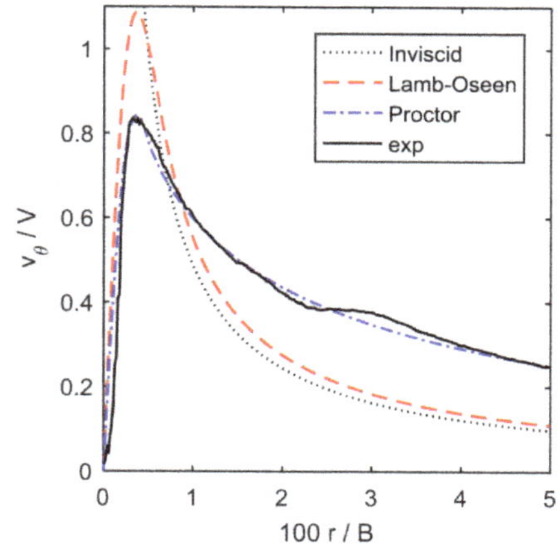

(**a**) Non-dimensional azimuthal velocity distribution

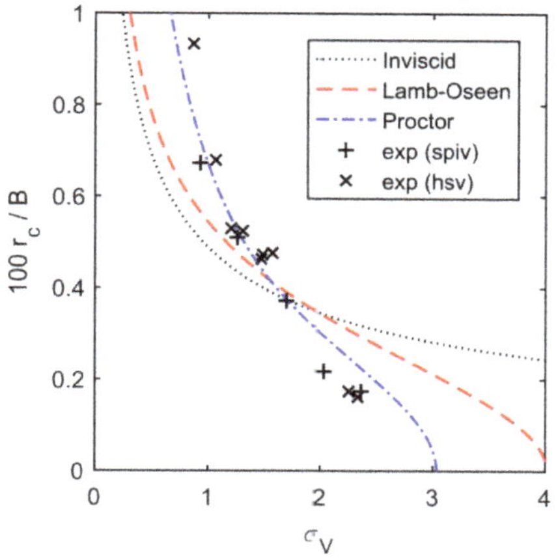

(**b**) Variation of cavity size with cavitation number

Figure 3. Measured and fitted azimuthal velocity distribution and cavity size variation for a wing of elliptical planform at 7 deg angle of attack. Experiments by Pennings et al. [29].

The cavity size was estimated using image analysis of photographs. Roughness was applied on two blades, except at the leading edge to investigate its effect and two blades were kept smooth. However, the geometry of one of the smooth blades differed from the other blades and results for this blade were excluded from the comparison. Overall, the change in cavity size due to the application of roughness, as well as due to variations of air content in the tunnel, was within the measurement uncertainty so the effect could not be determined.

For the present method, the tip-vortex strength was obtained from the circulation computed by PROCAL at 0.95R, with $R = D/2$, and the viscous core size was obtained from model-scale measurements by Jessup [31] scaled according to Equation (7), which is discussed later. Results are shown for the Proctor vortex model only. Initially, the values for β and p were kept identical to the values obtained for the wing, but these did not result in a good agreement with experimental data. This indicates that the vorticity roll-up of the wake of the propeller differs from that of the wing. A new tuning process was then applied for β and p, which resulted in a reasonable agreement with experimental data, as shown in Figure 4. The variation in Reynolds number in the experiments results in a variation in r_v in the vortex model. In the present method, this dependency was taken into account by using the Reynolds number scaling for cavitation inception of McCormick [32]. The viscous core size can then be computed from the measured value by Jessup [31] referred to as *ref*,

$$\frac{r_v}{c} = \left(\frac{r_v}{c}\right)_{ref} \left(\frac{\text{Re}}{\text{Re}_{ref}}\right)^{-m/2}, \quad (7)$$

with c the chord length and Re the Reynolds number for the chord length and the resultant velocity at 0.95R. The value for parameter m is computed using the formulation by Shen et al. [33] resulting in a value of 0.38 and 0.37 for the lower and higher Reynolds number, respectively. The effect of the change in Reynolds number on cavity size is well predicted by this method. The vortex model shows that the effect of the viscous core size on the cavity size decreases with increasing cavity size, as shown in Figure 4a.

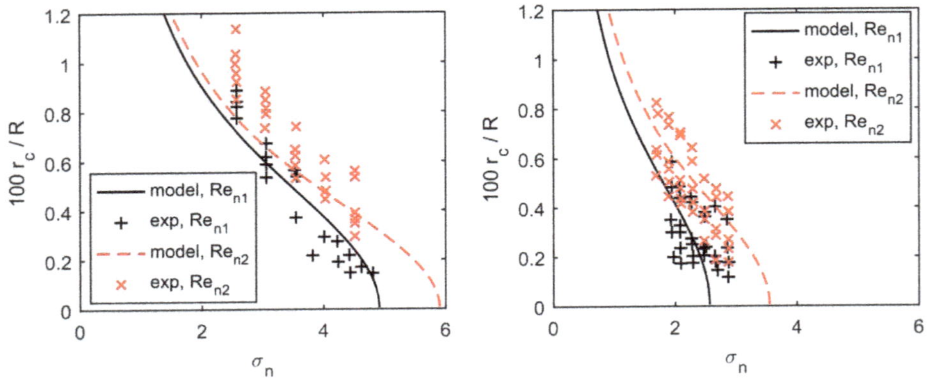

(**a**) $J = 0.4$

(**b**) $J = 0.5$

Figure 4. Measured and fitted cavity size variation for a propeller at two advance ratios. Experiments by Kuiper [30], $\mathrm{Re}_{n1} = 1.38 \times 10^6$, $\mathrm{Re}_{n2} = 2.76 \times 10^6$.

Results were also analyzed for other angles of attack for the wing and other advance ratios for the propeller. Good agreement was obtained for the other two angles of attack of the wing (5 and 9 deg) using the same settings for the vortex model (see also [26]). However, for the lowest advance ratio of the propeller, $J = 0.3$, and hence the case with highest vortex strength, the predicted cavity size was smaller than observed by experiments. A possible reason for this underprediction is that, in the experiment, a combination of sheet and vortex cavitation was present. Unfortunately, no detailed images are available for these conditions so this has not been further investigated.

2.3. Prediction of Source Levels

The semi-empirical formulation for the centre frequency and level of the hump was developed using non-dimensional parameters. In the present study, use has been made of a combination of model-scale, measured in the Depressurized Wave Basin (DWB), and full-scale experimental data for hull pressures obtained for twin-screw vessels. The rms-pressure amplitudes were converted into k_p-values according to Equation (8), where Δf corresponds to the resolution bandwidth of the amplitude spectrum and f_{bpf} to the blade passage frequency. The levels were converted to decibel values according to Equation (9). The frequencies were made non-dimensional with f_{bpf}:

$$k_p = \frac{p_{rms}}{\rho n^2 D^2 \sqrt{\Delta f / f_{bpf}}}, \tag{8}$$

$$k_p[dB] = 120 + 20 \log_{10}(k_p). \tag{9}$$

The resulting spectrum can then be interpreted as a non-dimensional power density spectrum or non-dimensional *rms* amplitude density spectrum. The hull-pressure spectra were converted to source levels by correcting for the propeller-hull clearance and the solid boundary factor. Only the centre pressure sensor located directly above the 12 o'clock position of the propeller was considered, as measurements at this location were available for all data sets.

To focus on the broadband character, the spectrum was converted to 1/6 octave band levels and scaled back to a power density spectrum, an example of which is given in Figure 5. The resulting spectrum then becomes a smoothened power density spectrum. Next, the centre frequency, the maximum level and the bandwidth of the hump were determined by a curve fit. In Figure 5, the centre frequency was determined at $f_c / f_{bpf} = 7.3$.

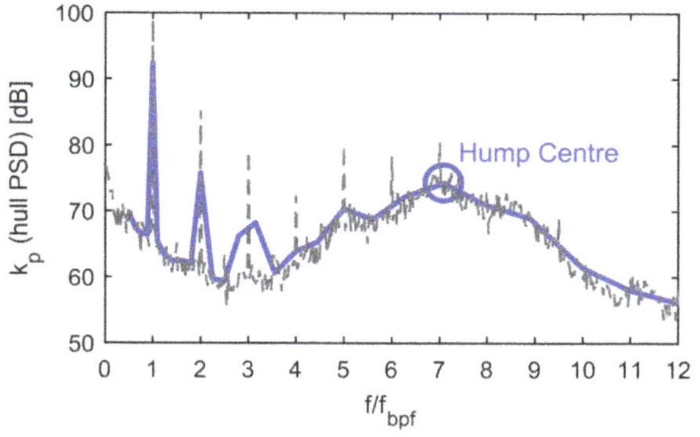

Figure 5. Example of a non-dimensional hull pressure power density spectrum of a two-bladed research propeller. The solid line is the 1/6 octave band smoothened spectrum.

The parameters for the semi-empirical model are the cavity size r_c made non-dimensional with diameter D, the cavitation number σ_n and the number of blades Z. The empirical relation for the maximum level, which equals the level of the hump centre, is given by

$$k_{p,\max} = a_p + 20\log_{10}\left\{\left(\frac{r_c}{D}\right)^{\kappa}\sqrt{Z}\right\}, \tag{10}$$

in which the non-dimensional empirical constants a_p and κ were obtained by curve fitting of Equation (10) to the experimental data. Raestad [21] has used $\kappa = 2$ in his formulation, but the value $\kappa = 3$ is used in the present method as this is closer to the trend seen in the datasets. The contribution from the total number of blades is summed as a set of incoherent sources. The formulation for the centre frequency is based on theoretical considerations and is given by

$$\frac{f_c}{f_{bpf}} = b_f \, \frac{1}{r_c/D} \frac{\sqrt{\sigma_n}}{Z}, \tag{11}$$

with b_f a non-dimensional empirical constant. Raestad derives this relation from the resonance frequency of a bubble, although the relation has also been derived from the dispersion relation for a cavitating vortex [3].

The vortex strength for the vortex models was obtained from the maximum circulation value at 0.95R of all blade positions computed by PROCAL using model-scale measured wake fields. For the analysis of sea trial conditions, these wake fields were scaled using a method developed in a Cooperative Research Ships (CRS) working group [34]. This method decomposes the ship wake into the contribution from the hull boundary layer and from the wake of struts and shaft, and scales these components separately using semi-empirical formulations.

The ships in the experimental dataset used to derive Equations (10) and (11), and to determine the empirical constants a_p, κ and b_f, consist of twin-screw vessels equipped with fixed-pitch podded propellers, and with fixed or controllable pitch propellers with an open-shaft arrangement. The test conditions were close to or in the design condition for varying ship speeds.

During the process of determining the empirical constants, it became apparent that the experimental datasets had to be divided in separate groups. A distinction was made between cases for which tip-vortex cavitation was dominant and cases for which, close to the tip, sheet cavitation was also present. Figure 6 shows examples of these two situations. This behaviour was also expected in the fit of the cavity size for the propeller tip vortex cavitation as discussed in Section 2.2.

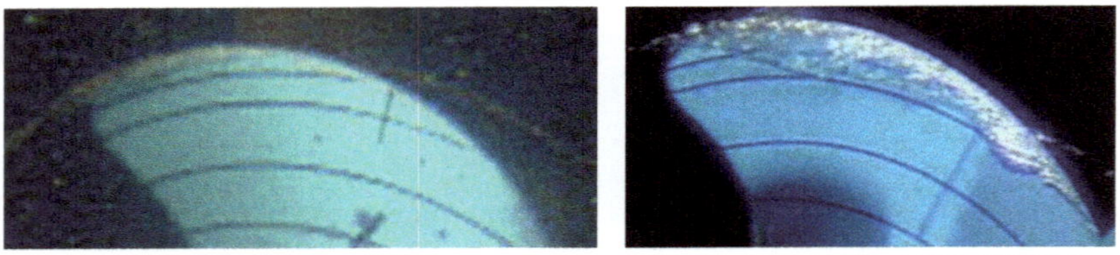

(a) Tip vortex cavitation (b) Tip-sheet and tip-vortex cavitation

Figure 6. Examples of cavitation patterns on the propellers used for the development of the method.

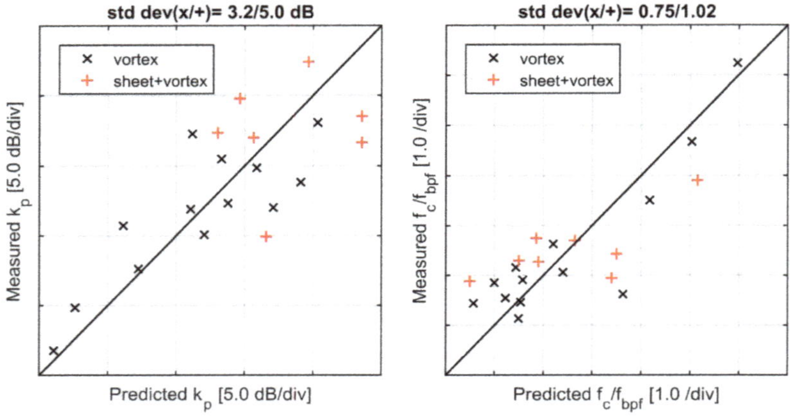

(a) Maximum level of hump (b) Centre frequency of hump

Figure 7. Comparison between measured and predicted levels and frequency of the centre of the hump for the cases used in the development of the method.

The accuracy of the fit for the maximum level of the hump and for the centre frequency for the two groups is presented in Figure 7a,b, respectively. An example of the fit of the noise level for an individual data set, not included in Figure 7, is presented in Figure 8. The propeller is a two-bladed research propeller with skew and tip-unloading tested for a range of cavitation numbers and thrust coefficients [3]. The cavitation pattern on this propeller is a cavitating vortex structure that is generated on the leading edge.

All results presented in Figures 7 and 8 are for the Proctor vortex model. Similar results could also be obtained with the other vortex models but with different empirical constants showing that, for the test-cases considered, the cavity size is in general much larger than the viscous core size. The results in Figure 7 are quite reasonable for the group with vortex cavitation although the standard deviation of the frequency is rather high. In particular, one condition is poorly predicted. The standard deviation for the group with tip-sheet and vortex cavitation is significantly higher than the group with tip-vortex cavitation only. However, Figure 8 shows that the trend for a single propeller with tip-sheet and vortex cavitation is well predicted by the used parameters.

The relation between maximum level, and centre frequency, and cavity size can easily be converted to a relation in terms of propeller thrust coefficient for the inviscid vortex model. The relation between cavity size, cavitation number and circulation is then given by

$$\frac{r_c}{D} = \frac{1}{2\pi}\frac{\Gamma_\infty}{nD}\frac{1}{\sqrt{\sigma_n}}. \tag{12}$$

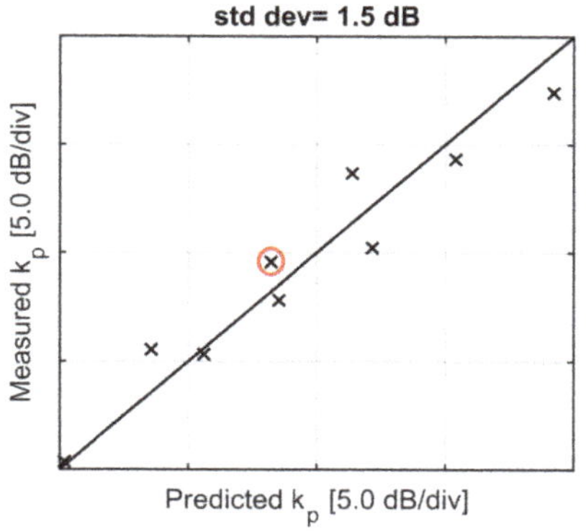

(**a**) Maximum level of hump

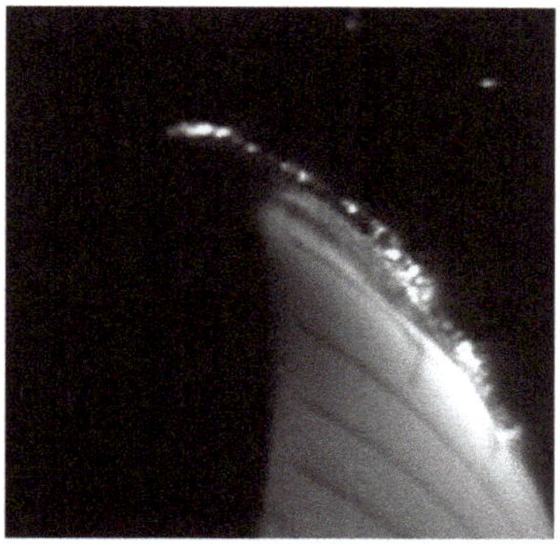

(**b**) Typical cavitation pattern

Figure 8. Comparison between measured and predicted maximum level of the hump for a two-bladed research propeller tested at various advance ratios and cavitation numbers. The cavitation pattern corresponding to the encircled symbol is shown on the right.

The non-dimensional vortex strength can be written as the product of the propeller blade thrust coefficient K_T/Z and a tip loading parameter τ. The relations for the maximum level and centre frequency then read

$$k_{p,\max} \propto \left(\frac{r_c}{D}\right)^{\kappa}\sqrt{Z} \propto \left(\frac{\tau K_T}{Z\sqrt{\sigma_n}}\right)^{\kappa}\sqrt{Z}, \tag{13}$$

$$\frac{f_c}{f_{bpf}} \propto \frac{1}{r_c/D}\frac{\sqrt{\sigma_n}}{Z} \propto \frac{\sigma_n}{\tau K_T}. \tag{14}$$

If it is assumed that K_T and J do not change with varying ship speed, the pressure amplitudes scale with ship speed V as $p_{rms} \propto V^{2+\kappa}$ and the centre frequency as $f_c \propto V^{-1}$.

3. Shape of the Spectrum

3.1. Source Level

The shape of the spectrum of cavitation noise is described by e.g., [35–37]. The spectrum can be divided in a low-frequency part and a high-frequency part. The low-frequency part is characterized by a hump due to the overall growth, collapse and rebounds of the cavity. The high frequency part is related to the final phase of the collapse process during which velocities may approach or exceed the speed of sound, and compressibility effects become important. The collapse and rebounds of the smaller size bubbles generated by the collapse of the large scale cavity also contribute to the high frequency part.

In the present method, it is assumed that f_c corresponds to a resonance frequency of the cavitating vortex. The related pressure signal can then be interpreted as a damped oscillatory signal, or, in a simplified way, as an oscillatory signal multiplied with a rectangular window. The Fourier transform of the latter is given by a sinc-function. Therefore, the shape of the spectrum, presented in decibel values, has been defined as

$$H_h(f) = 20\log_{10}\left\{\operatorname{sinc}\left(\frac{f-f_c}{0.830\,\Delta f_{-6dB}}\right)\right\}, \tag{15}$$

where Δf_{-6dB} corresponds to the bandwidth of the hump where the pressure amplitude is half the maximum amplitude. A small value for Δf_{-6dB} corresponds to a time trace with multiple rebounds (small damping) resulting in a narrow hump in the spectrum. A large value for Δf_{-6dB} corresponds to a highly damped system resulting in a wide hump in the spectrum. Analysis of experimental data suggested that Δf_{-6dB} is proportional to f_c, although the scatter is high.

The shape of the spectrum at much lower and higher frequencies than the centre frequency of the hump is modelled separately. The simple model used here consists of prescribed slopes at low and high frequency in a power density spectrum

$$H_s(f) = 10 \log_{10} \left\{ \frac{2(f/f_c)^{\alpha_l}}{1 + (f/f_c)^{\alpha_l - \alpha_h}} \right\}, \tag{16}$$

in which α_h corresponds to the slope for high frequency, with typical value $\alpha_h = -2$, and α_l corresponds to the slope for low frequency, for which a value $\alpha_l = 4$ is used as suggested in [35]. For different values of α_l and α_h, the maximum value of H_s is different from 0 dB, which needs to be corrected for.

The resulting spectrum is then taken as a weighted sum of powers of the two spectral functions

$$H(f) = 10 \log_{10} \left\{ \alpha 10^{H_h(f)/10} + (1-\alpha) 10^{H_s(f)/10} \right\}, \tag{17}$$

where α is a user defined parameter. For practical applications, this parameter is computed from

$$\Delta L_\alpha = 10 \log_{10}(1-\alpha), \tag{18}$$

with ΔL_α the difference in noise level between the maximum of the two-slope function and the maximum of the sinc function.

Examples of the shape of the spectrum for the source level are given in Figure 9 for three values of α, where the curve for $\alpha = 0.0$ corresponds to the spectrum for the two-slope function and the curve for $\alpha = 1.0$ corresponds to the spectrum by the sinc function. In its current version, the default value is $\alpha = 0.8$. For some cases, however, the default values do not give a good representation of the measured spectrum. It is remarked that the maximum absolute value of the argument of the sinc function has been limited to avoid the presence of 'sidelobes' in the spectrum.

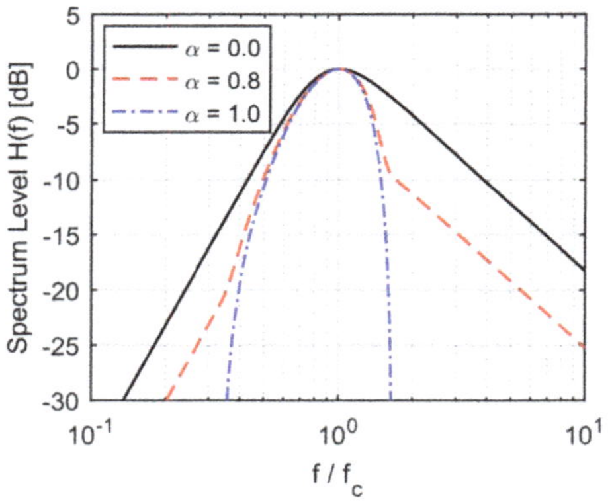

Figure 9. Examples of the shape of the spectrum for different values of α.

3.2. Hull-Pressure and URN Spectra

The hull-pressure spectrum is obtained from the source level spectrum by correcting for a solid boundary factor and for the propeller-hull clearance. The spectrum of the URN levels is obtained from the source level spectrum by correcting for the interference with the free surface (Lloyd's mirror). A simple formula to correct for the propagation loss (PL) due to Lloyd's mirror is given by Ainslie [38]

$$PL_{LM} = -10\log_{10}\left[\frac{1}{2} + \frac{1}{4k^2 d_s^2 \sin^2\theta}\right], \tag{19}$$

where k corresponds to the acoustic wave number, $k = 2\pi f/c$ with c the speed of sound, d_s to the submergence depth of the cavity collapse and θ to the hydrophone depression angle. The function is compared in Figure 10 to the theoretical formulation of the Lloyd's mirror effect for a flat free surface [39]. The frequency in Figure 10 is made non-dimensional with the critical frequency f_0,

$$f_0 = \frac{c}{4d_s \sin\theta}. \tag{20}$$

The interference patterns at frequencies above the critical frequency are usually not observed in ship noise data because of the presence of waves and bubbles near the free surface and the varying position between ship and hydrophone. For that reason, the Ainslie model assumes that, at high frequencies, the free surface leads to an uncorrelated image source.

The source level spectrum (SL) can now be defined as

$$SL(f) = L_{p,\max} + H(f)\left[\text{dB, re}\,1\,\mu\text{Pa}^2\text{m}^2/\text{Hz}\right], \tag{21}$$

with $L_{p,\max}$ the maximum level of the hump of the power density spectrum, which is the dimensional value of $k_{p,\max}$. The spectrum of the underwater radiated noise level (RNL) is computed from the SL by correcting for Lloyd's mirror

$$RNL(f) = SL(f) + PL_{LM}(f)\left[\text{dB, re}\,1\,\mu\text{Pa}^2\text{m}^2/\text{Hz}\right]. \tag{22}$$

The RNL levels, or URN levels in general, are often presented in one-third-octave band levels denoted by $RNL_{1/3}$ and $URN_{1/3}$, respectively. The present model is referred to as ETV-2, the 2nd version of the Empirical Tip Vortex cavity model.

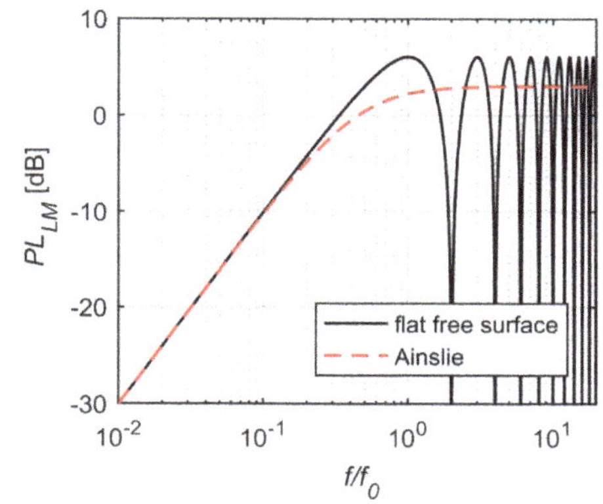

Figure 10. Propagation loss due to the Lloyd's mirror effect.

4. Comparison of Predicted and Measured Hull Pressures and URN

An example of the resulting spectrum of hull pressures predicted by the ETV model for the two-bladed research propeller is presented in Figure 11. Overall, the hump is well represented by the default spectral shape. The values of the empirical parameters for the maximum level and centre frequency were obtained from a dedicated fit to the test series for this propeller. The comparison between the resulting predicted and measured levels is given in Figure 8.

Examples of spectra for ship configurations used in the database are presented for model- and full scale in Figures 12 and 13, respectively. Both cases are part of the vortex-cavitation group for which results are presented in Figure 7. The model- test results show some disturbance at the eighth harmonic of the blade passage frequency generated by the propeller drive train. However, this frequency is located at a much higher frequency than the broadband hump due to the cavitating tip vortex. Both figures are examples of cases that are well predicted by the model. As shown in Figure 7, there are also some cases for which the maximum level and centre frequency are predicted less accurately.

The model- and full-scale data are all made non-dimensional using Equation (8). Since no significant differences in the trend of the maximum level and centre frequency with scale could be discerned, it is concluded that Equation (8) can be used to scale the broadband pressure spectra from model scale to full scale. In fact, the equation can also be obtained by rewriting the formulation that is used to scale the (low-frequency) URN by cavitation as presented by e.g., Strasberg [40] and Bark [41].

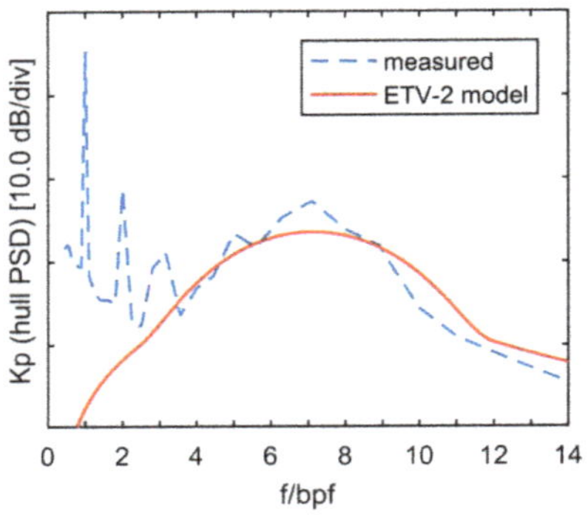

Figure 11. Example of the measured and predicted non-dimensional hull pressure spectrum for a two-bladed research propeller.

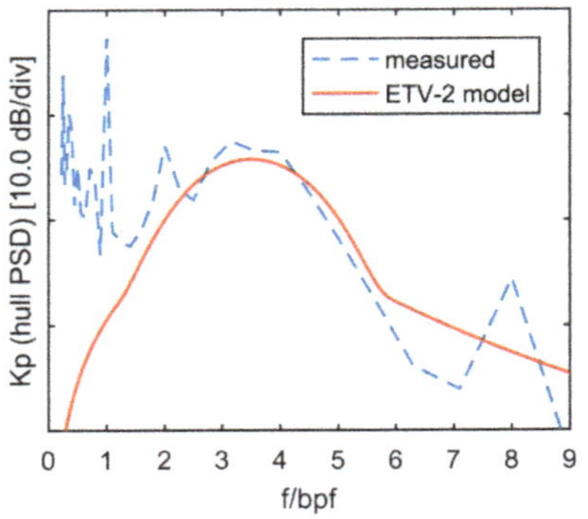

Figure 12. Example of the measured and predicted non-dimensional hull pressure spectrum for a ship propeller of the database tested at model scale.

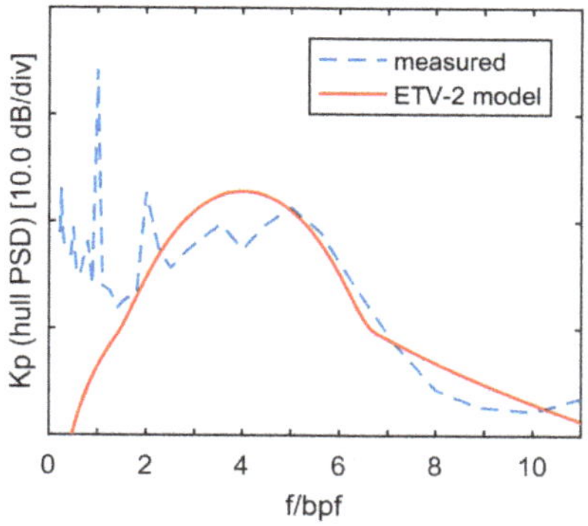

Figure 13. Example of the measured and predicted non-dimensional hull pressure spectrum for a ship propeller of the database tested at full scale.

The ETV-model has also been applied to predict the broadband hull-pressure levels on an 85 m combi freighter equipped with a single controllable pitch propeller. This ship was not in the database and, being a single screw vessel, it also does not resemble any ship in the data base. The sea trials for the ship, which included URN measurements in deep water, were performed by DAMEN, DNV-GL and MARIN and were financed by the CRS BROADBAND2 working group. The effective wake field for the PROCAL computation was obtained from a coupled RANS-BEM procedure [42,43]. Details of this data set are given in [44], which shows a comparison of the full-scale URN levels with predictions from model tests in the DWB. The cavitation pattern on the propeller was characterized by strong tip vortex cavitation. The values for the empirical constants derived using the complete database were used in the ETV method.

The comparison of broadband hull pressures as predicted by the ETV method and as measured during the sea trial at 10 knots is reasonable, as illustrated in Figure 14a. It is notable that the hump for this single screw ship is overpredicted by the ETV method. The URN of this vessel is dominated by a very pronounced hump that will be discussed later. This hump is not so obvious in the pressures measured by the centre transducer (P1) but can clearly be seen in the spectrum of the pressure sensor located closer to the collapse of the cavity (transducer P2).

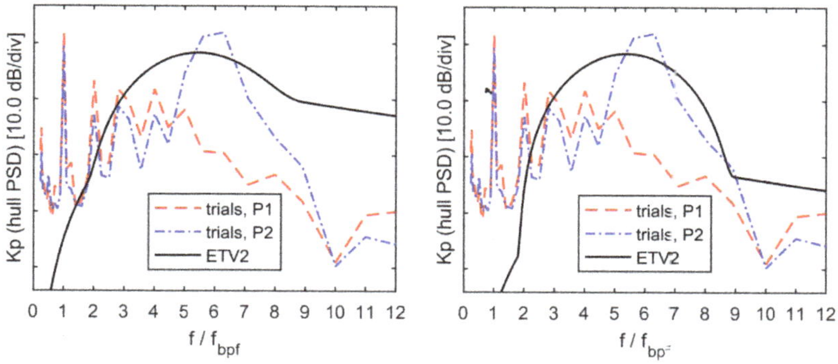

(**a**) Default values for the ETV-method (**b**) Adapted value for ΔL_α

Figure 14. Measured and predicted non-dimensional hull pressure spectrum for the combi-freighter at a ship speed of 10 knots. P1 corresponds to the centre transducer and P2 is located closer to the cavity collapse. Experimental data measured by DAMEN and MARIN.

In general, the results were considered to be quite acceptable for a range of pitch settings and shaft rotation rates as long as the cavitation was present on the back side of the blade. In its present form, the model was not capable of predicting the broadband pressure spectrum due to face side cavitation for a low pitch setting.

The model that was developed using data for hull pressures has also been applied to predict the URN due to propeller cavitation. The first test case considered for the prediction of URN is the single-screw combi freighter discussed above. The noise levels are presented in Figure 15a using the default values. The centre of the hump is well predicted in terms of level and frequency, but, at high frequencies, the predicted noise levels are approximately 10 dB too high. It is seen that the hump is very pronounced. The difference in noise levels between the hump and the high frequency region can easily be adjusted with the parameter ΔL_α. Results for $\Delta L_\alpha = -20$ dB are presented in Figure 15b. The default value for the slope of the spectrum at high frequencies is in good agreement with the sea-trial data. The hull-pressure prediction for this setting is also in better agreement with the experimental data (see Figure 14b).

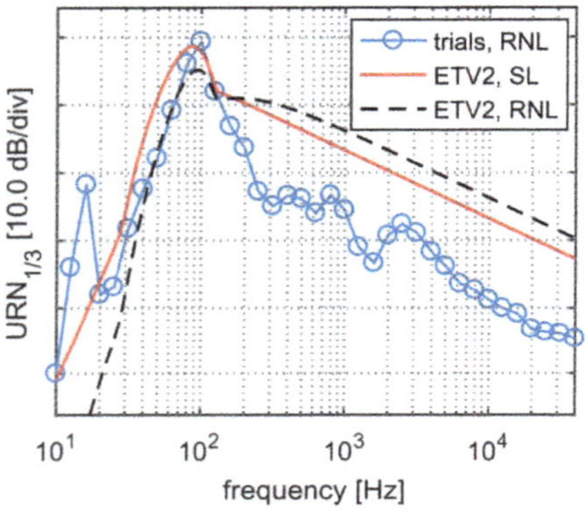

(a) Default values for the ETV-method

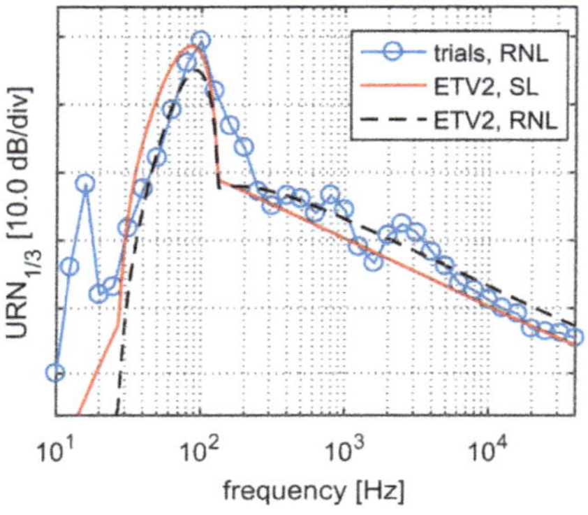

(b) Adapted value for ΔL_α

Figure 15. Measured and predicted radiated noise spectrum in one-third octave band levels for the combi-freighter at a ship speed of 10 knots. Experimental data by DNV-GL.

The second data set analyzed is that of the (twin-screw) cruise vessel Statendam as reported by Kipple [45]. The length between perpendiculars is 182 m and the ship is driven by controllable pitch propellers. The URN measurements were performed in deep water. The measured SL are directly taken from [45] and the RNL are computed from the received noise levels at 500 yards as reported in [45]. All measured levels are converted to levels at 1 m distance assuming spherical spreading loss. The applied formulation in [45] to compute the source levels from the measured radiated noise levels is unknown, but the applied propagation loss is similar to Equation (19), except for high frequencies where the applied correction is 0 dB instead of 3 dB.

Results with default values are presented in Figure 16a. The maximum level is very well predicted, but the measured values in that frequency region are reported to be mainly due to machinery noise. At high frequencies, the agreement is not as good because the assumed slope of the spectrum is smaller than in the experiments. Adjustment of the high frequency slope to $\alpha_h = -1.6$ gives good agreement (see Figure 16b).

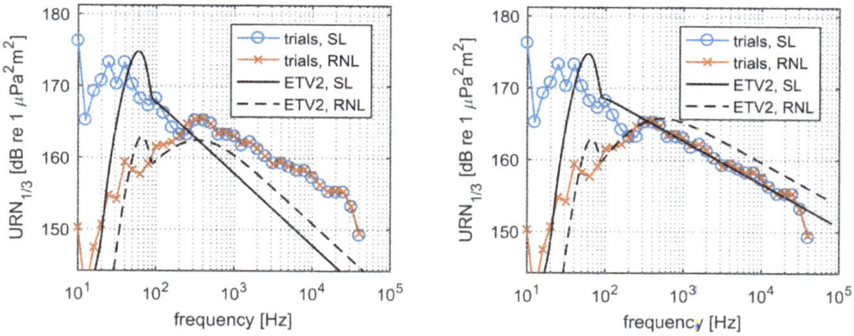

(**a**) Default values for the ETV-method (**b**) Adapted value for the high frequency slope α_h

Figure 16. Measured and predicted radiated noise spectrum in one-third octave band levels for the cruise vessel at a ship speed of 18 knots. Experimental data by Kipple [45].

5. Discussion

The ETV-method has been developed to predict the broadband noise by propeller tip-vortex cavitation as present on propellers of twin screw vessels. Both model- and full-scale hull-pressure data were used for this purpose. The tip-vortex cavity is attached to the blade and might also be interpreted as an attached leading-edge vortex cavity that is directly connected to the tip vortex.

The majority of the test conditions were at or close to the design condition of the propeller. This restriction in variability of conditions was deemed necessary to avoid the influence of sheet cavitation on the measured hull-pressure data. Nevertheless, the cavitation observations showed that, for some of the cases, a combination of tip vortex cavitation and sheet cavitation was present. This sheet cavity detaches from the leading edge upstream of the blade tip. Due to the skew of the blade, the re-entrant jet flow at the cavity closure has a strong radial component, which is why it is sometimes called a side-entrant jet [46]. This side-entrant jet forms a cavity closure vortex [47] that is chordwise oriented in the tip region and that collapses when the blade leaves the wake peak. The propellers with such a cavitation pattern have been considered as a separate group.

The broadband noise levels for the propellers with such sheet and vortex cavitation was significantly higher than for propellers with vortex cavitation alone, requiring adjustment of the empirical parameter for the maximum level of the hump. Surprisingly, the adopted methodology still seems to apply to such a cavitation pattern indicating that the noise due to the collapse of the sheet-cavity closure vortex follows a similar trend as for the tip-vortex cavity. Another possibility is that the collapse of the closure vortex excites the cavitating tip-vortex and that the tip-vortex cavity is still responsible for the broadband hump. The standard deviation of the fit for this small group of propellers with sheet cavitation was higher than for the group with tip-vortex cavitation alone. Due to the two groups of tuning parameters, some information on the expected propeller cavitation pattern is required in the application of the method. This information can for instance be obtained from a sheet cavitation computation by the applied boundary element method. Such information can probably also be used to improve the fit for the group of propellers with sheet cavitation.

The test-cases used for the development of the method were all twin-screw vessels. Despite the limited variability of wake fields for these ships, the broadband hump was also well predicted for a single-screw vessel equipped with skewed propeller blades. This gives some confidence that the method is applicable to a wider range of wake fields. The most important limitation of the method is that it only describes the noise due to tip-vortex cavitation and, as discussed above, due to a sheet-cavity closure vortex in the tip region. The method is therefore mostly suited for the analysis of tip-unloaded (and skewed) propeller blades and is not expected to work for heavy-loaded propellers with extensive sheet cavitation.

The shape of the spectrum was not studied in much detail in the present study and requires further research. This not only holds for the higher frequency part of the spectrum, as shown by the URN prediction of the two test-cases shown, but also for the width of the broadband hump.

6. Conclusions

A semi-empirical method to predict broadband hull-pressure fluctuations and radiated noise by cavitating tip vortices on tip-unloaded propeller blades has been presented. The method makes use of results obtained by a boundary element method and can therefore very easily be used for propeller noise evaluation, for instance in a propeller design process. The principal parameter in the method is the vortex cavity size from which the maximum noise level and the centre frequency of the broadband hump was computed.

The standard deviation of the model fit was about 3 dB for the maximum level and 0.75 for the non-dimensional centre frequency, as long as no sheet cavitation was present on the propeller. When sheet cavitation was present in the tip area, the methodology could still be used, but the empirical parameter to predict the maximum of the hump had to be adjusted. The standard deviation of the model fit was higher than for the group with only tip-vortex cavitation.

The method was also applied to predict hull pressure fluctuations and URN of two vessels that were not used to develop the empirical relations. A good prediction was obtained for the broadband hump, but the high frequency region of the URN required adjustment of the empirical parameters that describe the spectral shape.

During the development and application of the method, a number of issues were identified that require further investigation. An important aspect is the contribution of sheet cavitation and the interaction between the sheet-cavity closure vortex and the tip-vortex cavity in the generation of the broadband hump. Another aspect is the overall shape of the spectrum, in particular the relation between the levels of the hump and the levels in the high-frequency part. Physical aspects that influence the shape of the hump and the slope of the high frequency noise also require further investigation.

Acknowledgments: The present work has been performed within the Cooperative Research Ships (CRS) BROADBAND and BROADBAND2 working group, www.crships.org.

Conflicts of Interest: The author declares no conflict of interest.

References

1. Brubakk, E.; Smogeli, H. QE2 from turbine to diesel—Consequences for noise and vibration. In Proceedings of the IMAS Conference on the Design and Development of Passenger Ships, London, UK, 18–20 May 1988.
2. Carlton, J.S. Broadband cavitation excitation in ships. *Ships Offshore Struct.* **2015**, *10*, 302–307. [CrossRef]
3. Bosschers, J. Investigation of hull pressure fluctuations generated by cavitating vortices. In Proceedings of the First International Symposium on Marine Propulsors (SMP'09), Trondheim, Norway, 24–29 June 2009.
4. Pennings, P.C.; Bosschers, J.; Westerweel, J.; van Terwisga, T.J.C. Dynamics of isolated vortex cavitation. *J. Fluid Mech.* **2015**, *778*, 288–313. [CrossRef]
5. Bosschers, J. On the relation between tonal and broadband content of hull pressure spectra due to cavitating ship propellers. In Proceedings of the 9th International Conference on Cavitation (CAV2015), Lausanne, Switzerland, 6–10 December 2015.
6. Newman, M.; Abrahamsen, K. Measurement of underwater noise. In Proceedings of the Ship Noise and Vibration Conference, London, UK, 28–30 June 2007.
7. Foeth, E.J.; Bosschers, J. Localization and source-strength estimation of propeller cavitation noise using hull-mounted pressure transducers. In Proceedings of the 31st Symposium on Naval Hydrodynamics, Monterey, CA, USA, 11–16 September 2016.
8. Van Wijngaarden, H.C.J. *Prediction of Propeller-Induced Hull-Pressure Fluctuations*; Delft University of Technology: Delft, The Netherlands, 2011.

9. Johannsen, C.; van Wijngaarden, E.; Lücke, T.; Streckwall, H.; Bosschers, J. Investigation of hull pressure pulses, making use of two large scale cavitation test facilities. In Proceedings of the 8th International Symposium on Cavitation CAV2012, Singapore, 13–16 August 2012.
10. Bosschers, J.; Lafeber, F.H.; de Boer, J.; Bosman, R.; Bouvy, A. Underwater radiated noise measurements with a silent towing carriage in the Depressurized Wave Basin. In Proceedings of the 3rd International Conference on Advanced Measurement Technology for the maritime industry (AMT'13), Gdansk, Poland, 17–18 September 2013.
11. Tani, G.; Viviani, M.; Hallander, J.; Johansson, T.; Rizzuto, E. Propeller underwater radiated noise: A comparison between model scale measurements in two different facilities and full scale measurements. *Appl. Ocean Res.* **2016**, *56*, 48–66. [CrossRef]
12. Lafeber, F.H.; Bosschers, J. Validation of computational and experimental prediction methods for the underwater radiated noise of a small research vessel. In Proceedings of the PRADS2016, Copenhagen, Denmark, 4–8 September 2016.
13. Seol, H.; Suh, J.; Lee, S. Development of hybrid method for the prediction of underwater propeller noise. *J. Sound Vib.* **2005**, *288*, 345–360. [CrossRef]
14. Bosschers, J.; Vaz, G.; Starke, A.R.; van Wijngaarden, E. Computational analysis of propeller sheet cavitation and propeller-ship interaction. In Proceedings of the RINA MARINE CFD Conference, Southampton, UK, 26–27 March 2008.
15. Salvatore, F.; Testa, C.; Greco, L. Coupled hydrodynamics–Hydroacoustics BEM modelling of marine propellers operating in a Wakefield. In Proceedings of the First International Symposium on Marine Propulsors (SMP'09), Trondheim, Norway, 22–24 June 2009.
16. Matusiak, J. *Pressure and Noise Induced by a Cavitating Marine Screw Propeller*; Helsinki University of Technology: Espoo, Finland, 1992.
17. Brown, N.A. Thruster noise. In Proceedings of the Dynamic Positioning Conference of the Marine Technology Society, Houston, TX, USA, 12–13 October 1999.
18. Ligneul, P. Theory of tip vortex cavitation noise of a screw propeller operating in a wake. In Proceedings of the 17th Symposium on Naval Hydrodynamics, The Hague, The Netherlands, 13–14 June 1988.
19. Koronowicz, T.; Szantyr, J.A. Vortex cavitation as a source of high level acoustic pressure generated by ship propellers. *Acta Acust. United Acust.* **2006**, *92*, 175–177.
20. Berger, S.; Gosda, R.; Scharf, M.; Klose, R.; Greitsch, L.; Abdel-Maksoud, M. Efficient Numerical Investigation of Propeller Cavitation Phenomena causing Higher-Order Hull Pressure Fluctuations. In Proceedings of the 31st Symposium on Naval Hydrodynamics, Monterey, CA, USA, 11–16 September 2016.
21. Raestad, E. *Tip Vortex Index—An Engineering Approach to Propeller Noise Prediction*; The Naval Architect; The Royal Institution of Naval Architects: London, UK, July/August 1996, pp. 11–16.
22. Yamada, T.; Sato, K.; Kawakita, C.; Oshima, A. Study on prediction of underwater radiated noise from propeller tip vortex cavitation. In Proceedings of the 9th International Conference on Cavitation (CAV2015), Lausanne, Switzerland, 6–10 December 2015.
23. Vaz, G.; Bosschers, J. Modeling three dimensional sheet cavitation on marine propellers using a boundary element method. In Proceedings of the 6th International Symposium on Cavitation CAV2006, Wageningen, The Netherlands, 11–15 September 2006.
24. Bosschers, J.; Willemsen, C.; Peddle, A.; Rijpkema, D. Analysis of ducted propellers by combining potential flow and RANS methods. In Proceedings of the 4th International Symposium on Marine Propulsors (SMP'15), Austin, TX, USA, 31 May–4 June 2015.
25. Hommes, T.; Bosschers, J.; Hoeijmakers, H.W.M. Evaluation of the radial pressure distribution of vortex models and comparison with experimental data. In Proceedings of the 9th International Symposium on Cavitation (CAV2015), Lausanne, Switzerland, 6–10 December 2015.
26. Bosschers, J. An analytical and semi-empirical model for the viscous flow around a vortex cavity. *Int. J. Multiph. Flow* **2018**, in press. [CrossRef]
27. Lamb, H. *Hydrodynamics*, 6th ed.; Cambridge University Press: Cambridge, UK, 1932.
28. Proctor, F.; Ahmad, N.; Switzer, G.; Duparcmeur, F.L. Three-phased wake vortex decay. In Proceedings of the AIAA 2010-7991: AIAA Atmospheric and Space Environments Conference, Toronto, ON, Canada, 2–5 August 2010.

29. Pennings, P.C.; Westerweel, J.; van Terwisga, T.J.C. Flow field measurement around vortex cavitation. *Exp. Fluids* **2015**, *56*, 1–13. [CrossRef]
30. Kuiper, G. Cavitation Inception on Ship Propeller Models. Ph.D. Thesis, Delft University of Technology, Delft, The Netherlands, 1981.
31. Jessup, S.D. *An Experimental Investigation of Viscous Aspects of Propeller Blade Flow*; The Catholic University of America: Washington, DC, USA, 1989.
32. McCormick, B.W. On cavitation produced by a vortex trailing from a lifting surface. *J. Basic Eng.* **1962**, *84*, 369–379. [CrossRef]
33. Shen, Y.T.; Gowing, S.; Jessup, S. Tip vortex cavitation inception scaling for high Reynolds number applications. *J. Fluids Eng.* **2009**, *131*, 071301. [CrossRef]
34. Hally, D. *User's Guide for PIF-WAKE: The CRS PIF Wake Scaling Program for Single and Twin Screw Forms*; Technical Report ECR 2002-053; DRDC Atlantic: Ottawa, ON, Canada, 2002.
35. Fitzpatrick, H.M.; Strasberg, M. Hydrodynamic sources of sound. In Proceedings of the First Symposium on Naval Hydrodynamics, Washington, DC, USA, 24–28 September 1956; pp. 241–280.
36. Lövik, A. Scaling of propeller cavitation noise. In *Noise Sources in Ships*; Nordforsk: Stockholm, Sweden, 1981.
37. Blake, W.K. *Mechanics of Flow-Induced Sound and Vibration*; Academic Press Inc.: Cambridge, MA, USA, 1986.
38. Ainslie, M. *Principles of Sonar Performance*; Springer: Berlin, Germany, 2010.
39. Clay, C.C.; Medwin, H. *Acoustical Oceanography: Principles and Applications*; John Wiley & Sons Ltd.: Hoboken, NJ, USA, 1977.
40. Strasberg, M. Propeller cavitation noise after 35 years of study. In Proceedings of the ASME Noise and Fluids Engineering, Altanta, GA, USA, 27 November–2 December 1977.
41. Bark, G. Prediction of Propeller Cavitation Noise From Model Tests and Its Comparison With Full Scale Data. *J. Fluids Eng.* **1985**, *107*, 112–120. [CrossRef]
42. Starke, B.; Bosschers, J. Analysis of scale effects in ship powering performance using a hybrid RANS-BEM approach. In Proceedings of the 26th Symposium on Naval Hydrodynamics, Gothenburg, Sweden, 26–31 August 2012.
43. Rijpkema, D.; Starke, B.; Bosschers, J. Numerical simulation of propeller-hull interaction and determination of the effective wake field using a hybrid RANS-BEM approach. In Proceedings of the 3rd International Symposium on Marine Propulsors (SMP'13), Launceston, Australia, 5–8 May 2013.
44. Lloyd, T.; Lafeber, F.H.; Bosschers, J. Investigation and validation of procedures for cavitation noise prediction from model-scale measurements. In Proceedings of the 32nd Symposium on Naval Hydrodynamics, Hamburg, Germany, 5–10 August 2018.
45. Kipple, B. *Southeast Alaska Cruise Ship Underwater Acoustic Noise*; NSWCCD-71-TR-2002/S74; Naval Surface Warfare Center—Detachment Bremerton: Washington, DC, USA, 2002.
46. Foeth, E.-J.; van Terwisga, T.; van Doorne, C. On the Collapse Structure of an Attached Cavity on a Three-Dimensional Hydrofoil. *J. Fluids Eng.* **2008**, *130*, 071303. [CrossRef]
47. Bark, G.; Bensow, R.E. Hydrodynamic mechanisms controlling cavitation erosion. In Proceedings of the 29th Symposium on Naval Hydrodynamics, Gothenburg, Sweden, 26–31 August 2012.

Journal of
Marine Science and Engineering

MDPI

Article

DDES of Wetted and Cavitating Marine Propeller for CHA Underwater Noise Assessment

Ville M. Viitanen [1,*], Antti Hynninen [1], Tuomas Sipilä [1] and Timo Siikonen [2]

[1] VTT Technical Research Centre of Finland Ltd., 02150 Espoo, Finland; antti.hynninen@vtt.fi (A.H.); tuomas.sipila@vtt.fi (T.S.)

[2] Department of Mechanical Engineering, Aalto University, 02150 Espoo, Finland; timo.siikonen@aalto.fi

* Correspondence: ville.viitanen@vtt.fi

Received: 29 March 2018; Accepted: 3 May 2018; Published: 21 May 2018

Abstract: In this paper we present results of delayed detached eddy simulation (DDES) and computational hydroacoustics (CHA) simulations of a marine propeller operating in a cavitation tunnel. DDES is carried out in both wetted and cavitating conditions, and we perform the investigation at several propeller loadings. CHA analyses are done for one propeller loading both in wetted and cavitating conditions. The simulations are validated against experiments conducted in the cavitation tunnel. Propeller global forces, local flow phenomena, as well as cavitation patterns are compared to the cavitation tunnel tests. Hydroacoustic sources due to the propeller are evaluated from the flow solution, and corresponding acoustic simulations utilizing an acoustic analogy are made. The propeller wake flow structures are investigated for the wetted and cavitating operating conditions, and the acoustic excitation and output of the same cases are discussed.

Keywords: marine propeller; cavitation simulation; DDES; SST k-ω model; turbulence modelling; hydroacoustics

1. Introduction

The growing global shipping rates are generating increasing acoustic output in the underwater environment. The deep-ocean noise levels have grown over the past four decades, correlating with the observed increase in global shipping rates [1]. In addition, the adverse effects of shipping noise on marine mammals raised concern in the 1970s when the overlap between the main frequencies used by large baleen whales and the dominant components of noise from ships was noted [2]. Fish have also been observed to be disturbed by noise emitted from ships [3]. The sound emitted from naval and research vessels and submarines can interfere with measurement equipment, or can be used for detection. The noise emitting into the interior of the ship may disturb the crew as well as the passengers on board and increase hull vibration levels. The noise emission levels are especially important for cruise liners and yachts from the points of view of comfort and the mission of the ship.

Marine propellers are an important source of noise emitted from ships to the underwater environment and to the interior of the vessel. A non-cavitating propeller induces discrete peaks to the noise spectrum, which occur at the blade passing frequency and its multiples. These peaks are related to blade thickness and loading-induced pressure pulses. In addition, a propeller induces broadband noise, which is related to unsteadiness in the flow field. This corresponds to the turbulent fluctuations in the velocity and pressure fields. In the case of phase changes, the underwater noise from cavitation usually dominates other propeller-induced noise (excluding singing), and all other underwater noise from a ship [4]. Sheet and tip vortex cavitation generally increase the amplitude of the tonal pressure fluctuations, and cavitating vortices can also act as a source of broadband excitation [5–7]. Additionally, unsteady cavitation structures such as cloud and bubble cavities give rise to the propeller-induced broadband signature.

During the past decade, CFD (computational fluid dynamics) has been actively utilized to study propeller performance in wetted and cavitating conditions [8–20]. For instance, Turunen et al. [9], Asnaghi et al. [21], and Lidtke et al. [22] have investigated single- and two-phase propeller flows using the open-source CFD toolbox OpenFOAM. Flow structures in the wake and tip vortex of a propeller employing different RANS (Reynolds-averaged Navier–Stokes) or scale-resolving turbulence closures have been studied by Sipilä et al. [23], Guilmineau et al. [24], and Viitanen and Siikonen [25]. Additionally, higher-fidelity turbulence closures such as the LES or DES (large eddy simulation or detached eddy simulation, respectively) approaches have been used to compute the flow past marine propellers by Liefvendahl et al. [26], Lu et al. [10], Muscari et al. [11], Ji et al. [27], Chase and Carrica [28], and Balaras et al. [12]. For example, Ji et al. [27] simulated a cavitating marine propeller using a partially averaged Navier–Stokes (PANS) modelling approach, whereas Chase and Carrica [28] studied a submarine propeller with several turbulence modelling approaches including a coarse direct numerical simulation (DNS) method. Moreover, Balaras et al. [12] conducted investigations using LES in conjunction with an immersed boundary method. Specific attention towards the peculiar nuisance of propeller-induced noise has been given, for example, by Ianniello et al. [29], Lloyd et al. [30], and Lidtke et al. [22] via utilization of the Ffowcs Williams–Hawkings acoustic analogy, where the CFD results of the propeller flow dynamics were utilized as source terms for the acoustic simulations. Budich et al. [31] studied the Potsdam propeller test case (PPTC) in cavitating conditions, focusing on the shock wave dynamics and including an erosion assessment with a compressible flow solver.

The most general approach for acoustic simulations concerns the DNS, since the flow solution then includes both sound generation and its propagation. Unfortunately, DNS is computationally very expensive and limited to problems with low Reynolds numbers. Acoustic analogies may be utilized for the assessment of flow-induced noise based on an a priori flow solution, and the noise propagation is evaluated from the results of the flow simulation. Several different acoustic analogies are reviewed by Uosukainen [32]. Computationally-less-intensive integral methods can be used for external problems, but for practical cases (e.g., propeller in a cavitation tunnel), the assumption of a free-field space may not be valid. Then, one option is to use the variational formulation of Lighthill's analogy in the finite element method (FEM) context. We have chosen the latter approach for the investigations presented here. With this procedure, moving boundaries are naturally taken into account, and there are no limitations on the elasticity or geometry of the surrounding boundaries [33]. In the present acoustic solution, the different types of sound sources are not explicitly distinguished, and we do not utilize the pressure provided by the flow solution. Instead, the acoustic sources are evaluated from the so-called Lighthill tensor, which comprises the flow momentum components. On the other hand, our approach requires a volumetric discretization of the acoustic domain.

In this paper, we study a model-scale propeller, specifically the PPTC, in a uniform homogeneous inflow condition. We conducted wetted and cavitating DDES (delayed DES) of the propeller to investigate the influence of not only global blade loading effects but also transient wake characteristics and turbulent vortical flow structures, as well as sheet and tip vortex cavitation, on the harmonic and broadband noise. The combined hydrodynamic–hydroacoustic problem was solved via a hybrid approach. In the hybrid method, we assume that the flow solution based on DDES is decoupled from the acoustic propagation, which is composed of an acoustic FEM computation. The effective assumption is that the acoustic field does not modify the bulk flow solution. Consequently, we solve the flow problem independently of and prior to the solution of the acoustic problem.

We conducted the flow simulations with the general-purpose CFD solver FINFLO [8,34]. The code has been applied to both cavitating and non-cavitating propeller flows [23,25,35]. The propeller-induced sound pressure levels were obtained using ACTRAN [33]. We concentrated on one propeller operation point in wetted and cavitating conditions for the hydroacoustic analyses. Similar hybrid CFD–CHA (computational hydroacoustics) investigations were carried out for the PPTC propeller by Hynninen et al. [36] and Viitanen et al. [37].

In the next section, the hydrodynamic and hydroacoustic modelling approaches are described. Then, the test case is introduced, followed by an assessment and validation of the numerical results. Finally, conclusions are drawn from the presented results.

2. Flow Solution

2.1. Governing Equations

The flow model applied is based on a homogeneous flow assumption, which is a common assumption as far as cavitation is concerned [38]. The governing equations for the cavitation model are

$$\begin{aligned} \frac{\partial \alpha_k \rho_k}{\partial t} + \nabla \cdot \alpha_k \rho_k \mathbf{V} &= \Gamma_k, \\ \frac{\partial \rho \mathbf{V}}{\partial t} + \nabla \cdot \rho \mathbf{V}\mathbf{V} + \nabla p &= \nabla \cdot \boldsymbol{\tau}_{ij} + \rho \mathbf{g}, \end{aligned} \tag{1}$$

where p is the pressure, $\mathbf{V}$ the absolute velocity in a global non-rotating coordinate system, and $\boldsymbol{\tau}_{ij}$ the stress tensor, α_k is a void (volume) fraction of phase k, ρ_k the density, t the time. Γ_k the mass-transfer term, and $\mathbf{g}$ the gravity vector. The void fraction is defined as $\alpha_k = \mathcal{V}_k/\mathcal{V}$, where $\mathcal{V}_k$ denotes the volume occupied by phase k of the total volume, $\mathcal{V}$. For the mass transfer $\sum_k \Gamma_k = 0$ holds, and consequently only a single mass-transfer term is needed.

Although the phase temperatures do not play a significant role in cavitation, the energy equations are always solved in the present method. The aim is to apply a compressible form of the equations. In order to predict the correct acoustic signal speeds, a complete model is needed. The phase temperatures T_g and T_l also have some influence on the solution via the material properties that are calculated as functions of the pressure and phase temperatures. The calculation of the material properties is described in [39]. The sound speed c for a two-phase mixture is defined as

$$\frac{1}{\rho c^2} = \frac{\alpha}{\rho_g c_g^2} + \frac{1-\alpha}{\rho_l c_l^2} \quad \text{and} \quad \frac{1}{c_k^2} = \frac{\partial \rho_k}{\partial p} + \frac{1}{\rho_k}\frac{\partial \rho_k}{\partial h_k}. \tag{2}$$

In the expressions above, the indices g and l refer to gas and liquid phases, respectively, and h_k denotes the enthalpy of phase k.

The energy equations for phase $k = g$ or l are written as

$$\begin{aligned} &\frac{\partial \alpha_k \rho_k (e_k + \frac{V^2}{2})}{\partial t} + \nabla \cdot \alpha_k \rho_k (e_k + \frac{V^2}{2})\mathbf{V} = \\ -\nabla \cdot \alpha_k \mathbf{q}_k + \nabla \cdot \alpha_k \boldsymbol{\tau}_{ij} \cdot \mathbf{V} + q_{ik} &+ \Gamma_k (h_{k\text{sat}} + \frac{V^2}{2}) + \alpha_k \rho_k \mathbf{g} \cdot \mathbf{V}. \end{aligned} \tag{3}$$

Here e_k is the specific internal energy, $\mathbf{q}_k$ the heat flux, q_{ik} interfacial heat transfer from the interface to phase k, and $h_{k\text{sat}}$ saturation enthalpy. Since $\sum_k \Gamma_k = 0$, by adding the energy equations together, the following relationship is obtained between the interfacial heat and mass transfer

$$\Gamma_g = -\frac{q_{ig} + q_{il}}{h_{g\text{sat}} - h_{l\text{sat}}} \quad \text{and} \quad q_{ik} = h'_{ik}(T_{\text{sat}} - T_k). \tag{4}$$

Above, h'_{ik} is a heat transfer coefficient between the phase k and the interface. The interfacial heat transfer coefficients are based on the mass transfer, as shown in Section 2.5.

The momentum and total continuity equations in the homogeneous model do not change, except for the material properties like density and viscosity, which are calculated as

$$\rho = \sum_k \alpha_k \rho_k \quad \text{and} \quad \mu = \sum_k \alpha_k \mu_k, \tag{5}$$

where μ is the dynamic viscosity. The turbulence effects are currently handled using single-phase models for the mixture.

2.2. Finite-Volume Form

Equation (1) can be written in a general finite-volume form for a cell $\mathcal{V}_i$ as

$$\begin{aligned} \mathcal{V}_i \frac{d(\alpha_k \rho_k)_i}{dt} + \sum_j (S\alpha_k \rho_k \bar{u})_j = \mathcal{V}_i \Gamma_{k,i}, \\ \mathcal{V}_i \frac{d(\rho \mathbf{V})_i}{dt} + \sum_j \dot{m}_j \mathbf{V}_j + \sum_j S_j n_j p_j = \sum_j S_j (\boldsymbol{\tau}_{ij} \cdot \mathbf{n})_j + \mathcal{V}_i (\rho_i - \rho_0)\mathbf{g}, \end{aligned} \tag{6}$$

where the sum is taken over all cell surfaces j, $\mathbf{n}_j$ is a surface normal on a cell face, and S_j the cell face area. The mass flux $d\dot{m} = \rho \mathbf{V} \cdot \mathbf{n} dS = \rho \bar{u} dS$ can be identified in all field equations. In Equation (6), a Rhie–Chow-type damping term is added via the convective velocity $\bar{u}$. Instead of the commonly used scaling, the term is scaled using an artificial sound speed [40]. Pressure differences are applied and the flux terms are written in terms of the void fraction. However, the implicit solution is based on mass fractions:

$$\mathcal{V}_i \frac{d(\rho x)_i}{dt} + \sum_j \dot{m}_j x_j = \mathcal{V}_i \Gamma_i. \tag{7}$$

This form in the implicit stage is convenient, since the same mass flows can be used in the Jacobian matrices as in the case of the momentum equation.

Equation (6) can be applied for arbitrary cell shapes, although in the present solution a structured grid is applied. For the time derivatives, a second-order three-level fully implicit method is used. The viscous fluxes as well as the pressure terms are centrally differenced. For the convective part, the variables on the cell surfaces are evaluated using a third-order upwind-biased MUSCL (monotonic upstream-centred scheme for conservation laws) interpolation [41]. A flux limiter can be applied, but in this study this is done only for the void fraction. The application of a limiter function to the convective fluxes of the void fraction may be necessary, since it is essentially a discontinuous quantity through the phase boundary. This may lead to problems in a numerical solution, and the void fraction could locally obtain non-physical values amidst an iterative solver. Additionally, cavitation volumes can exhibit rapid temporal and spatial variation when, for instance, bursts of cloud cavities or fine cavitating vortices are present. Previously, it was shown that a compressive limiter with the void fraction equation especially improves the predicted tip vortex cavitation patterns [25]. Hence, a compressive "superbee" limiter of Roe [42] is employed for the cavitating cases considered in this study as well. A review of the high-resolution limiters for two-phase flows is given, for example, in [43].

2.3. Solution Algorithm

The solution method is a segregated pressure-based algorithm where the momentum equations are solved first, and then a pressure–velocity correction is made. The basic idea in the solution of all equations is that the mass balance is not forced at every iteration cycle, but rather the effect of the mass error is subtracted from the linearized conservation equations. A pressure correction equation was derived from the continuity Equation (1) linked with the linearized momentum equation. The method is based on the corresponding algorithm for a single-phase flow [40], and is described in detail in Ref. [25]. The resulting pressure-correction equation is

$$\left[\frac{0.01 \mathcal{V}_i}{\rho_i c_i^2 \Delta t} + \frac{\mathcal{V}_i |\Gamma_i|}{\Delta p_{\max}} \left(\frac{1}{\rho_{g,i}} - \frac{1}{\rho_{l,i}} \right) \right] p'_i - \sum_j \frac{S_j^2}{\bar{a}^u_{P,j}} (p'_{j+} - p'_i) = \sum_k \frac{\Delta \dot{m}_{k,i}}{\rho_{k,i}}, \tag{8}$$

where p'_{j+} is the pressure change in the cell on the other side of face j, $\bar{a}^u_{P,j}$ the diagonal term of the Jacobian matrix of the linearized momentum equation, and $\Delta \dot{m}_{k,i}$ an error in mass balance for phase k. The first term on the left-hand side is a result of the compressibility, but mainly serves as an under-relaxation for the pressure. An extra multiplier of 0.01 was added on the basis of test calculations. Another parameter $\Delta p_{\max}$ is used to control the pressure changes caused by the mass-transfer term [25].

Two different solution strategies were utilized for the simulations of the flow around the rotating propeller. The first one was to rotate the computational domain with the propeller rate of revolution, and integrate the governing equations in the physical time. Consequently, results obtained from this strategy are referred to as transient. The second approach exploited the fact that the governing equations can yield a steady-state solution when the equations are expressed in the coordinate system that is rotating with the propeller. This solution method is then referred to as quasi-steady.

In the transient simulations, a steady-state solution was sought within each physical time-step by iterating until the L_2 norms of the main variables decreased by a sufficient amount (i.e., at least 2–3 orders of magnitude). Approximately 100 inner iterations were usually required within each physical time-step for non-cavitating simulations. In the present study, between 150–200 inner iterations were made for the cavitating simulations.

In the quasi-steady approach, absolute velocities were used in the solution, and the rotational movement of the propeller was accounted for in the convection velocity and as source terms in the y- and z-momentum equations as the propeller was rotating around the x-axis. The equations were iterated until the global force coefficients and the L_2 norms of the main variables obtained a sufficiently steady level, with the L_2 norms having decreased to $10^{-5} \cdots 10^{-7}$.

2.4. Turbulence Modelling

Nominally a Reynolds-averaged form of Equations (1)–(3) was used, and the delayed detached-eddy simulation (DDES) [44] that combines RANS and LES was also applied in the same form. Usually, in cavitation modelling, turbulence is taken into account using single-phase closures. Also in this study, the turbulence modelling was applied for the homogeneous mixture. The choice of turbulence closure plays an essential role in the numerical prediction of the performance of a marine propeller. While the global forces or steady cavitation patterns near the blades generally might not considerably differ between the turbulence closures, the utilized model can have a significant influence on unsteady flow structures, or on the flow in the wake of the propeller. In the case of unsteady propeller cavitation, capturing the cavitation dynamics is crucial in order to assess not only the performance but also the erosive tendency of collapsing cavities, as well as the induced underwater noise. Moreover, accurate prediction of the wake flow is important when considering the propeller–rudder, propeller–pod, or multi-propeller interactions.

In this study, DDES is based on the shear stress transport (SST) $k-\omega$-model [45]. DDES is a slightly modified version of the detached-eddy simulation (DES). DES and DDES reduce to a RANS model in regions where the largest turbulent fluctuations are of a smaller size than the local grid spacing. Both are hybrid RANS/LES models, and function as an LES subgrid-scale model in regions where the local turbulent phenomena are of greater size than the local grid spacing [46]. A time-accurate solution is made to resolve turbulent fluctuations. In the present study, the calculations were performed up to the wall, and the height of the first cell was adjusted such that the non-dimensional wall distance $d^+ = \rho u_\tau d/\mu \lesssim 1$ for the first cell, with $u_\tau = \sqrt{\tau_w/\rho}$ being the friction velocity, τ_w the wall shear stress, and d is the normal distance from the solid surface to the centre point of the cell next to the surface.

In DES, the equation for the turbulent kinetic energy (k) can be written with a modified dissipation term as [47]:

$$\rho \frac{Dk}{Dt} = P - \frac{\rho k^{3/2}}{l_{DES}} + D, \tag{9}$$

where P is the production of turbulence, l_{DES} is the length scale, and D is the diffusion term. The DES length scale was computed as the minimum of the RANS length scale, $l_{RANS} = \sqrt{k}/\beta^*\omega$, and the local resolution Δ. Here $\beta^* = 0.09$ was a model constant. The parameter Δ was evaluated as the minimum of the local wall distance, and the grid resolution $\max(\Delta x_i)$, where Δx_i denotes the thickness of the cell in different index directions. The DES length scale was then

$$l_{DES} = \min(C_{DES}\Delta, l_{RANS}), \tag{10}$$

and the coefficient C_{DES} was computed from

$$C_{DES} = (1 - F_1)C_{DES}^{k-\varepsilon} + F_1 C_{DES}^{k-\omega}, \tag{11}$$

where the constants were $C_{DES}^{k-\varepsilon} = 0.61$, $C_{DES}^{k-\omega} = 0.78$, and F_1 is Menter's blending function [45]. Furthermore, when utilizing DDES, the length scale is replaced by the expression [44]

$$l_{DDES} = l_{RANS} - F_1 \max(0, l_{RANS} - C_{DES}\Delta). \tag{12}$$

Here $F_1 \to 1$ outside the boundary layer, and the length scale becomes $l_{DDES} = C_{DES}\Delta$ if the grid spacing permits. The DDES variant of DES aims to improve the accuracy compared to Equation (10), which has in some instances been observed to cause grid-induced separation.

2.5. Mass and Energy Transfer

A number of mass-transfer models have been suggested for cavitation [48]. Usually, the mass-transfer rate is proportional to a pressure difference from a saturated state or to a square root of that. In this study, the mass-transfer model is similar to that of Choi and Merkle [38]:

$$\Gamma_l = \frac{\rho_l \alpha_l \min\left[0, p - p_{\text{sat}}\right]}{\frac{1}{2}\rho_\infty V_\infty^2 (L_{\text{cav}}/V_{\text{cav}})\tau_l} + \frac{\rho_g \alpha_g \max\left[0, p - p_{\text{sat}}\right]}{\frac{1}{2}\rho_\infty V_\infty^2 (L_{\text{cav}}/V_{\text{cav}})\tau_g}, \tag{13}$$

where p_{sat} is the saturation pressure, ρ_∞ the reference (inlet) density, and V_∞ the corresponding velocity. The evaporation time constants were made non-dimensional using the reference length L_{cav} and the velocity related to cavitation (V_{cav}). In some cases, such as on a propeller blade, the cavitation length and velocity differ from the reference length L_{ref} and the reference velocity (V_∞). The time constants correspond to the parameters of the original model as $\tau_l = 1/C_{\text{dest}}$ and $\tau_g = 1/C_{\text{prod}}$. The empirical parameters of the cavitation model are calibrated in [49].

In the present method, the saturation pressure was based on the free-stream temperature, and the gas phase was assumed to be saturated (i.e., $T_g = T_{\text{sat}}$). Liquid temperature varies less because of the mass and energy transfer. Since the gas temperature was forced to be $T_g = T_{\text{sat}}$, $q_{ig} = 0$. From Equation (4), the interfacial heat transfer can be solved for the liquid phase

$$q_{il} = -(h_{g\text{sat}} - h_{l\text{sat}})\Gamma_g - q_{ig} = (h_{g\text{sat}} - h_{l\text{sat}})\Gamma_l. \tag{14}$$

Using Equations (13) and (14), the interfacial transfer terms in the continuity and energy equations can be solved.

In order to decrease the oscillations in the solution owing to the rapid changes in the mass transfer, the mass-transfer rate was under-relaxed between the iteration cycles as $\Gamma_l^{n+1} = \alpha_\Gamma \Gamma_l^* + (1 - \alpha_\Gamma)\Gamma_l^n$, where $\alpha_\Gamma = 0.5$ is an under-relaxation factor, n refers to the iteration cycle, and Γ_l^* is calculated from Equation (13). For small values $|\Gamma_l^n| < 0.1\,(\text{kg/m}^3\text{s})$, under-relaxation was not applied. The under-relaxation factor and the limit are quite arbitrary, although tested by numerous simulations.

3. Acoustic Solution

Instead of the direct solution of the compressible Navier–Stokes (N-S) equations, the hydrodynamic and hydroacoustic problems are treated separately. The former is obtained from compressible N-S equations (Equations (1)–(3)) using DDES, while the latter can be seen as a subsequent solution of the compressible N-S equations in isentropic conditions. The acoustic problem is analogous to a direct solution in the sense that the flow-generated noise propagates not through the mean flow, but according to a wave operator in a medium at rest. In this analogous problem, source terms are utilized to represent the flow. The effectively two-step approach requires one unsteady flow solution in the time domain and one subsequent acoustic analysis in the frequency domain.

The present solution to the acoustic problem is based on the acoustic analogy using the ACTRAN code [33]. The propagation problem, in which noise is generated by flow fluctuations, is replaced by an analogous problem, where the propagation is represented by a wave operator. In this study, the analogy of Lighthill [50] is used, which utilizes a wave equation for density. The wave equation can be derived by taking the time-derivative of the continuity equation, and the gradient of the momentum equations. Combining these two yields

$$\frac{\partial^2 \rho}{\partial t^2} - c_0^2 \frac{\partial^2 \rho}{\partial x_i^2} = \frac{\partial^2 T_{ij}}{\partial x_i \partial x_j}, \tag{15}$$

where c_0 is the speed of sound in the medium, and $T_{ij} = \rho u_i u_j + (p - c_0^2 \rho)\delta_{ij} + \tau_{ij}$ the Lighthill tensor. Furthermore, T_{ij} can be simplified by assuming isentropic, high-Re, and low-Ma flows to comprise only the fluid momentum components, or $T_{ij} \approx \rho u_i u_j$. The wave equation can be transformed to the frequency domain

$$-\omega^2 \tilde{\rho} - c_0^2 \frac{\partial^2 \tilde{\rho}}{\partial x_i^2} = \frac{\partial^2 \tilde{T}_{ij}}{\partial x_i \partial x_j}, \tag{16}$$

where ω is the angular frequency. In ACTRAN, the solution of the acoustic problem, Equation (15), is based on variational formulation of the wave equation for density propagation, transformed into the frequency domain, Equation (16). The code requires the Lighthill tensor in the frequency domain as an input. The source terms $\nabla \cdot (\nabla \cdot \tilde{T}_{ij})$ are related solely to the momentum components, which are provided by the flow solution.

In ACTRAN, instead of the density a transformed potential, ψ, defined through

$$\tilde{\rho} = -\frac{i\omega\psi}{c_0^2}, \tag{17}$$

is used. The alternative equation for the Lighthill analogy is then

$$\frac{\omega^2}{c_0^2}\psi + \frac{\partial^2 \psi}{\partial x_i \partial x_i} = \frac{1}{i\omega} \frac{\partial^2 \tilde{T}_{ij}}{\partial x_i \partial x_i}. \tag{18}$$

The weak variational formulation for the density propagation in the frequency domain involving stationary and moving geometry reads

$$-\int_\Omega \frac{\omega^2}{\rho_0 c^2} \psi \delta\psi d\Omega - \int_\Omega \frac{1}{\rho_0} \frac{\partial \psi}{\partial x_i} \frac{\partial \delta\psi}{\partial x_i} d\Omega = \int_\Omega \frac{1}{\rho_0 \omega} \frac{\partial \delta\psi}{\partial x_i} \frac{\partial T_{ij}}{\partial x_j} d\Omega - \int_\Gamma \frac{1}{\rho_0} \mathcal{F}(\tilde{\rho}\tilde{v}_i n_i) d\Gamma, \tag{19}$$

where Ω is the volume and Γ is the surface of the boundary enclosing the propeller. Using this formulation, the volume and surface source term influences can be evaluated. The quadrupole sources (e.g., due to turbulence) are included in the volume term, whereas the monopole and dipole sources are taken into account in the surface source term.

Two types of source terms are present in Equation (19). The surface source terms are evaluated on the surface surrounding the propeller blades and hub. Conformal surface enclosing the propeller is shown in Figure 1. The noise generation resulting from flow phenomena such as turbulence, blade loading, and cavitation take place inside the volume surrounding the blades and the hub, as shown in Figure 1, which is accounted for by mapping the sources as defined in Equation (19) onto the enclosing surface. In practice, in order to achieve the most accurate source description, the surface should be located close to the source (i.e., the unsteady flow). The volume mesh encloses the propeller and the wake region as shown in Figure 1, and the volume source terms in Equation (19) are evaluated in these volume elements. The hydroacoustic source terms are calculated in the time domain using oversampling by default at every half CFD time step to avoid aliasing effects during the Fourier transform, and are saved in an NFF database. The calculated hydroacoustic source terms are then transformed on the acoustic mesh by integrating over the CFD mesh using the shape functions of the acoustic mesh.

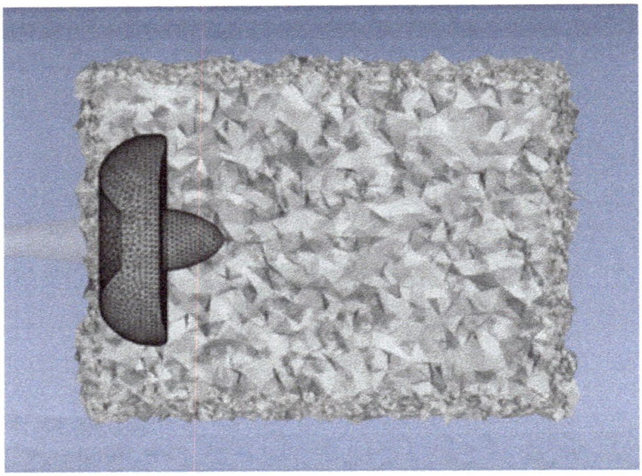

Figure 1. Lighthill surface mesh around the propeller blades and hub, and Lighthill volume mesh around the propeller, and in the wake of the propeller.

4. Computational Case

The computational case was the PPTC propeller [51]. The diameter of the model-size propeller is 0.250 m. The propeller has five blades, and a right-handed direction of rotation. The skew of the propeller is moderate. Table 1 summarizes the main geometrical parameters of the PPTC propeller. Photographs of the propeller are shown in Figure 2. A large database of experimental results has been made available by SVA Potsdam (http://www.sva-potsdam.de/pptc-smp11-workshop/).

Figure 2. Photographs of the Potsdam propeller test case (PPTC) propeller [51].

Table 1. Main geometric parameters of the PPTC propeller.

Diameter (m)	0.250
Pitch ratio at $r/R = 0.7$	1.635
Chord at $r/R = 0.7$	0.10417
Expanded area ratio	0.779
Skew (°)	18.837
Hub ratio	0.300
Number of blades	5
Rotation	Right-handed

We investigated the propeller operating in push configuration (i.e., the shaft was located in front of the propeller), and we considered three propeller points of operation. One operating point was simulated both in wetted and cavitating conditions, while the other two cases were simulated only in wetted conditions. The simulations were performed using a constant rate of revolution, $n = 20\ 1/\mathrm{s}$. The advance coefficient and the cavitation number are defined as

$$J = \frac{V_A}{nD} \quad \text{and} \quad \sigma_n = \frac{p - p_{\mathrm{sat}}}{\frac{1}{2}\rho(nD)^2}, \tag{20}$$

respectively, where V_A is the propeller speed of advance, n is the propeller rate of revolution, D is the propeller diameter, p is the pressure, p_{sat} is the saturation pressure, and ρ the fluid density. The first investigated operation point was $J = 1.019$ in wetted conditions as well as in cavitating conditions with $\sigma_n = 2.024$. The second non-cavitating case had the advance coefficient $J = 1.253$, and the third non-cavitating operating point was $J = 1.408$. The first propeller operating point corresponds to the PPTC Case 2.3.1 of smp'11 Workshop [52], while the cavitation tunnel LDV (laser Doppler velocimetry) measurements of the propeller were carried out for the second operating point [53]. The third operating point corresponds to Case 2.3.3 of the smp'11 Workshop. The thrust and torque of the propeller were non-dimensionalized as

$$K_T = \frac{T}{\rho n^2 D^4} \quad \text{and} \quad K_Q = \frac{Q}{\rho n^2 D^5}, \tag{21}$$

respectively, where T denotes the thrust and Q the torque of the propeller. Finally, the open water efficiency of the propeller is defined as

$$\eta_0 = \frac{J}{2\pi}\frac{K_T}{K_Q}. \tag{22}$$

Below, we describe first the CFD numerical setup utilized, the grid, and related boundary conditions, followed by a corresponding description of the numerical setup used in the hydroacoustic simulations.

4.1. CFD: Numerical Setup, Grid, and Boundary Conditions

The structured computational grid used consists of roughly 5.5 million cells in 28 grid blocks. The computational domain is shown in Figure 3. Due to the symmetric nature of the problem of a propeller operating in axial uniform inflow, only one blade was modelled. The blades, hub, and shaft were modelled as no-slip rotational surfaces, coloured black in Figure 3. Boundary-layer transition to turbulent flow was not taken into account in the present simulations. A velocity boundary condition was applied at the inlet, denoted as the red face, and a pressure boundary condition was applied at the outlet. A slip boundary condition was applied at the simplified tunnel walls, which are coloured green in Figure 3. Cyclic boundaries are denoted by the blue faces, and the whole computational domain was considered as rotating with the given rate of rotation. The inflow velocity was set based on the advance numbers of the propeller, and the background pressure level was set based on the cavitation number. The inlet was located five propeller diameters upstream of the propeller, and the outlet was located ten diameters downstream of the propeller. The rectangular cavitation tunnel of SVA Potsdam was here modelled as a circular duct of the same cross-sectional area, thus enabling also the quasi-steady computation of the problem. The radius of the computational domain was then 0.3385 m.

The surface grid on the suction side of the blade is shown in Figure 4a. The surface grid on the pressure side of the blade was similar. The grid had an O-O topology around the propeller blades. The grid resolution around the leading edge was fine, as shown in Figure 4b, and there were about 30 cells around the leading-edge radius. Due to the O-O topology, the same resolution was applied around the blade tip and the trailing edge as well. The grid was refined normal to the viscous surfaces such that $d^+ \approx 1$.

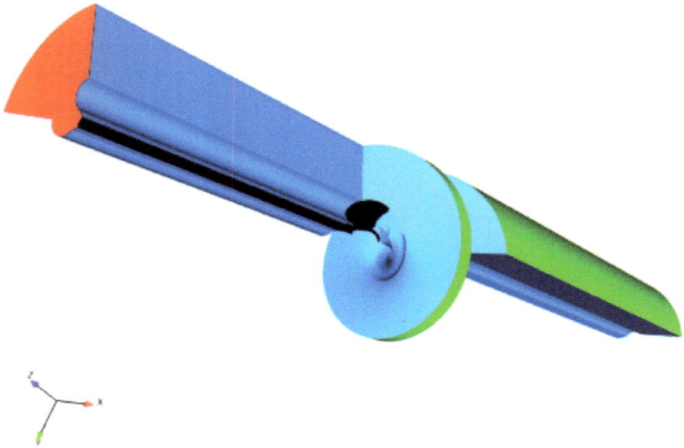

Figure 3. A perspective view of the grid topology used in the open-water computations.

The grid points in the helical blocks located downstream of the propeller were adaptively concentrated in the region of the tip vortex, based on the propeller loading. Figure 5 depicts the concentration of the grid points near the tip vortex induced by the rotating blades. In the figure, $|\Omega_i|$ denotes the absolute value of vorticity, and the propeller blades, hub, and shaft are coloured dark red. The figure shows exemplary views of the resolution on the finest grid, demonstrating that the tip vortex was well-maintained even beyond $x/D \approx 1$. There were roughly 18×14 grid points in the cross-section of the tip vortex on the finest grid on the plane $x/D = 1$. The helical blocks in the slipstream of the propeller were extended to a pitch corresponding to approximately 450° of rotation from the propeller plane.

The calculations were performed on three grid levels. On a coarse grid level, every second point in all directions was removed compared to a finer level grid. The fine grid had roughly 5.5 million cells, the medium grid roughly 0.7 million cells, and the coarse grid approximatively 0.08 million cells. The computations were initiated such that a quasi-steady solution was first obtained on the coarse grid, which was then used as an initial guess for the medium grid quasi-steady simulations. After a quasi-steady solution was obtained on the medium grid, time-accurate simulation was continued on the medium grid from the quasi-steady results for approximately 10 propeller revolutions. The quasi-steady solutions were obtained using Chien's low-Reynolds-number $k - \varepsilon$ turbulence model [54], and all transient approaches relied on DDES. The results of the time-accurate medium grid simulations were then used as an initial guess for the time-accurate simulations on the fine grid, and usually 5–10 propeller revolutions (or 25–50 propeller blade passages) were simulated on the fine grid. The presented approach is a relatively fast and overall efficient computational procedure for single and two-phase DDES of marine propellers. In the time-accurate simulations, a physical time-step of $\Delta t = 6.9444 \times 10^{-5}$ s was used, which corresponds to half a degree of propeller rotation. In this paper, we present the results that were obtained on the fine grid from time-accurate simulations.

(**a**) Grid resolution on the suction side. (**b**) Grid resolution near the leading edge (LE).

Figure 4. Grid resolution on the suction side and near the leading edge of the blade.

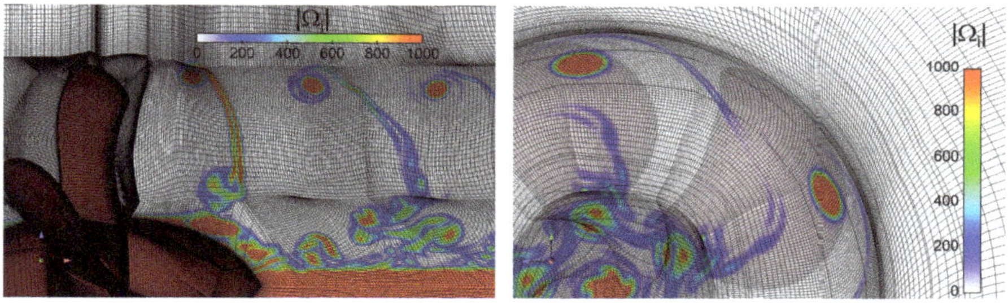

(**a**) Cut plane $y = 0$. (**b**) Cut plane $x/D = 1$.

Figure 5. Views of grid resolution on cut planes $y = 0$ and $x/D = 1$.

4.2. CHA: Numerical Setup, Grid, and Boundary Conditions

The acoustic FE mesh is shown in Figure 6, where the propeller is modelled inside the cavitation tunnel. There were more than six elements for the shortest wave length. In total, there were 450,000 quadratic tetrahedron elements in the mesh. At both ends of the cavitation tunnel, non-reflecting boundary conditions were used. The tunnel walls were modelled as rigid. The maximum frequency that can be resolved within the acoustic analyses is limited by the Nyqvist frequency and the time step used in the transient CFD computations, or $f_{max} = 0.5/\Delta t = 3.6$ kHz. In the CFD simulations, every second time-step was written for the acoustic simulation (i.e., with a time-step corresponding to one degree of propeller rotation). The lowest frequency that could be resolved, $f_{min} = 20$ Hz, was set by the physical simulation time chosen for the CFD–CHA coupling, which in this case was one propeller revolution. While the transient CFD simulations were conducted for up to ten propeller revolutions, the flow solution of the final propeller revolution was used as input for the CHA simulations.

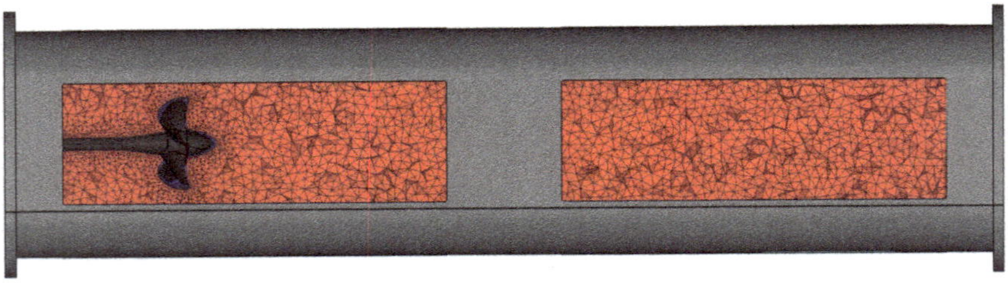

Figure 6. Acoustic finite element (FE) mesh inside the cavitation tunnel.

5. Results

Next, we report the results of DDES for the wetted and cavitating cases, followed by the CHA results. First, the global forces (i.e., the thrust, torque, and open water efficiency of the propeller) are given, followed by an assessment of flow and cavitation patterns near the propeller and in its wake. Finally, we show the results of the CHA simulations. In all figures depicting the propeller, unless stated otherwise, a snapshot of the simulations with the blade at the top dead centre position is shown, depicting instantaneous flow structures near the propeller. We compare the simulated and experimentally-determined propeller global performance characteristics at all studied advance numbers. The propeller wake was validated at $J = 1.253$, whereas we carried out detailed investigations of the wake flow structures and acoustic excitation at $J = 1.019$ both in wetted and cavitating conditions.

5.1. Validation and Cavitation Observations

A comparison of propeller performance characteristics is shown in Figure 7 in both wetted and cavitating conditions. The experimental results are given by Barkmann [55] for the wetted conditions, and by Heinke [51] for the cavitating conditions. In the figure, the CFD results are given for the wetted and cavitating cases at $J = 1.019$, and for the wetted cases at $J = 1.253$ and $J = 1.408$. Corresponding experimental results for the cavitating condition are shown as markers. The CFD-predicted open water characteristics were in good agreement with the experimental data, although the simulations tended to under-predict the thrust and torque of the propeller. For the lowest investigated propeller loading condition, the simulations agreed better with the experiments than for the two higher loadings. Overall, the thrust coefficient differed from the experimental results by 2–8% in the wetted case, whereas the difference was smaller for the investigated cavitating case, 3%. The torque coefficient agreed better with the experiments, and the differences were within 1–3% for both the wetted and cavitating cases. Deviations in the EFD (experimental fluid dynamics) and CFD propeller global forces could be due to various reasons, other than limitations in the numerical method. Possible confinement effects due to the geometry of the cavitation tunnel, which was simplified in the simulations, may be a source of the observed deviations. The driving mechanism, including the propeller shaft used in the experiments, could further add to the observed differences. However, the magnitudes of the deviations between the EFD and CFD global performance characteristics were relatively small and consistent with previous studies featuring more accurate turbulence modelling approaches [11,24,25,28].

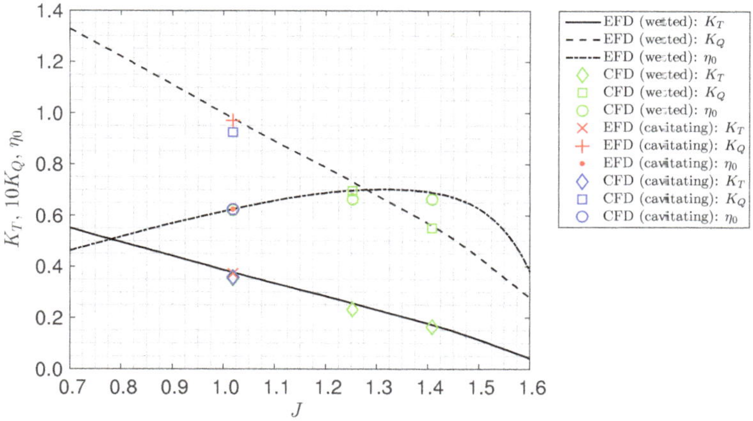

Figure 7. Comparison of global propeller performance characteristics. CFD: computational fluid dynamics; EFD: experimental fluid dynamics.

Next, we compare the axial wake of the propeller to the LDV measurements conducted in a cavitation tunnel. The measurements were made at non-cavitating conditions at $J = 1.253$, and the comprehensive LDV experimental results were reported by Mach [53]. The axial wakes at four axial distances from the propeller plane were compared, namely on the plane $x/D = 0.10$ in Figure 8, on the plane $x/D = 0.13$ in Figure 9, on the plane $x/D = 0.16$ in Figure 10, and on the plane $x/D = 0.20$ in Figure 11.

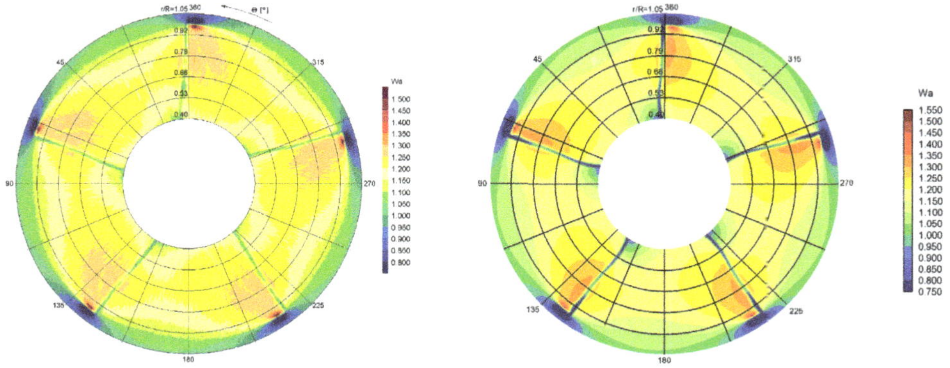

(**a**) EFD. (**b**) CFD.

Figure 8. Comparison of axial wake distributions at $x/D = 0.10$ behind the propeller.

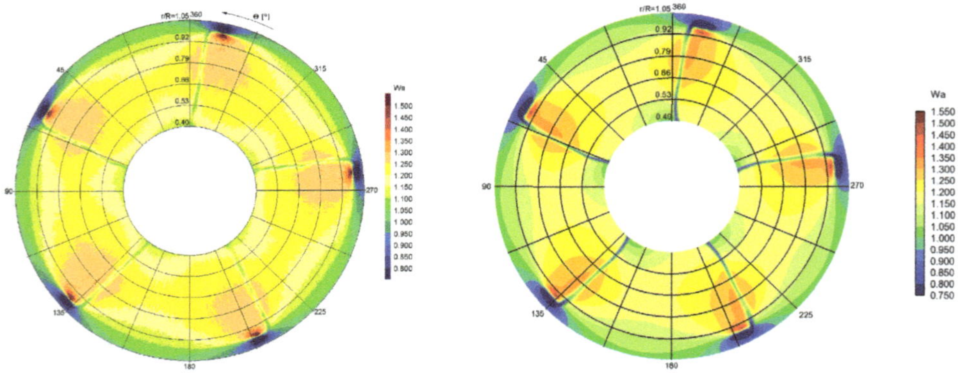

(**a**) EFD. (**b**) CFD.

Figure 9. Comparison of axial wake distributions at $x/D = 0.13$ behind the propeller.

Comparing the EFD and CFD axial wake distributions, it can be seen that overall the propeller wakes were mostly very similar between CFD and the experiments. The accelerated flow region on both sides of the thin decelerated blade wakes were also clearly present in the simulations. The spatial evolution of the blade wakes and associated flow patterns around them were also well captured in the simulations, including the form and strength of the tip vortices. The axial velocity distribution near the tip vortices was very similar, although the simulations predicted the maximum local wake at a slightly lower radius than was observed in the experiments. In addition, the simulations predicted a somewhat larger velocity deficit for the thin blade wakes. We note that the small deviations between the present simulations and the experiments were in line with observations made during the smp'11 workshop (http://www.marinepropulsors.com/proceedings-2011.php), where several CFD codes were compared with the LDV measurements.

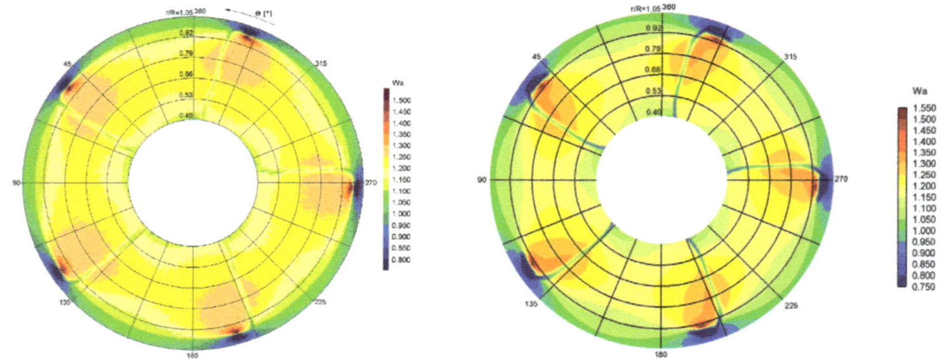

(**a**) EFD. (**b**) CFD.

Figure 10. Comparison of axial wake distributions at $x/D = 0.16$ behind the propeller.

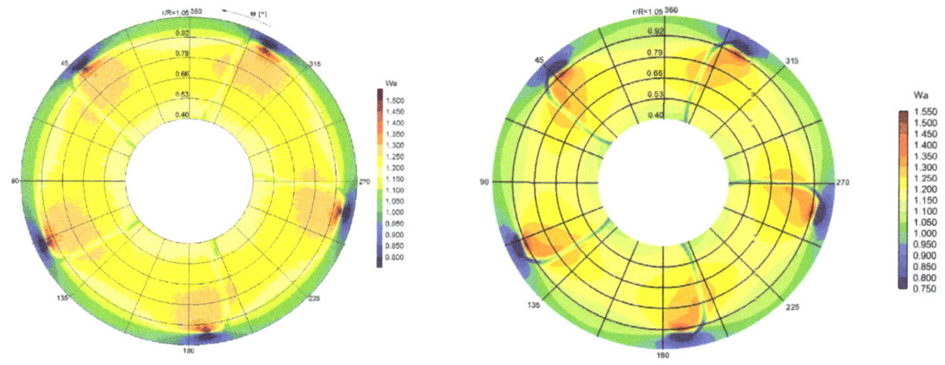

(**a**) EFD. (**b**) CFD.

Figure 11. Comparison of axial wake distributions at $x/D = 0.20$ behind the propeller.

Figures 12 and 13 show the cavitation patterns at $J = 1.019$ on the suction side of the propeller blades as well as in the propeller wake together with the observations made in the cavitation tunnel tests. The propeller has strong tip vortex and hub vortex cavitation, which was visible in the experiments and in the simulations. Shedding of root cavitation was observed in the experiments and in the simulations. It should be noted that the unstable behaviour was very mild in the simulations. Otherwise, the shape and extent of the root cavitation, as well as the tip and hub vortex cavitation, were captured well. The tip and hub vortex cavitation extending far behind the propeller were captured exceptionally well, as shown in Figure 13. Comparing the EFD and CFD results in Figure 13, it can be seen that the modal shapes of the cavitating tip vortex were also qualitatively well-captured. Furthermore, streak cavitation in the experiments at several radial locations near the leading edge of the blade was identified, while the simulation predicted sheet cavitation at the leading edge.

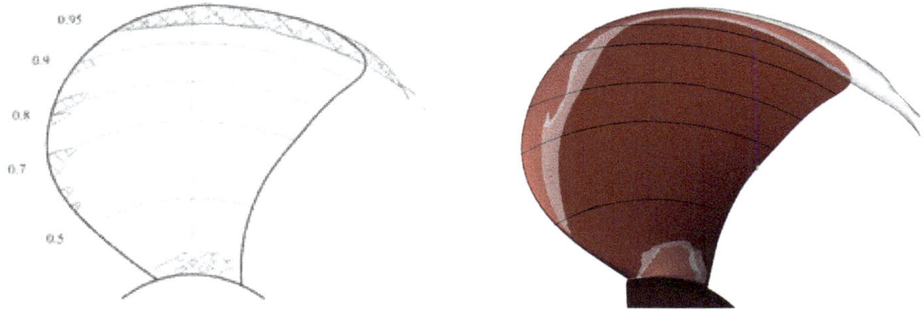

(**a**) EFD. (**b**) CFD.

Figure 12. Comparison of the cavitation patterns near the blade surfaces with the cavitation sketches according to observations made in the experiments, *cf.* Ref. [37].

(**a**) EFD.

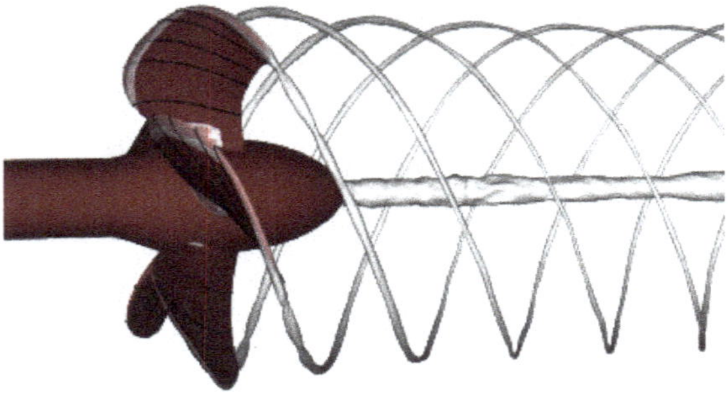

(**b**) CFD.

Figure 13. Comparison of the tip and hub vortex cavitation extents behind the propeller.

The surface restricted streamlines and pressure coefficients, $C_p = 2(p - p_\infty)/(\rho_\infty n^2 D^2)$, on the blade surface in wetted and cavitating conditions, are shown in Figure 14. The boundary layer flow was mostly circumferentially directed along the blade. The effect of cavitation on the surface restricted streamlines was significant. The re-entrant jets were directed towards the cavitating tip vortex at the closure line of the sheet cavitation, *cf.* Refs. [25,49]. In addition, flow separation was visible in the blade root region at cavitating conditions, caused by the blade root cavitation. Furthermore, the wetted case computation predicted a more radially extended although relatively fine separation region at the trailing edge of the blade.

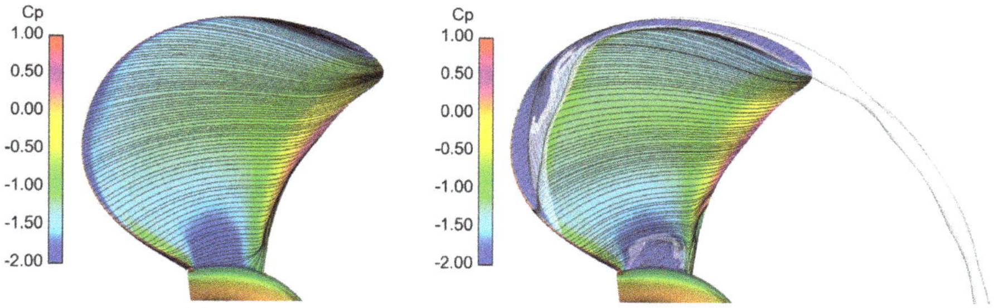

(a) Wetted conditions. (b) Cavitating conditions.

Figure 14. Surface restricted streamlines and non-dimensional pressure coefficients on the suction side of the blade surface. In the cavitating case, the iso-surface of void fraction value of 0.1 is shown as transparent grey.

5.2. Wake Flow Structures

The pressure coefficient near the propeller and in its wake is visualized in Figure 15. In the figure, we compare the wetted and cavitating conditions. A corresponding comparison of the magnitude of non-dimensional velocity ($V_{ref} = V_A$) is shown in Figure 16. It can be seen that the strong tip vortex was preserved well in the slipstream. In both the wetted and cavitating cases, the flow field near the tip vortex region was convected far in the wake, nearly unaffected by the distance it travelled. Dissipation of flow disturbances in the tip vortex region was low, and the tip vortices were well preserved up to the extent of the helical grid.

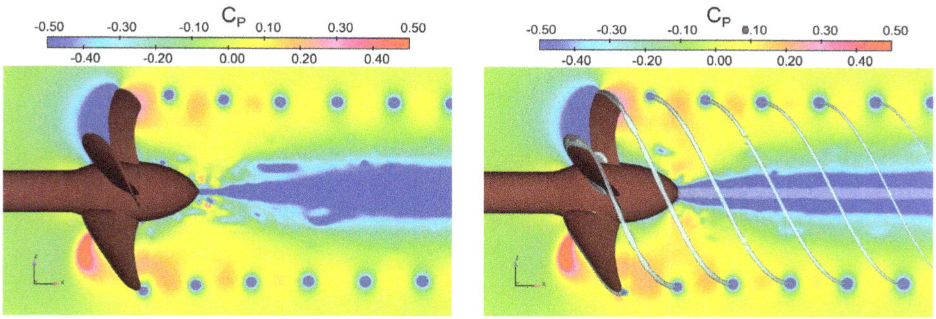

(a) Wetted conditions. (b) Cavitating conditions.

Figure 15. Distribution of the pressure coefficient near the propeller on the cut plane $y = 0$. In the cavitating case, the iso-surface of the void fraction $\alpha = 0.1$ is coloured light grey.

In Figure 17, the flow field near the propeller is visualized by an iso-surface of the second invariant of the velocity gradient tensor, or the Q criterion. The iso-surface had the value of $Q = 20,000$, and it was coloured by helicity $\mathrm{H} = \mathbf{V}_r \cdot \mathbf{\Omega}/(|\mathbf{V}_r||\mathbf{\Omega}|)$, where $\mathbf{V}_r$ is the relative velocity vector in the rotating reference frame. The helicity denotes the cosine of the angle between the relative velocity and the absolute vorticity vectors, and tended to ± 1 in the vortex cores, the sign indicating the direction of swirl. Several distinct areas characterized by different types of vortical flow structures were seen in the wake of the propeller in both the wetted and the cavitating cases. Distinct, regular, and strong vortical flow structures with dominant vorticity direction aligned with the flow were due to the tip vortices caused by the blades. As was noted above, these convect with little dissipation up to the extent of the helical grid. The modal shapes of the cavitating tip vortex are also seen in Figure 17b. At small radii behind the propeller hub, the wake was dominated by the vortical flow structures shed by the hub in both cases. This also had a relatively regular structure, in addition to being strong enough to maintain cavitation up to the extent of the helical grid, *cf.* also Figure 12. Flow structures that originate from the root of the blade then again differed more between the wetted and cavitating conditions. In the wetted case, several horseshoe-type vortical structures being shed in the wake existed, whereas in the cavitating case, the wake structures did not appear as clearly. Furthermore, distinct flow patterns originated from near the midspan of the blade. In the wetted case, as the blade wake departed the blades, the vortical flow structures seemed to converge to a distinct vortex filament which then coiled as a helical shape with its vorticity aligned with the flow (i.e., $H \approx 1$). Similar structures were present in the cavitating simulations, but with less clarity of the helical composition; closer observation near the trailing edges of the blades revealed that the vortical structures caused by the blade boundary layer destabilized just after the trailing edge (TE) for the cavitating case. These differences were due to the leading edge cavitation as well as the enhanced flow separation from the root section of the blades in the cavitating conditions, since the relatively stable root cavitation caused the flow separation (*cf.* Figure 14) as observed also in previous studies [25,37].

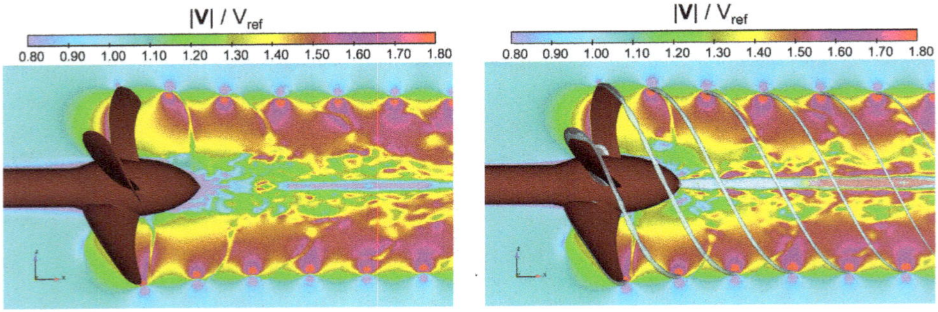

(**a**) Wetted conditions. (**b**) Cavitating conditions.

Figure 16. Distribution of non-dimensional velocity near the propeller on the cut plane $y = 0$. In the cavitating case, the iso-surface of the void fraction $\alpha = 0.1$ is coloured by light blue.

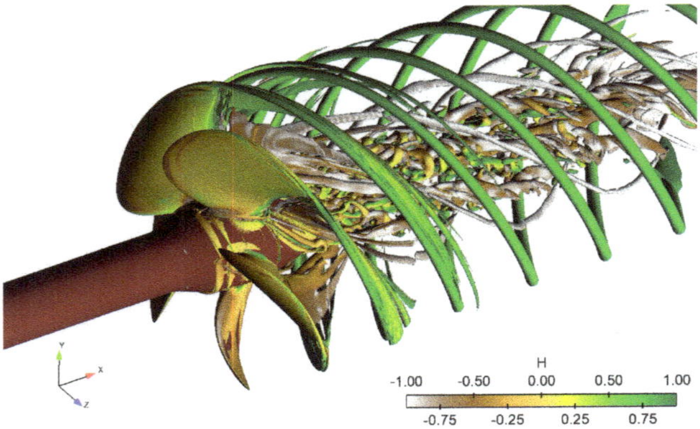

(**a**) Wetted conditions.

Figure 17. *Cont.*

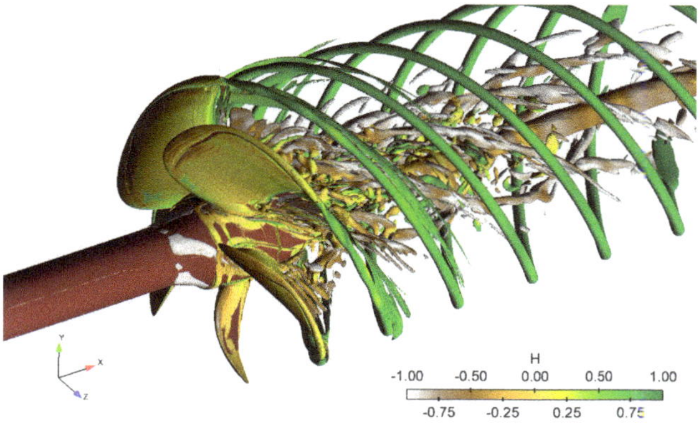

(**b**) Cavitating conditions.

Figure 17. Vortical flow structures visualized near the propeller by means of the Q criterion. The iso-surface of the Q criterion is coloured by helicity.

5.3. Acoustic Excitations

Figure 18 shows the acoustic source terms obtained from the CFD solution on the cut plane $y = 0$ utilizing different turbulence models in wetted conditions. As shown by Saarinen and Siikonen [56], the divergence of the divergence of the Lighthill tensor, or the source term of the acoustic wave equation, can in incompressible cases be simplified to $\nabla \cdot (\nabla \cdot \tilde{T}_{ij}) = S_{inc} = -2\rho Q$ (kg/m^{3}s^2), where Q is the second invariant of the velocity gradient tensor.

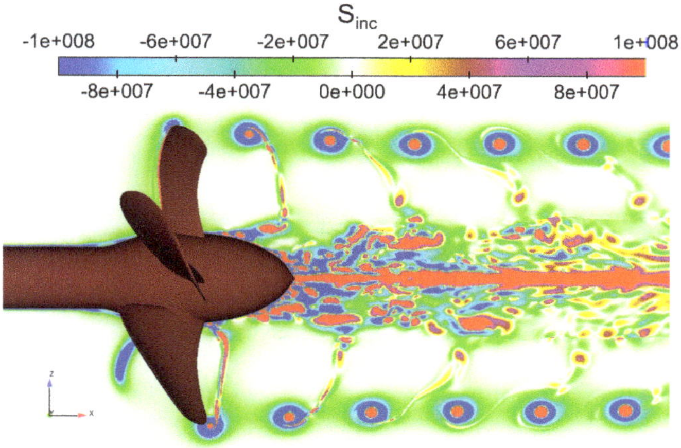

(**c**) Wetted conditions.

Figure 18. *Cont.*

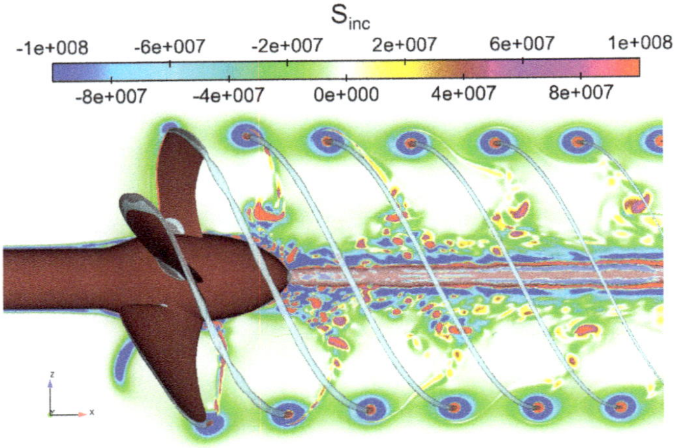

(**b**) Cavitating conditions.

Figure 18. Distribution of the acoustic source term based on delayed detached eddy simulation (DDES) near the propeller on the cut plane $y = 0$. In the cavitating case, the iso-surface of the void fraction $\alpha = 0.1$ is coloured by light blue.

Figure 19 furthermore shows the acoustic source terms, obtained from the CFD solution, on the plane $x/D = 0.5$ for the wetted and cavitating cases. The figure also shows the vortical flow structures in terms of the Q criterion, together with contours of the acoustic source term on the plane. We can see that in the wetted case, the acoustic source term distribution was more concentrated and regular near the hub vortex than in the cavitating case. In the cavitating case, the source distribution was spread to a slightly wider radius in the wake. For instance, the presence of the helical vortex filament that was shed behind the blades in the wetted case can be seen in the acoustic source term distributions. The tip vortex structure was slightly different between the wetted and cavitating conditions, and in the cavitating case around the elongated core of the tip vortex a larger area of negative source region was present.

The sound pressure can be very sensitive to the location of the investigation point in the propeller near-field. The sound field characteristics in a wave guide (e.g., a cavitation tunnel) depend considerably on the type of the source, in addition to its position and orientation, as concluded in Hynninen et al. [36]. To overcome this problem, in this paper, the mean square pressure over the cavitation tunnel domain was used to compare the results. A comparison of the mean square pressure inside the cavitation tunnel, in wetted and cavitating conditions, is shown in Figure 20. Additionally, the figure shows a corresponding comparison of the mean square pressure at the first ten blade passing frequencies. The results were obtained from the CHA simulations.

In both investigated cases, the acoustic excitation was high at discrete frequencies at the blade passing rate and its harmonics. A difference of more than 10 dB between the f_{bpf} and its second harmonic was observed in the wetted case, whereas in the cavitating case, we noted a corresponding difference of approximately 5 dB. We observed that cavitation resulted in greater acoustic excitation at the low-frequency range from $f > f_{bpf}$ and at the high-frequency end of the investigated range. Additionally, the sound pressure levels at the harmonics of the blade passing frequency were on average at a higher level for the cavitating case. Then again, at frequencies between tonals $8 < f/f_{bpf} < 10$, the non-cavitating case exhibited a greater acoustic excitation.

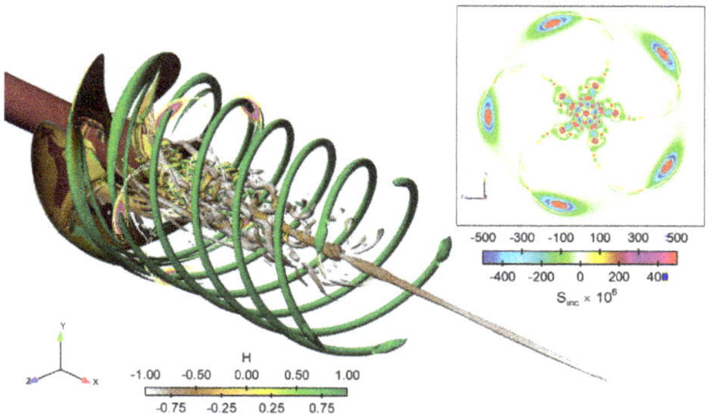

(a) Wetted conditions.

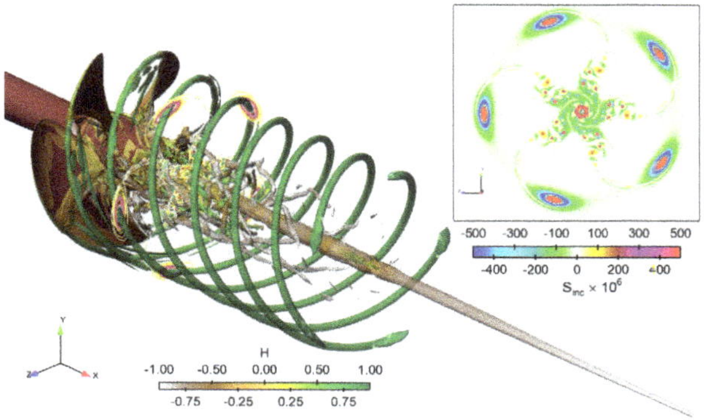

(b) Cavitating conditions.

Figure 19. Vortical flow structures visualized with the Q criterion and acoustic source term distributions in the propeller wake. The Q criterion is coloured by helicity. The boxed figure in the upper right corner shows the acoustic source term distribution on the plane $x/D = 0.5$.

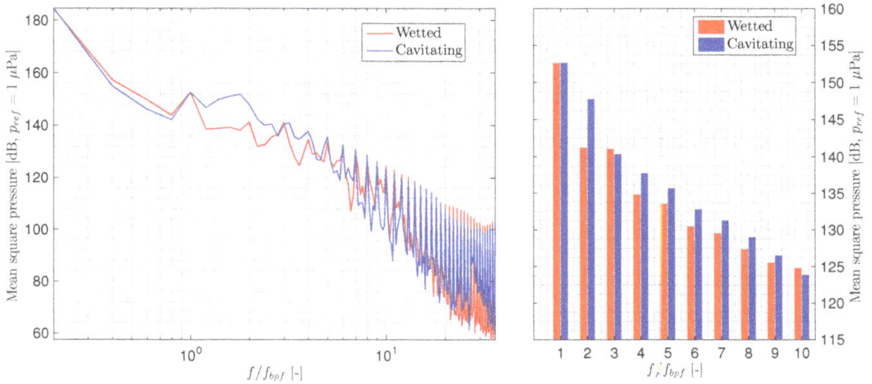

Figure 20. Simulated mean sound pressure levels inside the cavitation tunnel.

An increase in the predicted mean pressure levels was visible at around $f/f_{bpf} = 27$ in the cavitating case. The observed behaviour is due to the fact that an unsymmetrical tunnel mode was excited—a phenomenon which is discussed by Hynninen et al. [36]. For broadbanded frequencies higher than $f/f_{bpf} = 27$, the cavitating case had larger acoustic excitation. The relatively equal level of acoustic sound pressures at the blade passing rate then again presents a new issue, as previous simulations with the SST turbulence model coupled with an EARSM (explicit algebraic Reynolds stress model) [37] predicted higher acoustic excitations at the BPF for the cavitating case. This must be further investigated in the future.

In the studied wetted case, flow separation near the blade root or at the hub was not extensive. Flow around the blades behaved smoothly, and also the wake flow structures appeared somewhat clearer. In the cavitating case, the computations also predicted a rather stable root cavitation, which caused apparent changes to the flow geometry that led to separation near the trailing edge. This introduced instabilities in the wake as well, some being more prominent than those we observed in the wetted case. Consequently, an increased level of broadband noise was induced due to cavitation.

6. Conclusions

We have presented results of a hybrid CFD–CHA study of a marine propeller in a cavitation tunnel. DDES was used to obtain a transient solution of the propeller flow and wake dynamics, and an acoustic Lighthill analogy was used in the FEM context to simulate the propeller-induced acoustic propagation.

The predicted open water characteristics of the propeller were close to the experimental results, although the simulations had the tendency to under-predict the thrust and torque. Good agreement was observed between the numerically simulated propeller wake patterns and experimental LDV measurements, with the tip vortex evolution and other relevant wake flow details captured with the numerical simulations. Cavitation patterns were also well predicted, and the cavitating tip and hub vortices were excellently captured with the DDES.

However, if we are mainly investigating the global propeller performance characteristics, utilization of higher-fidelity turbulence closures is not always justified due to also the higher computational burden they impose. Then again, if one aims to resolve important vortical and cavitating flow structures and turbulent flow patterns also in the propeller wake, the choice of appropriate turbulence modelling approach becomes a key question. For such situations, hybrid RANS/LES type approaches offer attractive alternatives to the propeller two-phase flow simulations.

In the propeller noise simulations, a two-step hybrid approach was used. An examination of the CHA results with respect to the CFD-simulated flow field indicated that the sound pressure levels were reasonable, and effects due to cavitation were recognized. It was seen that the propeller wake could act as an acoustic source in a wide frequency range, while cavitating tip vortex enhanced the higher-order tonal signature. However, validation of the present acoustic simulations with experimental results is still needed. Yet, the comparison of single-point measurements conducted in a cavitation tunnel and the acoustic simulations is not straightforward. Sound pressure can be very sensitive to the location of the investigation point in the propeller near-field. The sound field characteristics in a wave guide (e.g., a cavitation tunnel) depend considerably on the type of the source, in addition to its position and orientation. The transformation of the results to corresponding free-field values is difficult. These issues are thoroughly discussed by Hynninen et al. [36].

A grid dependency study for the acoustic analyses is needed. In addition to evaluating the CFD sources from coarser and finer grids for the CHA, the sensitivity of the predicted noise levels to the numerical approximation used in the FEM should be investigated. To further improve the flow solution for smaller turbulent fluctuations and possible cavity instabilities, a finer temporal resolution could be utilized in the present DDES.

Cavitation also contributes to the acoustic excitation due to the source term related to the density variations. A scale-resolving turbulence modelling approach further enhances the predictions of the wetted and cavitating propeller-emitted noise as it aims to resolve, instead of to model, the turbulent flow fluctuations. In order to capture the possible broadband noise contribution due to rapidly varying cavities, special attention needs to be given to the cavitation modelling apart from the present compressible flow solution algorithm. Currently, mass-transfer models are used which are based on a pressure difference $p - p_{sat}$, on its square root or other similar relation. One option to improve the cavitation modelling would be a multi-scale two-phase flow model, such as that developed by Hsiao et al. [57]. The flow solution can be further developed by assuming unequal velocities for the phases [58]. This creates new challenges for the modelling of turbulence and interfacial transfer, which will be important research topics in the future.

Author Contributions: Ville M. Viitanen and Tuomas Sipilä constructed the CFD models used, and Ville M. Viitanen carried out the CFD simulations and analyses. Antti Hynninen generated the CHA model and conducted the acoustic simulations and analyses. Antti Hynninen, Tuomas Sipilä and Timo Siikonen contributed to the paper with valuable comments and suggestions. Ville M. Viitanen wrote most of the paper, and Timo Siikonen wrote parts of the section describing the flow solution algorithm used.

Conflicts of Interest: The authors declare no conflict of interest.

Abbreviations

The following abbreviations are used in this manuscript:

BPF	Blade passing frequency
CFD	Computational fluid dynamics
CHA	Computational hydroacoustics
DES	Detached eddy simulation
DDES	Delayed detached eddy simulation
DNS	Direct numerical simulation
EARSM	Explicit algebraic Reynolds stress model
EFD	Experimental fluid dynamics
FEM	Finite element method
LE	Leading edge
LES	Large eddy simulation
MUSCL	Monotonic upstream-centred scheme for conservation laws
PANS	Partially averaged Navier–Stokes
PPTC	Potsdam propeller test case
RANS	Reynolds averaged Navier–Stokes
SST	Shear stress transport
TE	Trailing edge

References

1. Andrew, R.K.; How, B.M.; Mercer, J.A.; Dzieciuch, M.A. Ocean ambient sound: Comparing the 1960s with the 1990s for a receiver off the California coast. *Acoust. Res. Lett. Online* **2002**, *3*, 65–70. [CrossRef]
2. Payne, R.; Webb, D. Orientation by means of long range acoustic signaling in baleen whales. *Ann. N. Y. Acad. Sci.* **1971**, *188*, 110–141. [CrossRef] [PubMed]
3. Mitson, R.B. Underwater noise of research vessels. *ICES Co-Oper. Res. Rep.* **1995**, *209*, 61.
4. Lightelijn, J.T. Advantages of Different Propellers for Minimising Noise Generation. In Proceedings of the 3rd International Ship Noise and Vibration Conference, London, UK, 26 September 2007.
5. Van Wijngaarden, E.; Bosschers, J.; Kuiper, G. Aspects of the cavitating propeller tip vortex as a source of inboard noise and vibration. In Proceedings of the ASME 2005 Fluids Engineering Division Summer Meeting, Houston, TX, USA, 19–23 June 2005; American Society of Mechanical Engineers: New York, NY, USA, 2005; pp. 539–544.
6. Bosschers, J. Investigation of Hull Pressure Fluctuations Generated by Cavitating Vortices. In Proceedings of the First International Symposium on Marine Propulsors (smp'09), Trondheim, Norway, 22–24 June 2009.

7. Pennings, P.; Westerweel, J.; van Terwisga, T. Sound signature of propeller tip vortex cavitation. *J. Phys. Conf. Ser.* **2015**, *656*, 012186. [CrossRef]
8. Sipilä, T.; Siikonen, T.; Saisto, I.; Martio, J.; Reksoprodjo, H. Cavitating propeller flows predicted by RANS solver with structured grid and small Reynolds number turbulence model approach. In Proceedings of the CAV2009-7th International Symposium on Cavitation, Ann Arbor, MI, USA, 16–20 August 2009; Volume 2, p. 1.
9. Turunen, T.; Siikonen, T.; Lundberg, J.; Bensow, R. Open-water computations of a marine propeller using OpenFOAM. In Proceedings of the ECFD VI-6th European Congress on Computational Fluid Dynamics, Barcelona, Spain, 20–25 July 2014; pp. 1123–1134.
10. Lu, N.X.; Bensow, R.E.; Bark, G. Large eddy simulation of cavitation development on highly skewed propellers. *J. Mar. Sci. Technol.* **2014**, *19*, 197–214. [CrossRef]
11. Muscari, R.; Di Mascio, A.; Verzicco, R. Modeling of vortex dynamics in the wake of a marine propeller. *Comp. Fluids* **2013**, *73*, 65–79. [CrossRef]
12. Balaras, E.; Schroeder, S.; Posa, A. Large-eddy simulations of submarine propellers. *J. Ship Res.* **2015**, *59*, 227–237. [CrossRef]
13. Lloyd, T.; Vaz, G.; Rijpkema, D.; Reverberi, A. Computational fluid dynamics prediction of marine propeller cavitation including solution verification. In Proceedings of the Fifth International Symposium on Marine Propulsors, smp'17, Helsinki, Finland, 12–15 June 2017.
14. Rijpkema, D.; Baltazar, J.; Falcão de Campos, J. Viscous flow simulations of propellers in different Reynolds number regimes. In Proceedings of the Fourth International Symposium on Marine Propulsors (smp'15), Austin, TX, USA, 31 May–4 June 2015.
15. Baltazar, J.; Rijpkema, D.; Falcão de Campos, J.A.C. On the Use of the $\gamma - \tilde{Re}_\theta$ Transition Model for the Prediction of the Propeller Performance at Model-Scale. In Proceedings of the Fifth International Symposium on Marine Propulsors (smp'17), Helsinki, Finland, 2–15 June 2017.
16. Rhee, S.H.; Joshi, S. CFD validation for a marine propeller using an unstructured mesh based RANS method. In Proceedings of the ASME/JSME 2003 4th Joint Fluids Summer Engineering Conference, Honolulu, HI, USA, 6–10 July 2003; American Society of Mechanical Engineers: New York, NY, USA, 2003; pp. 1157–1163.
17. Morgut, M.; Nobile, E. Numerical predictions of cavitating flow around model scale propellers by CFD and advanced model calibration. *Int. J. Rotating Mach.* **2012**, *2012*. [CrossRef]
18. Salvatore, F.; Streckwall, H.; van Terwisga, T. Propeller cavitation modelling by CFD-results from the VIRTUE 2008 Rome workshop. In Proceedings of the First International Symposium on Marine Propulsors, Trondheim, Norway, 22–24 June 2009.
19. Vaz, G.; Hally, D.; Huuva, T.; Bulten, N.; Muller, P.; Becchi, P.; Herrer, J.L.R.; Whitworth, S.; Macé, R.; Korsström, A. Cavitating flow calculations for the E779A propeller in open water and behind conditions: Code comparison and solution validation. In Proceedings of the Fourth International Symposium on Marine Propulsors, smp'15, Austin, TX, USA, 31 May–4 June 2015.
20. Gaggero, S.; Tani, G.; Viviani, M.; Conti, F. A study on the numerical prediction of propellers cavitating tip vortex. *Ocean Eng.* **2014**, *92*, 137–161. [CrossRef]
21. Asnaghi, A.; Feymark, A.; Bensow, R. Computational analysis of cavitating marine propeller performance using OpenFOAM. In Proceedings of the Fourth International Symposium on Marine Propulsors (smp'15), Austin, TX, USA, 31 May–4 June 2015.
22. Lidtke, A.; Turnock, S.; Humphrey, V. Use of acoustic analogy for marine propeller noise characterisation. In Proceedings of the Fourth International Symposium on Marine Propulsors (smp'15), Austin, TX, USA, 31 May–4 June 2015.
23. Sipilä, T.; Sanchez-Caja, A.; Siikonen, T. Eddy vorticity in cavitating tip vortices modelled by different turbulence models using the RANS approach. In Proceedings of the 11th World Congress on Computational Mechanics (WCCM XI), Barcelona, Spain, 20–25 July 2014; pp. 4741–4752.
24. Guilmineau, E.; Deng, G.; Leroyer, A.; Queutey, P.; Visonneau, M.; Wackers, J. Influence of the Turbulence Closures for the Wake Prediction of a Marine Propeller. In Proceedings of the Fourth International Symposium on Marine Propulsors, smp'15, Austin, TX, USA, 31 May–4 June 2015.
25. Viitanen, V.M.; Siikonen, T. Numerical simulation of cavitating marine propeller flows. In Proceedings of the 9th National Conference on Computational Mechanics (MekIT'17), Trondheim, Norway, 11–12 May 2017; International Center for Numerical Methods in Engineering (CIMNE): Trondheim, Norway, 2017; pp. 385–409, ISBN 978-84-947311-1-2.

26. Liefvendahl, M.; Felli, M.; Troëng, C. Investigation of wake dynamics of a submarine propeller. In Proceedings of the 28th Symposium on Naval Hydrodynamics, Pasadena, CA, USA, 12–17 September 2010.
27. Ji, B.; Luo, X.; Wu, Y.; Peng, X.; Xu, H. Partially-Averaged Navier–Stokes method with modified k–ε model for cavitating flow around a marine propeller in a non-uniform wake. *Int. J. Heat Mass Transf.* **2012**, *55*, 6582–6588. [CrossRef]
28. Chase, N.; Carrica, P.M. Submarine propeller computations and application to self-propulsion of DARPA Suboff. *Ocean Eng.* **2013**, *60*, 68–80. [CrossRef]
29. Ianniello, S.; Muscari, R.; Di Mascio, A. Ship underwater noise assessment by the acoustic analogy. Part I: Nonlinear analysis of a marine propeller in a uniform flow. *J. Mar. Sci. Technol.* **2013**, *18*, 547–570. [CrossRef]
30. Lloyd, T.; Rijpkema, D.; van Wijngaarden, E. Marine propeller acoustic modelling: Comparing CFD results with an acoustic analogy method. In Proceedings of the Fourth International Symposium on Marine Propulsors (smp'15), Austin, TX, USA, 31 May–4 June 2015.
31. Budich, B.; Schmidt, S.J.; Adams, N.A. Numerical investigation of a cavitating model propeller including compressible shock wave dynamics. In Proceedings of the Fourth International Symposium on Marine Propulsors (smp'15), Austin, TX, USA, 31 May–4 June 2015.
32. Uosukainen, S. *Foundations of Acoustic Analogies*; VTT Publications 757: Espoo, Finland, 2011.
33. Free Field Technologies SA. *Actran 16.1 User's Guide*; MSC Software Belgium: Newport Beach, CA, USA, 2016.
34. Sánchez-Caja, A.; Rautaheimo, P.; Siikonen, T. Computation of the incompressible viscous flow around a tractor thruster using a sliding-mesh technique. In Proceedings of the 7th International Conference on Numerical Ship Hydrodynamics, Nantes, France, 19–22 July 1999.
35. Viitanen, V.; Martio, J.; Sipilä, T. FINFLO two-phase URANS predictions of propeller performance in oblique flow. In Proceedings of the Fourth International Symposium on Marine Propulsors (smp'15), Austin, TX, USA, 31 May–4 June 2015.
36. Hynninen, A.; Tanttari, J.; Viitanen, V.M.; Sipilä, T. On predicting the sound from a cavitating marine propeller in a tunnel. In Proceedings of the Fifth International Symposium on Marine Propulsors (smp'17), Helsinki, Finland, 12–15 June 2017.
37. Viitanen, V.M.; Hynninen, A.; Lübke, L.; Klose, R.; Tanttari, J.; Sipilä, T.; Siikonen, T. CFD and CHA simulation of underwater noise induced by a marine propeller in two-phase flows. In Proceedings of the Fifth International Symposium on Marine Propulsors (smp'17), Helsinki, Finland, 12–15 June 2017.
38. Choi, Y.H.; Merkle, C.L. The application of preconditioning in viscous flows. *J. Comput. Phys.* **1993**, *105*, 207–230. [CrossRef]
39. Miettinen, A. *Simple Polynomial Fittings for Steam, CFD/THERMO-55-2007*; Report 55; Laboratory of Applied Thermodynamics, Aalto University: Espoo, Finland, 2007.
40. Miettinen, A.; Siikonen, T. Application of pressure- and density-based methods for different flow speeds. *Int. J. Numer. Methods Fluids* **2015**, *79*, 243–267, doi:10.1002/fld.4051. [CrossRef]
41. Van Leer, B. Flux-Vector Splitting for the Euler Equations. In Proceedings of the 8th International Conference on Numerical Methods in Fluid Dynamics, Aachen, Germany, 28 June–2 July 1982.
42. Roe, P.L. Some contributions to the modelling of discontinuous flows. In *Large-Scale Computations in Fluid Mechanics*; American Mathematical Society: Providence, RI, USA, 1985; pp. 163–193.
43. Viitanen, V. Verification of a Homogeneous Mixture Model for the Free Surface Problem. Master's Thesis, Aalto University, Espoo, Finland, 2015.
44. Spalart, P.R.; Deck, S.; Shur, M.; Squires, K.; Strelets, M.K.; Travin, A. A new version of detached-eddy simulation, resistant to ambiguous grid densities. *Theor. Comput. Fluid Dyn.* **2006**, *20*, 181–195. [CrossRef]
45. Menter, F. Influence of freestream values on $k-\omega$ turbulence model predictions. *AIAA J.* **1992**, *30*, 1657–1659. [CrossRef]
46. Shur, M.; Spalart, P.; Strelets, M.; Travin, A. Detached-Eddy Simulation of an Airfoil at High Angle of Attack. In *Engineering Turbulence Modelling and Experiments 4, Proceedings of the 4th International Symposium on Engineering Turbulence Modelling and Measurements, Ajaccio, Corsica, France, 24–26 May 1999*; Elsevier: New York, NY, USA, 1999; pp. 669–678.
47. Strelets, M. Detached eddy simulation of massively separated flows. In Proceedings of the 39th AIAA Aerospace Sciences Meeting and Exhibit, Reno, NV, USA, 8–11 January 2001.
48. Frikha, S.; Coutier-Delgosha, O.; Astolfi, J.A. Influence of the cavitation model on the simulation of cloud cavitation on 2D foil section. *Int. J. Rotating Mach.* **2009**, *2008*. [CrossRef]

49. Sipilä, T. RANS Analyses of Cavitating Propeller Flows. Licentiate Thesis, Aalto University, Espoo, Finland, 2012.
50. Lighthill, M.J. On Sound Generated Aerodynamically. I. *Gen. Theory Proc. R. Soc. Ser. A* **1952**, *211*, 564–587. [CrossRef]
51. Heinke, H.J. Potsdam Propeller Test Case (PPTC). Cavitation Tests with the Model Propeller VP1304. In *SVA Potsdam Model Basin Report No.3753*; Schiffbau-Versuchsanstalt Potsdam GmbH: Potsdam, Germany, 2011.
52. Barkmann, U.; Heinke, H.J.; Lübke, L. Potsdam propeller test case (PPTC). Test case description. In Proceedings of the Second International Symposium on Marine Propulsors (smp'11), Hamburg, Germany, 15–17 June 2011.
53. Mach, K.P. Potsdam Propeller Test Case (PPTC). LDV velocity measurements with the Model Propeller VP1304. In *SVA Potsdam Model Basin Report No.3754*; Schiffbau-Versuchsanstalt Potsdam GmbH: Potsdam, Germany, 2011.
54. Chien, K.Y. Predictions of Channel and Boundary-layer Flows with a Low-Reynolds-Number Turbulence Model. *AIAA J.* **1982**, *20*, 33–38. [CrossRef]
55. Barkmann, U. Potsdam Propeller Test Case (PPTC). Open water tests with the model propeller VP1304. In *SVA Potsdam Model Basin Report No.3752*; Schiffbau-Versuchsanstalt Potsdam GmbH: Potsdam, Germany, 2011.
56. Saarinen, P.; Siikonen, T. Simulation of HVAC flow noise sources with an exit vent as an example. *Int. J. Vent.* **2016**, *15*, 45–66. [CrossRef]
57. Hsiao, C.T.; Ma, J.; Chahine, G.L. Multiscale tow-phase flow modeling of sheet and cloud cavitation. *Int. J. Multiph. Flow* **2017**, *90*, 102–117. [CrossRef]
58. Siikonen, T. Numerical method for one-dimensional two-phase flow. *Numer. Heat Transf. Part A Appl.* **1987**, *12*, 1–18.

Journal of
Marine Science and Engineering

MDPI

Article

The Effect of Propeller Scaling Methodology on the Performance Prediction

Stephan Helma [1,*], Heinrich Streckwall [2] and Jan Richter [2]

1 Stone Marine Propulsion Ltd., SMM Business Park, Dock Rd, Birkenhead CH41 1DT, UK
2 Hamburgische Schiffbau-Versuchsanstalt GmbH, Bramfelder Straße 164, 22305 Hamburg, Germany; streckwall@hsva.de (H.S.); richter@hsva.de (J.R.)
* Correspondence: sh@smpropulsion.com; Tel.: +44-1255-420-005

Received: 28 February 2018; Accepted: 4 May 2018; Published: 24 May 2018

Abstract: In common model testing practise, the measured values of the self propulsion test are split into the characteristics of the hull, the propeller and into the interaction factors. These coefficients are scaled separately to the respective full scale values and subsequently reassembled to give the power prediction. The accuracy of this power prediction depends inter alia on the accuracy of the measured values and the scaling procedure. An inherent problem of this approach is that it is virtually impossible to verify each single step, because of the complex nature of the underlying problem. In recent years the scaling of the open-water characteristics of propeller model tests attracted a renewed interest, fuelled by competitive tests, which became the norm due to requests of the customer. This paper shows the influence of different scaling procedures on the predicted power. The prediction is compared to the measured trials data and the quality of the prediction is judged. The procedures examined are the standard ITTC 1978 procedure plus derivatives of it, the Meyne method, the strip method developed by the Hamburgische Schiffbau-Versuchsanstalt (HSVA) and the β_i-method by Helma.

Keywords: propeller scale effects; propeller open-water efficiency; surface roughness; equivalent profile; strip method; β_i-method

1. Introduction

The International Towing Tank Conference ITTC established the "1978 ITTC Performance Prediction Method" (ITTC 1978) [1], which is widely used to extrapolate the data collected during model test to full scale performance for trial or service condition. In recent years, it was suggested by more and more people—mainly designers of unconventional propellers—that the predictions made using this method often do not reflect the performance measured during ship trials, see for example Brown et al. [2]. Most authors believe that these deviations between prediction and measured performance is due to the scaling method for the open-water performance, which is needed by the ITTC 1978 power prediction method. Consequently, they either modified the ITTC method or came up with completely new methods (Praefke [3] and Helma [4,5]).

Since the performance of a full scale propeller is not easily available, Helma [4,5] suggested to scale the open-water data from tests performed at different Reynolds numbers to the full scale propeller, arguing that a good scaling method must give the same results for all model-test Reynolds numbers. It also mentions that the final validation should be done by comparing predicted with measured performance data, which is the topic of this paper.

2. Scaling Methods for Propellers

Currently, the following scaling methods are described in the literature:

1. Statistical methods
2. Analytical methods
3. CFD methods
4. Combinations of the above methods

2.1. Statistical Methods

Statistical methods try to match the measured data to the full scale performance by a relation derived by statistical analysis.

2.1.1. ITTC 1978 Method

The best known statistical propeller scaling method is described in ITTC's Performance Prediction Method (ITTC 1978). The origin of this method is described by Kuiper [6] "as based on statistics and the basis for the statistical values is very small". This method correlates the change in the thrust and torque coefficients K_T and K_Q to the change in the section drag Δc_D of a significant section profile, the chord length to diameter ratio c/D, the number Z of propeller blades and the pitch to diameter ratio P/D. The section drag again depends on the thickness to chord length ratio t/c and the Reynolds number Rn calculated with the section length c. According to the ITTC 1978 method, the integral characteristic of the propeller blade is substituted by a significant section located at a fractional radius of 0.75.

As long as the propeller to be scaled falls into the envelop of the propellers used in the statistical analysis, this method gives good results. Nevertheless it should be mentioned that this method introduces a dependency of the lift coefficient c_L of the significant profile on the pitch to diameter ratio P/D (see also Appendix A). The authors believe that this behaviour does not capture the underlying physics.

Another disadvantage can be seen in the fact that the method does not take the camber distribution of the sections into account, resulting in a lower correction for propellers with higher cambers.

2.1.2. Derivatives

Based on the same statistical approach, some authors tried to improve the accuracy of the ITTC 1978 method by using different form factors and friction lines [7].

2.2. Analytical Methods

Analytical methods derive the section's lift and drag coefficient c_L and c_D from the measured open-water data. There are two approaches described in the literature.

2.2.1. Meyne Method

The method of Lerbs/Meyne [8] addresses propeller performance scaling in combination with a propeller analysis step, which is lifting line based. It gives access to a hypothetical open-water performance of the propeller valid for a non-viscous fluid. In comparison with the experimental open-water results, specific friction corrections are obtained for model scale, while global friction adjustments are done for the full scale propeller. These are based on an equivalent profile assuming that the integral values of the whole blade are reasonably well reflected by the singular value of this profile. Meyne suggested to use the profile located at $0.75R$.

It should be mentioned that this method assumes that the propeller analysed has an optimum circulation distribution with minimum losses. An immediate result is that this method does not work as well for propellers with a non-optimum circulation distribution, such as propellers restricted in diameter or tip-unloaded propellers.

2.2.2. β_i-Method

Helma [4,5] showed that the mean hydrodynamic inflow angle $\bar{\beta}_i$ into an equivalent profile can be calculated from the open-water test as follows:

$$\tan \bar{\beta}_i(J) = -\gamma \frac{\mathrm{d}K_T(J)}{\mathrm{d}K_Q(J)}, \tag{1}$$

where the factor

$$\gamma = \frac{3}{8} \times \frac{1 - \left(\frac{d_h}{D}\right)^4}{1 - \left(\frac{d_h}{D}\right)^3} \tag{2}$$

is a purely geometric constant depending only on the ratio of the hub to propeller diameter d_h/D.

With this result, the measured thrust and torque can be split into the lift and drag of the equivalent profile, which can be scaled independently. In the last step of the calculation, they are combined again to the scaled thrust and torque figures.

The advantage of this method is the decomposing into the lift and drag coefficients, which are aligned with the hydrodynamic inflow and not the nose-tail pitch line. It also does not assume any special circulation distribution or a form drag of the section, but it still works on the assumption of an equivalent profile.

2.3. *CFD Methods*

The direct numerical simulation (DNS) represents the most accurate computational approach to describe surface friction effects, but is too expensive for scaling purposes. The large eddy simulation (LES) truncates the scales which are to be resolved to describe turbulent flow, but remains too costly when used with the exact propeller geometry. The RANS method reduces the required resolution in time and space further by introducing empirical turbulence models at all scales. RANS results are sensitive to grid qualities, turbulence modelling options and implementation details of the actual computer code. There is more work to be done until the RANS method gives repeatable results independent of the code used and the programme's operator.

2.4. *Combined Methods*

2.4.1. HSVA's Strip Method

The strip method was proposed as early as 1994 by Praefke [3] and recently deployed by HSVA [9]. It should cover any blade shape, because surface friction effects are treated in an integral manner by dividing the blade into strips covering all sections between the hub and the tip. A section drag coefficient is dedicated to each strip depending on the local Reynolds number.

2.4.2. Numerical Section Drag

A boundary layer solver might be linked with a potential flow solver to calculate section drag including effects ignored by the friction lines mentioned previously, see Thwaites [10], Head [11] or Drela [12].

3. Section Drag

With the exception of the CFD-based and the Meyne methods, all methods rely on the a priori knowledge of the drag coefficient c_D of the propeller section profiles, either for a significant section or for all sections, which are subsequently integrated over the blade. According to Abbot and Doenhoff [13], the drag of a two-dimensional section can be composed of the frictional drag c_F

of one side of a flat plate and a drag increase c_{2d}, due to the shape of the section, which can be written in coefficient form as

$$c_D = c_{F,u} + c_{F,l} + c_{2d}, \tag{3}$$

where the indices u and l denotes the upper and lower faces of the section.

Whereas the frictional resistance coefficient c_F depends solely on the Reynolds number Rn and the relative roughness k/c (with k the finished roughness of the propeller), the form drag c_{2d} shows a complicated relationship to the thickness to chord ratio t_{max}/c, the Reynolds number, the relative roughness and also c_F.

3.1. Viscous Drag of a Flat Plate

Generally speaking, two flow regimes can be observed:

1. Laminar flow and
2. turbulent flow.

Depending on the Reynolds number, the inflow turbulence and the surface condition of the flat plate, the flow might become turbulent while travelling along the flat plate, leading to a laminar-turbulent transition zone. The following cases can be observed:

Laminar case: The flow is laminar over the entire plate. The viscous drag decreases with increasing Reynolds number.

Transition governed case: The flow starts to become turbulent. The higher the Reynolds number, the earlier this transition occurs. Since the viscous resistance of a turbulent flow is higher than that of a laminar flow, the frictional drag increases with higher Reynolds numbers.

"Fully" turbulent case: If the transition point moves close to the leading edge, the influence of the laminar flow on the overall drag diminishes and the viscous drag decreases again with an increase in the Reynolds number.

The frictional resistance of the laminar flow follows a simple correlation to the Reynolds number. However, open-water tests of propellers are conducted around Reynolds numbers of 10^5 and 10^6, thus putting them into the transition region. Real size propellers work well above a Reynolds number of 10^7, generally around 5×10^7, subjecting them to fully turbulent flow.

It is theoretically possible to calculate the frictional resistance of a hydrodynamically smooth flat plate for laminar and fully turbulent flow. For one side of the plate these are given as local skin friction coefficients c_f as follows:

Laminar region:

$$c_f = \frac{0.664}{\sqrt{\mathrm{Rn}_x}}, \tag{4}$$

Turbulent region:

$$\sqrt{\frac{2}{c_f}} = \frac{1}{\varkappa} \ln \left(\mathrm{Rn}_\delta \sqrt{\frac{c_f}{2}} \right) + B + \frac{2\Pi}{\varkappa}, \tag{5}$$

where Rn_x is the Reynolds number based on the distance x from the leading edge; Rn_δ is the Reynolds number based on the boundary layer thickness at x; and $\varkappa$, B and Π are constant factors suitably selected.

The behaviours of the friction lines discussed are shown in Figure 1.

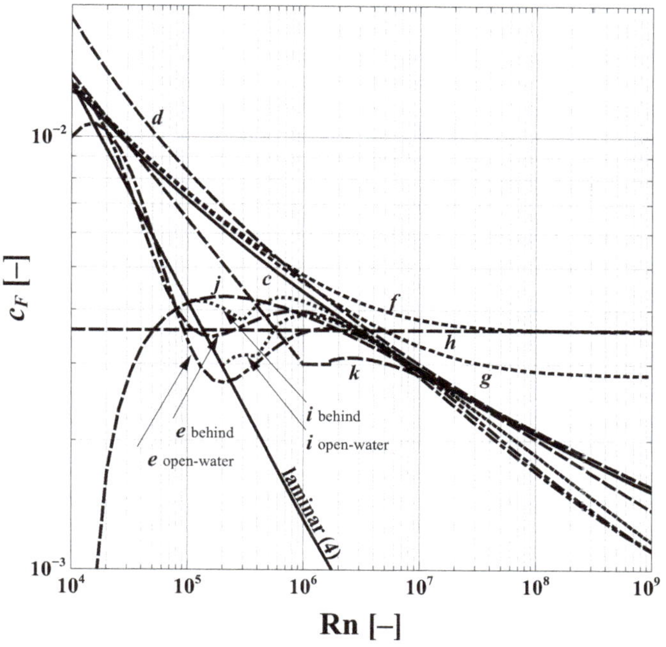

Figure 1. Values of the friction coefficients for the whole plate. The relative roughness k/c is taken as 2×10^{-5}. The capital letters reference Tables 1 and 2.

3.1.1. Friction Line for Laminar Flow

Equation (4) can be readily integrated to give the normalized frictional drag c_F for one side of the whole plate:

$$c_F = \frac{2 \times 0.664}{\sqrt{\text{Rn}}} \quad (6)$$

3.1.2. Friction Line for Fully Turbulent Flow

For the turbulent flow, Equation (5) gives c_f implicitly in terms of Rn_δ. To make matters worse, the Reynolds number used in this context depends on the local boundary layer thickness δ, which is not known. There were many approximations given for both coefficients c_f and c_F, some are listed in Table 1. Some of them are based on a theoretical model and some are curves fitted to experimental data, which might explain the differences in these lines. Five relationships do not explicitly include the relative roughness k/c, whereas the ITTC 1978 and two more general friction lines take the surface finish into account.

It must be noted that the Streckwall et al. friction line e includes the form drag Δc_{2d} (see Section 3.2).

Table 1. Exemplary values mentioned in the literature for the viscous drags c_f and c_F for one side of a flat plate with fully turbulent flow. The first five relationships assume a hydrodynamically smooth surface, the others take the relative roughness k/c of the surface into account.

Local skin friction coefficients:
a Schlichting [14]: $c_f = 0.0592/\sqrt[5]{\mathrm{Rn}_x}$
b von Kármán [15]: $1/\sqrt{c_f} = 4.15\log_{10}\left(c_f\mathrm{Rn}_x\right) + 1.7$
Friction coefficients for the whole plate:
c Schönherr [16]: $0.242/\sqrt{c_F} = \log_{10}\left(c_F\mathrm{Rn}\right)$
d ITTC 1957 [17]: $c_F = 0.075/\left(\log_{10}\mathrm{Rn} - 2\right)^2$
e Streckwall et al. [9] [a]: $c_F = a\left[d/\left(1+r^2\right) + (1-d)\,e^{-\frac{1}{2}r^2}\right]$ with $r = \left(\log_{10}\mathrm{Rn} - b\right)/c$ and suitable constants a, b, c and d
f Schlichting [14] [b,c]: $\sqrt{2/c_F} = \frac{1}{\varkappa}\ln\left(\mathrm{Rn}\times c_F/2\right) + 2 + \frac{1}{\varkappa}\ln 3.4 - \frac{1}{\varkappa}\ln\left(3.4 + \mathrm{Rn}\sqrt{c_F/2}\times k/c\right)$
g Schlichting, according to Schulze [7] [b,c,d]: $\sqrt{2/c_F} = \frac{1}{\varkappa}\ln\left(\mathrm{Rn}\times c_F/2\right) + 5 - \frac{1}{\varkappa}\ln\left(3.4 + k^+_{tech}\right)$ with $k^+_{tech} \approx 0.01\mathrm{Rn}\frac{k}{c}$
h ITTC 1978 [1]: $c_F = \left[1.89 + 1.62\log_{10}\left(c/k\right)\right]^{-2.5}$

[a] Factors: a, b, c, and d = 0.02145435039201, 4.174741254548, 0.9112701673967, and 3.029492755962 (open-water condition), respectively. a, b, c, and d = 0.02546689917582, 3.980310869827, 0.9163209307599, and 2.704789857162 (behind condition), respectively.
[b] Factor $\varkappa = 0.41$.
[c] Note that $2 + \ln 3.4/\varkappa \approx 5$ for $\varkappa = 0.41$.
[d] There is a discrepancy between the presented formula $\sqrt{1/c_F} = \log\left(c_F/2\times\mathrm{Rn}\right)/\varkappa + 5 - \log\left(3.4 + k^+_{tech}\right)$ with $k^+_{tech} \approx 0.001\mathrm{Rn}\times k/c$ and the corresponding diagram in [7]. Using the natural logarithm and $k^+_{tech} \approx 0.01\mathrm{Rn}\times k/c$ resolves this inconsistency.

3.1.3. Friction Lines for Transition Region

The two friction lines presented in Equations (4) and (5) are only valid if the flow is either laminar or turbulent over the whole length of the flat plate. Strictly speaking, this is only possible for laminar flow, since each turbulent flow starts near the leading edge as laminar flow and trips to turbulent instantaneously at the transition point. With the Reynolds number increasing, this transition occurs relatively closer to the leading edge. For high enough Reynolds numbers, this contribution of the laminar region gets so small that it can be neglected; the flow can be considered fully turbulent and calculated accordingly.

The distance x_t of the transition point from the leading edge depends not only on the local Reynolds number, but also on the surface quality of the plate, the turbulence of the inflow and—in the case of profiles—the chord wise pressure gradient. These factors pose a formidable challenge to calculate a universal friction line for the transition region.

Some friction lines found in the literature are given in Table 2. It is noteworthy that none of these takes the surface roughness of the plate or the turbulence of the inflow into account, which depend on the common practice followed by the testing facilities. It can be assumed that the relationships given by ITTC 1978 and Schulze (*d*, *j* and *k*) include these effects in a general way (for the ITTC 1978 line) or for one special towing tank (the Schulze line). The transition line *i* as used by Streckwall et al.

integrates the local friction coefficients $c_{f,lam}$ and $c_{f,turb}$ for the laminar and turbulent flow over the length of the plate. It assumes that the transition occurs at the position x_t from the trailing edge. This formulation would result in a discontinuity of the boundary layer thickness at the transition point. For this paper, this formulation was improved in such a way that the local Reynolds number $\mathrm{Rn}_{x,\delta}$ for the turbulent flow was adapted, so that in the transition point the impulse loss thickness δ_2 is the same for the laminar and the turbulent boundary layer (Schlichting [14]).

Table 2. Exemplary values for the viscous drag c_F of a flat plate mentioned in the literature. x_t is the position from the leading edge along the plate where the flow trips from laminar to turbulent [a].

i Transition (with δ_2-continuity):
$c_F = \int_0^{x_t} c_{f,lam} \mathrm{d}x + \int_{x_t}^{c} c_{f,turb} \mathrm{d}x$ $= \int_0^{x_t} 0.664/\sqrt{\mathrm{Rn}_x} \mathrm{d}x + \int_{x_t}^{c} 0.0592/\sqrt[5]{\mathrm{Rn}_{x,\delta}} \mathrm{d}x$ with suitable x_t and $\mathrm{Rn}_{x,\delta}$
j ITTC 1978 [1]: $c_F = 0.044/\mathrm{Rn}^{1/6} - 5/\mathrm{Rn}^{2/3}$
k Schulze [7] [b]: $c_F = \begin{cases} 0.3/\sqrt[3]{\mathrm{Rn}} & \text{for } \mathrm{Rn} < 10^6 \\ 0.003 & \text{for } 10^6 \leq \mathrm{Rn} \leq 1.7 \times 10^6 \\ 3.913/(\ln \mathrm{Rn})^{2.58} - 1700/\mathrm{Rn} & \text{for } \mathrm{Rn} > 1.7 \times 10^6 \end{cases}$

[a] Streckwall et al. specified the Reynolds numbers of the transition point x_t for the open-water condition as 4×10^5 (suction side) and 2×10^5 (pressure side), and for the behind condition as 3×10^5 (suction side) and 1×10^5 (pressure side) [9].
[b] In his presentation, Schulze has an obvious misprint when giving the value of 0.03 for the middle part [7]. It should be 0.003 to assure continuity at its boundaries.

3.2. Form Drag

The form drag c_{2d} is the increase of the drag of a two-dimensional section when compared to the purely viscous drag of a flat plate. The reason for this increase is two-fold: Firstly, the flow velocity over the surface is higher due to the thickness of the profile increasing the viscous drag. Secondly, the pressure does not recover entirely due to losses in the boundary layer:

$$c_{2d} = \Delta c_F + c_P, \tag{7}$$

where Δc_F is the relative increase of the friction due to the speed increase and c_P is the drag coefficient due to pressure losses.

The increase of the friction due to the higher velocity depends on the relative speed increase, the position of the transition point and the pressure gradient along the profile. For symmetrical sections, the speed increase is the same for both sides and only depends on the thickness to chord ratio t_{max}/c. The position of the transition point depends on the Reynolds number, the turbulence of the inflow and the section shape. Values found in the literature for symmetrical profiles are given in Table 3.

The pressure drag coefficient c_P can only be determined by analysing experimental results. It can be argued that it also depends on the position of the transition point, the pressure gradient and the occurrence of separation. Some values found in the literature for symmetrical profiles without flow separation are given in Table 4.

Table 3. Values for the relative form drag $\Delta c_F/(2c_F)$ for symmetrical section profiles mentioned in the literature. For the NACA 64 and 65 laminar profiles, the transition point was fixed at $0.09c$. Hoerner also emphasised that the given relationship for these sections are only valid for rough surfaces [18].

A ITTC 1978 [1]: $2t_{max}/c$
B Hoerner, t_{max} at $0.3c$ [18]: $2t_{max}/c$ for $10^6 < \mathrm{Rn} < 10^7$
C Hoerner, NACA 64 and 65 [18]: $1.2t_{max}/c$
D Torenbeek [19]: $2.7t_{max}/c$

Table 4. Values for the relative pressure drag $c_P/(2c_F)$ for symmetrical section profiles without flow separation mentioned in the literature. See also explanations given for Table 3.

A ITTC 1978 [1]: 0
B Hoerner, t_{max} at $0.3c$ [18]: $60\,(t_{max}/c)^4$ for $10^6 < \mathrm{Rn} < 10^7$
C Hoerner, NACA 64 and 65 [18]: $70\,(t_{max}/c)^4$
D Torenbeek [19]: $100\,(t_{max}/c)^4$

4. Section Lift

Abott and von Doenhoff remarked in their book that the turbulent lift coefficient of aerofoils changes only minimally with the Reynolds number [13]. There are claims that this observation does not hold for propellers, e.g., Bugalski et al. [20], who also gave a formula for the lift coefficient as a function of the Reynolds number. This possible influence of the Reynolds number is not included in the current investigation.

5. Methodology

Since full scale open-water performance data for propellers are not easily attainable, the most straight-forward approach is to compare the power and shaft revolutions predicted from model tests with the values measured during full scale trials:

$$C_P = \frac{P_{DT}}{P_{DS}}, \tag{8}$$

$$C_N = \frac{n_T}{n_S}, \tag{9}$$

where C_P and C_N are model–ship correlation factors for the power and shaft speed, respectively; P_{DT} and P_{DS} are measured and predicted delivered power, respectively; and n_T and n_S are measured and predicted shaft revolutions, respectively.

If the prediction methods were perfect, these two model–ship correlation factors would be 1 for every model test analysed However, it is to be expected that these factors scatter around a mean value, which is not necessarily 1. This shift in the mean value can be corrected by applying the model–ship correlation factor as a final step in the power prediction, as stated in the ITTC 1978 method. A measure of the scatter is the standard deviation, which assumes that the distribution of values forms the bell shaped curve. The smaller the standard deviation the closer are the scattered values to the mean value, hence the better is the scaling method. Note that, before calculating the standard deviation, the values must be normalized to give a mean value of 1.

The analysis was run by the Hamburgische Schiffbau-Versuchsanstalt (HSVA).[1]

HSVA currently uses the strip method and previously used the Lerbs/Meyne [8] and the standard ITTC 1978 [1] methods, so these were already available. The β_i-method was implemented by Stone Marine Propulsion (SMP), just as the ITTC method. This was necessary, so that the ITTC method can be used with different friction lines and form factors, which were implemented by SMP as well. Using this newly developed program, it was possible to calculate the open-water characteristics for the self-propulsion and full scale conditions for all possible combinations of scaling methods, friction lines and form factors.

Based on HSVA's databases of performance predictions and sea trials, the intersection of both sets was identified. The expected power was calculated according to HSVA's standard performance prediction method [9].

The 25 scaling methods used are summarized in Table 5. The finished roughness k of the propeller in full scale was assumed to be 20 μm.

Table 5. Scaling methods λ investigated. A mark in the $\int$-columns denotes that the scaling procedure integrates the sectional friction over the whole blade. The capital letters specifying the friction lines, form and pressure drag are references to the Tables 1–4, respectively (OW, open-water; SP, self-propulsion; and FS, full scale). The finished roughness k of the propeller in full scale was assumed to be 20 μm.

			Friction Lines			Drag	
	Method	$\int$	OW	SP	FS	Form~	Pressure~
A	ITTC		*j*	*j*	*h*	*A*	*A*
B	ITTC		*j*	—	*h*	*A*	*A*
C	ITTC	×	*j*	*j*	*h*	*A*	*A*
D	ITTC	×	*j*	—	*h*	*A*	*A*
E	ITTC		*j*	*j*	*f*	*A*	*A*
F	ITTC		*j*	—	*f*	*A*	*A*
G	ITTC	×	*j*	*j*	*f*	*A*	*A*
H	ITTC	×	*j*	—	*f*	*A*	*A*
I	ITTC		*e*	*e*	*e*	incl.	*A*
J	ITTC		*e*	—	*e*	incl.	*A*
K	ITTC		*i*	*i*	*i*	*A*	*A*
L	ITTC		*k*	*k*	*g*	*D*	*D*
M	Meyne		—	—	—	—	—
N	Strip	×	*e*	*e*	*e*	incl.	*A*
O	Strip	×	*e*	—	*e*	incl.	*A*
P	β_i		*j*	*j*	*h*	*A*	*A*
Q	β_i		*j*	—	*h*	*A*	*A*
R	β_i	×	*j*	*j*	*h*	*A*	*A*
S	β_i	×	*j*	—	*h*	*A*	*A*
T	β_i		*j*	*j*	*f*	*A*	*A*
U	β_i		*j*	—	*f*	*A*	*A*
V	β_i	×	*j*	*j*	*f*	*A*	*A*
W	β_i	×	*j*	—	*f*	*A*	*A*
X	β_i		*i*	*i*	*i*	*A*	*A*
Y	β_i	×	*i*	*i*	*i*	*A*	*A*

1 Data processing was done solely at HSVA and results on trial predictions related to the various propeller scaling approaches were kept anonymous. The anonymous results on trial prediction were exclusively stored in normalized form, meaning that the quality of power and shaft speed predictions were expressed by the two model–ship correlation factors C_P and C_N.

5.1. Performance Prediction Method

At HSVA, self-propulsion is simulated via the so-called Continental Method. Model speeds are to be converted to full scale following Froude's similarity law. Under self-propulsion, the hull model is additionally towed by a force F_D to compensate for increased surface friction effects present in model scale. The 1957 ITTC friction line is used to prescribe a hull surface friction coefficient C_F (to be applied to the wetted surface) in model and full scale [17]. F_D depends on the difference in surface friction and the evaluation process solely involves the difference $C_{FM} - C_{FS}$ without any form factor entering. A correlation allowance C_A (a function of the vessel's length and its block coefficient) is added instead to evaluate F_D. C_A covers added resistance of the zinc anodes, standard hull roughness and small openings. Assuming a complete geometrical similarity and disregarding any superstructure, the model scale thrust would be directly convertible to full scale. However, the full scale prediction of the shaft speed needs a prediction of the actual full scale effective wake fraction and the Yazaki's method [21] is applied for this purpose. As usual, to allow for a trial prediction, the normalized propeller thrust measured behind the model is enlarged to account for air resistance of the superstructure and (if necessary) to include appendages and hull openings not present during the model tests. The thrust correction causes adequate power and shaft speed adjustments for the sea trials, achievable by the aid of the propeller open-water diagram.

5.2. Analysis

The intersection of performance predictions and sea trials available at HSVA consisted of 360 data records (see Table 6). For 183 records, the open-water characteristics could not be calculated for all scaling methods considered. By visual inspection it was found that three sets have an obvious mistake in either the available trial or prediction data. The remaining 174 datasets consist of 38 unique propeller-hull configurations. The mean values of C_P and C_N were calculated for each of these configurations using each of the 25 propeller scaling methods listed in Table 5. Finally, the resulted distribution was filtered according to Tukey's range test: If for one dataset more then half of the C_P values are outside Tukey's range calculated with the typical value of $k = 1.5$, this dataset was disregarded. The Tukey's range for valid data points was calculated with:

$$Q_{1,\lambda} - k\,(Q_{3,\lambda} - Q_{1,\lambda}) < C_{P,i,\lambda} < Q_{3,\lambda} + k\,(Q_{3,\lambda} - Q_{1,\lambda})\,, \tag{10}$$

where $Q_{1,\lambda}$ and $Q_{3,\lambda}$ are lower and upper quartiles of all C_P values for one scaling method λ, respectively. Applying this test yields a total of 35 ships to be included in the evaluation. The range of ships are shown in Table **??**.

Table 6. Number of used datasets.

	Discarded	**Remaining**
Available datasets	—	360
Open-water data could not be calculated	183	177
Errors in reference data	3	174
Unique hull–propeller combinations	38	
Outliers according to Tukey	3	35
	Total	35

Table 7. Range covered by the 35 ships included in the final analysis.

Ship type	Mainly bulk carriers and container vessels	
Ship length	L_{pp}	140–340 m [a]
Ship speed	V_s	14–26 kn
Propeller diameter	D_P	4.5–9.1 m [b]
Pitch to diameter ratio	P/D	0.76–1.11
Number of blades	N	4–6
Blade area ratio	A_e/A_0	0.4–1.02
Number of propellers	Single screw	

[a] Two vessels have a length below 100 m, [b] For the range of L_{PP} given above.

Using these 35 valid datasets, the following values were calculated for each scaling method λ:

- Mean value of all datasets for each scaling method λ:

$$\overline{C}_{P,\lambda} = \frac{1}{N}\sum_{i=1}^{N} C_{P,i,\lambda}, \tag{11}$$

- Normalized model–ship correlation factors:

$$C^*_{P,i,\lambda} = \frac{C_{P,i,\lambda}}{\overline{C}_{P,\lambda}}, \tag{12}$$

- Standard deviation S_P of all normalized datasets for each scaling method λ:

$$S^*_{P,\lambda} = \sqrt{\frac{1}{N}\sum_{i=1}^{N}\left(C^*_{P,i,\lambda} - 1\right)^2}, \tag{13}$$

where N is the number of valid datasets and $C_{P,i,\lambda}$ is the model–ship power correlation factor for the i-th dataset. Note that the normalized mean values $\overline{C}^*_{P,\lambda}$ are always 1.

The normalization (Equation (12)) of the model–ship correlation factors is necessary, because the ITTC power prediction method defines the model–ship correlation factors as multiplication factors (compared to an offset to be added) and the standard deviation changes, whenever the underlying data-set is multiplied by a constant factor. Without the normalization, it would favour scaling methods with smaller $\overline{C}_{P,\lambda}$ values.

6. Results

The mean values $\overline{C}_{P,\lambda}$ of the model–ship power correlation factors $C_{P,i}$ for each propeller scaling method λ (Equation (11)) are shown in Figure 2. It should be noted that a mean value of 1 is not necessarily an indicator for the quality of the propeller scaling method.

The following patterns can be seen in Figures 2 and 3:

1. The mean values of the model–ship power correlation factor is about 1 for most investigated scaling methods.
2. The scaling methods which do not scale down to the Reynolds number of the self-propulsion test typically perform better than the same method using the scaled down open-water characteristics to analyse the self-propulsion test (*B–A*, *D–C*, *F–E*, *H–G*, *J–I*, *O–N*, *Q–P*, *U–T* and *W–V*, but not *S–R*).
3. The methods using the original Schlichting friction line *f* for the full scale propeller tend to perform better (*E–A*, *F–B*, *G–C*, *H–D*, *T–P* and *W–S*, but not *U–Q* and *V–R*).
4. All methods using the local surface friction *i* trying to capture the transition from laminar to turbulent flow do not perform very well (*K*, *X* and *Y*).

5. The β_i-methods integrating the friction forces over the whole blade perform better than the same method using only the friction force of a significant profile (*R–P*, *S–Q*, and *W–U*, but not *V–T* and the notable exception *Y–X*). For the ITTC 1978 methods, this trend is reversed (*C–A*, *D–B*, *G–E* and *H–F*).
6. The most recent methods perform better than the original ITTC 1978 method *A*.
7. The β_i-methods *R* and *W* integrating the ITTC 1978 and Schlichting friction lines *h* and *f* over the whole blade perform best, closely followed by other β_i-methods using different approaches regarding the handling of the viscous resistance. The next best method is the HSVA strip method in a version *O*, which does not scale down to the Reynolds number range of the self-propulsion tests.

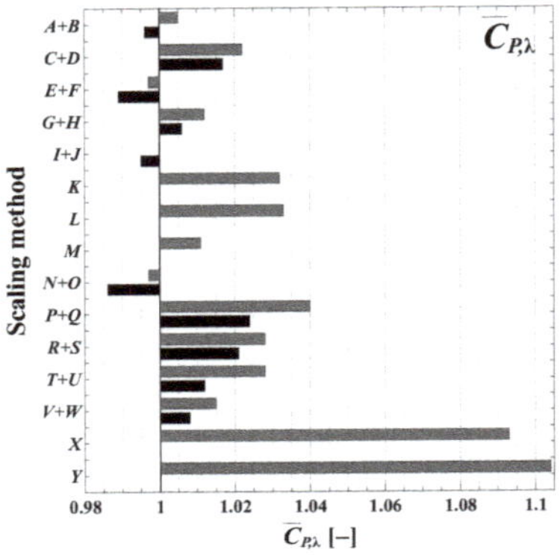

Figure 2. Mean values $\overline{C}_{P,\lambda}$ of the model–ship power correlation factors $C_{P,\lambda}$ for all scaling methods λ. If the same scaling method was investigated with and without scaling to the Reynolds number of the self-propulsion test, the results are grouped together; the black bar stands for the no scaling to the self-propulsion test. The capital letters reference Table 5.

Figure 3 shows the standard deviation $S^*_{P,\lambda}$ of the normalized model–ship power correlation factors $C^*_{P,i,\lambda}$ (Equation (13)).

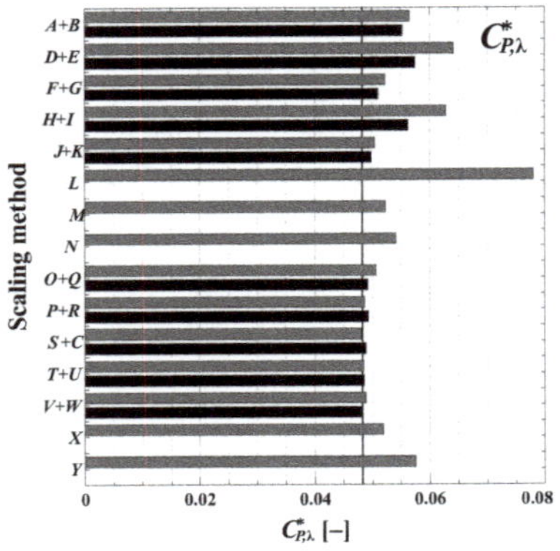

Figure 3. Standard deviation $S^*_{P,\lambda}$ of the normalized model–ship power correlation factor $C^*_{P,i}$ for all scaling methods λ. If the same scaling method was investigated with and without scaling to the Reynolds number of the self-propulsion test, the results are grouped together, the black bar stands for no scaling to the self-propulsion test. The capital letters reference Table 5.

7. Discussion

The scaling approaches compared in this paper rely on the estimates of the normalized drag forces experienced by either a single section representing the blade or—in the integral cases, such as the strip method—by individual sections building up the blade. All methods but the Meyne and β_i-methods consider the resistance force vector to be orientated in the direction of the nose-tail line; both the Meyne and the β_i-method calculate the direction of the hydrodynamic inflow and aligns the resistance force parallel to it. In terms of a favourable standard deviation of the normalized power correction factor C_P^*, the quality of a specific approach is also linked to the qualities of the drag estimates achieved in model or full scale. The ability to account correctly for local Reynolds number variations, either in the model or full scale Reynolds number region, would be reflected in a lower standard deviation.

Generally, it can be considered more challenging to capture Reynolds number sensitivities in the model scale case due to the extended presence of laminar flow over the model propeller blade. Referencing item 2 from the list presented in Section 5, it could be concluded that none of the used friction lines is able to predict the friction forces in model scale accurately. Better results are achieved by using the open-water data without any correction applied to analyse the self-propulsion test. One possible reason for this behaviour might lie in the fact that the inflow into the propeller during the self-propulsion test is more turbulent due to the boundary layer of the ship model than during the open-water test. This turbulent inflow would trigger an earlier transition from laminar to turbulent flow similar to the open-water test in undisturbed inflow.

The data presented indicate that for full scale the original Schlichting friction line f might be a good choice for any scaling procedure.

One has to be aware that not only the propeller scaling method has an influence on the predicted power and shaft revolutions: The favourable settings for arriving at the target $\overline{C}_P = 1$ will depend on other corrections entering the set-up and evaluation of propulsion tests. In particular the scaling of hull resistance and effective wake are to be mentioned in this context. It should be noted that more important than a $\overline{C}_P$ value of unity is a small standard deviation. The offset of the mean value can be compensated for, whereas the standard deviation is a measure of the spread of predicted values, which should be kept small.

It should be mentioned that the strip method N has the advantage to have been confirmed previously by quite the same dataset used in this investigation to analyse all scaling methods.

All propeller scaling methods investigated in this paper are supposed to show negligible differences in view of their influence on the full scale shaft speed prediction. Consequently, the isolated sensitivity of the shaft speed forecast on the traditional propeller scaling approaches is to be considered minor. This can be shown by calculating the ratios

$$C_{N,i,\lambda}^{*(\mathrm{ITTC})} = \frac{C_{N,i,\lambda}}{C_{N,i,\lambda=\mathrm{ITTC}}}, \tag{14}$$

where $C_{N,i,\lambda=\mathrm{ITTC}}$ is the model–ship correlation factor for the shaft speed calculated using the ITTC 1978 propeller scaling method A. A ratio close to 1 for all propeller scaling methods indicates a minor sensitivity of $C_{N,i,\lambda}$, which is shown in Figure 4 for all 177 valid datasets used in this investigation. This confirms that the predictions for the shaft speed are hardly a matter of propeller scaling contrary to the power predictions. The causes of this behaviour would justify an investigation on its own.

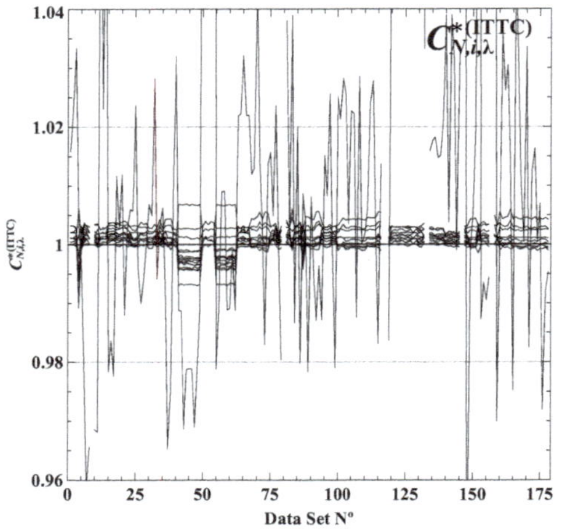

Figure 4. $C^{*(ITTC)}_{N,i,\lambda}$ of the model–ship shaft speed correlation factors $C_{N,i,\lambda}$ normalized with the ITTC 1978 propeller scaling methods. The small scatter around the value of 1 indicates a small influence of the propeller scaling method on the prediction of the shaft speed.

8. Conclusions

The power and shaft revolutions were predicted by the standard HSVA performance prediction method but with 25 different propeller scaling methods. These predictions were compared to the measured trials data to quantify the quality of each propeller scaling method. The standard deviations of the normalized model–ship power correlation factors were calculated as a measure of the quality of the prediction. All investigated methods showed a mean value of this correlation factor of about 1.

Typically, better results can be expected if the open-water propeller characteristics are not scaled down to the Reynolds number of the self-propulsion test. A possible reason is explained in Section 6, but its cause should be investigated thoroughly, e.g., by paint tests and measuring the turbulence of the inflow into the propeller in open-water and behind condition. Finally, either a Reynolds number for the open-water test should be established, which results in the same flow pattern as in the self-propulsion test, or a friction line should be developed, which takes into account the increased turbulence of the inflow behind the ship model.

From the data available, the most promising friction line for the full scale propeller is the original Schlichting line *f*.

From the propeller scaling methods investigated, the β_i-method in its variants *R* and *W*, where the friction forces are integrated over the whole blade, showed the best results. The most likely reason is that this method aligns the drag forces with the actual hydrodynamic inflow angle experienced by the propeller blade and that it does not need to know the sectional form drag.

It might be worthwhile to investigate numerical methods to calculate the section drag, such as outlined by Thwaites [10] and Head [11]. Drela implemented the transition from laminar to turbulent flow in the software XFoil [12]. The values for the section drag calculated numerically can be used by any scaling method investigated in this paper.

It must be mentioned that the original 360 trial datasets were reduced to just 35 unique propeller–hull combinations due to rigorous data checks and averaging over sister-ship cases. HSVA is currently investing in a maintenance program for the database to allow for an enlargement of datasets, which would pass the rigorous consistency checks.

With more data available, it becomes feasible to run an optimization on a parametrized friction line with the target to further minimize the scatter of the model–ship power correlation factors.

In the current investigation, the influence of different formulations of the form and pressure drags was not investigated.

Finally, it must be noted that no unconventional propellers, such as end-plate, tip-raked propellers or propellers with unconventional section shapes, such as the NPT propeller, were present in the datasets available to the current investigation. Because of the underlying physical principles, it can be assumed that the β_i-methods integrating the friction forces over the whole blade will also perform best for these propellers. When unconventional propellers are included in such investigations, one should be aware that their number is very small compared to more conventional designs, hence their influence on the overall outcome is most likely negligible. Since most newly developed propeller scaling methods claim to give more accurate results for unconventional propellers, this group of propellers must be looked at separately to isolate the effect of different propeller scaling methods.

9. Final Note

The software developed to calculate the scaled open-water characteristics is published with an open license on the principal author's GitLab website https://gitlab.com/sphh/PyOWscaling. All functionality are realized as plug-ins written in Python. It is easy to write your own plug-in to implement other propeller scaling methods, friction lines and in- and output formats. It would be appreciated if you made your plug-ins available to the public.

Author Contributions: The idea of this paper emerged during discussions between S.H. and H.S. J.R. set up and ran the power predictions. H.S. contributed the computer code for the strip method, parts of the manuscript, preliminary post processing and evaluated the data filtering. Stephan Helma contributed the computer code for the other scaling methods. He did the data filtering, the final post processing and wrote most of the manuscript.

Conflicts of Interest: The authors declare no conflict of interest.

Abbreviations

The following abbreviations are used in this manuscript:

CFD	Computational Fluid Dynamics
DNS	Direct Numerical Simulation
FS	full scale
HSVA	Hamburgische Schiffbau-Versuchsanstalt GmbH
ITTC	International Towing Tank Conference
ITTC 1978	ITTC Performance Prediction Method [1]
LES	Large Eddy Simulation
OW	open-water
SMP	Stone Marine Propulsion Ltd.
SP	self-propulsion
RANS	Reynolds-averaged Navier-Stokes [equations]

Appendix A Review of ITTC 1978 Scaling Procedure

The total propeller force F can be composed from the propeller thrust T and torque Q

$$F^2 = T^2 + \left(\frac{Q}{x\frac{D}{2}}\right)^2,$$

where x = fractional lever of the torque, which does not change with the scaling.

This relation can be made dimensionless by dividing by $\left(\rho n^2 D^4\right)^2$:

$$K_F^2 = K_T^2 + \frac{4}{x^2} K_Q^2.$$

Using the ITTC 1978 [1] scaling for K_T and K_Q

$$\Delta K_T = 0.3 \Delta c_D \frac{P}{D} \frac{c}{D} Z,$$

$$\Delta K_Q = -0.25 \Delta c_D \frac{c}{D} Z,$$

it is possible to scale the propeller force coefficient:

$$\begin{aligned} {K'_F}^2 &= {K'_T}^2 + \frac{4}{x^2} {K'_Q}^2 \\ &= (K_T + \Delta K_T)^2 + \frac{4}{x^2} (K_Q + \Delta K_Q)^2 \\ &= \left(K_T + 0.3 \Delta c_D \frac{P}{D} \frac{c}{D} Z\right)^2 + \\ &\quad + \frac{4}{x^2} \left(K_Q - 0.25 \Delta c_D \frac{c}{D} Z\right)^2. \end{aligned}$$

Hence

$$K'_F = \mathrm{f}(K_T, K_Q, \Delta c_D \frac{c}{D} Z, \frac{P}{D}),$$

which shows that the scaled propeller force is not only a function of the thrust and torque figures and the increase in section drag $\Delta c_D Z c/D$, but also of the pitch to diameter ration P/D. In the opinion of the authors this dependency cannot be explained with first principles. This surprising result is a property of all scaling methods which are based on the ITTC procedure.

References

1. Performance Prediction Method. *ITTC—Recommended Procedures and Guidelines*; Effective Date 2014, Revision 03; ITTC: The Hague, The Netherlands, 1978.
2. Brown, M.; Sanchez-Caja, A.; Adalid, J.G.; Black, S.; Sobrino, M.P.; Duerr, P.; Schroeder, S.; Saisto, I. Improving Propeller Efficiency through Tip Loading. In Proceedings of the 30th Symposium on Naval Hydrodynamics, Hobart, Tasmania, Australia, 2–7 November 2014.
3. Praefke, E. Multi-Component Propulsors for Merchant Ships—Design Considerations and Model Test Results. In Proceedings of the SNAME Symposium (Propeller/Shafting'94), Virginia Beach, VA, USA, 20–21 September 1994.
4. Helma, S. An Extrapolation Method Suitable for Scaling of Propellers of any Design. In Proceedings of the Fourth International Symposium on Marine Propulsors (Smp'15), Austin, TX, USA, 4 June 2015.
5. Helma, S. A scaling procedure for modern propeller designs. *Ocean Eng.* **2016**, *120*, 165–174. [CrossRef]
6. Kuiper, G. *The Wageningen Propeller Series*; MARIN: Wageningen, The Netherlands, 1992.
7. Schulze, R. Neue Verfahren zur Reynoldszahlkorrektur für Freifahrtmessungen an Modellpropellern. Presented at the STG-Sprechtag, Moderne Propulsionskonzepte, Hamburg, Germany, 16 June 2016.
8. Meyne, K. *Untersuchung der Propellergrenzschichtströmung und der Einfluss der Reibung auf die Propellerkenngrößen*; STG-Jahrbuch: Hamburg, Germany, 1972.
9. Streckwall, H.; Greitsch, L.; Müller, J.; Scharf, M.; Bugalski, T. Development of a Strip Method Proposed as New Standard for Propeller Performance Scaling. *Ship Technol. Res.* **2013**, *60*, 58–69. [CrossRef]
10. Thwaites, B. Approximate Calculation of the Laminar Boundary Layer. *Aeronaut. Q.* **1949**, *1*, 245–280. [CrossRef]
11. Head, M.R. Entrainment in the Turbulent Boundary Layer. Aeronautical Research Council Reports, 1958. Available online: https://pdfs.semanticscholar.org/16b6/5a95963109996d23ce1f3e953fc493939f9e.pdf (accessed on 28 February 2018).
12. Drela, M. XFoil Subsonic Airfoil Development System, 2013. Available online: http://web.mit.edu/drela/Public/web/xfoil/ (accessed on 28 February 2018).

13. Abbott, I.H.; von Doenhoff, A.E. *Theory of Wing Sections*; Dover Publications, Inc.: New York, NY, USA, 1959.
14. Schlichting, H.; Gersten, K. *Grenzschicht-Theorie*; Springer Verlag: Berlin, Germany, 2006.
15. von Kármán, T. Turbulence and Skin Friction. *J. Aeronaut. Sci.* **1934**, *1*, 1–20. [CrossRef]
16. Schönherr, K.E. Resistance of Flat Surfaces Moving Through a Fluid. *Trans. Soc Nav. Archit. Mar. Eng.* **1932**, *40*, 279–313.
17. Resistance Test. *ITTC—Recommended Procedures and Guidelines*; Effective Date 2011, Revision 03; ITTC: Kongens Lyngby, Denmark, 1957.
18. Hoerner, S.F. *Fluid-Dynamic Drag, Theoretical, Experimental and Statistical Information*; Hoerner Fluid Dynamics: Bakersfield, CA, USA, 1965.
19. Torenbeek, E. Synthesis of Subsonic Airplane Design. *The Netherlands and Martinus Nijhoff*; Delft University Press: Delft, The Netherlands, 1992.
20. Bugalski, T.; Streckwall, H.; Szantyr, J. Critical Review of Propeller Performance Scaling Methods, Based on Model Experiments and Numerical Calculations. *Polish Marit. Res.* **2013**, *20*, 71–79. [CrossRef]
21. Yazaki, A. A Diagram to Estimate the Wake Fraction for a Actual Ship from a Model Tank Test. In Proceedings of the 12th International Towing Tank Conference (ITTC), Roma, Italy, September 1969.

Journal of
Marine Science and Engineering

MDPI

Article

Coupling Numerical Methods and Analytical Models for Ducted Turbines to Evaluate Designs

Bradford Knight [1,2,*,†], Robert Freda [2], Yin Lu Young [1] and Kevin Maki [1]

1 Department of Naval Architecture and Marine Engineering University of Michigan, Ann Arbor, MI 48109, USA; ylyoung@umich.edu (Y.L.Y.); kjmaki@umich.edu (K.M.)
2 V[2] Wind Inc., Boston, MA 02116, USA; robfreda@v2wind.com
* Correspondence: bgknight@umich.edu; Tel.: +1-734-764-6470
† Current address: 2600 Draper Dr, Ann Arbor, MI 48109, USA.

Received: 27 February 2018; Accepted: 11 April 2018; Published: 16 April 2018

Abstract: Hydrokinetic turbines extract energy from currents in oceans, rivers, and streams. Ducts can be used to accelerate the flow across the turbine to improve performance. The objective of this work is to couple an analytical model with a Reynolds averaged Navier–Stokes (RANS) computational fluid dynamics (CFD) solver to evaluate designs. An analytical model is derived for ducted turbines. A steady-state moving reference frame solver is used to analyze both the freestream and ducted turbine. A sliding mesh solver is examined for the freestream turbine. An efficient duct is introduced to accelerate the flow at the turbine. Since the turbine is optimized for operation in the freestream and not within the duct, there is a decrease in efficiency due to duct-turbine interaction. Despite the decrease in efficiency, the power extracted by the turbine is increased. The analytical model under-predicts the flow rejection from the duct that is predicted by CFD since the CFD predicts separation but the analytical model does not. Once the mass flow rate is corrected, the model can be used as a design tool to evaluate how the turbine-duct pair reduces mass flow efficiency. To better understand this phenomenon, the turbine is also analyzed within a tube with the analytical model and CFD. The analytical model shows that the duct's mass flow efficiency reduces as a function of loading, showing that the system will be more efficient when lightly loaded. Using the conclusions of the analytical model, a more efficient ducted turbine system is designed. The turbine is pitched more heavily and the twist profile is adapted to the radial throat velocity profile.

Keywords: numerical methods; ducted turbine; computational fluid dynamics; CFD

---

1. Introduction

Hydrokinetic turbines are a means to extract energy from the world's oceans, rivers, and tidal streams. Energy extraction from renewable sources can reduce human dependence on fossil fuels and mitigate global warming. Ducts have been used to augment hydrokinetic turbines, wind turbines, and propellers for decades. Understanding the fluid dynamics effects of energy extraction within the duct will allow for the development of more efficient hydrokinetic turbines, which can in turn reduce the overall generation cost of marine renewable energy.

Figure 1 shows a duct which passively accelerates the flow at the throat. The duct is comprised of two sections: an intake (from inlet to throat) and a diffuser (from throat to exit).

Ducts to increase the power of a wind or hydrokinetic turbine have been researched for decades [1–4]. Grumman Aerospace Corporation performed research on ducted wind turbines in the 1970s and 1980s by analyzing ducts with screens to represent the pressure drop caused by a turbine [2,3]. These ducts are characterized by high angle diffusers that result in an area ratio between the diffuser exit and throat of 2.78 and mass flow efficiencies that are less than 60% when the duct is empty [2]. This research has been used to evaluate other models, like van Bussel's, which uses back pressure at the diffuser exit as a calibration variable [2,5]. The ducted turbine, the Vortec 7, was the culmination of the research performed by Grumman Aerospace Corporation [6]. Ducts have also been applied to other wind turbines like the Windlens turbines [7] and to hydrokinetic turbines designed by OpenHydro, Solon, and Lunar Energy [8]. The OpenHydro turbine is characterized by an open center turbine with a rim-drive generator, but produces only an estimated 0.34 coefficient of power [8,9].

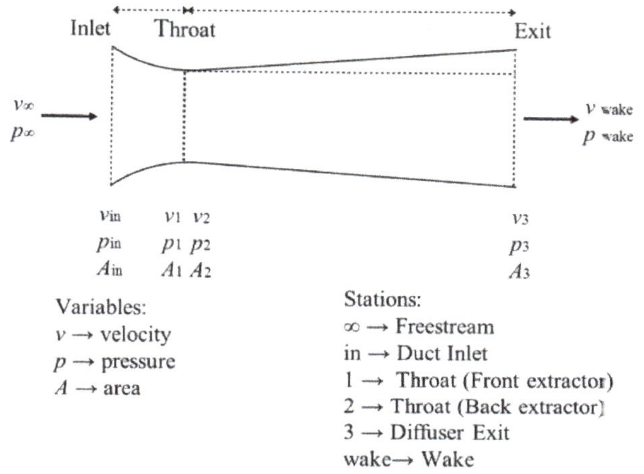

Figure 1. Schematic of a ducted turbine.

Power extraction with ducted turbines creates complicated flow effects around both the turbine and the duct. Numerical methods like Reynolds averaged Navier–Stokes (RANS) computational fluid dynamics (CFD) can be useful for detailed analysis of the ducted turbine system, but analytical models can provide a useful preliminary analysis. The objective of this study is to evaluate the differences between a RANS CFD prediction and an analytical model prediction for ducted turbine performance.

Prior to examining the ducted turbine, two RANS CFD methods are examined for a freestream turbine: a steady state moving reference frame (MRF) and an unsteady RANS (URANS) rotating sliding mesh. The MRF approach is less expensive than the URANS rotating sliding mesh approach, but the rotating sliding mesh approach can lead to more accurate predictions [10,11]. The MRF method is broadly used for analysis of axisymmetric hydrokinetic turbine analysis, especially when the focus is upon blade loading [12–14]. Both the MRF and URANS rotating sliding mesh techniques are evaluated using the geometry of a marine current turbine which was tested in a cavitation tunnel and towing tank [15]. The experiments have been widely used to evaluate other numerical tools ranging from Boundary Element Methods [16] to RANS coupled with FEA [17].

This study examines how the performance of the turbine tested by Bahaj et al. [15] changes when placed inside of a high-efficiency duct. The study expands upon prior work [18]. An analytical model is derived and the results are compared to the CFD predictions of the turbine in the duct. The predictions of the analytical model are also compared with other analytical models. The analytical models are compared to two data sets. The first data set is the screen tests presented by Gilbert et al. [2]. The second data set is the RANS CFD results of the Bahaj et al. [15] turbine inside of a tube. The new analytical model differs from other models via the assumption that the pressure drop occurs at the accelerated throat velocity, instead of as a function of the average of wake and freestream velocities. The analytical model, once corrected for viscous effects, is used to improve the design of the ducted turbine system.

2. Materials and Methods

The methodology is comprised of four parts. First, an analytical model is derived. Second, the CFD model for the freestream turbine is described. Third, the CFD model for the ducted turbine model setup is described. Fourth, the numerical methods and turbulence modeling for both the freestream and ducted CFD models are discussed. To compare the analytical model to the RANS CFD results, each must be validated independently. The freestream CFD model is compared to the tow tank results by Bahaj et al. [15], and the analytical model is compared to ducted screen tests [2].

OpenFOAM version 2.4.x is used for both freestream and ducted analyses. The steady state MRF solutions are completed using the OpenFOAM solver simpleFOAM with MRF, referred to as MRFSimpleFOAM. The rotating mesh URANS study is completed using the OpenFOAM solver pimpleDyMFOAM with an Arbitrary Mesh Interface (AMI). MRFSimpleFOAM and pimpleDyMFOAM are used to calculate the steady state and transient solutions, respectively, for the freestream turbines. MRFSimpleFOAM is used to calculate the steady state solution for the ducted turbine.

MRFSimpleFOAM is a steady state solver that uses an MRF to simulate a spinning turbine. MRFSimpleFOAM solves the steady state, incompressible Navier–Stokes Equations. The MRF specifies that the blades and the surrounding MRF domain are in a 'spinning frame' at the specified rate. The rotating zone is a circular cylinder.

This same rotating zone is used for the pimpleDyMFOAM analysis. The difference between the pimpleDyMFOAM (transient) analysis and the MRFSimpleFOAM (steady state) analysis is that the rotating zone physically rotates in the transient case. The maximum Courant number for the reported transient cases is 1.0.

MRFSimpleFOAM utilizes a stationary mesh and the flow is assumed steady, so the solution time is less than the transient rotating mesh URANS approach. This simplification can lead to less accurate results than the rotating mesh URANS approach [10,11]. Both methods are examined in the freestream case to ensure that the MRF method provides accurate results for this study.

The flow is assumed incompressible and gravity is neglected. The ducted turbine results assume steady state.

The freestream model is created to validate the model against Bahaj et al. [15] turbine tests in a tow tank at 1.4 m/s. Fresh water at 15° Celsius is used for both the freestream and ducted models. This correlates to a kinematic viscosity, $\nu = 1.1386 \times 10^{-6}$ m^2/s, and a density of $\rho = 999.1$ kg/m^3 [19]. The 0.8 m diameter turbine is a three-bladed horizontal axis turbine, with a twenty degree root pitch, a five degree tip pitch, and NACA 63-8xx sections [15]. The effects of the hub and the upright support are assumed to be small and are ignored. The turbulent intensity for the freestream turbine is 2.9%. The ducted model uses the same turbine geometry and water parameters with the addition of an efficient duct design provided by V^2 Wind Inc. This duct is designed for a ducted wind turbine operating in near ground conditions. Therefore, ambient turbulent intensity is assumed 10% for the inlet boundary condition of the ducted CFD cases. The specifics of the duct geometry are discussed in Section 2.3.

The mesh is created using snappyHexMesh. SnappyHexMesh is an octree mesher which applies refinement levels to geometry by dividing each cell evenly into eight smaller cells. The mesh domain size and background grid is the same for the freestream and ducted models. The rectangular domain is set to extend 13 m upstream of the turbine, 37 m downstream of the turbine, and 8 m in both the vertical and lateral directions. Three meshes are used: coarse, base, and fine. The OpenFOAM utility blockMesh is used to create a structured grid domain which is input to snappyHexMesh. The blockMesh for the base mesh is set to have a uniform block of 12 cells × 12 cells in the vertical and lateral directions for a square zone of 2 m × 2 m. From the edge of this uniform zone to the edge of the domain, 8 stretched cells are used. There are a total of 160 uniformly distributed cells in the longitudinal direction. The coarse mesh has half the number of cells in each direction as the base mesh, while the fine mesh has double the cells specified in each direction.

The analytical model is derived and compared to the ducted screen tests [2], the CFD predictions of a turbine inside of a tube, and the ducted turbine CFD predictions.

2.1. Derivation of the Analytical Model for Ducted Turbines

The analytical model uses a control volume analysis to determine the effective power extracted inside of a duct. The fundamental equations are Bernoulli's Equation, the conservation of mass equation, and the effective power equation shown in Equations (1)–(3), respectively. Equation (1) shows Bernoulli's Equation for an empty duct. During extraction a pressure drop occurs between Stations 1 and 2 as shown by Equation (3). Figure 1 defines the station numbers and variables where p is pressure, v is velocity, $\dot{m}$ is mass-flow rate, ρ is density, A is area, P_e is effective power, and $\Delta p_{1,2}$ is the pressure drop. The thrust, T, is the product of $\Delta p_{1,2}$ and A_1 as shown by Equation (3). The fundamental assumption of the analytical model, based on the momentum equation, is that the pressure drop occurs as a function of the initial throat velocity without extraction, v_{1o}, and the throat velocity under extraction, v_1, as shown in Equation (4). Therefore, the model assumes that the pressure drop is the difference between the dynamic pressure at the throat without extraction and the dynamic pressure at the throat with extraction. Thus, dynamic pressure at the throat only decreases as a function of the pressure drop. v_{1o} must be determined by experiments or CFD analysis since most ducts do not have ideal mass flow rates. In this study, v_{1o} is the empty throat velocity predicted by the steady state RANS CFD using the base mesh. An ideal mass flow rate is when v_{1o} is equal to the product of v_∞ and the ratio of the maximum frontal area, $A_{\max}$, to A_1. The analytical model also assumes that the mass flow efficiency does not change during extraction. Therefore, reduction in velocity is purely a function of power extraction. Mass flow efficiency, η, defined by Equation (5), is the ratio of the initial velocity in the duct compared to the ideal velocity in the duct.

$$\begin{aligned} p_{\text{tot}} &= p_\infty + \frac{1}{2}\rho v_\infty^2 = p_{\text{in}} + \frac{1}{2}\rho v_{\text{in}}^2 = p_1 + \frac{1}{2}\rho v_1^2 \\ &= p_2 + \frac{1}{2}\rho v_2^2 = p_3 + \frac{1}{2}\rho v_3^2 = p_{\text{wake}} + \frac{1}{2}\rho v_{\text{wake}}^2 \end{aligned} \quad (1)$$

$$\begin{aligned} \dot{m} &= \rho A_\infty v_\infty = \rho A_{\text{in}} v_{\text{in}} = \rho A_1 v_1 = \rho A_2 v_2 \\ &= \rho A_3 v_3 = \rho A_{\text{wake}} v_{\text{wake}} \end{aligned} \quad (2)$$

$$P_e = \Delta p_{1,2} A_1 v_1 = T v_1 \quad (3)$$

$$\Delta p_{1,2} = \frac{1}{2}\rho(v_{1o}^2 - v_1^2) \quad (4)$$

$$\eta = \frac{v_{1o}}{v_\infty A_{\max}/A_1}. \quad (5)$$

The analytical model is compared to the steady CFD results by comparing the velocity and power to the throat coefficient of thrust, C_{T,v_1}, defined in Equation (6).

$$C_{T,v_1} = \frac{T}{\frac{1}{2}\rho A_1 v_1^2}. \quad (6)$$

α, defined in Equation (7), is the percentage of the initial dynamic pressure that is converted into thrust. By iterating α we can calculate v_1 for each level of conversion of dynamic pressure to thrust as shown in Equation (8).

$$\alpha = \frac{T}{\frac{1}{2}\rho A_1 v_{1_o}^2} \tag{7}$$

$$v_1 = v_{1_o}\sqrt{1-\alpha}. \tag{8}$$

2.2. Freestream CFD Model

SnappyHexMesh applies nine refinements levels to the turbine surface, creates a wake refinement zone with five refinement levels, and creates the rotating zone with five refinement levels. The turbine diameter, D, is 0.8 m. The wake refinement zone has a diameter of $1.1D$ that ranges from the blade plane upstream to $2D$ (1.6 m) downstream. The rotating zone is $1.1D$ wide and ranges from $0.156D$ (0.125 m) upstream to $0.312D$ (0.25 m) downstream. The size of the rotating zone is the same for both the MRF and sliding mesh simulations. Analysis is done to ensure that the blades are independent of the rotating zone size. The rotating zone diameter dependence is examined by doubling the selected rotating zone diameter and ensuring that the forces on the turbine are negligibly affected. Similarly, the computational domain is doubled in size to ensure that the domain size negligibly affects the results. The coarse mesh has 0.73 million cells, the base mesh has 3.11 million cells, and the fine mesh has 16.16 million cells. Figure 2 shows the base mesh for the freestream model. The left image shows a cut plane of the whole domain, where the flow travels from left to right. The left boundary is a velocity inlet with zero-gradient pressure, the turbine blades are no-slip walls, the sides of the domain are symmetry planes, and the outlet is a fixed pressure velocity inlet-outlet. The right image shows a cut-plane of the mesh, the mesh on the AMI, and the turbine blades.

(**a**)

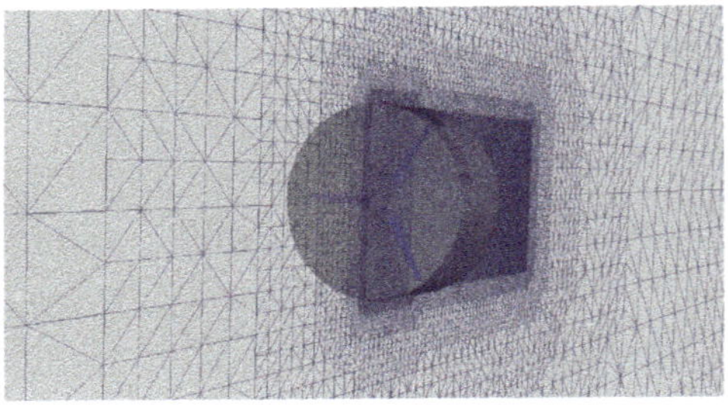

(**b**)

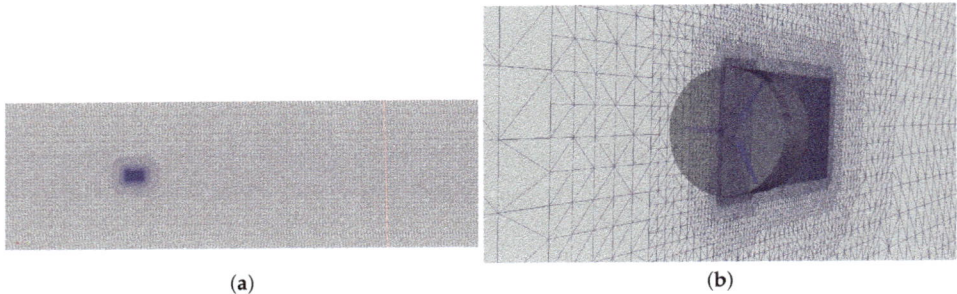

Figure 2. Freestream mesh. (**a**) The whole domain. (**b**) The mesh near the freestream turbine with the wake refinement and the cylindrical AMI.

2.3. Ducted CFD Model

The main difference between the ducted CFD study and the freestream study is the addition of the duct. The duct length, L, is $2.63D$ (2.107 m). The exit plane of the diffuser is $0.736L$ (1.55 m) downstream of the blade plane and the inlet is $0.264L$ (0.557 m) upstream of the blade plane. The duct has an inlet diameter of $1.59D$ (1.27 m), a throat diameter of $1.2D$ (0.96 m), and a maximum diameter at the duct exit of $1.92D$ (1.536 m). This correlates to a 20% tip gap, based on turbine radius, between the turbine blade and throat wall. For a centrally mounted turbine, it is easier to manufacture a system with a tip gap. Furthermore, the tip gap provides a space where marine life can safely pass through the duct to outside the swept area of the rotor and the turbine tip is less likely to hit marine growth on the duct, thus reducing maintenance costs. The area ratio between diffuser exit and throat is $A_3/A_1 = 2.56$. The rotating zone is set to have a radius of $1.1D$ (0.44 m), so that there is equal distance between the duct wall and the turbine tip. The wake refinement zone ranges from $0.57L$ (1.2 m) upstream to $1.73L$

(3.65 m) downstream. This size of a refinement zone correlates to two diffuser lengths downstream from the turbine plane. Due to the higher mesh count of the ducted turbine, only 4 refinements are used for the wake and rotating zone. Five refinements are used on the duct and nine refinements are applied to the turbine blades.

The boundary condition on the duct is a no-slip wall with wall functions. The Reynolds number on the blades is on the same order as the Reynolds number for the freestream blades because the apparent velocity at the blades is dominated by the spin rate at high TSRs. The diffuser is in the turbulent regime. All other settings remain constant for the ducted turbine study. More TSRs are examined than in the freestream case. Grid dependence is examined with coarse mesh and base mesh simulations. The coarse mesh has 1.35 million cells and the base mesh has 7.35 million cells. Figure 3 shows the base mesh for the ducted model. The left image shows a cut plane of the whole domain, where the flow travels from left to right. The left boundary is a velocity inlet with zero-gradient pressure, the turbine and duct are no-slip walls, the sides of the domain are symmetry planes, and the outlet is a fixed pressure velocity inlet-outlet. The duct is shown in the image for reference. The right image depicts a meshed cut-plane of the volume through the duct and one of the blades. The bottom image shows a cross sectional view of the duct.

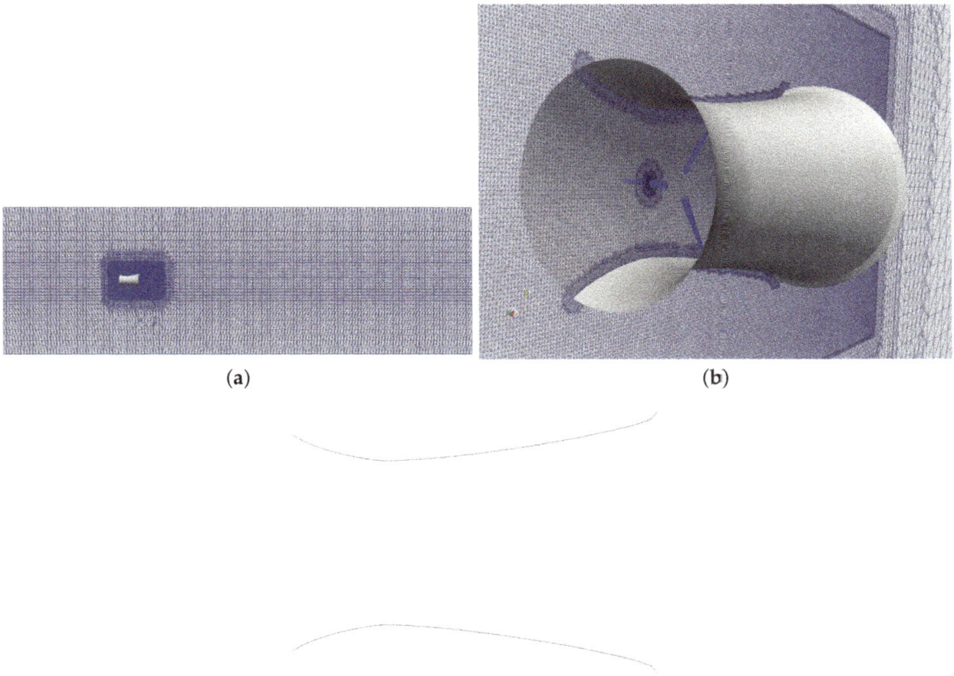

Figure 3. Ducted turbine mesh. (**a**) The whole meshed domain sliced in the middle. (**b**) The mesh near the ducted turbine. (**c**) Cross section of the duct.

2.4. Numerical Methods and Turbulence Modeling

The Reynolds number, Re, of the turbine ranges from 1.0×10^5 at the lowest tip speed ratio (TSR) to 2.5×10^5 at the highest TSR. This is in the transitional regime. The equations for TSR, blade Reynolds number, Re_{blade}, and duct Reynolds number, Re_{duct}, are respectively shown in Equations (9)–(11). TSR is calculated as a function of the turbine radius, r, the rotational speed,

ω, and freestream velocity, v_∞. Re_{blade} is calculated as a function of the chord at 70% of the radius, c_{r_o} = 0.03 m, v_∞, ν, the radius at 70%, r_0 = 0.28 m, and ω. Re_{duct}, calculated with duct maximum diameter, D_{duct} = 1.536 m, at freestream velocity, is 1.9×10^6. The power is calculated as the product of the torque and rotational speed.

$$\text{TSR} = \frac{\omega r}{v_\infty} \tag{9}$$

$$Re_{\text{blade}} = \frac{\sqrt{v_\infty^2 + (\omega r_o)^2} c_{r_o}}{\nu} \tag{10}$$

$$Re_{\text{duct}} = \frac{v_\infty D_{\text{duct}}}{\nu}. \tag{11}$$

The $k - \omega$ SST turbulence model is used with wall functions. This turbulence model assumes that the flow is turbulent. Re_{blade} is transitional but Re_{duct} is turbulent. A transitional turbulence model could be used in the future to potentially improve results. Due to the very thin boundary layer thickness, base grids are used with an average y^+ of around 20 on the freestream blades and around 30 on the ducted turbine blades for computational efficiency, especially for the transient simulations. The y^+ on the duct is 135. The forces on the turbine are dominated by the inertia of the fluid. For a well-designed turbine like this, with little separation, the differences between laminar and turbulent boundary layers on peak performance will likely be small. Potential flow methods have been used to accurately estimate the blade loading for the turbine examined herein [20], thus, while on-design it is expected that laminar and turbulent predictions will also be accurate. To illustrate this, the freestream turbine is run with both laminar and a $k - \omega$-SST turbulence model. Wall functions are applied to the walls (turbine and duct). In this study, RANS is used to depict separation at the inlet of the duct. Methods such as potential flow codes would not be able to depict separation. However, wall resolved grids with $y^+ = 1$ could be used to better capture viscous effects.

3. Results

The results are described in three subsections. The first subsection discusses the freestream CFD predictions. The second subsection discusses the CFD results for the ducted turbine. The third subsection evaluates the accuracy of the analytical model by comparing it to experimental and CFD results.

3.1. Freestream CFD Predictions versus Experiments

Equations (12) and (13) define the coefficient of thrust, C_T, and the coefficient of power, C_P, respectively.

$$C_T = \frac{T}{\frac{1}{2}\rho A_1 v_\infty^2} \tag{12}$$

$$C_P = \frac{P}{\frac{1}{2}\rho A_1 v_\infty^3}. \tag{13}$$

The CFD predictions of the base and fine mesh match well with the experimental tow tank results [15], with the exception of high TSRs. At the highest TSR of 11.4, only the fine mesh matches the experimental results. Figure 4 shows C_T and C_P as a function of TSR for the experiment and the steady state CFD.

The coarse mesh over-predicts C_T at high TSRs and under-predicts the C_P. This is likely caused by the coarseness of the mesh not accurately depicting the finest features of the body surface. For the base and fine meshes, the laminar and $k - \omega$-SST models both match well with the experimental results for C_P. For the base mesh, the difference between the laminar and $k - \omega$-SST model is less than a percent, for all but the last two TSRs, with 1.32% and 2.19% difference, respectively. Even though both the laminar and turbulent simulations match the C_P results well, the laminar simulations match the results better, especially in the off-peak range of the highest TSR. The laminar model also does better at predicting the C_T at higher TSRs.

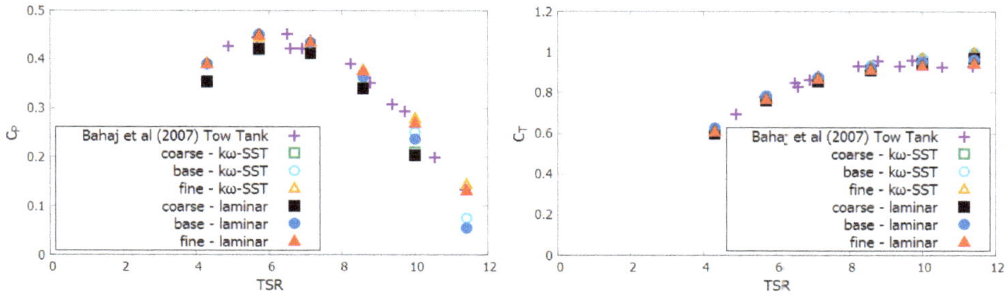

(a) (b)

Figure 4. Freestream C_P and C_T as a function of TSR. (**a**) C_P; (**b**) C_T.

Figure 5 compares the results of the freestream predictions for the transient and steady-state methods. The coarse transient solution is within 1.5% of the coarse MRF solution. The freestream transient base mesh predicts higher C_P than the coarse mesh. The transient base mesh predicts a similar C_P to the steady state fine mesh at TSR = 10.00. For the calculation of mean blade loading, especially on-design, there is little benefit to using the transient method since it is much more computationally expensive than the steady state method.

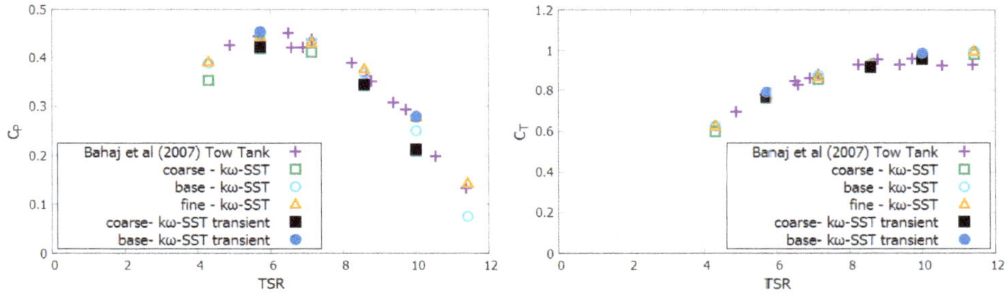

(a) (b)

Figure 5. Steady state and transient results for freestream C_P and C_T as a function of TSR. (**a**) C_P vs. TSR; (**b**) C_T vs. TSR.

3.2. CFD Results of the Ducted Turbine

Equation (14) depicts the power coefficient, $C_{P,A_{\max}}$ as a function of the maximum frontal area of the duct, $A_{\max}$.

$$C_{P,A_{\max}} = \frac{P}{\frac{1}{2}\rho A_{\max} v_\infty^3}. \tag{14}$$

The addition of the duct leads to a decrease in performance from the freestream turbine based on $C_{P,A_{\max}}$. If the power is examined based on the frontal area of the turbine alone, then the C_P would be higher. However, this is an unfair comparison due to the additional structure and frontal area necessary for the duct. Figure 6 shows the ducted turbine results for the coarse and base meshes. The $C_{P,A_{\max}}$ is shown as a function of both TSR and C_{T,v_1}. This figure also shows the velocity ratio v_1/v_∞ as a function of C_{T,v_1} and C_P as a function of TSR. The plot of C_P as a function of TSR demonstrates that the power relative to the turbine area is increased compared to the freestream turbine, but a more fair comparison would examine $C_{P,A_{max}}$ instead. This figure demonstrates that the coarse mesh and the base mesh for the MRF results are in good agreement, except for the second lowest speed, TSR = 5.

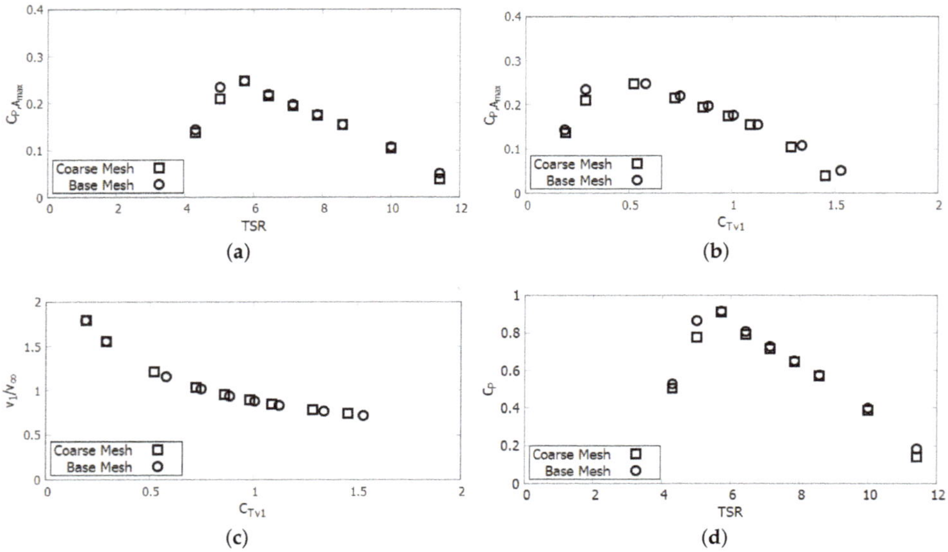

Figure 6. Ducted results. (**a**) $C_{P,A_{max}}$ vs. TSR; (**b**) $C_{P,A_{max}}$ vs. C_{Tv_1}; (**c**) v_1/v_∞ vs. C_{Tv_1}; (**d**) C_P vs. TSR.

The left side of Figure 7 shows a cut plane through the center of the turbine and the resultant velocity field for TSRs of 4.29, 5.71, and 11.43. Asymmetry due to blade count is noted. Partial separation occurs at the inlet at the lowest TSR and more significant separation occurs as the TSR increases. As the separation increases, the efficiency of the duct decreases. The low velocity region behind the duct is not convected downstream. This phenomenon should be compared to particle image velocimetry (PIV) results or to large eddy simulation (LES) predictions which will not suffer from the temporal averaging of RANS predictions. LES predictions may predict more accurate wake structures as well as better resolve the flow features both inside and outside the duct. As the loading increases the angle of attack of the flow entering the intake changes. The intake could be shortened or rounded to reduce the separation at higher loadings. The right hand side of Figure 7 shows a cut plane through the center of the turbine and the resultant velocity field for a TSR of 5.71 for the coarse and base mesh. Overall, the flow structures are similar between the two meshes, but more numerical diffusion is seen with the coarse mesh especially in the wake and separation region to the outside of the duct.

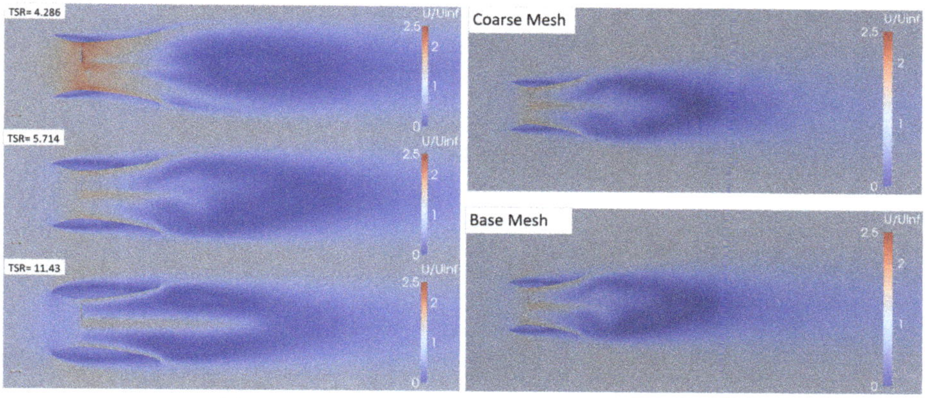

(a) (b)

Figure 7. Flow visualization of ducted turbine. (**a**) Velocity field of ducted turbine at different TSRs. (**b**) Coarse mesh flow and base mesh flow at TSR = 5.714.

3.3. *Comparison of the Analytical Model to Experimental and CFD Results of Ducted Turbine Cases*

The accuracy of an analytical model for ducted turbine performance can be determined by its ability to predict throat velocity, power, and thrust. This study compares the analytical model to experimental screen results, the turbine inside of a tube, as well as the CFD simulations previously described.

3.3.1. Analytical Model Prediction of Screen Tests for a Ducted Turbine

The analytical model is compared to screen tests of the 30° Gilbert et al. diffuser [2] . The diffuser has an area ratio, $A_{\max}/A_1$ of 2.78 and η less than 60%. These screen tests are a means of measuring the effective power extracted by a turbine, without the non-uniformity or complexity that a real turbine may pose. Figure 8 shows the velocity ratio, $C_{P,A_{\max}}$, and C_T as a function of C_{T,v_1}. Other analytical models like those proposed by Jamieson [21] as well as Werle and Presz [22], share the similar assumption that the pressure drop occurs as a function of the freestream and wake velocities, instead of the throat velocity.

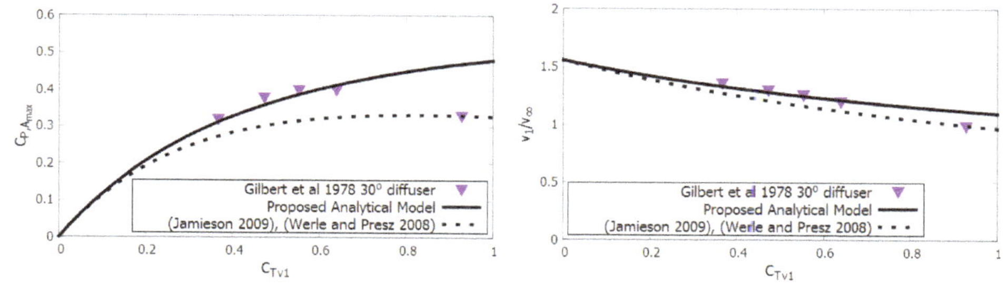

(a) (b)

Figure 8. Analytical predictions along with the screen tests of Gilbert et al. [2]. (**a**) $C_{P,A_{\max}}$ vs. C_{Tv_1}; (**b**) v_1/v_∞ vs. C_{Tv_1}.

The analytical models that assume a pressure drop is a function of freestream and wake velocities are unable to accurately predict the results of the screen test. In comparison, the proposed analytical model predicts each data point with more accuracy, with the exception of the highest C_{T,v_1} value, as shown in Figure 8. At the highest C_{T,v_1}, the over-prediction of power and thrust likely occurs because the accelerative efficiency is assumed to be constant. The mass flow efficiency of the duct may reduce at higher values of C_{T,v_1} since the stagnation point on the leading edge likely translated inward, leading to separation. The analytical model therefore could be improved with the ability to predict loss in mass flow efficiency at higher loadings.

3.3.2. Analytical Model Prediction of the RANS CFD Results for a Turbine in a Uniform Tube

A tube with an infinitely thin wall thickness represents the simplest duct with $A_{in} = A_1 = A_2 = A_3$. The tube used was the same length as the duct and had a diameter equal to the throat diameter of the duct. Werle and Presz state that their model arrives at the freestream result when a turbine is placed in a tube [22]. However, a turbine does not perform at the same efficiency in a tube as it does in the freestream. Figure 9 shows the results of the Bahaj et al. turbine [15] inside of the tube. The $C_{P,A_{max}}$ is over-predicted by a considerable margin, but the proposed analytical model only slightly over-predicts the velocity and the effective power. This could be caused by viscous effects in the CFD calculation. The coefficient of effective power at the throat is defined in Equation (15).

$$C_{P,\text{eff}} = \frac{T v_1}{\frac{1}{2}\rho A_{\max} v_\infty^3}. \tag{15}$$

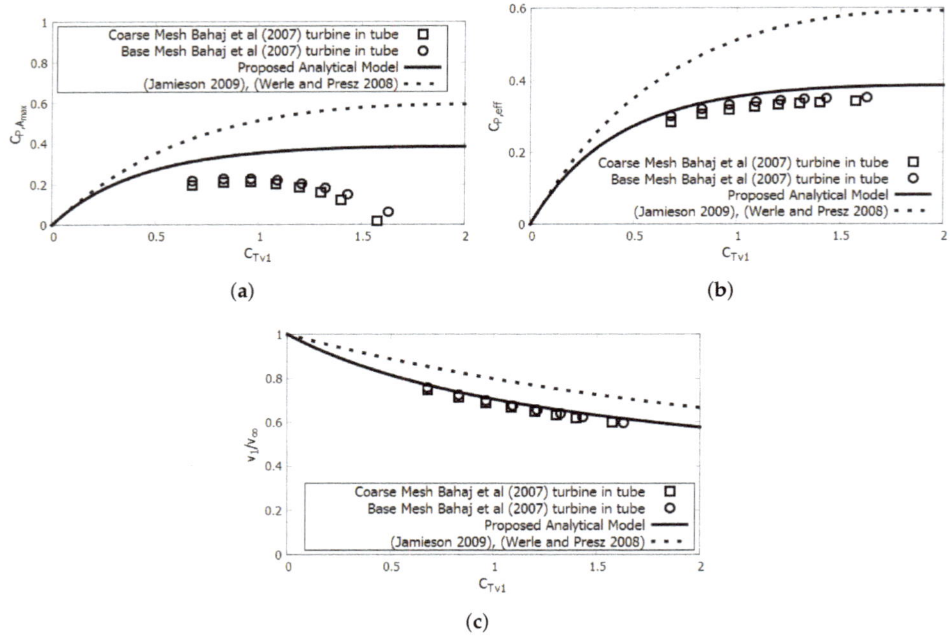

Figure 9. Analytical models along with the CFD of Bahaj et al. turbine [15] in a tube with diameter equal to duct throat diameter. (**a**) $C_{P,A_{max}}$ vs. C_{Tv_1}; (**b**) $C_{P,\text{eff}}$ vs. C_{Tv_1}; (**c**) v_1/v_∞ vs. C_{Tv_1}.

3.3.3. The Analytical Model's Ability to Predict RANS CFD Results for a Ducted Turbine

Since the analytical model assumes uniform and ideal extraction, the performance of the ducted turbine will be over-predicted due to viscous effects and non-uniformity. The non-effective power related axial induction is referred to as blockage. Figure 10 shows that the analytical model over-predicts both the CFD predictions of velocity and the coefficient of power.

This over prediction is related to the large-scale separation at the leading edge of the duct under extraction, shown in Figure 7. This large-scale separation leads to a decrease in η. Since the analytical model does not account for blockage other than that caused by uniform power extraction, it is important to correct the throat velocity for each data point. By iterating α and η, we can calibrate v_1 for each level of C_{T,v_1}. Figure 11 depicts the results when the analytical velocity is corrected to match the CFD. The velocity ratio is shown on the left and the $C_{P,A_{\max}}$ is shown on the right. Despite the velocity having been corrected, the power is still over-predicted since the turbine does not ideally convert effective power to power. The analytical model predicts effective power, and not the actual power produced by the turbine which is the product of torque and rotational velocity. Therefore, the variable of importance is the effective power, which is also shown on the right of Figure 11. Based on this calibration, we can determine the value of η as a function of C_{T,v_1}, shown in Figure 12. This demonstrates how the interaction between the turbine and duct leads to decreased performance. Uniform extraction, like that simulated by a screen may cause a lower rate of rejection than a real turbine. η drops from nearly ideal to below 50% with high loading.

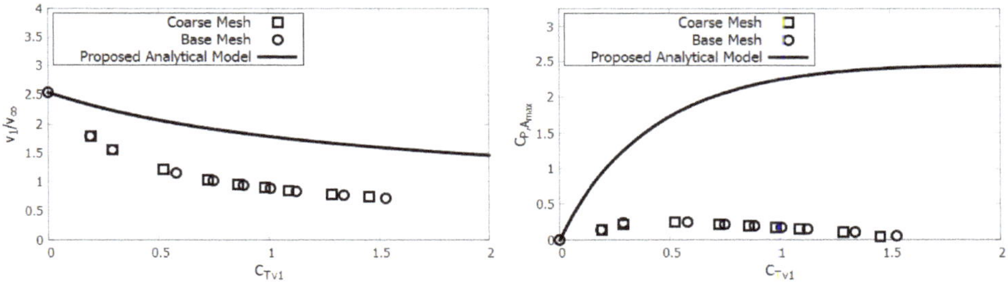

(a) (b)

Figure 10. Analytical model and ducted CFD predictions. (**a**) v_1/v_∞ vs. C_{Tv_1}; (**b**) $C_{P,A_{\max}}$ vs. C_{Tv_1}.

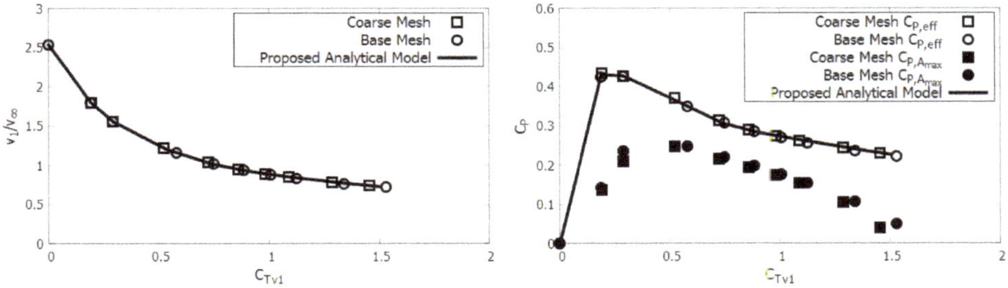

(a) (b)

Figure 11. Analytical model and ducted CFD predictions with calibrated velocity. (**a**) v_1/v_∞ vs. C_{Tv_1}; (**b**) C_P vs. C_{Tv_1}.

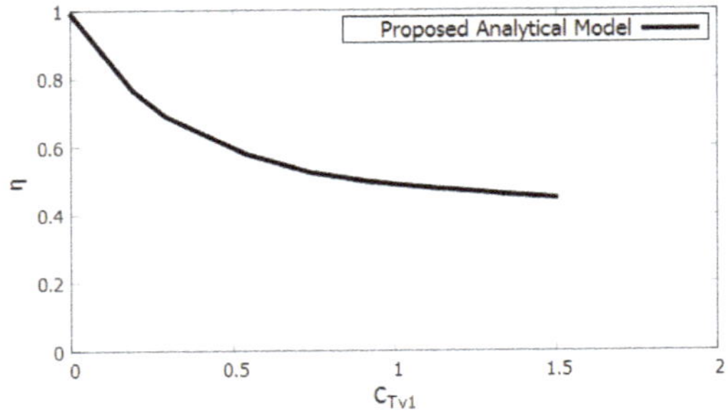

Figure 12. η as a function of loading.

4. Discussion

The analytical model demonstrates that the duct's η reduces significantly with high loading. Therefore, the turbine inside the duct must have a high $C_{P,A_{max}}$ to $C_{P,eff}$ ratio. One method to improve this ratio is to pitch the turbine so that the lift vector is directed more in the direction to produce torque instead of thrust. Design implications of coupling the algorithm with the numerical solutions are subsequently discussed.

Design Implications

The design of the ducted turbine can be improved by adapting the turbine to operate better within the duct. The pitch is systematically varied from the original 20° root pitch to a 50° root pitch in 7.5° increments. TSR is varied at each pitch to determine which TSR produces the maximum power for each pitch. Figure 13 shows the results for the maximum power for each pitch of the ducted turbine. The $C_{P,A_{max}}$ is shown on the left axis and the corresponding C_{T,v_1} is shown on the right axis. C_{T,v_1} decreases as pitch is increased. The optimum $C_{P,A_{max}}$ occurs at a root pitch of 42.5° with nearly a 40% improvement in power over the original pitch. Appendix A shows a similar study for pitch variation inside of a tube with the same length as the duct and the diameter of the throat. It shows how the ratio of $C_{P,A_{max}}$ to $C_{P,eff}$ varies with TSR, correlating streamlines to illustrate why the power ratio changes as a function of TSR, and the pitch distribution changes as a function of radial position.

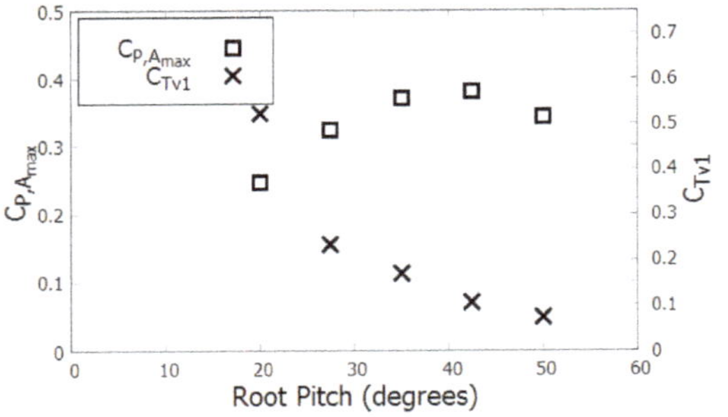

Figure 13. Effect of pitch on $C_{P,A_{max}}$ and C_{T,v_1} for a ducted turbine.

Adding pitch improves design considerably but does not account for the fact that the duct induces a radial variation in velocity not present in the freestream case, nor that a heavily pitched blade must have more twist than a less pitched blade. To account for this, the twist is increased so that each radial section has the same angle of attack as the freestream turbine. For simplicity, this calculation ignores the difference in induced tangential velocity, but uses the radial velocity distribution, rotation speed of each radial section, and the geometric pitch angle at each radial section to determine the angle of attack. The blade is designed for a root pitch of 44.6°. The twist profile and the method to determine the twist profile is shown in Appendix B. The twisted blade is pitched further to 50°, 55°, and 60° root pitch. For this twist, it is found that a 50° root pitch is most efficient. At a 50° root pitch, the effect of reducing the tip gap is examined by scaling the blades up by ten percent. Theory suggests that reducing the tip gap should reduce the unloading at the tip and improve performance. This again leads to an increase in performance. As a function of pitch angle, Figure 14 shows $C_{P,A_{\max}}$ for the best TSR for each pitch angle.

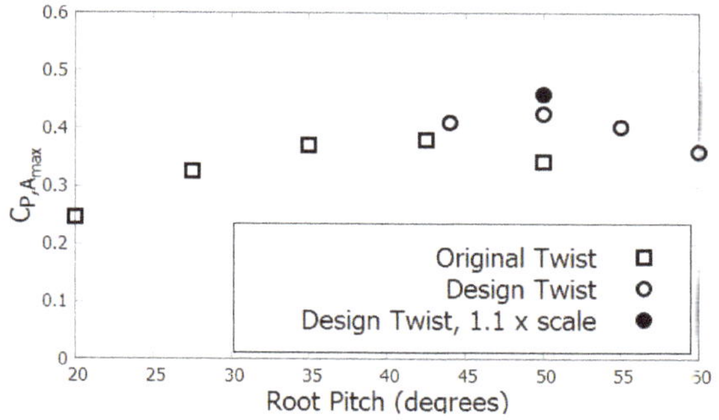

Figure 14. Effect of pitch, twist, and scale for a ducted turbine.

5. Conclusions

The freestream CFD results match the tow tank results. A fine mesh resolution is required to properly calculate the off-performance loading for the highest TSRs.

While the results for the freestream turbine closely agree with the experimental results, good correlation, especially near peak performance, is also achievable with less computationally demanding methods like panel codes. RANS CFD is useful for pinpointing flow features that panel codes are not able to model, such as separation. Separation at the leading edge and exit of the duct are critical design features that affect performance. RANS is a useful tool for an initial analysis of these design features and is less computationally expensive than alternatives like LES.

It is evident that the underlying analytical system of equations for a ducted turbine is different from that of a freestream turbine. The thrust not associated with power production must be reduced for a ducted turbine to be efficient, since the effective power at the throat, $C_{P,\text{eff}}$ causes a reduction in the momentum and thus mass flow through the device. Therefore, the turbine should be optimized to increase the ratio of $C_{P,A_{\max}}$ to $C_{P,\text{eff}}$.

The RANS model has shown that when a freestream turbine is placed inside of a duct, the performance decreases. However, this performance can be improved by reducing the non-power thrust. An analytical model was developed to help analyze these results. The analytical model accurately predicts the data from Gilbert et al. [2]. This suggests that the duct that Gilbert et al. examined [2] maintained a constant level of η over a range of C_{T,v_1} that is longer than that of the duct examined with CFD. This also demonstrates that the pressure drop across the turbine occurs at the accelerated velocity v_1. The analytical model over-predicts the velocity calculated by the CFD models of an initially efficient duct. This is because RANS CFD accounts for viscous effects and the blockage, but the analytical model does not. This blockage is likely the result of complicated viscous flow interaction, non-ideal extraction, and the reduction in η as the stagnation point moves inwards on the duct. The analytical model could be improved by incorporating a method to predict loss of mass flow efficiency during extraction. This study shows that the analytical model can be used to determine whether a duct design is efficient enough to pursue with higher order tools and that it can determine the degree to which the turbine causes losses within the duct.

Using the conclusions from the analytical model a more efficient ducted turbine was designed. Increasing the pitch of the blade significantly improved the $C_{P,A_{\max}}$ to $C_{P,\mathrm{eff}}$ ratio. By increasing the pitch, adapting the twist, and increasing the scale of the blade, the ducted turbine's $C_{P,A_{\max}}$ exceeded the original freestream turbine's C_P. This design procedure could be continued to create an even more efficient ducted turbine. While not explored in depth in this study, the effect of tip gap should be used to further optimize the turbine. Furthermore, the duct geometry could be altered to better match the characteristics of a specific turbine design.

This work demonstrates that the duct and turbine should be designed together to create an optimal system. An efficient freestream turbine cannot simply be placed in a duct and expected to perform with similar efficiency as it did in the freestream. By coupling the analytical model with RANS CFD results we are able to determine how to optimize the ducted turbine system.

More research is needed to better understand how to efficiently extract power from ducted turbines. It will be important to further analyze the effects of turbulent intensity on the system design and the effects of transition from laminar to turbulent. A transitional turbulence model with wall resolved grids could be used to better evaluate viscous effects. It is also important to understand and quantify what causes the blockage and rapid decrease in η. More detailed CFD models should be created to analyze the effects of the turbine wake interaction with the duct and the translation of the stagnation point inwards leading to separation at the inlet. To do this, detached eddy simulation or LES should be used.

6. Patents

The duct used in this study is based off of V^2 Wind Inc.'s patent application: US20160305247A1.

Acknowledgments: This study was made possible by two sources of funding. Funding for the authors from the University of Michigan was made possible by the Office of Naval Research Grant N00014-16-1-2969 and program manager Kelly Cooper. Funding for V^2 Wind Inc.'s contributions was made possible by the United States Army Small Business Innovation Research Contract W911QY-13-C-0054.

Author Contributions: Bradford Knight conducted the CFD experiments. Robert Freda discovered predictive errors with existing analytical tools for ducted turbines. All authors provided input for an improved analytical model. Yin Lu Young and Kevin Maki provided guidance for the project. Bradford Knight wrote the paper, and all authors contributed to editing the paper.

Conflicts of Interest: The authors declare no conflict of interest. Robert Freda and Bradford Knight are employed by V^2 Wind Inc., which is a wind turbine company. This research was motivated by a desire to better understand how power is extracted from ducts and to develop a better analytical method to analyze ducted turbines.

Abbreviations

The following abbreviations are used in this manuscript:

AMI	arbitrary mesh interface
CFD	computational fluid dynamics
LES	large eddy simulation
MRF	moving reference frame
RANS	Reynolds averaged Navier–Stokes
URANS	unsteady Reynolds averaged Navier–Stokes

Appendix A. Effect of Modifying the Pitch of the Turbine in a Tube

Figure A1 shows the power ratio as a function of both TSR and C_{T,v_1}, as well as $C_{P,A_{\max}}$ as a function of C_{T,v_1} for the turbine inside of a tube at different root pitch angles. As the pitch is increased, the optimum TSR decreases and the maximum $C_{P,A_{\max}}$ decreases. Conversely, the $C_{P,A_{\max}}$ to $C_{P,\mathrm{eff}}$ ratio increases. As noted in the body of the paper, by improving this ratio, the efficiency of the ducted turbine will increase when this ratio is improved and the turbine is paired with a high efficiency duct.

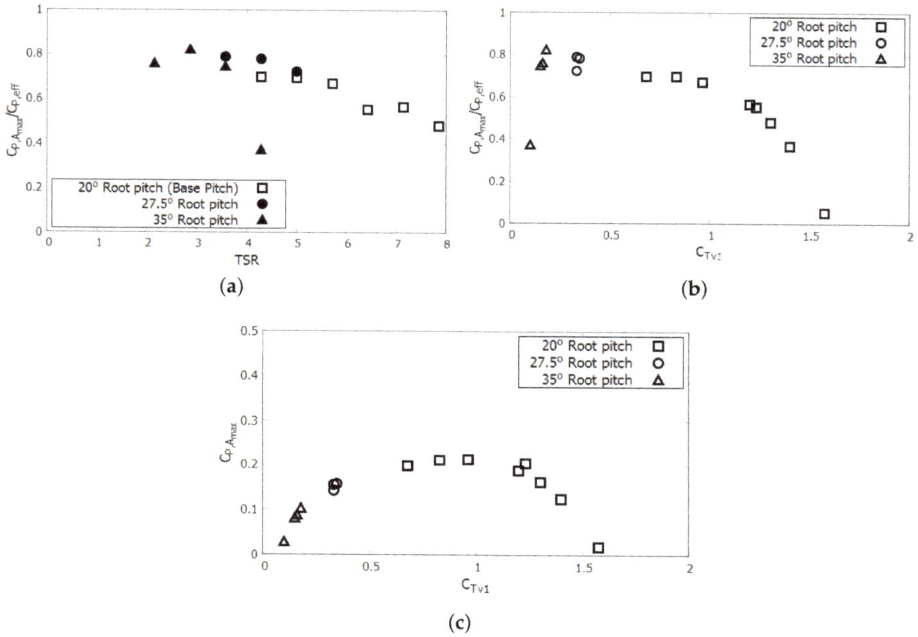

Figure A1. Effect of pitch on the power ratio and $C_{P,A_{max}}$. (**a**) $C_{P,A_{max}}/C_{P,eff}$ vs. TSR; (**b**) $C_{P,A_{max}}/C_{P,eff}$ vs. C_{Tv_1}; (**c**) $C_{P,A_{max}}$ vs. C_{Tv_1}.

Figure A2 shows the streamlines at 70% span for the 35° root pitch case in the tube. The left hand image shows the streamlines for the TSR that produces the maximum power and the image on the right shows the over spin case. As pitch increases, the operating TSR decreases since the foil will begin to operate at a negative angle of attack at a lower TSR earlier. As the blade begins to over-run the flow, the performance rapidly drops off as shown by the power ratio in Figure A1 for the 35° root pitch at a TSR of 4.28. On the other hand, peak performance occurs when the flow has a shock free entry as shown in the streamlines for the TSR of 2.85 case.

Figure A2. Streamlines at 70% span for the 35° root pitch turbine in the tube. (**a**) Peak TSR; (**b**) Over spin.

Figure A3 shows the pitch distribution. The 20° root pitch case is the original pitch distribution specified by Bahaj et al. (2007). The other pitch distributions maintain the same twist profile but have an increased pitch.

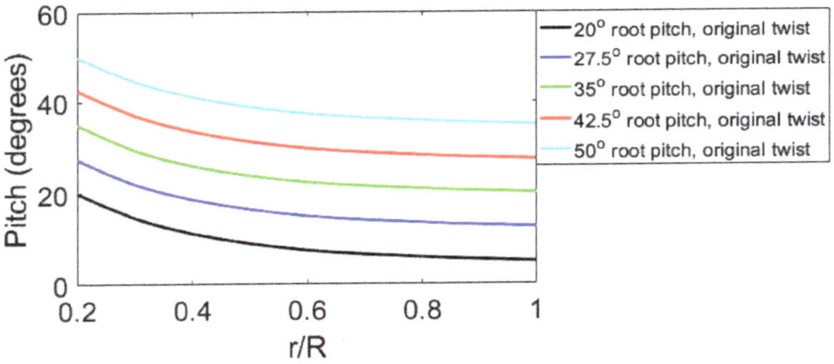

Figure A3. Pitch distribution for original blade at different root pitch angles.

Appendix B. Design Twist and Pitched Blade Details for the Ducted Turbine

The pitch profile was determined using Equation (A1), where $\phi(r)$ is the pitch at radial location r, $u(r)$ is the velocity at the blade plane at radial location r, and ω is the rotational speed. The subscripts, f and d, stand for freestream and ducted, respectively. Thus, ignoring differences in induced velocity as the pitch angle change, we are able to estimate the pitch necessary so that the blades sections operate at the same angle of attack. The velocities used are the blade plane velocity under extraction for the optimum freestream case, and the ducted case nearest the design point. Therefore, the induced velocity profile for the ducted turbine was calculated based on the 42.5° root pitch case. The method could be refined by using the CFD to determine the induced velocity near the leading edge of the blade and therefore better determine the optimal pitch.

$$\phi(r)_f - tan^{-1}(\frac{u_f(r)}{\omega_f r}) = \phi(r)_d - tan^{-1}(\frac{u_d(r)}{\omega_d r}). \quad \text{(A1)}$$

The modified blade with designed twist is much more heavily twisted than the original blade. The blade twist is nearly 30°, compared to the original freestream blade which had 15° of twist. The left plot of Figure A4 shows the pitch distribution for the blade with design twist at different root pitches. The right plot of Figure A4 below shows the pitch distribution for each of the blades tested. The optimal pitch for the blade with designed twist was 50°, but the optimal pitch for the blade with the original twist was 42.5°. When the right plot of Figure A4 is examined, this result makes sense since the pitch is similar to the original twist blade with a 50° pitch near the hub, but has nearly the same pitch as the original twist blade with a 35° root pitch at the tip. Therefore, the blade with original twist at 42.5° is much flatter, but somewhat averages the pitch distribution of the more optimal twist profile.

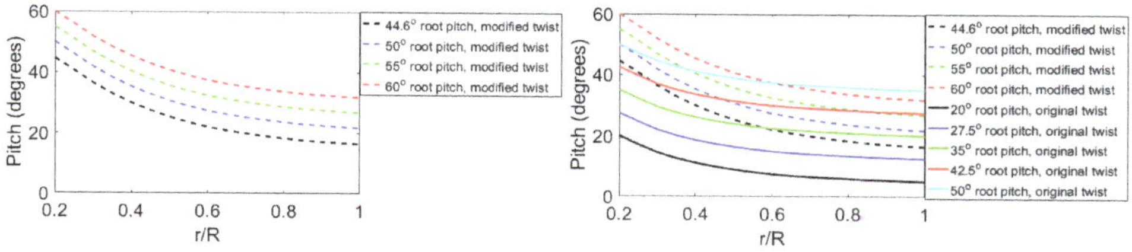

(a) (b)

Figure A4. Pitch distribution for the original blade at different root pitch angles and for the design twist blade at different root pitch angles. (**a**) Design twist; (**b**) all blades.

References

1. Igra, O. Shrouds for Aerogenerators. *AIAA J.* **1976**, *14*, 1481–1483.
2. Gilbert, B.L.; Oman, R.A.; Foreman, K.M. Fluid dynamics of diffuser-augmented wind turbines. *J. Energy* **1978**, *2*, 368–374, doi:10.2514/3.47988.
3. Gilbert, B.L.; Foreman, K.M. Experiments With a Diffuser-Augmented Model Wind Turbine. *J. Energy Resour. Technol.* **1983**, *105*, 46–53, doi:10.1115/1.3230875.
4. Lilley, G.; Rainbird, W. *A Preliminary Report on the Design and Performance of a Ducted Windmill*; Report No. 102; College of Aeronautics Cranfield: Cranfield, UK, 1956.
5. Van Bussel, G. The science of making more torque from wind: Diffuser experiments and theory revisited. *J. Phys. Conf. Ser.* **2007**, *75*, 012010.
6. Phillips, D.; Flay, R.; Nash, T. Aerodynamic analysis and monitoring of the Vortec 7 Diffuser-Augmented wind turbine. In *Transactions of the Institution of Professional Engineers New Zealand: Electrical/Mechanical/Chemical Engineering Section*; The Institution: Wellington, New Zealand, 1999; Volume 26.
7. Ohya, Y.; Karasudani, T. A Shrouded Wind Turbine Generating High Output Power with Wind-lens Technology. *Energies* **2010**, *3*, 634–649, doi:10.3390/en3040634.
8. Rourke, F.O.; Boyle, F.; Reynolds, A. Tidal energy update 2009. *Appl. Energy* **2010**, *87*, 398–409, doi:10.1016/j.apenergy.2009.08.014.
9. Zhou, Z.; Benbouzid, M.; Charpentier, J.F.; Scuiller, F.; Tang, T. Developments in large marine current turbine technologies—A review. *Renew. Sustain. Energy Rev.* **2017**, *71*, 852–858, doi:10.1016/j.rser.2016.12.113.
10. Liu, J.; Lin, H.; Purimitla, S.R. Wake field studies of tidal current turbines with different numerical methods. *Ocean Eng.* **2016**, *117*, 383–397, doi:10.1016/j.oceaneng.2016.03.061.
11. Siddiqui, M.S.; Rasheed, A.; Tabib, M.; Kvamsdal, T. Numerical Analysis of NREL 5MW Wind Turbine: A Study Towards a Better Understanding of Wake Characteristic and Torque Generation Mechanism. *J. Phys. Conf. Ser.* **2016**, *753*, 032059.
12. Ren, Y.; Liu, B.; Zhang, T.; Fang, Q. Design and hydrodynamic analysis of horizontal axis tidal stream turbines with winglets. *Ocean Eng.* **2017**, *144*, 374–383, doi:10.1016/j.oceaneng.2017.09.038.
13. Song, M.; Kim, M.C.; Do, I.R.; Rhee, S.H.; Lee, J.H.; Hyun, B.S. Numerical and experimental investigation on the performance of three newly designed 100 kW-class tidal current turbines. *Int. J. Nav. Arch. Ocean Eng.* **2012**, *4*, 241–255, doi:10.2478/IJNAOE-2013-0093.

14. Morris, C.; O'Doherty, D.; O'Doherty, T.; Mason-Jones, A. Kinetic energy extraction of a tidal stream turbine and its sensitivity to structural stiffness attenuation. *Renew. Energy* **2016**, *88*, 30–39, doi:10.1016/j.renene.2015.10.037.
15. Bahaj, A.; Molland, A.; Chaplin, J.; Batten, W. Power and thrust measurements of marine current turbines under various hydrodynamic flow conditions in a cavitation tunnel and a towing tank. *Renew. Energy* **2007**, *32*, 407–426, doi:10.1016/j.renene.2006.01.012.
16. Liu, P.; Bose, N. Parametric Analysis of Horizontal Axis Tidal Turbine Hydrodynamics for Optimum Energy Generation. In Proceedings of the Third International Symposium on Marine Propulsors, Australian Maritime College, University of Tasmania, At Launceston, Australia, 5–8 May 2013; pp. 242–256.
17. Park, S.; Park, S.; Rhee, S.H. Influence of blade deformation and yawed inflow on performance of a horizontal axis tidal stream turbine. *Renew. Energy* **2016**, *92*, 321–332, doi:10.1016/j.renene.2016.02.025.
18. Knight, B.; Freda, R.; Young, Y.L.; Maki, K. Evaluation of Different Numerical and Analytical Strategies to Analyze a Ducted Marine Current Turbine. In Proceedings of the Fifth International Symposium on Marine Propulsors, VTT Technical Research Center of Finland Ltd., Helsinki, Finland, 12–15 June 2017.
19. Fresh Water and Seawater Properties. In Proceedings of the 26th International Towing Tank Conference Specialist Committee on Uncertainty Analysis (ITTC), Rio de Janeiro, Brail, 28 August–3 September 2011.
20. Young, Y.L.; Motley, M.R.; Yeung, R.W. Three-Dimensional Numerical Modeling of the Transient Fluid-Structural Interaction Response of Tidal Turbines. *J. Offshore Mech. Arct. Eng.* **2009**, *132*, doi:10.1115/1.3160536.
21. Jamieson, P.M. Beating Betz: Energy Extraction Limits in a Constrained Flow Field. *J. Sol. Energy Eng.* **2009**, *131*, doi:0.1115/1.3139143.
22. Werle, M.J.; Presz, W.M. Ducted Wind/Water Turbines and Propellers Revisited. *J. Propul. Power* **2008**, *24*, 1146–1150, doi:10.2514/1.37134.

Journal of
Marine Science and Engineering

MDPI

Article

Marine Turbine Hydrodynamics by a Boundary Element Method with Viscous Flow Correction

Francesco Salvatore *, Zohreh Sarichloo and Danilo Calcagni

CNR-INSEAN, National Research Council, Marine Technology Research Institute, Via di Vallerano 139, 00128 Rome, Italy; zohreh.sarichloo@insean.cnr.it (Z.S.); danilo.calcagni@cnr.it (D.C.)

* Correspondence: francesco.salvatore@cnr.it; Tel.: +39-06-5029-9313

Received: 30 March 2018; Accepted: 1 May 2018; Published: 8 May 2018

Abstract: A computational methodology for the hydrodynamic analysis of horizontal axis marine current turbines is presented. The approach is based on a boundary integral equation method for inviscid flows originally developed for marine propellers and adapted here to describe the flow features that characterize hydrokinetic turbines. For this purpose, semi-analytical trailing wake and viscous flow correction models are introduced. A validation study is performed by comparing hydrodynamic performance predictions with two experimental test cases and with results from other numerical models in the literature. The capability of the proposed methodology to correctly describe turbine thrust and power over a wide range of operating conditions is discussed. Viscosity effects associated to blade flow separation and stall are taken into account and predicted thrust and power are comparable with results of blade element methods that are largely used in the design of marine current turbines. The accuracy of numerical predictions tends to reduce in cases where turbine blades operate in off-design conditions.

Keywords: marine current turbine; hydrodynamics; boundary element methods; trailing wake models; viscous flow correction

1. Introduction

Marine or hydrokinetic turbines for the production of renewable energy from tidal and ocean currents is a rapidly growing technology. Large scale installations mainly consist of horizontal axis turbines installed on structures fixed to the seabed or supported by floating platforms.

The relatively fast maturation of hydrokinetic turbine technology as compared to other ocean energy harvesting systems is partly due to experience gained over the last decades in the wind energy sector. In most cases the shape of marine turbine blades resembles wind rotor blades except for the aspect ratio that is substantially smaller to resist hydrodynamic loads in water. It is thus not surprising that Blade Element Momentum methods (simply, BEM) originally developed for wind turbines are extensively used for analysis and design of tidal and ocean current turbines, see e.g., [1]. BEM provides fast and reliable estimates of turbine performance if suitable tuning is applied to overcome important methodology weaknesses [2,3]. Specifically, blade loading is derived by prescribed lift and drag properties of two-dimensional profiles and semi-empirical three-dimensional flow corrections are necessary to account for blade tip effects, blade/hub interaction, number of blades.

In contrast to this, the hydrodynamic design of marine propellers is typically based on boundary element or panel methods that, under inviscid-flow assumptions, provide a consistent representation of the three-dimensional flow around rotors in steady or unsteady flow. To avoid confusion with blade element (momentum) methods, the terminology Boundary Integral Equation Method (BIEM) is used here. In spite of that, only few examples of applications of BIEMs to hydrokinetic turbines exist, see e.g., Young et al. [4], He et al. [5]. Results in the literature highlight the difficulty of boundary element methods to correctly describe the hydrodynamic performance of turbines designed to extract energy from an onset flow. A major difficulty is that turbine blades frequently operate at high angle of attack and viscosity induced separation and stall significantly affect generated thrust and power. Baltazar and Falcão de Campos [6,7] address the problem by proposing models to correct inviscid-flow predictions by a BIEM for three-dimensional steady flows. The lift force contribution evaluated at each blade section by BIEM is corrected by comparing viscous and inviscid flow lift coefficients of two-dimensional profiles representatives of blade sections, while blade section drag is taken from two-dimensional polar curves. The Kutta-Joukowski law is used to determine the incoming velocity to blade sections, and the viscous flow code X-Foil is used to determine polar curves. The methodology includes an iterative wake alignment model to adjust wake pitch to the local flow, while radial expansion is not modelled.

The problem is tackled in the present work by revisiting the approach in [6] and developing original viscous flow and trailing wake models that are integrated into a BIEM originally developed for marine propellers, see e.g., Salvatore et al. [8,9] and Pereira et al. [10]. In the proposed methodology, the trailing wake geometry is determined by a semi-analytical model with wake pitch alignment consistent with turbine-induced velocity perturbation calculated by BIEM and an experimental-based definition of the expansion rate of the streamtube downstream of the rotor plane. Next, a viscous flow correction is determined by comparing distributions of blade loads by BIEM and lift and drag properties of representative blade sections under viscous flow conditions obtained from available experimental data or from numerical predictions by two-dimensional viscous flow solvers. Turbine wake-induced velocity by BIEM is used to determine the effective inflow to blade sections.

The resulting methodology with Viscous-Flow Correction (VFC) is referred to here as BIEM-VFC. A validation study for the proposed computational model is addressed by considering two case studies taken from the literature with experimental results for three-bladed model turbines. Specifically, Gaurier et al. [11] present results from the first round-robin test on tidal turbines carried out in the framework of the EU-FP7 MaRINET Project [12], with turbine performance measurements from two towing tanks (Strathclyde University and CNR-INSEAN) and two flume tanks (IFREMER and CNR-INSEAN). Next, Bahaj et al. [13] present a detailed characterization of marine current turbine performance by considering the effects of blade pitch variations. For this case study, BIEM-VFC is also compared with other numerical models based on BEM and BIEM. Results of the comparative analyis provide a clear overview of the accuracy of the proposed BIEM-VFC methodology and its range of applicability as a marine current turbine analysis and design tool.

The paper is organised as follows. The theoretical and computational BIEM-VFC methodology is outlined in Section 2, with details of viscous flow and trailing wake models. The validation study is addressed in Sections 3–5, while strenghts and weaknesses of the methodology are discussed in Section 6.

2. Theoretical Model

The computational model proposed here for the hydrodynamic analysis of marine current turbines is based on a Boundary Integral Equation Method (BIEM) that is valid under inviscid flow assumptions. The methodology has been originally developed at CNR-INSEAN to study marine propulsors, see e.g., Salvatore et al. [8,9], and Pereira et al. [10].

The extension of the methodology to analyse marine turbine flows requires the introduction of suitable models to describe trailing vorticity dynamics and to correct blade loads when turbine blades undergo flow separation and stall. In this section, the original BIEM is briefly reviewed and models specifically developed for turbine trailing vorticity and viscosity effects are described in detail.

Assuming the onset flow is incompressible and inviscid, the perturbation velocity $\mathbf{v}$ induced by the turbine may be described by a scalar potential as $\mathbf{v} = \nabla\varphi$, and general mass and momentum equations are dramatically simplified. Mass conservation yields the Laplace equation for the perturbation velocity potential, $\nabla^2 \varphi = 0$, while the momentum equation reduces to the Bernoulli equation for pressure p

$$\frac{\partial \varphi}{\partial t} + \frac{1}{2}\|\nabla \varphi + \mathbf{v}_I\|^2 + \frac{p}{\rho} + g z_0 = \frac{1}{2}\|\mathbf{v}_I\|^2 + \frac{p_0}{\rho}, \tag{1}$$

where p_0 is the free-stream reference pressure, $\mathbf{v}_I = \mathbf{w} + \boldsymbol{\Omega} \times \mathbf{x}$ is the inflow velocity as seen from an observer fixed with blades rotating at angular velocity $\boldsymbol{\Omega}$ while $\mathbf{w}$ is the onset flow velocity. In case of uniform inflow aligned to turbine axis x, one has $\boldsymbol{\Omega} = \Omega \mathbf{e}_x, \mathbf{w} = V \mathbf{e}_x$ with $\mathbf{e}_x$ unit vector along x, see Figure 1. Finally, $g z_0$ is the hydrostatic head term referred to a reference vertical position $z = 0$.

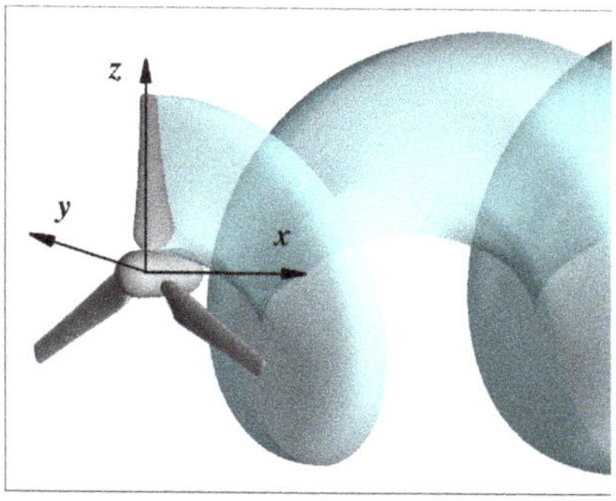

Figure 1. Sketch of the frame of reference associated to the solid boundary describing an isolated turbine and the surface of trailing wake shed by one blade.

The Laplace equation for φ is solved via a boundary integral formulation where problem unknowns are distributed on the body surface $\mathcal{S}_B$ and on the *trailing wake* surface $\mathcal{S}_W$. According to potential flow theory for lifting bodies, the trailing wake denotes a zero-thickness layer where vorticity generated by lifting surfaces is shed into the downstream flow. Through a classical derivation (see, e.g., Morino [14]) the following boundary integral representation for φ at an arbitrary field point $\mathbf{x}$ is obtained

$$E(\mathbf{x})\,\varphi(\mathbf{x}) = \oint_{\mathcal{S}_B} \left(\frac{\partial \varphi}{\partial n} G - \varphi \frac{\partial G}{\partial n} \right) d\mathcal{S}(\mathbf{y}) - \int_{\mathcal{S}_W} \Delta\varphi \frac{\partial G}{\partial n} d\mathcal{S}(\mathbf{y}). \tag{2}$$

Dealing with isolated turbines modelling by BIEM, a typical schematization is to represent the device as an assembly of blades attached to a nacelle of finite length immersed in an unbounded flow, as sketched in Figure 1. As a consequence, $\mathcal{S}_B$ denotes the solid surface combining blades and nacelle. Vector $\mathbf{n}$ is the unit normal to body and wake surfaces, pointing outward on solid boundaries and from pressure to suction sides at blade trailing edges to define the orientation on the wake. The symbol Δ in Equation (2) is used to denote discontinuity of velocity potential across the trailing wake surface defined according to the convention used for the unit normal to $\mathcal{S}_W$, whereas $E(\mathbf{x})$ is a function that makes the same equation to be valid for points $\mathbf{x}$ on the body surface ($E = 1/2$) or inside the fluid domain, $E = 1$. Moreover, quantities $G, \partial G/\partial n$ are unit source and dipoles in the unbounded three-dimensional space and depend only from the mutual position between the collocation point $\mathbf{x}$ and the influencing point $\mathbf{y}$ on the boundary surfaces. A distinguishing feature of the present formulation is that analytical expressions are used to evaluate the exact contributions of source and dipole terms on hyperboloidal quadrilateral surface elements, see [14] for details.

Boundary conditions for the velocity potential are imposed at infinity (vanishing perturbation φ), on solid surfaces (impermeability, $\partial\varphi/\partial n = -\mathbf{v}_I \cdot \mathbf{n}$) and on the trailing wake, where convection of vorticity generated on blades is imposed and a Kutta-Morino condition is used to impose identity between velocity potential difference at blade trailing edge pressure and suction sides and $\Delta\varphi$ on the wake.

Equation (2) with $E = 1/2$ and related boundary conditions represents a boundary integral equation whose solution determines φ on the body surface. By discretizing boundaries $\mathcal{S}_B$ and $\mathcal{S}_W$ into surface panels, and enforcing Equation (2) at centroids of body panels, a linear set of algebraic equations is obtained. The wake surface $\mathcal{S}_W$ can be determined as a part of the solution by a wake-alignment iterative procedure, as described in [15]. A faster and more robust approach is used in the present study as described in Section 2.1 below.

Once Equation (2) is numerically solved, the velocity potential and its gradient are known on the body surface and pressure can be evaluated using the Bernoulli Equation (1). Hydrodynamic loads generated by the turbine are then obtained by integrating pressure and tangential stress τ over the blades surface. In particular, the force contribution of a blade element of radial extension $\mathrm{d}r$ and chord c can be written as

$$\mathbf{df}(r) = \mathbf{df}_p(r) + \mathbf{df}_\tau(r) = (-p\,\mathbf{n} + \tau\,\mathbf{t})\; d\mathcal{S}, \tag{3}$$

where $d\mathcal{S} = c\,\mathrm{d}r$, $\mathbf{t}$ is the unit vector tangent to the surface and aligned to the local flow and quantities $\mathbf{f}_p, \mathbf{f}_\tau$ denote, respectively, contributions by normal (pressure) and tangential (friction) stress. Integrating elementary forces on all blades, turbine thrust T and torque Q follow

$$T = \int_{S_B} \mathbf{f} \cdot \mathbf{e}_x \, d\mathcal{S}, \qquad Q = \int_{S_B} (\mathbf{x} \times \mathbf{f}) \cdot \mathbf{e}_x \, d\mathcal{S}. \tag{4}$$

Surface stress τ is not part of the inviscid-flow solution and could be evaluated by a coupled viscous/inviscid model in which BIEM is combined with a boundary-layer model, as described in Salvatore et al. [8]. A simplified approach popular in marine propeller models consists of estimating quantity τ from formulas valid for attached laminar and turbulent flow over a flat plate, see e.g., [16]

$$\begin{aligned} C_F &= 1.328/\sqrt{Re_r} \qquad (Re_r < 10^5) \\ C_F &= 0.075/\,(log_{10}(Re_r) - 2)^2 \quad (Re_r \geq 10^5) \end{aligned} \tag{5}$$

where $C_F = \tau/\frac{1}{2}\rho V_I^2(r)$ is the friction coefficient, and

$$Re_r = c(r)V_I(r)/\nu = c(r)\sqrt{V^2 + (\Omega r)^2}/\nu, \tag{6}$$

defines the Reynolds number characterizing the flow around the blade section at radius r, where ν is the water kinematic viscosity.

The accuracy of this approximated viscosity correction to hydrodynamic loads by BIEM is typically limited to attached flows on blade sections at low angle of attack. The VFC approach proposed here aims at coping with a wider range of conditions including flow separation and stall, as outlined in Section 2.2.

2.1. Trailing Wake Model

In the present study, a semi-analytical model is used to determine the wake surface $\mathcal{S}_W$ in Equation (2). The wake is defined as a generalised helicoidal surface with distributions of axial pitch and radial expansion of the streamtube downstream of the rotor that are consistent with the operating mode of hydrokinetic turbines.

For the axial pitch, two regions are considered: the *tip-vortex* region and the *blade wake* extending spanwise between vortices released at blade root and tip. In the blade wake, trailing vortices are convected downstream with velocity given as the average of the onset flow speed and of the velocity perturbation induced by the wake itself, $\mathbf{v}_w$. A boundary integral representation of $\mathbf{v}_w$ is obtained by taking the gradient of the velocity potential Equation (2). Here, an approximated representation of this velocity field across the fluid region of interest is obtained by using BIEM to evaluate $\mathbf{v}_w$ at the rotor plane and imposing a linear variation downstream to match a given farfield distribution.

Then, the axial component of the wake-induced velocity, $v_{x,w} = \mathbf{v}_w \cdot \mathbf{e}_x$, may be written as

$$v_{x,w} = (1 - \xi_x) \frac{\partial \tilde{\varphi}}{\partial x}|_{RP} + \xi_x \, v_{x,w}|_{FF}, \tag{7}$$

where ξ_x is a normalised abscissa with $\xi_x = 0$ at rotor trailing edge and $\xi_x = 1$ at the downstream end of the discretised wake surface. Consistent with Betz theory [17], the axial induced velocity in the farfield $v_{x,w}|_{FF}$ is twice the intensity at the rotor plane. Symbol ($\tilde{}$) denotes the wake-induced velocity potential obtained from the gradient of the wake contribution in Equation (2), while subscript RP refers to the rotor plane axial position.

In the tip-vortex region, Okulov and Sørensen [18] describe a trailing vortex shedding model with axial velocity given as the average between velocity in the blade wake, Equation (7), and the unperturbed axial velocity V outside the streamtube. Thus, denoting by $\phi_{w,0}$ the hydrodynamic pitch associated to the unperturbed flow, one obtains the following expressions for the wake pitch ϕ_w

$$\phi_{w,bla}(x,r) = \left(1 + \frac{v_{x,w}(x,r)}{V}\right) \phi_{w,0}; \qquad \phi_{w,tip}(x) = \frac{1}{2}\left(\phi_{w,bla}(x,\hat{r}) + \phi_{w,0}\right), \tag{8}$$
$$\phi_w(x,r) = \xi_r \phi_{w,tip}(x) + (1 - \xi_r)\phi_{w,bla}(x,r),$$

where pedices bla and tip denote, respectively, blade wake and tip vortex, and ξ_r is a radial weight function (in the present analysis, $\xi_r = (r/R)^3$ has been used, where R is the turbine radius). In the evaluation of $\phi_{w,tip}$, the blade wake pitch $\phi_{w,bla}$ is evaluated at a representative radial station $\hat{r}$.

Next, the radial expansion of the wake streamtube downstream of the rotor plane is determined as

$$r = R + r_0 \left(1 - e^{-\xi_x / C_2}\right), \tag{9}$$

where constants r_0, C_2 are derived from experimental data describing the wake evolution of hydrokinetic turbines over a range of operating conditions. In the present study, two datasets are considered: Micek et al. [19] with wake flow measurements up to 10 diameters downstream of a three-bladed turbine with 3% onset flow turbulence, and Del Frate et al. [20] fitting measurements in the axial range $0 < x/D < 0.8$ and the limit at infinite distance downstream of the rotor from Betz theory, $r/D = \sqrt{2}/2$. Results extracted from the two datasets and the analytical expansion law from Equation (9) with $r_0 = 0.35, C_2 = 2$ are illustrated in Figure 2. It should be noted that at large axial distance from the rotor, the proposed law determines an expansion rate that is intermediate between idealised conditions from the Betz theory and real conditions affected by non negligible background turbulence.

Combining Equations (7) to (9), the generalised helicoidal surface defining the trailing wake $\mathcal{S}_W$ is obtained. In fact, the evaluation of the velocity potential $\tilde{\varphi}$ in Equation (7) depends on the definition of surface $\mathcal{S}_W$ in Equation (2) and hence an iterative procedure is required. In the numerical analysis addressed in the present work, it has been found that the iteration converges after few steps.

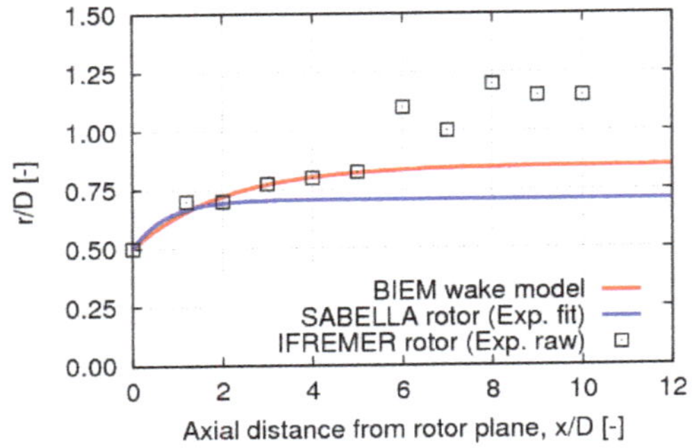

Figure 2. Streamtube radius downstream of rotor plane from Equation (9) and comparative data from experiments.

2.2. Viscous-Flow Correction Model

Assumptions of inviscid, irrotational flow underlying BIEM yield that turbine hydrodynamics is studied by fast numerical solutions of a linear problem with unknowns distributed only on the solid surface of the turbine. Unfortunately, turbine performance is dramatically affected by blade flow separation and stall and hence neglecting viscosity effects may result into completely unreliable predictions of turbine hydrodynamic loads and power output.

A methodology is proposed here to correct blade loads predicted under inviscid-flow assumptions by a procedure that preserves the reduced computing effort typical of BIEM. The idea is to *(i)* identify conditions where blade flow is subject to boundary layer separation and stall and *(ii)* estimate the effect of viscosity on blade loads under such conditions. The BIEM model including this viscous flow correction is hereafter referred to as BIEM-VFC.

For this purpose, sectional loads along blade span evaluated by BIEM are compared to lift and drag properties of two-dimensional (2D) profiles describing blade sections. Equivalence between operating conditions of three-dimensional rotating blade sections and corresponding 2D profiles is enforced in terms of local Reynolds number Re_r (see Section above) and of the *effective* angle of attack α_e.

Quantity α_e defines the angle of attack where wake-induced velocity contributions are accounted for to evaluate the total velocity incoming to blade sections. A graphical definition of α_e is given in Figure 3, where inflow velocity components and hydrodynamic force components for a turbine blade section at radius r are sketched. Axial and tangential induced velocity components, respectively Δu_i and Δv_i, represent three-dimensional flow effects induced by trailing vortices shed by blades. These quantitites are zero in case of 2D flow around a lifting surface of infinite span and the effective and nominal angle of attack α coincide.

Lift and drag properties representative of blade section shape and operating conditions (α_e, Re_r) are deduced from 2D foil polar curves, as sketched in Figure 4. Flow separation occurs when the lift curve departs from linear dependence with incidence α (points labelled as SE+, SE-), while stall occurs when lift drops as α increases in absolute value and drag rises abruptly (point ST).

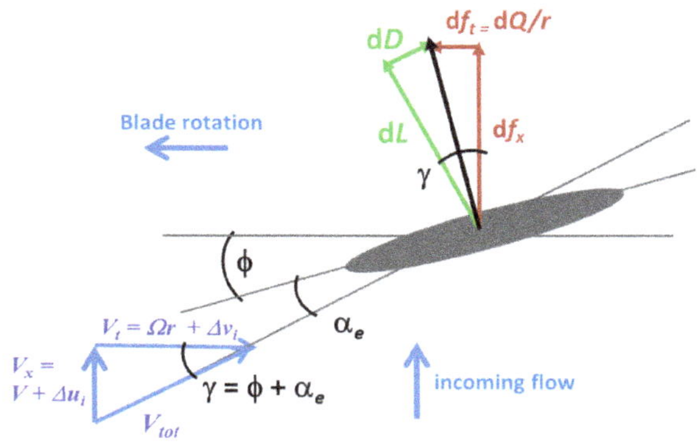

Figure 3. Inflow velocity components and hydrodynamic force components on turbine blade section at radius r.

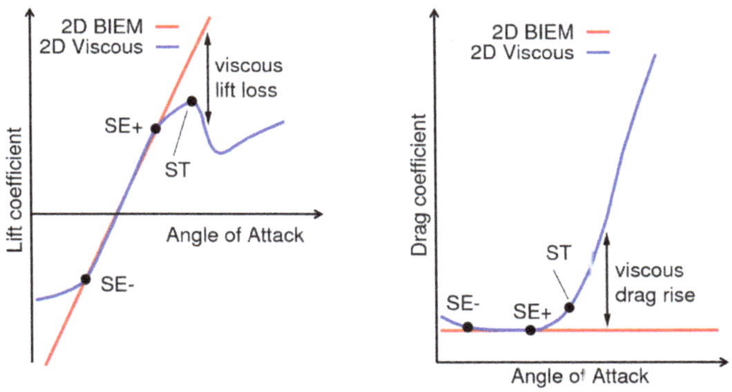

Figure 4. Notional lift and drag curves of a two-dimensional profile: viscous flow and inviscid flow conditions with flat plate drag correction compared.

Inviscid-flow solutions by BIEM determine blade sectional loads that are consistent with linear relationship between lift and angle of attack and, using the flat-plate analogy in Equation (5) with minimum drag reflecting attached flow conditions (curves in red in Figure 4). The comparison between sectional lift and drag properties motivates the following definition of factors to correct sectional loads by BIEM to represent both attached and separated flow conditions:

$$\begin{aligned} \mathcal{K}_D(\alpha_e, Re_r) &= \mathrm{d}D_{2D}/\mathrm{d}D_{2D}^{inv} \\ \mathcal{K}_L(\alpha_e, Re_r) &= \mathrm{d}L_{2D}/\mathrm{d}L_{2D}^{inv} \end{aligned} \tag{10}$$

where D_{2D}^{inv} and L_{2D}^{inv} are, respectively, drag and lift per unit length determined under inviscid 2D flow conditions (i.e., by a 2D BIEM) at angle of attack α_e, while D_{2D} and L_{2D} are profile drag and lift under 2D viscous flow conditions.

Once quantities $\mathcal{K}_D, \mathcal{K}_L$ are known, blade loads correction is obtained through the following procedure. From the BIEM solution, sectional contributions to axial force $\mathrm{d}f_x$ and tangential force $\mathrm{d}f_t$ are determined from Equation (3). Next, wake-induced velocity along blade span is determined by taking the gradient of Equation (2) (with $E = 1$), and the radial distribution of the effective angle of attack $\alpha_e(r)$ is evaluated. Radial distributions of sectional drag and lift dD, dL follow by projecting force in direction normal and tangent to the effective inflow, as sketched in Figure 3, where Φ is the angular pitch of the blade section at radius r.

Separating pressure-induced and friction-induced contributions to force $\mathbf{df}$ as defined in Equation (3), lift and drag contributions are also split into pressure-induced and friction-induced terms. Correction factors from Equation (10), yield

$$d\hat{L}_p = \mathcal{K}_L dL_p, \qquad d\hat{D}_p = \mathcal{K}_L^2 dD_p$$
$$d\hat{L}_\tau = \mathcal{K}_D dL_\tau, \qquad d\hat{D}_\tau = \mathcal{K}_D dD_\tau \tag{11}$$

where symbol $(\hat{\ })$ labels viscous-flow corrected quantities. While corrections for pressure-induced lift L_p and friction-induced drag D_τ are obvious, the assumption made here is that correction factor for drag $\mathcal{K}_D$ can be used to account for flow separation and stall effects on friction-induced lift L_τ. Pressure-induced drag D_p correction by $\mathcal{K}_L^2$ stems from the approximated relationship between induced drag and lift that is broadly valid for lifting surfaces. Numerical studies prove that contributions from *diagonal* terms L_τ and D_p are very small as compared to, respectively, contributions D_τ, L_p.

Converting lift and drag back to respective axial and tangential load components yields quantity $d\hat{f}_x$ that integrated along blade span returns corrected blade axial force, while quantity $d\hat{Q} = d\hat{f}_t r$ returns corrected blade torque. Summing over all blades, turbine corrected thrust $\hat{T}$ and torque $\hat{Q}$ are obtained (formally, Equation (4) with $\mathbf{f}$ replaced by $\hat{\mathbf{f}}$).

A full exploitation of the viscosity correction model described above implies that an iterative procedure is enforced to make the potential flow solution consistent with the modified loading on blades. No iteration has been considered in the present analysis and the subject is briefly address later in Section 6.

3. Case Studies for Validation of Computational Model

Numerical applications of the proposed computational model are discussed by considering two case studies taken from the literature. Both cases address three-bladed model turbines designed for research activity.

For a turbine with radius R, swept area $A = \pi R^2$, rotating at angular speed $\Omega = 2\pi n$ in a current with nominal freestream velocity V, the Tip Speed Ratio (TSR) is defined as

$$TSR = \Omega R / V. \tag{12}$$

Turbine performance is described through thrust, torque and power coefficients, respectively C_T, C_Q, C_P, defined as

$$C_T = \frac{T}{\frac{1}{2}\rho A V^2}, \qquad C_Q = \frac{Q}{\frac{1}{2}\rho A V^2 R}, \qquad C_P = \frac{\Omega Q}{\frac{1}{2}\rho A V^3} = C_Q \times \text{TSR}$$

where $P = \Omega Q$ is the power generated by the turbine.

3.1. Fixed Pitch Turbine

Gaurier et al. [11] describe a 700 mm diameter, fixed-pitch model turbine developed at IFREMER, France. The model has been the subject of the first round-robin test on tidal turbines carried out in the framework of the EU-FP7 MaRINET Project [12], with turbine performance measurements from two towing tanks (Strathclyde University and CNR-INSEAN) and two flume tanks (IFREMER and CNR-INSEAN). Turbine performance curves measured at inflow speed between 0.6 and 1.2 m/s are presented as mean values and standard deviations.

Main turbine geometry parameters are summarized in Table 1. A full description is given in [11]. This testcase is referred to here as the IFREMER-FP turbine.

Table 1. IFREMER-FP turbine main geometry parameters.

Rotor diameter, D	700 [mm]
Blades number, Z	3
Pitch angle at 70% span, Φ	7.3 [deg]
Thickness ratio, 75% span, t/c	0.21
Hub/rotor diameter ratio	0.131
Blade section profile	NACA 63-4xx

3.2. *Variable Pitch Turbine*

Bahaj et al. [13] investigate a 800 mm diameter, variable-pitch model turbine developed at the University of Southampton (U.K.). This model was analysed by extensive towing tank and cavitation tunnel tests. Experimental data provide turbine performance at different blade pitch settings, with blades rotated about the spanwise axis over a range of 15 degrees, from $\Phi = 15°$ to $30°$, while $\Phi = 20°$ is taken as the design condition. This pitch definition refers to the nose-tail angle of the blade at radius $r/R = 0.2$. Turbine performance curves are available for inflow speed between 0.8 and 2.0 m/s (cavitation tunnel tests) and between 0.8 and 1.5 m/s (towing tank tests).

Main turbine geometry parameters are summarized in Table 2, while a complete description can be found in [13]. This testcase is referred to here as the UoS-VP turbine.

Table 2. UoS-VP turbine geometry parameters.

Rotor diameter, D	800 [mm]
Blades number, Z	3
Pitch angle at 20% span, Φ	15, 20, 25, 27, 30 [deg]
Thickn. ratio, 75% span, t/c	0.151
Hub/rotor diameter ratio	0.125
Blade section profile	NACA 63-8xx

4. Fixed Pitch Turbine Study

The IFREMER-FP turbine performance is analysed with reference to experimental conditions corresponding to the highest inflow speed, $V = 1.2$ m/s. This choice is motivated to avoid a too small Reynolds number characterizing the blade flow. The tip speed ratio is varied between zero and 8. Comparing with the physical model in [11], it may be noted that the stanchion supporting the turbine is not considered in numerical simulations. Similarly, the nacelle portion downstream of the turbine hub is not present in the computational model, see Figure 5.

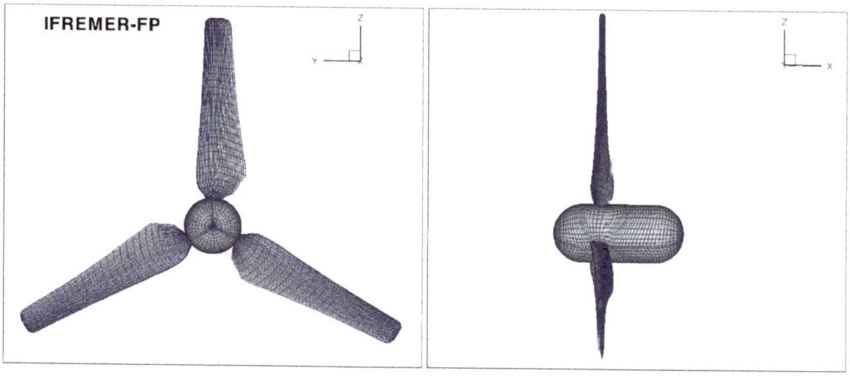

Figure 5. IFREMER-FP turbine. Computational grid used for calculations by BIEM.

Blade and hub surface discretization parameters are determined as the result of a grid sensitivity study. Each blade is discretized into M_B elements chordwise from leading edge to trailing edge and N_B elements spanwise. Four blade grid levels with $M_B = 24, 36, 48, 60$ and $N_B = 20, 30, 40, 50$ are considered. Hub and wake surface discretizations are built according to blade grid refinement level. Figure 6 presents inviscid-flow thrust and torque predicted by BIEM using the four grids. Three representative TSR values are considered. The torque coefficient evaluated by including the viscous flow correction is also presented to highlight that the VFC model has no effect on the sensitivity of results to grid refinement. From these results it is concluded that a discretization with $M_B = 36$, $N_B = 30$ is adequate to minimise the effect of further grid refinement on results. In this case, the hub surface is divided into 42 and 54 elements, respectively, in circumferential and longitudinal directions, and the wake is discretized into 36 elements along radius and 60 elements streamwise per revolution. The wake portion considered in the numerical solution of Equation (2) extends for 10 revolutions.

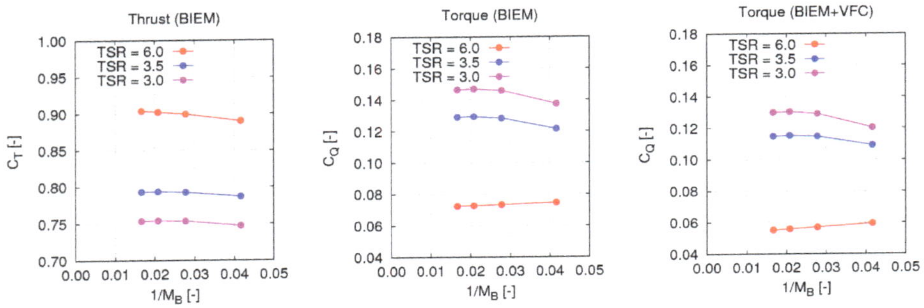

Figure 6. IFREMER-FP turbine. Calculated thrust and torque coefficients as a function of discretization parameter M_B. **Left** and **center**: thrust and torque by non-corrected BIEM; **right**: torque by BIEM-VFC.

The trailing wake model described in Section 2.1 is used to determine the turbine wake surface. Figure 7 maps the intensity of wake-induced axial velocity evaluated by BIEM at axial locations corresponding to rotor blade trailing edge and at different radial positions over a range of operating conditions identified by the parameter TSR.

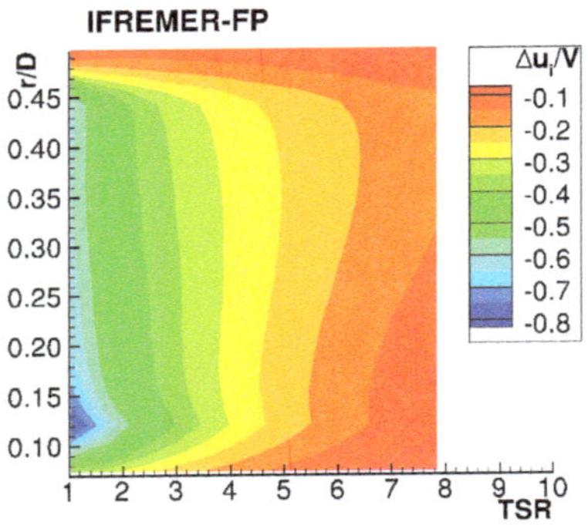

Figure 7. IFREMER-FP turbine. Calculated axial induced velocity distribution at rotor plane.

The resulting surfaces for three representative values of *TSR* are shown in Figure 8. The effect of *TSR* on wake pitch is clear: trailing vortices are rapidly shed away from the rotor at low *TSR*, while wake spirals pack-up close to the rotor as *TSR* increases. In all cases, wake pitch increases from inner radii to the tip vortex.

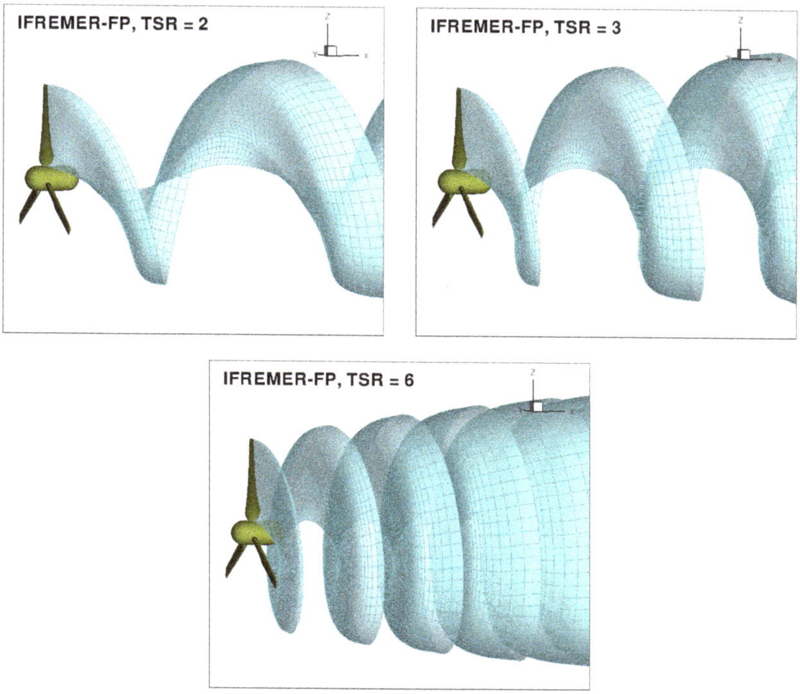

Figure 8. IFREMER-FP turbine. Trailing wake geometry of BIEM model at different operating conditions: $TSR = 2$ (**top left**), $TSR = 3$ (**top right**), $TSR = 6$ (**bottom**).

Viscous flow effects on blade loads are determined by applying the VFC model described in Section 2.2. The evaluation of correction factors in Equation (11) requires that blade section lift and drag properties are known. For this purpose, the inviscid-flow BIEM solution is used to estimate the local Reynolds number Re_r from Equation (6), and the effective angle of attack α_e in Figure 3, at all blade sections for the *TSR* range from zero to 8 considered in model tests. Results in left Figure 9 show that the local Reynolds number varies between 1×10^5 and 3.5×10^5 over most of the *TSR* range of interest. Right Figure 9 depicts a positive effective angle of attack α_e that increases from values close to zero for the highest *TSR* to 20–25 degrees for *TSR* between 1 and 2. At given *TSR*, both Reynolds and angle of attack present limited variations over a wide blade portion between 30% and 90% of span.

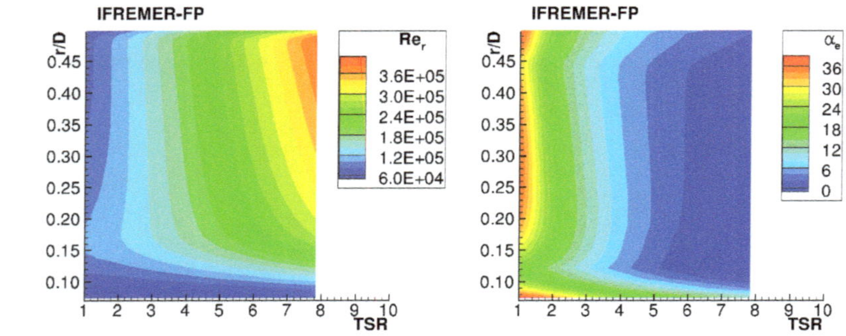

Figure 9. IFREMER-FP turbine. Reynolds number Re_r (**left**) and effective angle of attack α_e (**right**) as a function of radius r and of TSR.

Experimental data of lift and drag curves of NACA 63-4xx profiles are available in the literature only at Reynolds number of 10^6 and higher, which is outside the range of interest in the present analysis as shown above. Lack of experimental data is overcome here by using numerical predictions of polar curves by the X-Foil code [21]. This solver integrates a BIEM for two-dimensional, steady flow with the solution of boundary layer equations in integral form and is largely used in combination with blade element (momentum) methods. The NACA 63-421 profile corresponding to 21% thick blade sections at 70% of span is taken as representative of all blade sections. At angle of attack beyond stall, X-Foil predictions are not reliable and polar curves are completed by the following extrapolation procedure. At very high incidence angles (here, $\alpha \geq 30°$), lift and drag values are taken from experimental data for the NACA 0015 profile by Sheldahl and Klimas [22]. The assumption is that for very high angles, hydrodynamic loads are weakly sensitive to Reynolds number and to profile shape details. A polynomial fit is used to merge NACA 63-421 data from X-Foil and high angle of attack NACA 0015 data at angle of attack between stall and 30 degrees.

Lift and drag curves calculated by X-Foil are presented in Figure 10. In particular, results for 5 values of Reynolds number are considered in order to adequately describe lift and drag properties over the Re_r range of interest (Figure 9 Left). Lift and drag curves from experimental data at $Re = 3E6$ in [23] are also given for comparison.

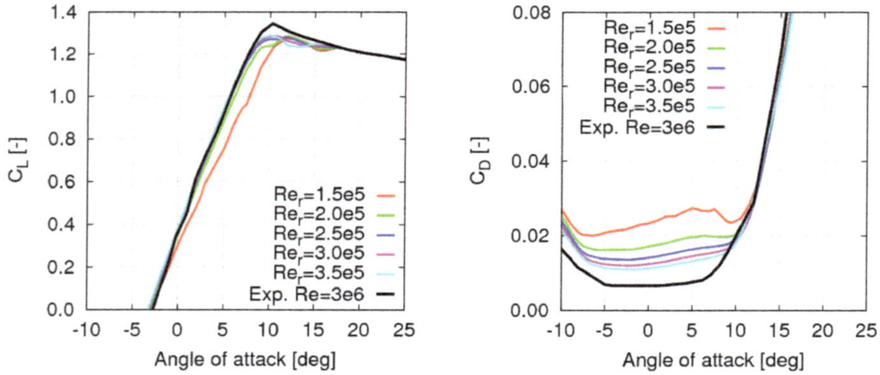

Figure 10. NACA 63-421 2D foil: lift (**left**) and drag (**right**) coefficients calculated by X-Foil and from experiments [23].

Figure 11 maps correction factors $\mathcal{K}_D, \mathcal{K}_L$ along blade span and over the *TSR* range of interest. Values close to one denote conditions where blade flow is attached or weakly separated and no correction of sectional loads by BIEM is needed. This occurs at $TSR \simeq 2.5$ and higher, which corresponds to effective angle of attack below 10–12 degrees, as shown in Figure 9. At lower *TSR*, the effective angle of attack increases up to stall, as apparent from polar curves in Figure 10. As expected, the lift factor $\mathcal{K}_L$ drops below 1, while the drag factor $\mathcal{K}_D$ rapidly grows, to simulate, respectively, stall-induced lift loss and drag rise.

Predicted turbine thrust, torque and power curves are presented in Figure 12 and compared with experimental data at $V = 1.2$ m/s from three facilities involved in the round-robin test: CNR-INSEAN towing tank (INSEAN), IFREMER flume tank (IFREMER) and Kelvin Hydrodynamics Laboratory at University of Strathclyde (KHL). Results from the fourth facility participating to the round robin test, CNR-INSEAN flume tank, are omitted here since they fall within the range given by those considered in plots. For the sake of precision, measured thrust and power coefficients only are presented in [11]. Here, also the torque coefficient is considered because this quantity provides a direct indication of the accuracy of blade tangential forces evaluated by the numerical model.

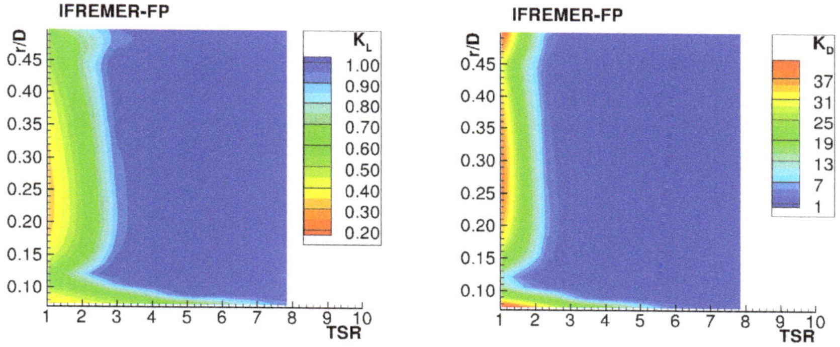

Figure 11. IFREMER-FP turbine. Correction factors for radial contributions to lift (**left**) and drag (**right**) as a function of radius *r* and of *TSR*.

It is important to observe that present experimental and numerical results use different definitions of thrust and torque. Numerical thrust and torque are determined by integrating hydrodynamic loads on blade surfaces, while in the experimental set-up, turbine torque denotes the axial moment measured by a torque sensor placed between the rotating hub and the fixed nacelle. Assuming the contribution to torque of the rotating hub is negligible, numerical and experimental data are consistent. Both numerical and experimental power are evaluated from the hydrodynamic torque Q as $P = 2\pi n Q$.

Less direct is the comparison between numerical and measured thrust. Turbine thrust reported in [11] denotes the axial force at the top of the mast supporting the turbine. This quantity combines blades thrust with a non negligible resistance contribution D_{HDM} from hub, nacelle and the mast piercing the free surface. Tests performed at IFREMER of a dummy IFREMER-FP rotor with no blades determined $D_{HDM} = 16.89N$ at $V = 0.8$ m/s (not reported in [11]). For the sake of completeness, top left Figure 12 also presents measured axial force with the D_{HDM} contribution subtracted. This result is referred to as 'Exp IFREMER Corr.'.

Numerical results in Figure 12 include both BIEM without viscosity correction and corrected values by Equation (11) (label BIEM-VFC). As expected from the discussion above, viscosity effects are negligible at $TSR = 5$ and higher, while small differences between standard BIEM (that is, with non viscous flow corrections) and BIEM-VFC predictions are noted for $3 < TSR < 5$. In this range, numerical and experimental results for torque and power are in good agreement, while thrust is underestimated in numerical results. The reason for this difference in thrust is not clear and could be related to hub, nacelle and mast resistance contributions that are only approximately subtracted from axial force measurements.

At *TSR* lower than 3, massive flow separation and stall determine a dramatic reduction of thrust, torque and power that is missed in standard BIEM results, while BIEM-VFC results capture the correct trend. In particular, measured peak values of C_Q and C_p are matched at the correct *TSR* values. At very low *TSR*, where deep stall conditions occur on blades, the BIEM-VFC model overpredicts both torque and power, but the viscous flow correction allows to recover most of the error affecting inviscid-flow predictions by non-corrected BIEM.

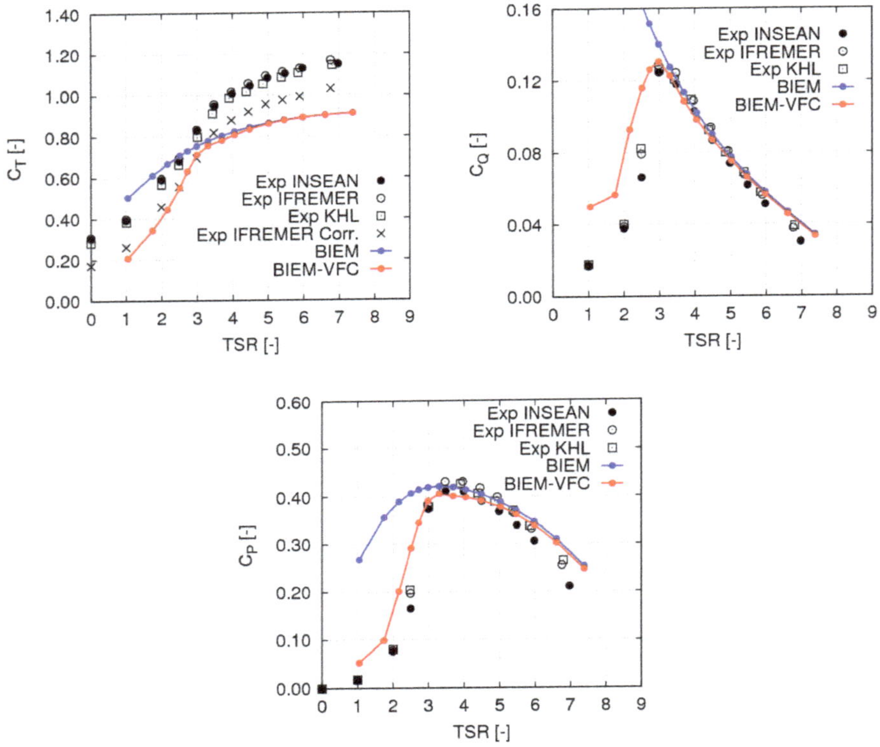

Figure 12. IFREMER-FP turbine performance predictions by BIEM and BIEM-VFC compared to experimental data in [11]: thrust (**top left**); torque (**top right**) and power (**bottom**) coefficients.

Figures 13 and 14 address blade pressure distributions evaluated by BIEM. Specifically, the pressure coefficient is defined as

$$C_p = \frac{p - p_0}{\frac{1}{2}\rho V_I^2}, \tag{13}$$

where the pressure p is evaluated by BIEM and $V_I(r) = [V^2 + (\Omega r)^2]^{1/2}$ is the velocity of the flow incoming to the blade section at radius r. Recalling that the VFC model applies only to global loads and not to the pressure distribution, calculated C_p is representative only in the *TSR* range where viscosity correction is not significant. For the present case, this approximately holds for $TSR > 3$. Figure 13 depicts pressure distributions on blades pressure and suction sides at *TSR* = 3.3 (peak power condition, see Figure 12). The effect of *TSR* on blade pressure distribution is illustrated in Figure 14, where C_p along the blade section at $r/R = 0.7$ for four values of *TSR* is plotted. As expected, the pressure jump between pressure and suction sides tends to reduce as *TSR* is increased from the peak power condition.

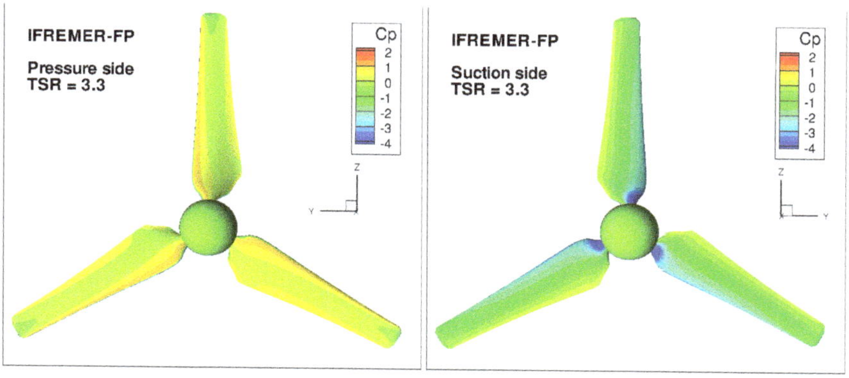

Figure 13. IFREMER-FP turbine. Pressure distribution evaluated by inviscid-flow BIEM, *TSR* = 3.3 (peak power condition). **Left**: pressure side; **right**: suction side.

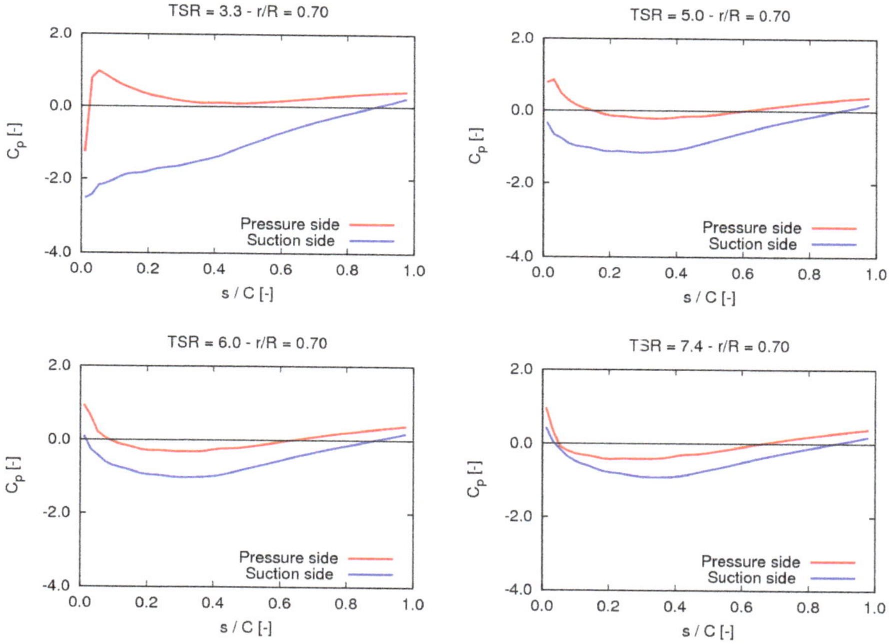

Figure 14. IFREMER-FP turbine. Pressure distribution evaluated by inviscid-flow BIEM at radial section at 70% of blade span. From **top left** to **bottom right**: *TSR* = 3.3, 5, 6, 7.4.

5. Variable Pitch Turbine Study

The variable-pitch UoS-VP turbine described in Bahaj et al. [13] represents a valuable benchmark to investigate the capability of a computational model to capture the effect of blade pitch variations on turbine loads and in particular to correctly describe performance in off-design conditions. As for the fixed-pitch IFREMER-FP turbine discussed above, a simplified three-dimensional model is used in which the aft portion of the nacelle and the supporting stanchion are omitted. Another difference exists at blade root where NACA 63-8xx sections are used in the computational model, while the physical model presents cylindrical sections to actuate pitch variations. Figure 15 shows the computational grid built for BIEM calculations. Discretization parameters are similar to the IFREMER-FP case.

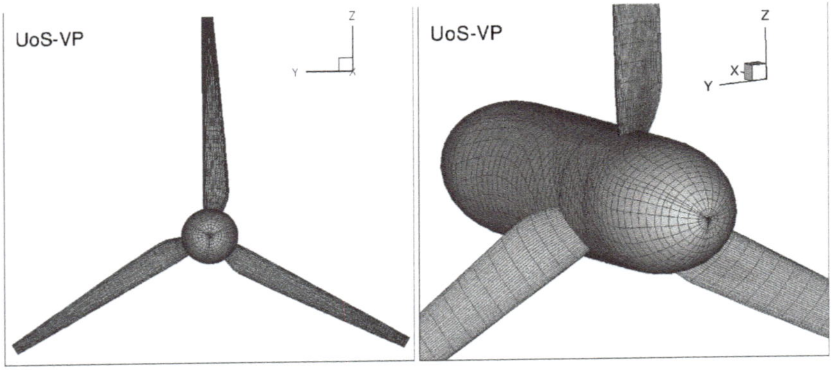

Figure 15. UoS-VP turbine. Three-dimensional model and computational grid for BIEM analysis. **Left**: front view; **right**: details of hub and blade roots.

Figure 16 depicts the intensity of wake-induced velocity $v_{x,w}$ from Equation (7) evaluated by BIEM at axial locations corresponding to rotor blade trailing edge and 70% of blade span. Different blade pitch settings and a range of operating conditions corresponding to model tests are plotted. For a given value of *TSR* the intensity of the induced velocity increases with the blade pitch setting. It should be noted that for the lowest pitch angle case the calculated value of the normalised induced velocity tends to exceed the theoretical limit $\Delta u_i / V = 0.5$ at the rotor plane given by the Betz theory.

The resulting trailing wake surfaces for the design condition $\Phi = 20°$ and for three representative values of *TSR* are shown in Figure 17.

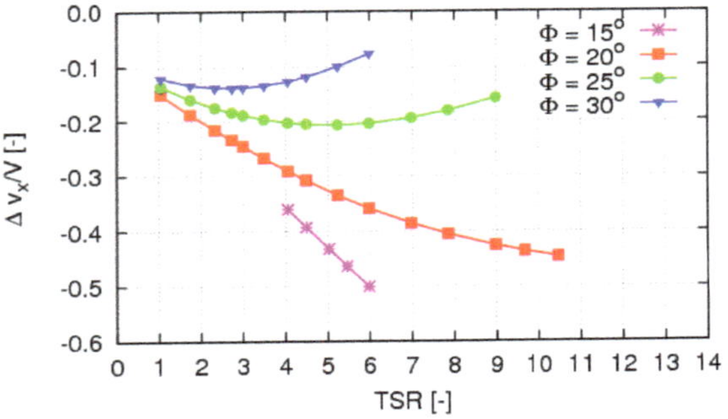

Figure 16. UoS-VP turbine. Axial induced velocity distribution at axial position corresponding to blade trailing edge and 70% of blade span. Different pitch settings Φ compared.

Turbine operating conditions considered in the present analysis refer to selected cavitation tunnel test conditions from [13] as summarized in Table 3.

Recalling Equation (6), Reynolds number Re_r characterizing blade section flow at radius r depends on the inflow velocity V. Figure 18 maps its distribution as a function of radius and *TSR* for pitch setting $\Phi = 20°$, while Figure 19 compares Re_r at 70% of blade span for the highest and lowest inflow speed cases from Table 3. Results indicate that Re_r approximately varies between 1×10^5 and 3.5×10^5 over most of the operating range of interest here.

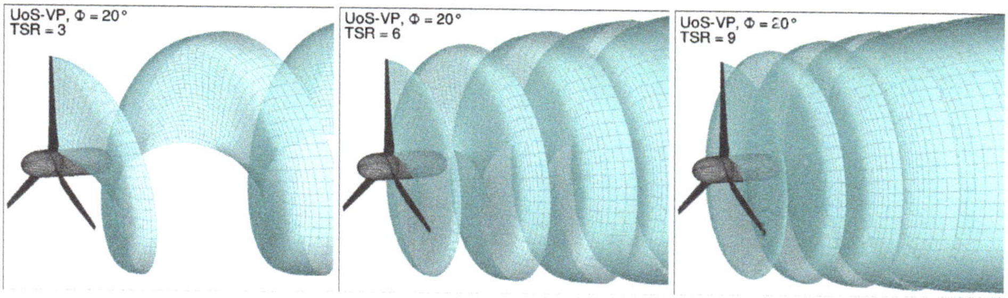

Figure 17. UoS-VP turbine. Wake geometry of BIEM model at different operating conditions. From **left** to **right**, *TSR* = 3, 6, 9. Design pitch setting, $\Phi = 20°$.

Table 3. UoS-VP turbine. Inflow speed conditions.

Blade pitch setting, Φ [deg]	15	20	25	27	30
Inflow speed, V [m/s]	1.40	1.73	1.54	1.30	1.54

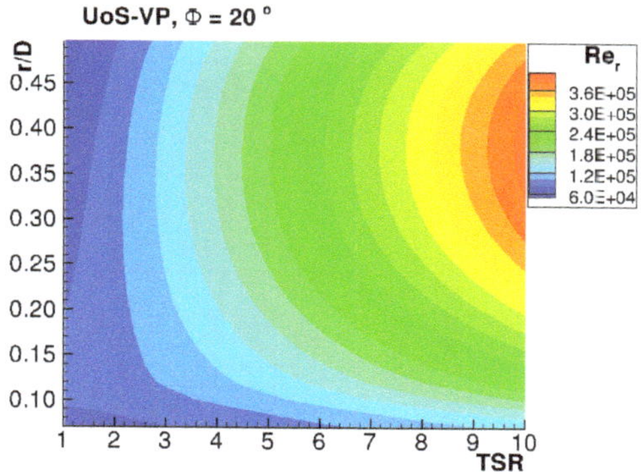

Figure 18. UoS-VP turbine. Reynolds number Re_r as a function of radius r and of turbine operating condition (*TSR*).

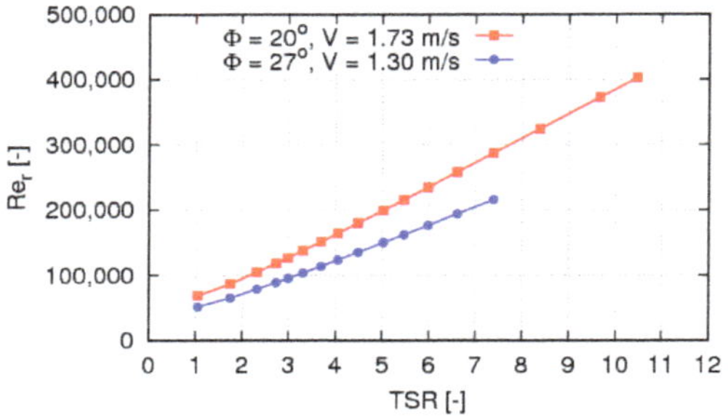

Figure 19. UoS-VP turbine. Reynolds number Re_r at radius $r/R = 0.7$ for pitch settings corresponding to the highest inflow speed ($V = 1.73$ m/s, $\Phi = 20°$), and for the lowest inflow speed ($V = 1.3$ m/s, $\Phi = 27°$).

The effective angle of attack α_e evaluated by standard BIEM is presented in Figure 20. Specifically, α_e distributions along blade span at variable *TSR* are presented for design pitch setting, $\Phi = 20°$, and Figure 21 presents the variability of this quantity at different pitch settings at 70% of blade span. Case $\Phi = 20°$ shows blade sections mostly operating in the range $-5° < \alpha_e < 25°$ with higher values only at $TSR < 2$. As expected, larger pitch angles determine lower α_e values.

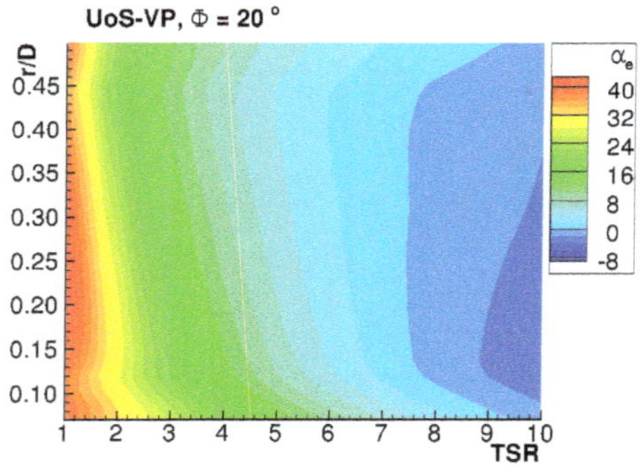

Figure 20. UoS-VP turbine. Effective angle of attack α_e as a function of radius r and of turbine operating condition (*TSR*). Design pitch setting $\Phi = 20°$.

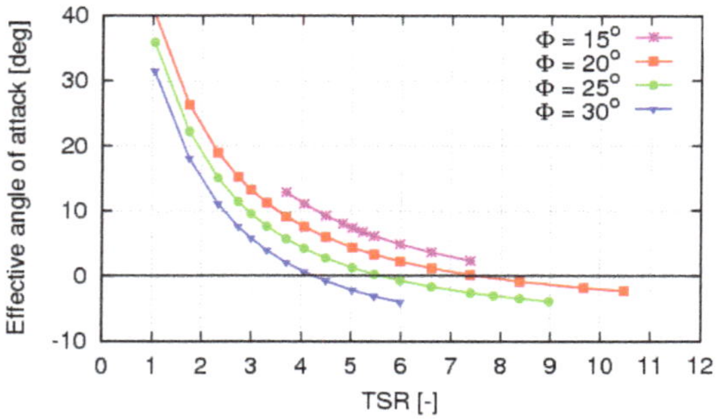

Figure 21. UoS-VP turbine. Effective angle of attack α_e at radius $r/R = 0.7$ for different pitch settings and *TSR*.

Reference [24] provides lift and drag curves of the NACA 63-815 foil at $Re = 8 \times 10^5$. This 15% thick foil is taken as representative of UoS-VP turbine sections whose thickness ratio varies from 0.176 at 50% of span to 0.126 at tip. Recalling Figures 18 and 19, $Re = 8 \times 10^5$ is higher than the range of interest in the present analysis. In order to obtain lift and drag data in the actual Re_r and α_e ranges, the X-Foil code is used and 6 polar curves for $1.5 < Re_r < 4.0 \cdot 10^5$ are evaluated. Polar data are completed at very high angle of attack using NACA 0015 profile data and polynomial interpolation as described in Section 3.2 for the IFREMER-FP turbine. Resulting lift and drag curves are plotted in Figure 22 and experimental data from [24] are also shown for comparison. Lift curves show that stall conditions are predicted by X-Foil at about 10–12 degrees, while experimental data show a more gradual transition to stall between 8 and 12 degrees. Results for drag are in agreement only at negative angle of attack, while experimental data present quite larger C_D values than X-Foil between 2 and 12 degrees. These discrepancies cannot be explained because of the different *Re* numbers in the two datasets. Furthermore, drag measurements also show a large scatter.

Lift and drag properties predicted by X-Foil are used to feed the viscous flow correction model described in Section 2.2. Contour maps in Figure 23 show $\mathcal{K}_L$ and $\mathcal{K}_D$ factors for design pitch setting $\Phi = 20°$, while Figure 24 depicts the variation of these quantities at $r/R = 0.7$ over the pitch settings range. In case $\Phi = 20°$, viscosity effects on blade section lift and drag are negligible at *TSR* of about 3.5–4 and higher, which corresponds to non-separated flow conditions at angle of attack below 8–10 degrees, see Figures 20 and 22. At lower *TSR*, the lift correction factor $\mathcal{K}_L$ gradually decreases to about 0.3 (lift loss under stall) while the drag correction factor $\mathcal{K}_D$ suddenly increases to values of 30 and more (drag crisis).

Consistent with sectional angle of attack values commented above, pitch settings $\Phi > 20°$ limit flow separation and stall effects to very low values of *TSR*, while in case $\Phi = 15°$, most of the addressed operating range is under the effect of flow separation and stall.

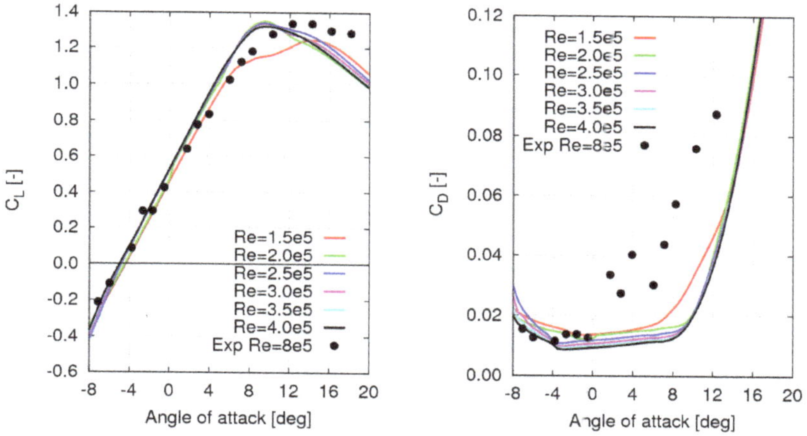

Figure 22. NACA 63-815 2D foil: lift (**left**) and drag (**right**) coefficients used for the viscous flow correction of BIEM.

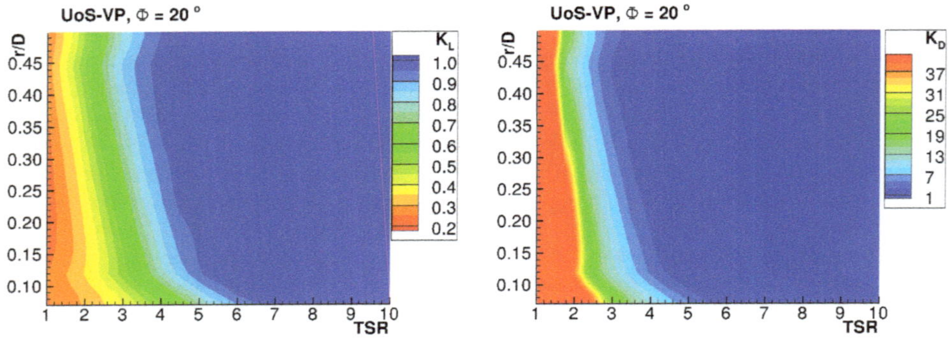

Figure 23. UoS-VP turbine. Correction factors for radial contributions to lift (**left**) and drag (**right**) as a function of radius r and of operating condition (*TSR*). Pitch setting $\Phi = 20°$.

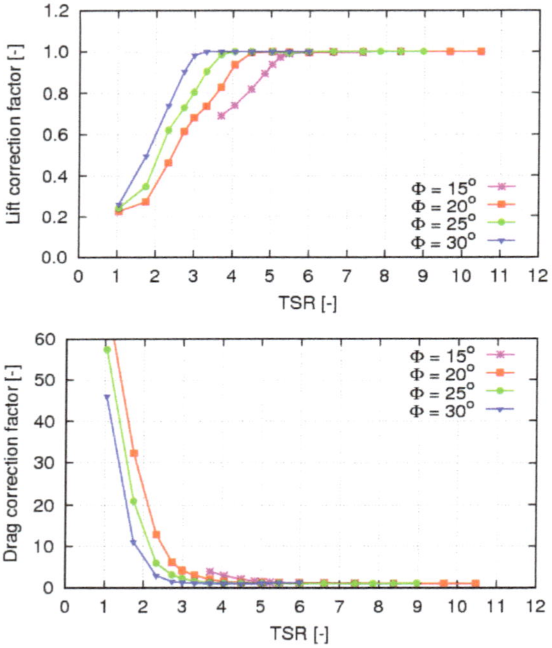

Figure 24. UoS-VP turbine. Lift correction factor $\mathcal{K}_L$ (**top**) and drag correction factor $\mathcal{K}_D$ (**bottom**) at radius $r/R = 0.7$ for different Pitch settings.

Turbine thrust and power predictions by BIEM and by BIEM-VFC using blade section polar data from X-Foil are compared with model test measurements from [13] in Figures 25 and 26. In general, it may be noted that standard BIEM predictions of both thrust and power fairly reproduce experimental data only in the high *TSR* range for pitch setting cases $20° < \Phi < 27°$. For operating conditions corresponding to peak power *TSR* and lower values of *TSR* , standard BIEM results overpredict both thrust and power, since the effects of blade flow separation and stall are not captured. When the VFC model is applied to correct BIEM, predicted thrust and power of cases $20° < \Phi < 27°$ are in fair agreement with measured data over a full *TSR* range. For extreme off-design cases $\Phi = 15^\circ$ and $\Phi = 30^\circ$, large differences between numerical and experimental results are observed even if the viscous flow correction is applied. In particular, at $\Phi = 15^\circ$ predicted C_T presents an unphysical trend with increasing *TSR* . A possible explanation for this result is that under extreme off-design conditions, blade flow is affected by a complex separated flow phenomenology that is beyond the limits of the proposed VFC model, and a detailed CFD analysis would be necessary. Unfortunately, experimental data do not give information at very low *TSR* where deep-stall conditions are expected. Large scattering of measured thrust and power is also noted in extreme off-design conditions.

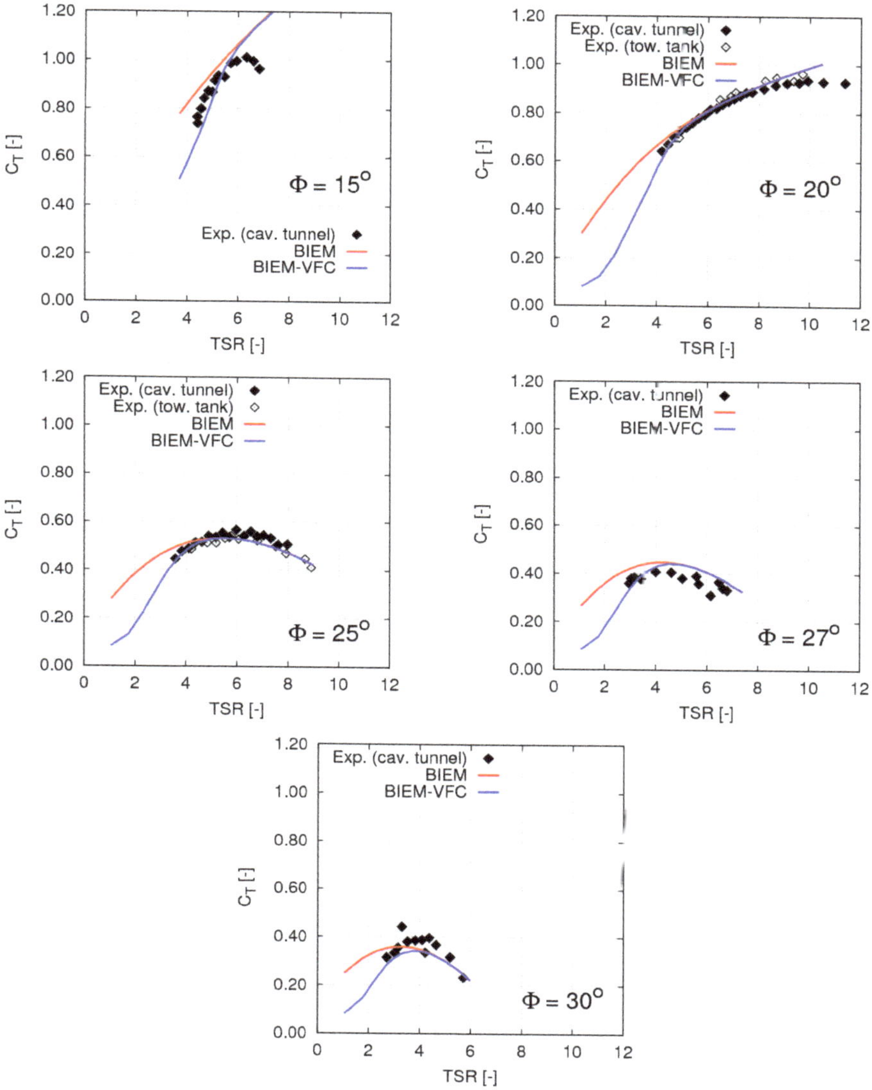

Figure 25. UoS-VP turbine performance predictions by BIEM and BIEM-VFC compared to experimental data in [13]. Thrust coefficient at pitch settings $\Phi = 15^\circ, 20^\circ, 25^\circ, 27^\circ, 30^\circ$.

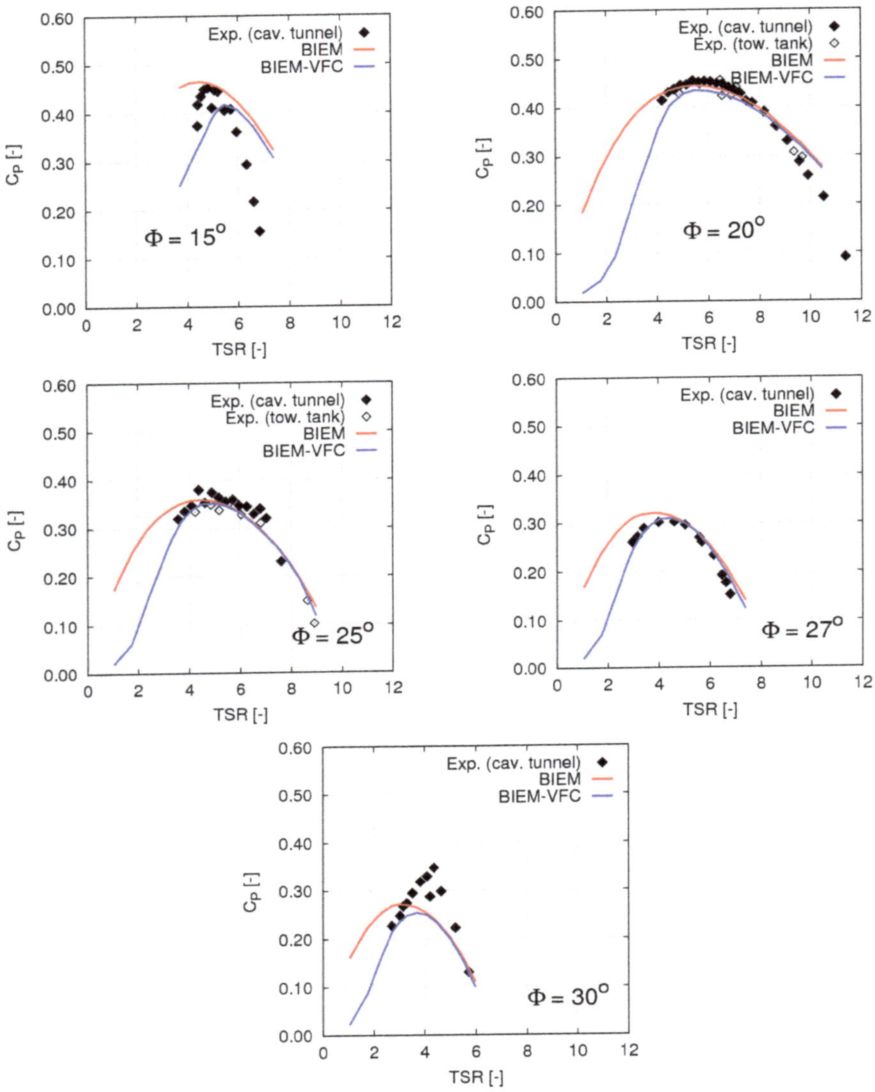

Figure 26. UoS-VP turbine performance predictions by BIEM and BIEM-VFC compared to experimental data in [13]. Power coefficient at pitch settings $\Phi = 15°, 20°, 25°, 27°, 30°$.

The capability of the BIEM-VFC model to correctly describe turbine performance trends at different pitch settings can be discussed considering results in Figure 27, where four performance indicators are considered: maximum value of thrust coefficient $C_{T_{max}}$, maximum value of power coefficient $C_{P_{max}}$, and corresponding values of *TSR* where maxima are established (labelled, respectively, as TSR@$C_{T_{max}}$, and TSR@$C_{P_{max}}$). Numerical predictions by BIEM-VFC and polar data from X-Foil (label: VFC, XFOIL Polar) are compared with polynomial fits of experimental data from cavitation tunnel tests as presented in Figure 7 of [13] (label: Model tests). Numerical results by BIEM-VFC using experimental data for blade section lift and drag taken from measurements in [24] are also presented (label: VFC, Exp. Polar).

Quantity $C_{T_{max}}$ by BIEM-VFC and X-Foil and the corresponding *TSR* values fairly reproduce the trend observed in experiments over the range $20^\circ < \Phi < 30^\circ$. Similar comments can be made for the maximum power except for case $\Phi = 30^\circ$, where predicted $C_{P_{max}}$ is some 20% lower than measured. However, fitted experimental data at $\Phi = 30^\circ$ show a rather inconsistent trend with Φ. At off-design pitch $\Phi = 15^\circ$, BIEM-VFC and X-Foil results match experimental data for $C_{P_{max}}$ and the corresponding *TSR*, while the $C_{T_{max}}$ prediction is not plotted since numerical results show an unphysical trend with increasing *TSR* as already discussed.

Figure 27 allows to compare BIEM-VFC results based on X-Foil predictions of sectional lift and drag properties with those obtained by considering measured lift and drag in [24]. As already commented in Figure 22, measured drag is much higher than X-Foil predictions over a significant range of angle of attack. As expected, this results in underestimation of turbine power coefficient, bottom left Figure 27. Smaller differences between measured and X-Foil lift observed in left Figure 22, have a negligible effect on predicted turbine thrust coefficient, as shown in top Figure 27.

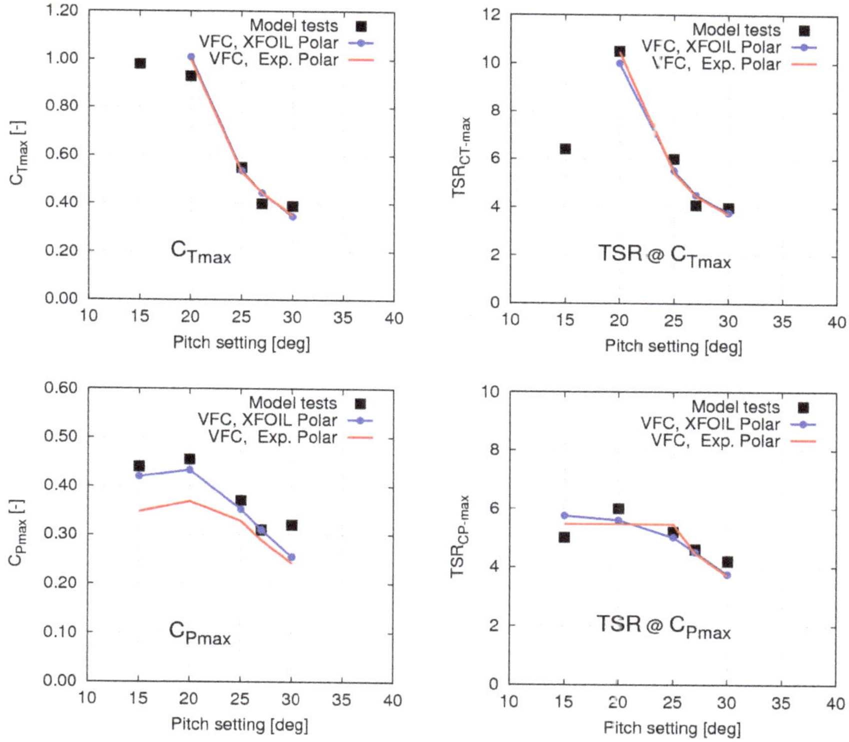

Figure 27. UoS-VP turbine. Effect of pitch setting Φ: from **top** to **bottom**, **left** to **right**, $C_{T_{max}}$, $C_{P_{max}}$ and corresponding *TSR* values TSR@$C_{T_{max}}$, TSR@$C_{P_{max}}$.

In order to complete the present validation study, it is also interesting to compare results by the proposed BIEM-VFC approach with data from the literature obtained using different computational models. Two cases are considered here: Bahaj et al. [25] present results by two solvers based on Blade Element Method (BEM) for pitch settings from $\Phi = 15^\circ$ to 27°. Next, Baltazar and Falcão de Campos [7] present results by BIEM with viscous flow corrections for cases $\Phi = 20^\circ, 25^\circ$ and 27°.

For the sake of clarity, comparisons with BEM and BIEM results from the literature are presented in different figures. Thrust and torque coefficient predictions by the present BIEM-VFC model are compared with results by BEM [25] in Figures 28 and 29. Results by BEM solvers GH-Tidal and SERG-Tidal show an accuracy with respect to model test results in [13] that is broadly comparable to what is obtained using the present BIEM-VFC. Predictions by BIEM-VFC and GH-Tidal are closer to experiments than SERG-Tidal for pitch settings $\Phi = 20°, 25°, 27°$, while the opposite holds for $\Phi = 27°$. This trend is only in part confirmed in Figure 29 where power coefficient results are shown. While BIEM-VFC and SERG-Tidal show a comparable accuracy for C_p at $\Phi = 20°, 25°, 27°$, results by GH-Tidal overestimate experimental data. It is also noted that off-design case $\Phi = 15°$ shows similar results between BEM and the present BIEM-VFC for the thrust coefficient, while power coefficient results are very different, with BEM models fairly capturing power peak and failing to predict results at higher *TSR* .

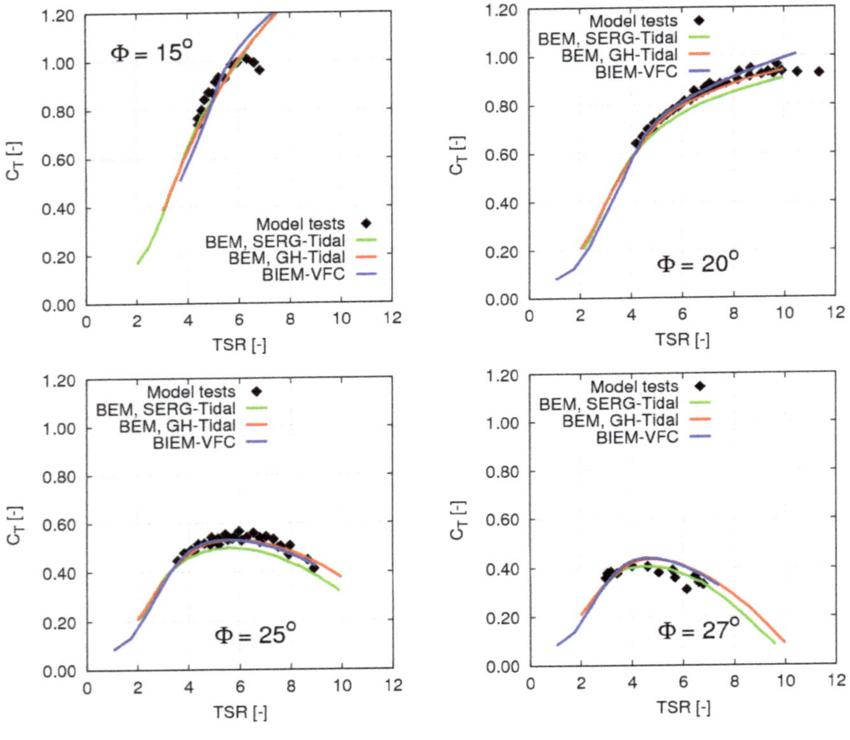

Figure 28. UoS-VP turbine performance predictions by BIEM-VFC compared to results of BEM models from [25]: thrust coefficient C_T. Pitch settings $\Phi = 15°$ to $27°$.

Comparisons between the present BIEM-VFC model and the BIEM with viscous flow correction proposed in [7] are presented in Figure 30. In general, the agreement of the two numerical models with experimental data is comparable with some exceptions. Results by the present BIEM-VFC model better reproduce measured turbine power in case $\Phi = 20°$ and thrust in case $\Phi = 25°$, while results from [7] better predict thrust in case $\Phi = 27°$. In these cases, differences occur throughout the *TSR* range and hence they can be explained as a combined effect of different viscosity correction and trailing wake models in the two formulations.

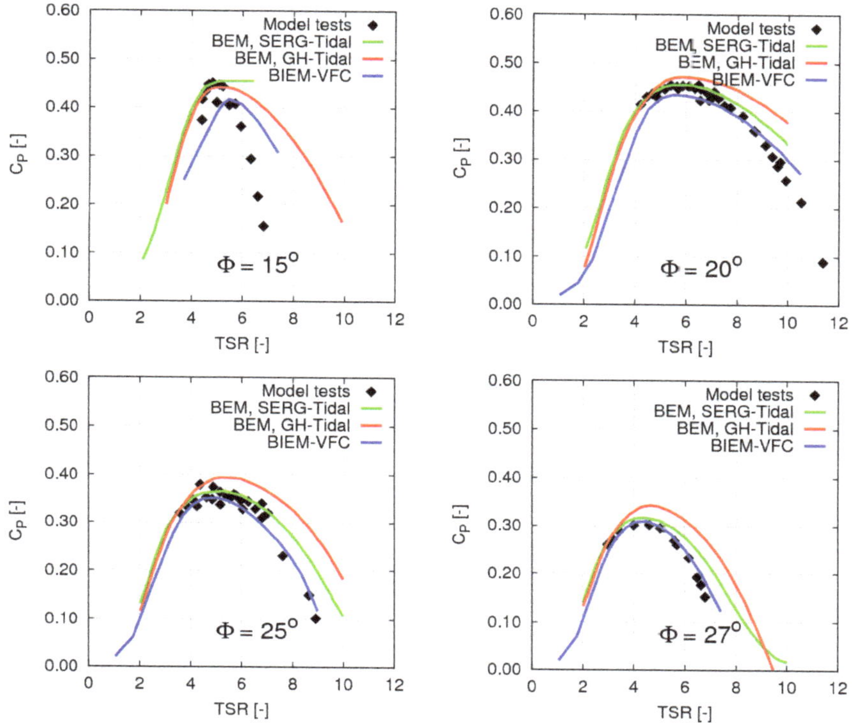

Figure 29. UoS-VP turbine performance predictions by BIEM-VFC compared to results of BEM models from [25]: power coefficient C_p. Pitch settings $\Phi = 15°$ to $27°$.

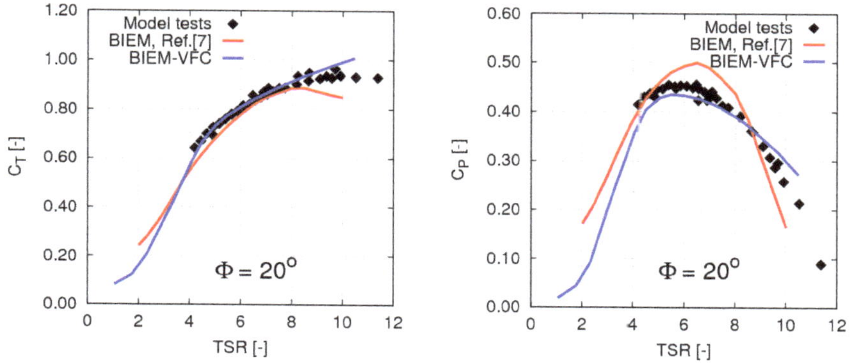

Figure 30. *Cont.*

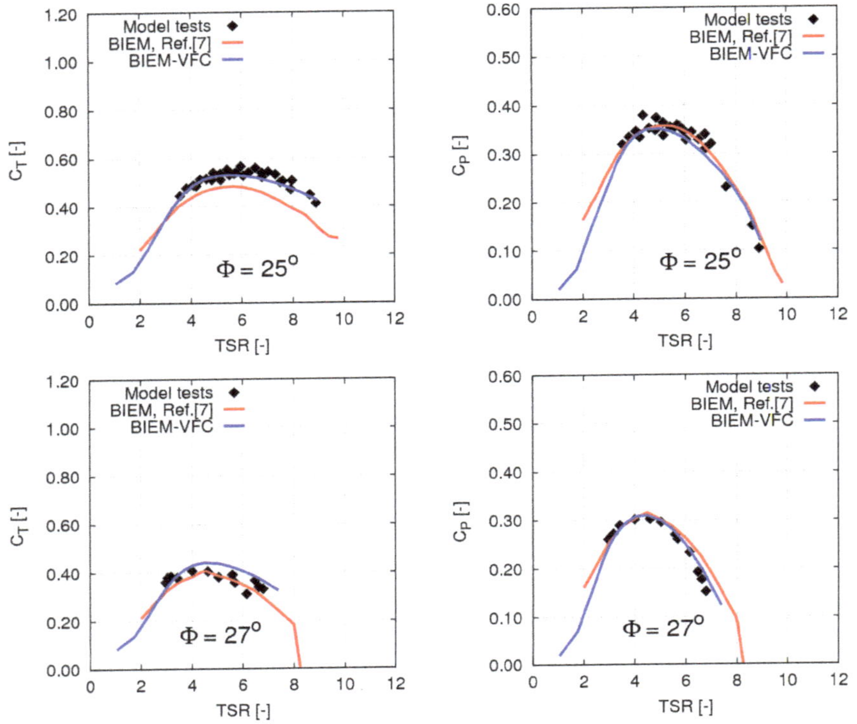

Figure 30. UoS-VP turbine performance predictions by BIEM-VFC compared to results from the BIEM model in [7]: thrust coefficient C_T (**left**); and power coefficient C_P (**right**). Pitch settings $\Phi = 20°$ to $27°$.

6. Discussion

The analysis of present results compared to reference data from the literature highlights the importance of adding a viscous flow correction to standard BIEM results obtained under inviscid-flow assumptions. Although simple and partially based on semi-empirical corrections, the proposed BIEM-VFC approach allows to significantly improve the reliability of turbine performance predictions by BIEM over a full range of operating conditions, including design and off-design blade pitch settings.

The main advantage of BIEM modelling compared to blade element momentum methods routinely used for marine turbines is the possibility to determine a consistent representation of the three-dimensional flow around rotor blades and hub in steady or unsteady flow. Pressure distributions on the blade surface can be used as input to predict the occurrence of cavitation and to estimate its detrimental effects in terms of vibrations, noise, erosion.

Nonetheless, a major weakness of the present BIEM-VFC approach is that viscosity correction applies only to blade loads and not to the potential flow solution as a whole. In particular, the correction does not apply to the intensity of the vortex sheet shed at blade trailing edge, nor to the induced velocity field necessary to evaluate the effective angle of attack. Neglecting these effects is expected to be the source of errors in performance predictions when the turbine operates at *TSR* lower than the peak power condition.

To overcome this limitation, a generalization of the present VFC model is the subject of work underway. Specifically, trailing vorticity distributions that are compatible with the correction of blade loads determined by the VFC scheme described in Section 2.2 can be obtained through an iterative procedure in which the direct relationship between axial and tangential force contributions by a blade element at radius r and turbine-induced velocity perturbation is derived by momentum and moment of momentum balance. Furthermore, the boundary integral representation (2) is generalised to include additional source terms according to the viscous/inviscid coupling methodology proposed in [26].

In addition to single turbine performance studies, the BIEM-VFC methodology is also applied to study the hydrodynamic behaviour of turbines operating in arrays. In this case, the inviscid-flow BIEM with VFC model is combined with a viscous flow solver (Reynolds-Averaged Navier-Stokes, RANS) to correctly describe the turbulent, vortical stream that characterizes the inflow to a turbine in the wake of similar devices placed upstream. An application of this combined BIEM-VFC and RANS computational methodology has been discussed by the authors in [27], where the interaction between two three-bladed turbines axially aligned with the upstream flow is analysed and numerical results are compared with experimental data from [28].

7. Conclusions

A computational methodology for the hydrodynamic analysis of horizontal axis marine current turbines has been presented, and results of a validation study have been discussed. The approach is based on a Boundary Integral Equation Model (BIEM) for inviscid flows that is combined with a trailing wake model specific for hydrokinetic turbines and with a viscous flow correction model (VFC) to include blade flow separation and stall effects on predicted hydrodynamic loads. The latter is derived by a semi-empirical approach in which inviscid-flow blade loads by BIEM are corrected on the basis of lift and drag properties of two-dimensional foils describing blade sections under equivalent three-dimensional flow conditions.

Numerical predictions by BIEM-VFC have been validated through comparisons with experimental data and with numerical results from the literature. The analysis highlights the capability of the proposed methodology to correctly describe turbine performance over a full range of operating conditions. Specifically, reliable predictions of turbine thrust, torque and power are obtained at medium/high tip speed ratio regimes, where blade flow is mostly attached, but also at relatively low tip speed ratio, where blade flow separation and stall determine thrust loss and drag crisis. More in details, good predictions of turbine performance are obtained for blade pitch settings close to design, while discrepancies for both thrust and torque (power) are observed in off-design conditions.

Comparing BIEM-VFC with other computational models in the literature, a key finding is that the accuracy of the proposed approach is aligned with blade element methods that are routinely used for the analysis and design of marine as well as wind turbines. Such a result is particularly important in that the present methodology based on BIEM provides a physically consistent description of the three-dimensional flow around a turbine in arbitrary onset flow, while blade element methods rely on taylored, case-dependent corrections for blade tip effects, for blade/hub interaction, number of blades. Well known limitations of blade element methods to analyse non-uniform flow conditions as well as to study turbine cavitation are also overcome through the more general description of turbine flow obtained by a BIEM approach.

Future work will address the generalization of the present VFC scheme to achieve trailing vorticity distributions and induced velocity distributions that are fully consistent with the viscosity correction applied on blade loads. Further validation studies will focus on the capability of the generalised BIEM-VFC model to predict turbine performance at low *TSR* and when turbine blades operate in off-design conditions.

Author Contributions: F.S. and Z.S. developed the original viscous flow correction and trailing wake models while D.C. and F.S. adapted the existing BIEM model. Z.S. and F.S. were responsible for computational model validation studies. All the authors contributed to results analysis and discussion.

Acknowledgments: The work described has been funded under the CNR-INSEAN Project ULYSSES (Underpinning LaboratorY for Studies on Sea Energy Systems). The authors wish to thank Benoit Gaurier for his kind support in the analysis of validation data from experiments at IFREMER. Part of validation data have been developed under the EU-FP7 MaRINET Project (Grant 262552).

Conflicts of Interest: The authors declare no conflict of interest.

Nomenclature

Symbol	Description	Units
c	Turbine blade chord	[m]
C_p	Pressure coefficient	[-]
C_P	Power coefficient	[-]
C_F	Friction coefficient	[-]
C_Q	Torque coefficient	[-]
C_T	Thrust coefficient	[-]
D	Turbine diameter, $2R$	[m]
D	Drag	[N]
$\mathcal{K}_D$	Drag correction factor	[-]
$\mathcal{K}_L$	Lift correction factor	[-]
L	Lift	[N]
n	Turbine rotational speed	[s^{-1}]
P	Turbine power	[W]
p_0	Reference pressure	[Pa]
Q	Turbine torque	[Nm]
R	Turbine radius	[m]
Re_r	Reynolds number, Equation (6)	[-]
T	Turbine thrust	[N]
TSR	Tip Speed Ratio	[-]
V	Freestream velocity	[ms^{-1}]
α	angle of attack	[deg]
ν	Kynematic viscosity	[m^2s^{-1}]
φ	Velocity scalar potential	[m^2s^{-1}]
Ω	Turbine rotational speed	[$rads^{-1}$]
ϕ	Wake (linear) pitch	[m]
Φ	Blade pitch	[deg]
ρ	Water density	[kgm^{-3}]

References

1. Hansen, M.O.L. *Aerodynamics of Wind Turbines*, 2nd ed.; Earthscan: London, UK, 2008.
2. Buhl, M.L., Jr. *A New Empirical Relationship between Thrust Coefficient and Induction Factor for the Turbulent Windmill State*; Technical Report NREL/TP-500-36834; National Renewable Energy Laboratory: Golden, CO, USA, 2005.
3. Shen, W.Z.; Mikkelsen, R.; Sœrensen, J.N.; Bak, C. Tip Loss Corrections for Wind Turbine Computations. *Wind Energy* **2005**, *8*, 457–475. [CrossRef]
4. Young, Y.L.; Motley, M.R.; Yeung, R.W. Three-Dimensional Numerical Modeling of the Transient Fluid-Structural Interaction Response of Tidal Turbines. *J. Offshore Mech. Arct. Eng.* **2010**, *132*, 1–12. [CrossRef]
5. He, L.; Xu, W.; Kinnas, S.A. Numerical Methods for the Prediction of Unsteady Performance of Marine Propellers and Turbines. In Proceedings of the ISOPE 2011 Twenty-First International Offshore and Polar Engineering Conference, Maui, HI, USA, 19–24 June 2011.

6. Baltazar, J.; Falcão de Campos, J.A.C. Unsteady Analysis of a Horizontal Axis Marine Current Turbine in Yawed Inflow Conditions With a Panel Method. In Proceedings of the SMP'09 First International Symposium on Marine Propulsors, Trondheim, Norway, 22–24 June 2009; Koushan, K., Steen, S., Eds.; MARINTEK (Norwegian Marine Technology Research Institute): Trondheim, Norway, 2009.
7. Baltazar, M.J.; Falcão de Campos, J.A.C. Hydrodynamic design and analysis of horizontal axis marine current turbines with lifting line and panel methods. In Proceedings of the OMAE2011 ASME 30th International Conference on Ocean, Offshore and Arctic Engineering, Rotterdam, The Netherlands, 19–24 June 2011; The American Society of Mechanical Engineers: Rotterdam, The Netherlands, 2011; Volume 5, pp. 453–465.
8. Salvatore, F.; Testa, C.; Greco, L. A Viscous/Inviscid Coupled Formulation for Unsteady Sheet Cavitation Modelling of Marine Propellers. In Proceedings of the CAV 2003 Fifth International Symposium on Cavitation, Osaka, Japan, 1–4 November 2003.
9. Salvatore, F.; Greco, L.; Calcagni, D. Computational analysis of marine propeller performance and cavitation by using an inviscid-flow BEM model. In Proceedings of the SMP'11 Second International Symposium on Marine Propulsion, Hamburg, Germany, 15–17 June 2011.
10. Pereira, F.; Salvatore, F.; Di Felice, F. Measurement and Modelling of Propeller Cavitation in Uniform Inflow. *J. Fluids Eng.* **2004**, *126*, 671–679. [CrossRef]
11. Gaurier, B.; Germain, G.; Facq, J.-V.; Johnstone, C.M.; Grant, A.D.; Day, A.H.; Nixon, E.; Di Felice, F.; Costanzo, M. Tidal energy Round Robin tests: Comparisons between towing tank and circulating tank results. *Int. J. Mar. Energy* **2015**, *12*, 87–109. [CrossRef]
12. EU-FP MaRINET Project. Available online: https://cordis.europa.eu/project/rcn/98372_en.html (accessed on 29 March 2018).
13. Bahaj, A.S.; Molland, A.F.; Chaplin, J.R.; Batten, W.M.J. Power and thrust measurements of marine current turbines under various hydrodynamic flow conditions in a cavitation tunnel and a towing tank. *Renew. Energy* **2007**, *32*, 407–426. [CrossRef]
14. Morino, L. Boundary Integral Equations in Aerodynamics. *Appl. Mech. Rev.* **1993**, *46*, 445–466. [CrossRef]
15. Greco, L.; Salvatore, F.; Di Felice, F. Validation of a Quasi—Potential Flow Model for the Analysis of Marine Propellers Wake. In Proceedings of the ONR-2004 Twenty-fifth ONR Symposium on Naval Hydrodynamics, St. John's, NL, Canada, 8–13 August 2004; National Academies Press: Washington, DC, USA, 2004.
16. Carlton, J.S. *Marine Propellers and Propulsion*, 1st ed.; Butterworth–Heinemann: Oxford, UK, 1994.
17. Betz, A. The Maximum of the Theoretically Possible Exploitation of Wind by Means of a Wind Motor. *Wind Eng.* **2013**, *37*, 441–446. (Translation by Hamann et al. of original paper published in 1920 by the author in German) [CrossRef]
18. Okulov, V.L.; Sørensen, J.N. Maximum efficiency of wind turbine rotors using Joukowsky and Betz approaches. *J. Fluid Mech.* **2010**, *649*, 497–508. [CrossRef]
19. Micek, P.; Gaurier, B.; Germain, G.; Pinon, G.; Rivoalen, E. Experimental study of the turbulence intensity effects on marine current turbines behaviour. Part I: One single turbine. *Renew. Energy* **2014**, *66*, 729–746. [CrossRef]
20. Del Frate, C.; Di Felice, F.; Alves Pereira, F.; Romano, G.P.; Dhomé, D.; Allo, J.-C. Experimental Investigation of the turbulent flow behind a horizontal axis tidal turbine. In *Progress in Renewable Energies Offshore, Proceedings of the RENEW 2016 2nd International Conference on Renewable Energies Offshore, Lisbon, Portugal, 24–26 October 2016*; CRC Press: Boca Raton, FL, USA, 2016.
21. Drela, M. XFOIL: An Analysis and Design System for Low Reynolds Number Airfoils. In *Low Reynolds Number Aerodynamics, Proceedings of the Conference Notre Dame, IN, USA, 5–7 June 1989*; Lecture Notes in Engineering, Bd. 54; Springer: Berlin/Heidelberg, Germany, 1989; pp. 1–12; ISBN 978-3-540-51884-6.
22. Sheldahl, R.E.; Klimas, P.C. *Aerodynamic Characteristics of Seven Symmetrical Airfoil Sections Through 180-Deg. Angle of Attack for Use in Aerodynamic Analysis of Vertical Axis Wind Turbines*; Technical Report SAND80-2114; SANDIA: Albuquerque, NM, USA, 1981,
23. Abbott, I.R.; von Doenhoff, A.E. *Summary of Airfoil Data*; NACA Report 824; NACA: Boston, MA, USA, 1945.
24. Molland, A.F.; Bahaj, A.S.; Chaplin, J.R.; Batten, W.M.J. Measurements and predictions of forces, pressures and cavitation on 2D sections suitable for marine current turbines. *J. Eng. Marit. Environ.* **2004**, *218*, 127–138.
25. Bahaj, A.; Batten, W.; McCann, G. Experimental verifications of numerical predictions for the hydrodynamic performance of horizontal axis marine current turbines. *Renew. Energy* **2007**, *32*, 2479–2490. [CrossRef]

26. Morino, L.; Salvatore, F.; Gennaretti, M. A New Velocity Decomposition for Viscous Flows: Lighthill's Equivalent-Source Method Revisited. *Comput. Methods Appl. Mech. Eng.* **1999**, *173*, 317–336. [CrossRef]
27. Salvatore, F.; Calcagni, D.; Sarichloo, Z. Development of a Viscous/Inviscid Hydrodynamics Model for Single Turbines and Arrays. In Proceedings of the EWTEC 2017 Twelfth European Wave and Tidal Energy Conference, Cork, Ireland, 27 August–1 September 2017.
28. Mycek, P.; Gaurier, B.; Germain, G.; Pinon, G.; Rivolaen, E. Experimental Study of the Turbulence Intensity Effects on Marine Current Turbines Behaviour. Part II: Two Interacting Turbines. *Renew. Energy* **2014**, *68*, 876–892. [CrossRef]

Journal of
*Marine Science
and Engineering*

MDPI

Article

Numerical Simulation and Uncertainty Analysis of an Axial-Flow Waterjet Pump

Ji-Tao Qiu [1], Chen-Jun Yang [1,*], Xiao-Qian Dong [1], Zong-Long Wang [2], Wei Li [1] and Francis Noblesse [1]

1 State Key Laboratory of Ocean Engineering, Collaborative Innovation Center for Advanced Ship and Deep-Sea Exploration, Shanghai Jiao Tong University, Shanghai 200240, China; qiujitao@sjtu.edu.cn (J.-T.Q.); xiaoqiandong0330@163.com (X.-Q.D.); wli@sjtu.edu.cn (W.L.); noblfranc@gmail.com (F.N.)
2 Marine Design and Research Institute of China (MARIC), Shanghai 200011, China; wzonglong@126.com
* Correspondence: cjyang@sjtu.edu.cn

Received: 30 March 2018; Accepted: 5 June 2018; Published: 11 June 2018

Abstract: Unsteady Reynolds-averaged Navier–Stokes simulations of an axial-flow pump for waterjet propulsion are carried out at model scale, and the numerical uncertainties are analyzed mainly according to the procedure recommended by the twenty-eighth International Towing Tank Conference. The two-layer realizable k-ε model is adopted for turbulence closure, and the flow in viscous sub-layer is resolved. The governing equations are discretized with second-order schemes in space and first-order scheme in time and solved by the semi-implicit method for pressure-linked equations. The computational domain is discretized into block-structured hexahedral cells. For an axial-flow pump consisting of a seven-bladed rotor and a nine-bladed stator, the uncertainty analysis is conducted by using three sets of successively refined grids and time steps. In terms of the head and power over a range of flow rates, it is verified that the simulation uncertainty is less than 4.3%, and the validation is successfully achieved at an uncertainty level of 4.4% except for the lowest flow rate. Besides this, the simulated flow features around rotor blade tips and between the stator and rotor blade rows are investigated.

Keywords: waterjet propulsion; axial-flow pump; unsteady; RANS; uncertainty analysis; tip clearance flow

1. Introduction

Due to high propulsive efficiency and superior cavitation performance, waterjet propulsion is particularly suited for high-speed crafts. In recent years, it has been gradually applied to large-scale ships; in such cases, it becomes important to accurately evaluate the hydraulic performance of the pump in order to predict the ship's powering performance correctly. The waterjet propulsion pumps have been designed by means of potential flow-based theoretical methods since the last century. To improve the cavitation performance of the rotor, experimental and numerical investigations are often conducted to analyze the tip-clearance flow apart from optimizing blade loading distribution and section profiles. With the rapid advances in computer hardware capability and CFD software technology, viscous flow simulation has increasingly become preferred to assist in designing the waterjet propulsion pumps, as it is capable of providing more detailed and realistic flow information in addition to more accurate hydraulic performance predictions than those by potential flow methods. Meanwhile, however, the reliability of CFD simulation has become an issue that has drawn increased attention from both research and design communities. Starting with the convergence study of discretization scheme and grid size, the theories and methodologies are gradually being developed for analyzing the numerical uncertainty in CFD simulation.

In 1998, the American Institute of Aeronautics and Astronautics (AIAA) presented the first CFD uncertainty analysis procedure [1], which consists of two processes: verification and validation (V&V); Roache [2,3] published a monograph on CFD uncertainty analysis, and proposed the grid convergence index (GCI) method for verification with at least two sets of grids. In 1999, the 22nd International Towing Tank Conference (ITTC) [4,5] introduced a preliminary procedure for CFD uncertainty analysis based on the work of Stern et al. [6] and Coleman et al. [7], where the generalized Richardson extrapolation (RE) approach [8] was employed to evaluate the uncertainty of numerical simulations. In 2002, according to the GCI method, the 23rd ITTC revised the procedure by including the safety factor approach in the existing ITTC procedure [9]. In 2017, the 28th ITTC introduced the least squares root (LSR) method for error estimation [10] based on the studies by Eça et al. [11] and Larsson et al. [12]. Besides this, Shen et al. [13] discussed the probability distribution of CFD simulation results, analyzed the sources of numerical uncertainty, and proposed an uncertainty analysis method based on orthogonal experiment design and variance analysis.

Simonsen et al. [14] conducted an uncertainty analysis of steady Reynolds-averaged Navier–Stokes (RANS) simulations for the tanker Esso Osaka in straight-ahead, pure drift, and static rudder conditions according to the procedure proposed by Stern et al. [15]. As far as the resistance and hydrodynamic moments are concerned, the analysis results for the appended hull indicate that the validation is at a high level but not achieved for all the cases investigated. Zhang et al. [16] simulated the SUBOFF submarine model by using three sets of successively refined grids and analyzed the numerical uncertainty of hull surface pressure distributions according to the ITTC procedures. Yang et al. [17] analyzed the numerical uncertainty due to grid size in the CFD simulation of an open-water propeller according to the ITTC procedures and investigated the effects of turbulence models. It is noted that, as ships and propulsors have a complex geometry, extremely fine grids (and time steps for unsteady flows) are needed in order for the solution to fall in the asymptotic range. Limited by hardware resources, three sets of grids are typically used for the RE-based analysis, and it seems difficult for the solution to reach the asymptotic range in complicated three-dimensional flows.

The uncertainty analysis can be done more rigorously for two-dimensional flows around simple geometries. Rosetti et al. [18] simulated the unsteady flow around a two-dimensional circular cylinder over a large range of the Reynolds number using five sets of grids and five time-step sizes and analyzed the simulation uncertainties. It was shown that the differences between numerical and experimental results stemmed from modeling errors instead of numerical ones. Diskin et al. [19] conducted a grid-convergence study of RANS simulations for the NACA 0012 airfoil and a flat plate by using three in-house codes and concluded that establishing an asymptotic converge order was more difficult than expected even with very fine grids. The authors also found that the grid resolution in the vicinity of geometric singularities, such as pointed leading/trailing edge, was essential to improve the accuracy and convergence property of CFD simulation.

The internal flow of the axial-flow pump is complicated due to the clearance flow around rotor tips and the unsteady interaction between rotor and stator blade rows, and the fidelity of CFD simulation results may influence the designer's judgment of the pump's cavitation performance. Therefore, the analysis of numerical uncertainties is very important, but the related research work is insufficient to date. In this work, unsteady flow simulations of an axial-flow pump for waterjet propulsion are carried out at model scale by solving the unsteady Reynolds-averaged Navier–Stokes (RANS) equations, and the numerical uncertainties are analyzed mainly based on the procedure recommended by the 28th ITTC [10]. Typical operating conditions are selected for the evaluation of the numerical uncertainties arising from spatial and temporal discretization. Besides this, the simulated flow features around rotor blade tips and between the stator and rotor blade rows are investigated.

2. CFD Uncertainty Analysis Procedure

The uncertainty analysis procedure recommended by the ITTC assesses CFD uncertainty at a 95% confidence level, i.e., the probability that the interval $(-U, +U)$ contains the error δ is 95%, where U is the magnitude of uncertainty. Mainly based on the ITTC procedure [10], the V&V processes that are specific to the present problem are detailed in this section.

2.1. Verification

Verification is a process for estimating the uncertainty and error in numerical simulations which assesses whether the modeling equations have been solved correctly [20]. Being defined as the difference between a simulation result S and the truth T, the simulation error δ_S is composed of the numerical error δ_{SN} and the modeling error δ_{SM}, viz, $\delta_S = S - T = \delta_{SN} + \delta_{SM}$. When a numerical solution falls in the asymptotic range, the δ_{SN} can be estimated together with the error contained in that estimate via the corrected approach; otherwise, only the numerical uncertainty U_{SN} can be estimated to provide a boundary for the δ_{SN} via the uncorrected approach. In this work, only the uncorrected approach is used; details about the corrected approach can be found in [10].

The numerical uncertainty is expressed as

$$U_{SN}^2 = U_I^2 + U_G^2 + U_T^2 + U_P^2 \tag{1}$$

where the subscripts I, G, T and P denote the uncertainties due to iteration, grid size, time step size, and other parameters, respectively. For an axial-flow pump, it is more accurate to simulate the unsteady flow arising from rotor/stator interactions. Then, the uncertainties due to the U_G and U_T in (1) are replaced by a combined discretization uncertainty U_{GT} that is evaluated by refining the grids and time-step simultaneously at uniform refinement ratios. The U_{SN} is alternatively expressed as

$$U_{SN}^2 = U_I^2 + U_{GT}^2 + U_P^2 \tag{2}$$

In this work, the uncertainty due to other parameters, U_P, is not investigated and will be neglected in the present analysis.

The iteration uncertainty U_I is defined as

$$U_I = \frac{1}{2}(\widetilde{S}_{\max} - \widetilde{S}_{\min}) \tag{3}$$

where $\widetilde{S}_{\max}$ and $\widetilde{S}_{\min}$ denote the maximum and minimum values of simulated results in the last two oscillation periods according to [10]. In this work, however, the iteration uncertainty is estimated conservatively by using the data over a complete revolution of the rotor, since the simulated integral quantities oscillate with time randomly and at high frequencies. Besides, it is necessary to exclude the contributions from the simulated physical oscillations in the integral quantities, although they are usually quite small in magnitude.

For the verification based on three solutions, the convergence ratio R is defined as

$$R = \varepsilon_{21}/\varepsilon_{32} = (S_2 - S_1)/(S_3 - S_2) \tag{4}$$

where S_1, S_2 and S_3 denote the solutions yielded from fine, medium, and coarse grids and time steps, respectively. The convergence is monotonic when $0 < R < 1$ but oscillatory when $-1 < R < 0$. If $|R| > 1$ the solutions diverge and the uncertainty cannot be estimated. For monotonic convergence, the RE approach is used to estimate the U_{GT}. For oscillatory convergence, the U_{GT} is estimated by

$$U_{GT} = (S_U - S_L)/2 \tag{5}$$

where S_U and S_L denote, respectively, the maximum and minimum values in the simulation results. Note that the U_{GT} yielded from (5) is just a rough estimate for the boundary of δ_{SN}.

According to the RE approach, the estimated error δ^*_{RE} is calculated by

$$\delta^*_{RE} = \frac{\varepsilon_{21}}{r^p - 1} \tag{6}$$

where r and p denote the parameter refinement ratio and the observed order of accuracy, respectively. The observed order of accuracy is determined by

$$p = \frac{\ln(\varepsilon_{32}/\varepsilon_{21})}{\ln(r)} \tag{7}$$

The correction factor, C, is defined as

$$C = \frac{r^p - 1}{r^{p_{est}} - 1} \tag{8}$$

where p_{est} is an estimate for the limiting order of accuracy of the first term as the grid and time-step sizes go to zero and the asymptotic range is reached, i.e., $C \to 1$. The uncorrected approach is used to estimate the U_{SN} only. The U_{GT} is estimated by the following formulae that are proposed by Wilson et al. [21] and recommended in [10],

$$U_{GT} = \begin{cases} [9.6(1-C)^2 + 1.1]|\delta^*_{RE}|, & |1-C| < 0.125 \\ (2|1-C| + 1)|\delta^*_{RE}|, & |1-C| \geq 0.125 \end{cases} \tag{9}$$

As pointed out in the ITTC procedure [10], (9) makes a conservative estimation when $C \to 1$ or $C < 1$.

2.2. Validation

Validation is defined as a process for assessing simulation modeling uncertainty U_{SM} by using benchmark experimental data and, when conditions permit, estimating the modeling error δ_{SM} itself. This process investigates whether the correct equations are solved [20]. The comparison error E is defined as the difference between experimental data D and simulation result S, i.e.,

$$E = D - S = T + \delta_D - (T + \delta_{SM} + \delta_{SN}) = \delta_D - (\delta_{SM} + \delta_{SN}) \tag{10}$$

The validation uncertainty U_V is defined as

$$U_V^2 = U_D^2 + U_{SN}^2 \tag{11}$$

where U_D denotes experimental uncertainty. To determine whether the validation has been achieved, the comparison error E is compared to the validation uncertainty U_V and the programmatic validation requirement U_{reqd}. If the three variables are unequal to each other, one of the following cases must be true:

(1) $|E| < U_V < U_{reqd}$
(2) $|E| < U_{reqd} < U_V$
(3) $U_{reqd} < |E| < U_V$
(4) $U_V < |E| < U_{reqd}$
(5) $U_V < U_{reqd} < |E|$
(6) $U_{reqd} < U_V < |E|$

In the first three cases, the validation is achieved at the U_V level, but the modeling error cannot be estimated as the comparison error is below the uncertainty (noise) level. Particularly, in the first case the validation is successfully achieved at a level below U_{reqd}.

In the last three cases, the validation is not achieved but the modeling error can be estimated. If $|E| >> U_V$, the modeling error dominates and is approximately equal to E. In case (4) the validation is successful at the $|E|$ level.

3. Unsteady RANS Simulation of an Axial-Flow Pump

Numerical simulations are conducted for a model-scale axial-flow pump by solving the unsteady RANS equations for incompressible fluids. The pump consists of a seven-bladed rotor and a nine-bladed pre-swirl stator. The shroud diameter d is 300 mm. The tip clearance of the rotor is 0.3 mm. The axial spacing between the rotor and the stator is 30 mm. Figure 1 shows the geometric model of the rotor and the stator.

Figure 2 shows the computational domain, which is bounded by a circular cylindrical shroud 16 times d in length. The rotor is located at the longitudinal center of the domain. The rectangular area marked out in Figure 2 is a sub-domain containing the rotor (as well as a portion of the shroud surface and the hub), which is defined in a coordinate system attached to the rotor. The remaining portions of the computational domain are defined in an earth-fixed coordinate system.

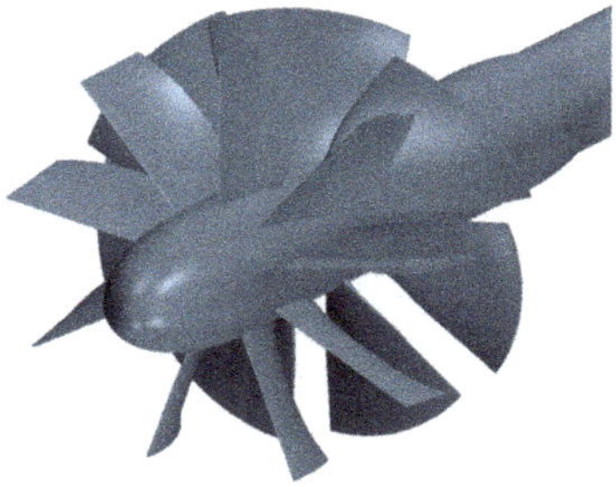

Figure 1. The geometric model of the seven-bladed rotor and the nine-bladed pre-swirl stator. The stator blades are connected to the shroud surface without clearance. The shroud surface is omitted.

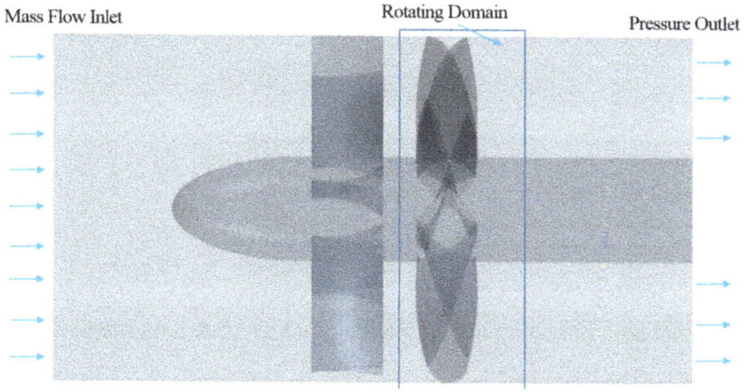

Figure 2. The computational domain for the axial-flow pump. The domain is a circular cylinder bounded by the shroud surface, 16 times the shroud diameter in length. The rotor is located at the longitudinal center of the domain, while the stator is located upstream of the rotor.

The computational domain is discretized into block-structured hexahedral cells by means of the grid generator ICEM CFD 16.0. Figure 3 illustrates the grid topology around rotor blade sections. The thin area around blade surfaces are discretized with C-grid layers, the area in between the C-grid areas of adjacent blades are discretized with L grids, and the small area upstream of the leading edge is discretized with H grids. The area around stator blade sections are discretized with H grids. In a spanwise direction, H grids are used for both the rotor and the stator. Figure 4 shows the grid structure in the tip clearance of a rotor, which is the same in topology along the chord.

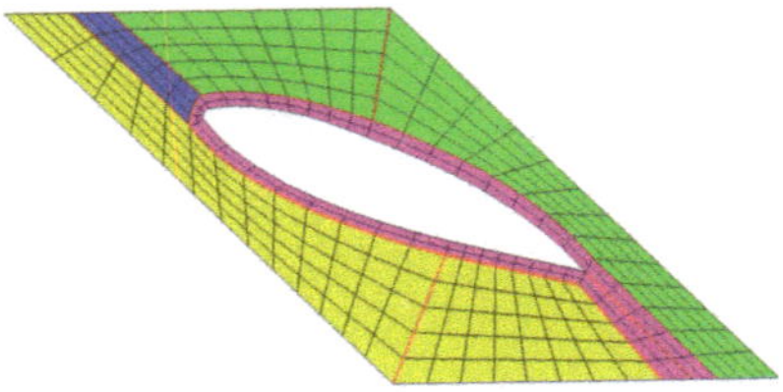

Figure 3. Grid topology around rotor blade sections.

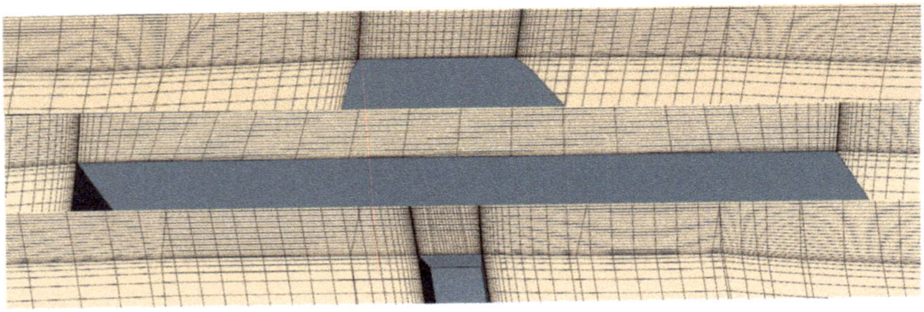

Figure 4. The grid structure in the tip clearance of a rotor. The sections perpendicular to the nose-tail chord of the tip are shown at 10% (**top**), 50% (**middle**), and 90% (**bottom**) chord length.

Three sets of successively refined grids are generated with a uniform grid refinement ratio, $r_G = \Delta h_2/\Delta h_1 = \Delta h_3/\Delta h_2 = \sqrt{2}$, where h_3, h_2 and h_1 denote, respectively, the sizes of coarse grid G_3, medium grid G_2, and fine grid G_1. Figure 5 shows the surface grids for rotor and stator blades. The key parameters of the computational grids are listed in Table 1.

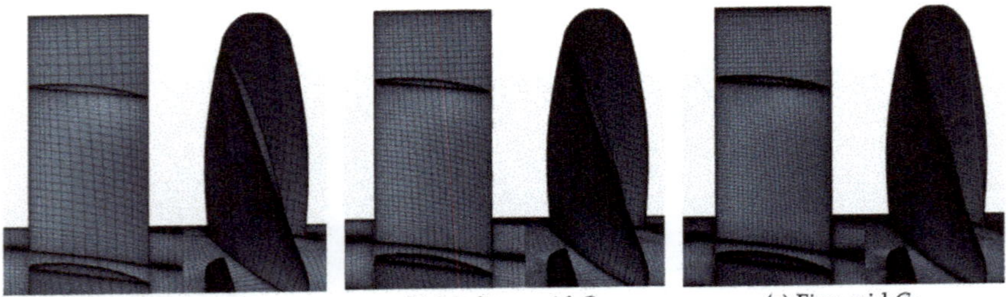

(**a**) Coarse grid G_3 (**b**) Medium grid G_2 (**c**) Fine grid G_1

Figure 5. The surface grids for rotor and stator blades.

Table 1. Key parameters of the computational grids.

Grid ID	Maximum Cell Size on Blade Surface (mm)		First-Layer Cell Height from Blade Surface (mm)		Number of Cells in Tip-Clearance		Total Number of Cells (Million)
	Stator	Rotor	Stator	Rotor	Radial	Circumferential	
G_3	5	4	0.02	0.01	10	16	1.79
G_2	3.54	2.83	0.014	0.007	15	22	4.87
G_1	2.5	2	0.01	0.005	20	30	13.17

As illustrated in Figure 2, the upstream boundary condition is set as the mass flow inlet, where the flow rate is prescribed; the downstream boundary condition is set as the pressure inlet, where the pressure relative to the reference pressure of the flow domain (atmospheric pressure) is set to zero. All the body surfaces are set as non-slip boundaries. The part of the shroud surface in the rotating sub-domain containing the rotor is set as stationary relative to the fixed coordinate system. At the inlet and the outlet, the turbulence intensity and the turbulent viscosity ratio are set to 2% and 2, respectively.

The inlet boundary condition is an issue that can influence the modeling uncertainty in present simulations since the computational domain is just a part of the closed loop used in the pump tests. In our modeling approach, the computational domain is made long enough (16 times the shroud diameter) to allow the boundary layer on the shroud surface to develop fully before the flow arrives at the stator. Figure 6 shows profiles of axial flow velocity V_a at three streamwise stations when $Q = 0.42\ m^3/s$. It seems that the boundary layer becomes fully developed from 1.7 m downstream of the inlet (0.45 m to the nose of shaft cap), and the inlet velocity profile prescribed by the software would have little influence on the modeling uncertainty.

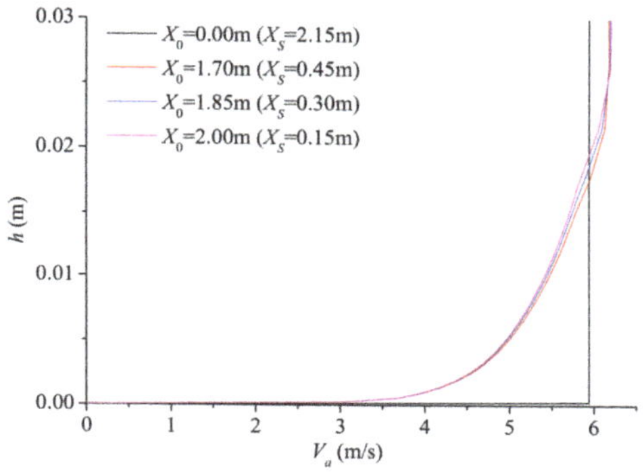

Figure 6. Simulated axial velocity profiles in shroud-surface boundary layer, where h is the normal distance from the shroud surface, X_0 is the distance measured from the inlet, and X_S is the distance to the nose of shaft cap.

The unsteady RANS simulations are carried out using the CFD software STAR-CCM+ 12.02. The two-layer realizable k-ε model is used for turbulence closure, and the flow in viscous sub-layer is resolved. The two-layer approach was first proposed by Rodi [22]. As another method of applying the k-ε model in the viscous sub-layer and the buffer layer, it works with either low-Reynolds number type meshes ($y^+ \sim 1$) or wall-function type meshes ($y^+ > 30$). In the layer next to the wall, functions of the y^+ are used to specify the turbulent dissipation rate ε and the turbulent viscosity μ_t. Otherwise, the transport equation for ε is solved. The equation for k is solved across the entire flow domain. The realizable k-ε model was proposed by Shih et al. [23], which consists of a new model equation for the turbulent dissipation rate and a new eddy viscosity formulation. The model was validated for different flow types, including rotating homogeneous shear flows, and found to perform much better than the standard k-ε model.

The governing equations are discretized with second-order schemes in space and a first-order scheme in time and solved by the semi-implicit method for pressure-linked equations (SIMPLE). It would be more accurate to use a second-order scheme for the discretization in time, but convergence problems were encountered when the flow rate was low. To be consistent with the order of discretization accuracy in space [11,24], three time-step sizes having a uniform refinement ratio, $r_T = \Delta t_2/\Delta t_1 = \Delta t_3/\Delta t_2 = 2$, are used in the unsteady simulations, where Δt_3, Δt_2, and Δt_1 denote coarse, medium, and fine time-step sizes, respectively. Note that the grids and time steps must be coarsened or refined simultaneously. To expedite convergence, steady-flow simulations are used to initialize unsteady simulations. In each time step, 20 iterations are performed to reduce the residuals to an acceptable level.

4. Numerical Uncertainty Analysis for the Axial-Flow Pump

The analysis of numerical uncertainty in the present RANS simulations is conducted for the axial-flow pump described in Section 3. Three volumetric flow rates ($Q = 0.35\ \text{m}^3/\text{s}$–$0.471\ \text{m}^3/\text{s}$) are considered which cover a wide range of the pump's operating conditions. The uncertainties in the simulated head and power are evaluated.

The head H and power P of a pump are defined as

$$\left.\begin{aligned} H &= \frac{p_d - p_u}{\rho g} + \frac{V_d^2 - V_u^2}{2g} \\ P &= \frac{2\pi N M}{60} \end{aligned}\right\} \tag{12}$$

where p_d and p_u denote shroud-surface pressures averaged over the circumferences at $2d$ downstream and $2d$ upstream of the rotor, respectively; V_d and V_u denote the axial velocities averaged over the shroud cross sections at $2d$ downstream and $2d$ upstream of the rotor, respectively. The locations where the pressures and axial velocities are numerically evaluated are the same as those in the experiments for the pump model considered here. In (12) the gravitational acceleration and the density of water are denoted by g and ρ, respectively; the torque and the rotational speed (r/min) of the rotor are denoted by M and N, respectively.

4.1. Verification

The rotational speed of the rotor is set to 1450 r/min for the three flow rates considered, which is same as that in the model experiments [25]. The Reynolds number based on the relative inflow speed and the chord length at $0.7R_{tip}$ is 2.3×10^6, where R_{tip} is the tip radius of the rotor. The coarse time step size Δt_3 is set to 1.1494×10^{-4} s, which corresponds to a blade angular displacement of 1° per time step. The medium and fine time step sizes, Δt_2 and Δt_1, correspond to 0.5° and 0.25° blade angular displacement per time step, respectively. To resolve the flow in viscous sub-layer, the near-wall grid layers need to satisfy $y^+ \sim 1$. Corresponding to the grid sizes as listed in Table 1, the ranges of surface-averaged y^+ are given in Table 2 for the flow rates simulated. Figure 7 shows the limiting streamlines on rotor and stator blade surfaces when $Q = 0.42\ \text{m}^3/\text{s}$. On rotor blade surfaces, flow separation occurs mainly on the suction side, in inner radii and close to the trailing edge. On stator blade surfaces, the flow is converging on the suction side but diverging on the pressure side; flow separation occurs on the suction side only, in close proximity to the trailing edge, and from about 40% span to the tip. It seems that the stator blade geometry may need to be improved. The flow patterns simulated with the coarse grid G_3 and the fine grid G_1 are quite similar to each other.

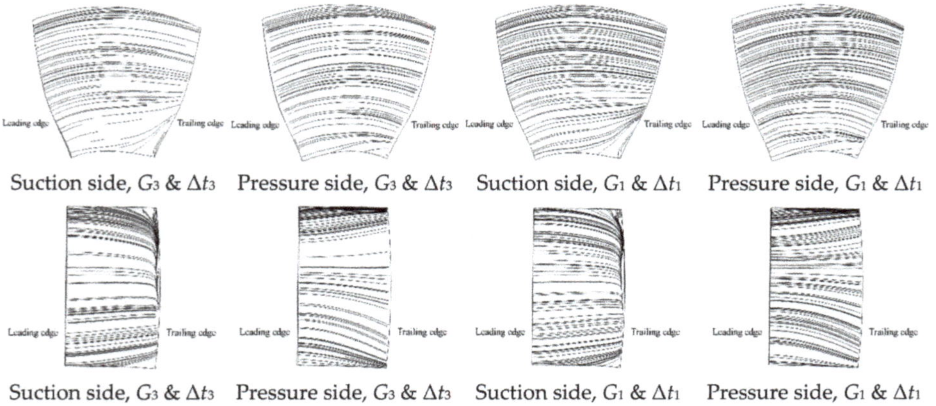

Figure 7. Limiting streamlines on rotor (**top row**) and stator (**bottom row**) blade surfaces, $Q = 0.42\ \text{m}^3/\text{s}$.

Table 2. Range of the surface-averaged y^+, $Q = 0.35$ m^3/s–0.471 m^3/s.

Grid ID	Stator	Rotor	Tip Clearance
G_3	3.2–4.2	1.9–2.3	5.7–5.8
G_2	2.3–3.0	1.4–1.6	4.7–4.8
G_1	1.7–2.3	1.0–1.3	3.3–3.4

Figure 8 shows an example of the residuals in the last few time steps before convergence. The residual in continuity drops to 10^{-5}–10^{-6}, those in velocity components to 10^{-5}–10^{-7}, and those in k and ε to 10^{-6}–10^{-8} when the solution converges. The speed of convergence is slow at low flow rates.

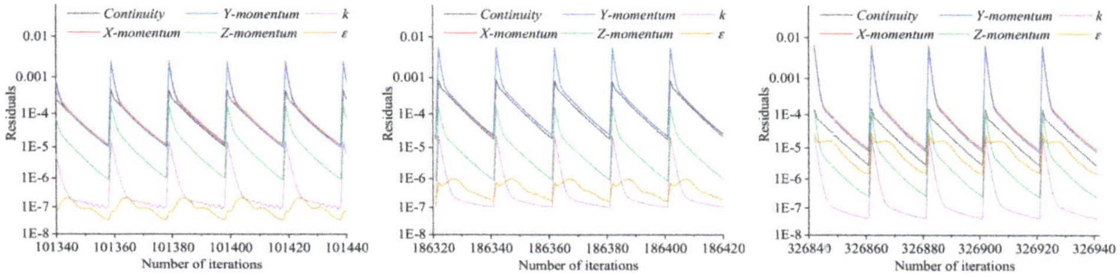

(a) G_3 & Δt_3 **(b)** G_2 & Δt_2 **(c)** G_1 & Δt_1

Figure 8. Example of residuals in the last few time steps before convergence, $Q = 0.42$ m^3/s.

Apart from the residuals, the head and power averaged over a complete revolution of the rotor are used to decide whether the simulation has converged, since the interactions between the rotor and the stator are time-dependent despite the fact that the inflow is uniform. Specifically, the time-averaged head and power of two consecutive revolutions are compared, and when the relative differences are less than 0.1%, results of the last revolution are taken as the final solution. Figure 9 shows the convergence history of the head H and the power P for $Q = 0.42$ m^3/s as an example. The heads differ by less than 0.05% in the last two revolutions for the three solutions, but speed of convergence is much lower than that of the power.

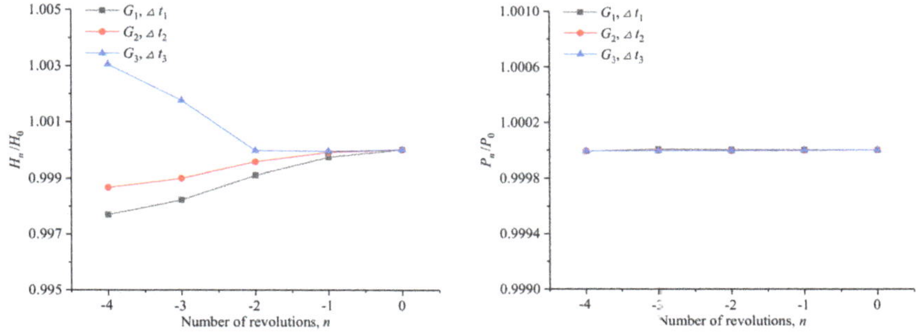

Figure 9. The convergence history of the averaged head H (**left**) and power P (**right**) in the last five revolutions before convergence, $Q = 0.42$ m^3/s. The abscissa, n, denotes the number of revolutions relative to the last revolution ($n = 0$). The ordinates H_n and P_n denote the head and power of the nth revolution, respectively.

Table 3 shows a comparison of the simulation results and the experimental data [25]. The head and power are mostly under-predicted, and the comparison error ranges between −3.5% and −0.6% for the head and −6.0% and 0.2% for the power.

Table 3. Comparison of simulation results and experimental data [25] for the axial-flow pump.

Q (m³/s)	Grid, Time-Step Size	Simulation Result S		Experimental Data D		Comparison Error E (%D)	
		H (m)	P(kW)	H (m)	P (kW)	H	P
0.35	G_3, Δt_3	7.320	29.473	7.450	31.341	−1.47	−5.96
	G_2, Δt_2	7.373	29.541			−1.04	−5.74
	G_1, Δt_1	7.406	29.584			−0.59	−5.61
0.42	G_3, Δt_3	5.221	25.343	5.400	25.957	−3.31	−2.36
	G_2, Δt_2	5.247	25.414			−2.83	−2.09
	G_1, Δt_1	5.267	25.473			−2.45	−1.87
0.471	G_3, Δt_3	3.476	20.341	3.600	20.305	−3.44	0.18
	G_2, Δt_2	3.521	20.456			−2.18	0.75
	G_1, Δt_1	3.546	20.513			−1.49	1.02

The iteration uncertainty is evaluated by (3) according to the time-domain results in the last revolution of the rotor. Note that, due to the interactions between the rotor and the stator, the simulated unsteady head and power contain the components at a number of frequencies. The interaction components should be removed when evaluating the iteration uncertainty. According to the relation given by Strasberg et al. [26] for contra-rotating propellers, the interaction components exist at the following frequencies:

$$f = (kZ_R + mZ_S)N/2 \tag{13}$$

where $kZ_R = mZ_S$, k and m are positive integers; Z_R and Z_S are blade numbers of the rotor and the stator, respectively; N is the rotational speed (shaft frequency) of the rotor. For the axial-flow pump considered here, $Z_R = 7$, $Z_S = 9$, the lowest frequency of the interaction components is 63 times the shaft frequency. To remove the interaction components, Fourier analyses are performed for the simulated unsteady head and power first; then, the amplitudes at the interaction frequencies are set to zero and the time series are reconstructed. Figure 10 shows a comparison of the time series before and after removing the interaction components. For the combination of rotor and stator blade numbers considered here, the interaction components have little influence on the simulated total fluctuations.

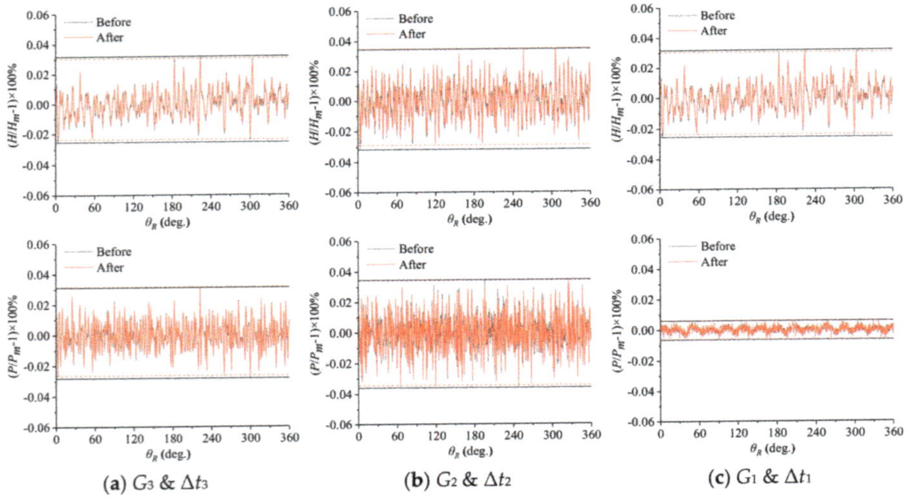

Figure 10. Comparison of the unsteady head H (**top row**) and power P (**bottom row**) before (black lines) and after (red lines) removing the components arising from stator/rotor interaction. The abscissa, θ_R, denotes the angular position of a rotor blade. The head and power are expressed respectively in percents relative to H_m and P_m, the averages over a complete revolution. $Q = 0.42$ m^3/s.

The reconstructed time-series are then used to estimate the iteration uncertainties. Table 4 shows the results, where S denotes the simulation results as listed in Table 3.

Table 4. Iteration uncertainties in the simulations.

Q (m^3/s)		$G_3, \Delta t_3$ (%S)	$G_2, \Delta t_2$ (%S)	$G_1, \Delta t_1$ (%S)
0.35		0.03	0.02	0.02
0.42	H	0.03	0.03	0.01
0.471		0.04	0.04	0.01
0.35		0.03	0.02	0.01
0.42	P	0.03	0.03	0.01
0.471		0.04	0.04	0.01

Table 5 shows the analysis results for U_{GT}, the uncertainties due to grid and time-step sizes. The following can be observed:

(1) The convergence ratio is $0 < R < 1$, which indicates the simulation results converge monotonically when the discretization in time and space is refined simultaneously and consistently. This result justifies the use of the generalized Richardson extrapolation (RE) for evaluating the observed order of accuracy p and the estimated error δ^*_{RE}.

(2) The estimated limiting order of accuracy p_{est} is set to 2, since the governing equations are discretized with second-order schemes in space, and $r_T = r_G^2$ although a first-order scheme is used for the discretization in time.

(3) The correction factor C is sufficiently far from 1 in most cases. Therefore, only the uncertainty U_{GT} is evaluated to give a boundary of the simulation error.

(4) The iteration uncertainties as shown in Table 4 are negligibly smaller than U_{GT}, hence $U_{SN} \approx U_{GT}$.

(5) The numerical uncertainties are less than 4.3%; and the uncertainties in simulated head are higher than those in simulated power, especially at the low are high flow rates.

Table 5. Results of the numerical uncertainties at different flow rates.

Q (m^3/s)		ε_{21}	ε_{32}	R	p	δ^*_{RE}	C	U_{GT}	U_{SN} (%S)
0.35		−0.033	−0.053	0.642	1.280	−0.060	0.558	0.113	1.52
0.42	H	−0.021	−0.025	0.814	0.593	−0.091	0.228	0.231	4.27
0.471		−0.025	−0.045	0.564	1.651	−0.033	0.772	0.048	1.33
0.35		−0.043	−0.068	0.625	1.355	−0.071	0.599	0.129	0.41
0.42	P	−0.059	−0.071	0.836	0.516	−0.302	0.196	0.787	3.03
0.471		−0.057	−0.115	0.496	2.025	−0.056	1.017	0.062	0.30

4.2. Validation

The absolute value of comparison error $|E|$ is evaluated according to the experimental data D [25] and the simulation results S based on the fine grid G_1 and fine time step size Δt_1. The validation uncertainty U_V is calculated according to (11). Due to the lack of experimental uncertainty data, the experimental uncertainty U_D is assumed to be 0.8% by summing up the system accuracies in measuring the flow rate Q, the pressure difference $p_d - p_u$, and the torque M, etc. The results in Table 6 indicate that

(1) The validation is successfully achieved at the U_V level of 1–4%, except for the power P at Q = 0.35 m^3/s and Q = 0.471 m^3/s.

(2) For the power P at Q = 0.471 m^3/s, the validation is successful at the $|E|$ level of 1%, although the comparison error is larger than the validation uncertainty.

(3) For the power P at Q = 0.35 m^3/s, the comparison error is much larger than the validation uncertainty, which indicates that the modeling error is large, and the validation is not achieved.

(4) In most cases investigated here, the principal source of error is unidentifiable since the comparison errors are quite close to the validation uncertainties.

Table 6. The comparison error $|E|$ and validation uncertainty U_V at different flow rates.

Q (m^3/s)	$\|E\|$ (%D)		U_V (%D)	
	H	P	H	P
0.35	0.59	5.61	1.72	0.90
0.42	2.46	1.86	4.35	3.14
0.471	1.50	1.02	1.55	0.86

5. Simulated Flow Features

In this section, a detailed investigation is carried out for simulated flow features around rotor blade tips and between the stator and rotor blade rows based on the simulation results with fine grids G_1 at the flow rate $Q = 0.42$ m^3/s.

5.1. Tip Clearance Flow

The tip clearance flow is an important factor that influences the cavitation performance of the rotor. Figure 11 shows the streamlines around a blade tip. The flow separates at locations near the leading edge, travels downstream towards the suction side of the blade, and merges together to form the leakage vortices. The flow separation is relatively small at other locations of the tip section. In Figure 12 it is shown that a vortex forms at the leading edge and detaches from the tip surface due to the strong adverse pressure gradients close to the leading edge.

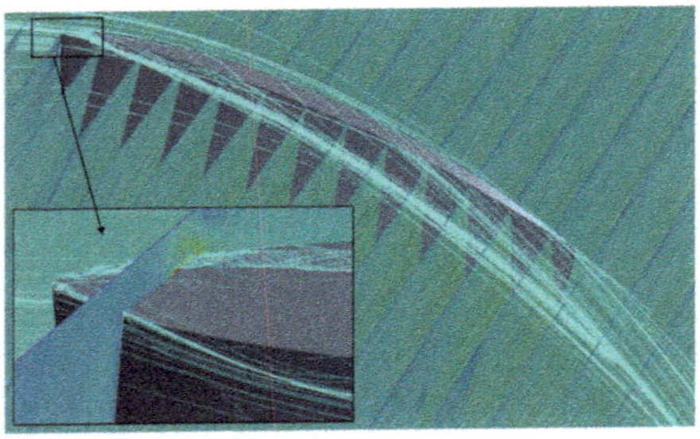

Figure 11. The streamlines around a blade tip. $Q = 0.42$ m^3/s.

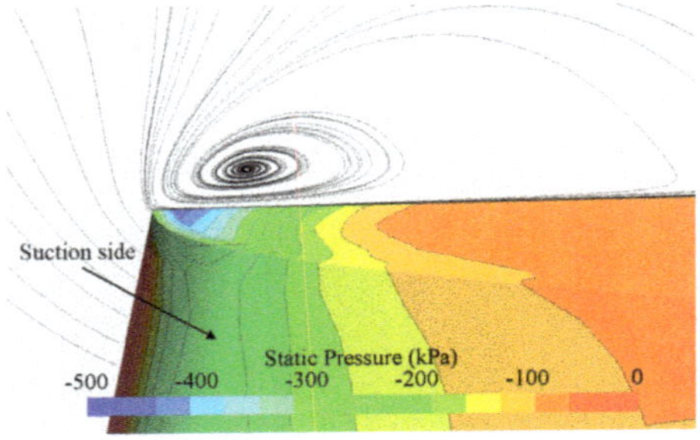

Figure 12. Pressure contours on the tip surface and streamlines in the tip clearance. $Q = 0.42$ m^3/s.

Figure 13 shows in more detail the streamlines in sections across the tip surface. In Figure 13a–c, it is clearly seen that the secondary flow from the pressure side to the suction side of the tip drives the vortical flow that detaches from the vicinity of the leading edge to move towards the suction side as it travels downstream. Meanwhile, the secondary leakage vortices are formed by the separated vortices from the pressure-side corner of the tip at other chordwise locations and travel downstream towards the pressure side of the tip just like the main leakage vortices. The phenomenon of flow separation is evident in the area where there is a large pressure difference between the pressure and suction sides and attenuates gradually behind the mid-chord. The main and secondary leakage vortices may merge at the mid-chord. Finally, they merge near the trailing edge to form a strong vortex. Unlike the open propellers, the tip leakage vortices contract little as they travel downstream.

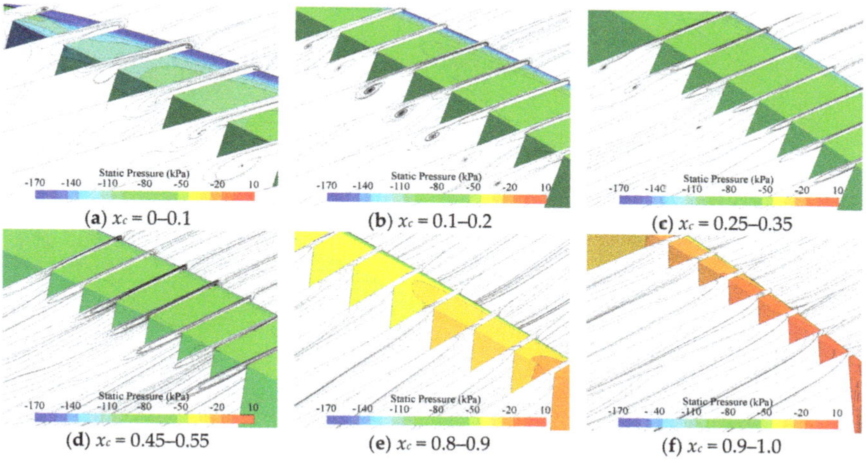

Figure 13. The streamlines (colored by static pressures) in sections across the tip surface, where x_c is the chordwise distance from the leading edge in fractions of the chord length. $Q = 0.42\ m^3/s$.

Figure 14 shows the in-plane velocity profiles in the tip clearance, at 10% and 50% chord length. Close to the tip surface, there is a jet-like flow due to the rotation of the blade, which becomes stronger as it goes from the suction side to the pressure side. Over the major part of the tip clearance, the cross flow is clearly driven by the pressure difference between the pressure and the suction sides, although it slows down due to the boundary layer on the shroud surface. Such features are generally similar at the two chordwise locations investigated. However, in the vicinity of the leading edge, the cross flow oscillates at a period of 40 degrees, which is just the angular spacing between adjacent blades of the stator, but becomes almost steady at the mid-chord. The oscillation is probably due to the impingement of the vortices shed from the pre-swirl stator blades.

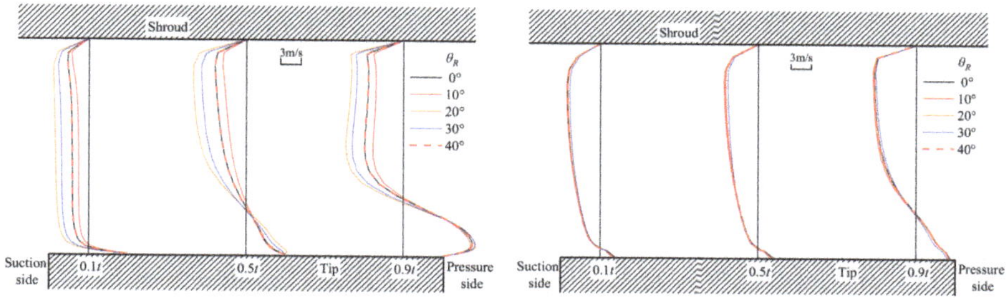

Figure 14. In-plane absolute velocity profiles in the tip clearance at 10% (**left**) and 50% (**right**) chord length. The velocity profiles shown at five instantaneous positions of the rotor θ_R cover the angular spacing between adjacent stator blades. From suction side to pressure side, three sections are taken at 10%, 50% and 90% of the local thickness t, respectively. The velocity magnitude is zero at the vertical straight lines. The shroud is stationary. The rotor tip rotates from left to right. $Q = 0.42\ m^3/s$.

5.2. Interactions between Rotor and Stator Blades

Figures 15 and 16 show the pressure and vorticity contours at typical time instants when the rotor blades sweep across adjacent stator blades. From the variations in the pressure contours around rotor blade sections, it is inferred that the angle of attack changes periodically due to the impingement of shed vortices from stator blades. Meanwhile, the rotor blades counter-act upon the stator blades by changing the pressures on the latter periodically, especially in the downstream part of blade sections. Due to the sliding interfaces between stator and rotor blade rows, the shed vortices behind stator blade sections dissipate abruptly when entering the rotor zone, which is clearly non-physical. It is not impossible to alleviate the grid dissipation by densifying the grids in between the upstream interface and the leading edges of rotor blades, but it would be very expensive computationally.

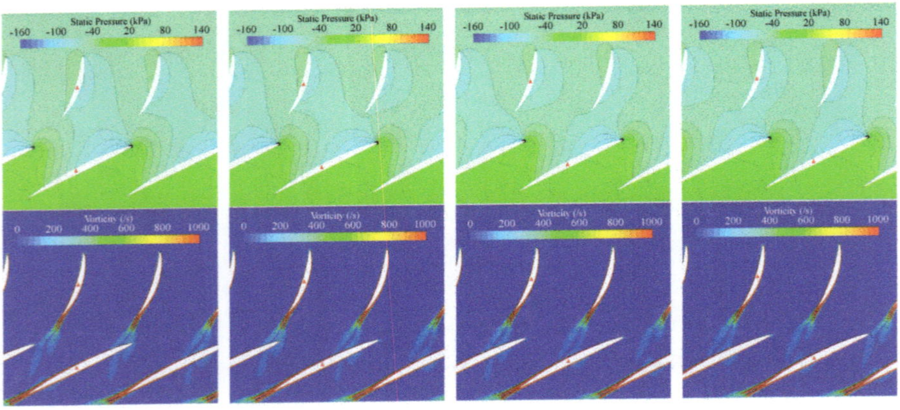

Figure 15. The pressure (**top row**) and vorticity (**bottom row**) around stator and rotor blade sections at $0.7R_{tip}$. The rotor blade angle θ_R = 0°, 10°, 20°, 30° (from **left** to **right**). Q = 0.42 m^3/s.

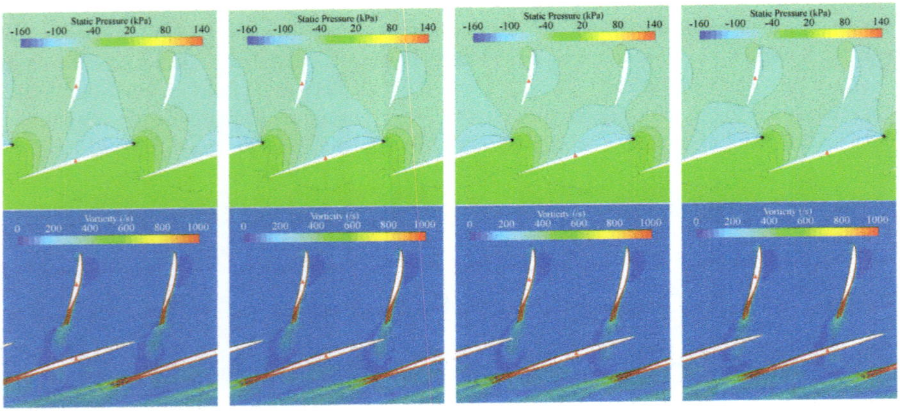

Figure 16. The pressure (**top row**) and vorticity (**bottom row**) around stator and rotor blade sections at $0.95R_{tip}$. The rotor blade angle θ_R = 0°, 10°, 20°, 30° (from **left** to **right**). Q = 0.42 m^3/s.

To investigate the rotor/stator interactions quantitatively, the oscillating pressures on rotor and stator blade sections are shown in Figure 17 at $0.7R_{tip}$ and $0.95R_{tip}$. The pressures on a rotor blade oscillate nine periods when the rotor completes a revolution, because the rotor blade sweeps across all the nine blades of the stator. For the same reason, the pressures on a stator blade oscillate seven periods since the rotor is seven-bladed. The oscillation amplitudes at the two radii shown in Figure 17 are quite close to each other, for both the rotor and the stator.

On rotor blades, the pressure oscillations are strong close to the leading edge due to the impingement of the vortices shed from the stator but attenuate towards the trailing edge. The oscillations on the pressure side are stronger than those on the suction side. On stator blades, however, the pressure oscillations are strong close to the trailing edge but attenuate towards the leading edge. Apparently, this is due to the counter-action from the rotor blades downstream. Besides, the oscillation amplitudes on the suction and pressure sides are close to each other.

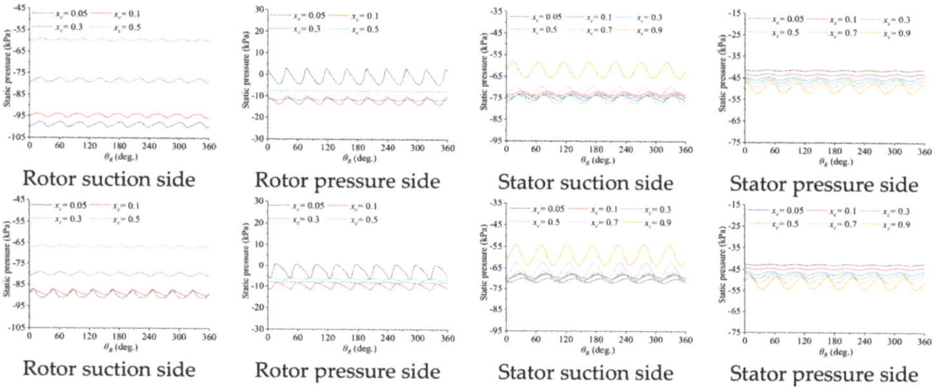

Figure 17. Blade-surface pressure oscillations in a complete revolution of the rotor at $0.7R_{tip}$ (**top row**) and $0.95R_{tip}$ (**bottom row**). x_c denotes the chordwise location from section nose in fractions of the chord length. θ_R denotes the angular position of a rotor blade. $Q = 0.42$ m^3/s.

Figure 18 shows the unsteady torques of a rotor blade and a stator blade, respectively. The amplitude of torque oscillations on the stator blade is about three times larger than that on the rotor blade. As shown in Figure 17, the oscillating pressures on the suction and pressure sides of the rotor blade are almost out of phase, but those of the stator blade are more or less in phase. This is the reason why the resultant torque oscillations on a rotor blade are weaker. It is noted that, for either the rotor or the stator, the oscillations in the total force of all the blades are much weaker than that of one blade (see Figure 10), and they only occur at the frequencies subject to the theoretical relation (13) when the inflow is uniform or axisymmetrical.

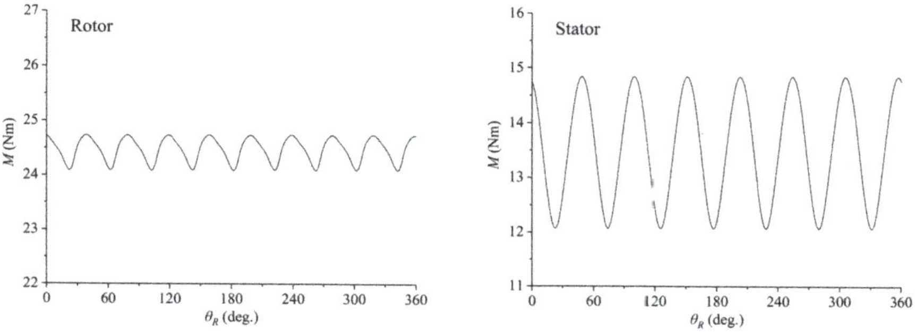

Figure 18. Oscillations of the torques on a rotor blade (**left**) and a stator blade (**right**). θ_R denotes the angular position of the rotor blade. $Q = 0.42$ m^3/s.

6. Concluding Remarks

In this work, unsteady RANS simulations of an axial-flow pump for waterjet propulsion are carried out and the numerical simulation uncertainties are analyzed. Unlike the ITTC procedure, the grid uncertainty U_G and time step uncertainty U_T are replaced by U_{GT}, the numerical uncertainty when grid and time-step sizes are refined simultaneously and consistently. For complex three-dimensional flow problems, the parameter refinement ratios need to be chosen appropriately so that the number of grids is not excessively large, and the time-step size is not too small. However, by doing so, it is almost impossible for the solutions to reach the asymptotic range. The analysis results indicate that the numerical uncertainties in present simulations are less than 4.3%. The validation is successfully achieved in most cases, except for the power at the lowest flow rate considered due to large modeling errors. For the flow rates considered, it is impossible to identify the principal source of error because the comparison errors are quite close to the validation uncertainties. So far as the head and power are concerned, it seems that the present simulation method based on block-structured grids works well from a practical point of view. However, from an uncertainty point of view, the grid and time-step sizes are still not small enough although further refinement would be very challenging computationally and even impractical.

Based on the simulation data using fine grids, the flows in tip-clearance and between rotor and stator blades are investigated. The formation and evolution of the tip leakage vortex is shown by visualizing the flow in the transverse sections along the chord of the tip section. The interactions between rotor and stator blades result in oscillating pressures on each blade of the rotor and the stator. But the oscillations are possibly under-predicted since the vortices shed from stator blades are dissipated artificially when entering the rotor zone.

The simulation method can be improved in two aspects at least. One is to use a second-order scheme for temporal discretization, the other is to reduce the artificial dissipation of the vorticity shed from stator blades due to insufficient grid density in the rotor zone. The latter seems, again, to be quite challenging.

Author Contributions: The numerical computation and analysis were done by J.-T.Q. and X.-Q.D.; C.-J.Y., W.L., and F.N. contributed to the simulation and analysis methodology, as well as composition of the manuscript; Z.-L.W. provided the model test data and discussions about the test setup and measuring methods.

Acknowledgments: The present work was funded by the Waterjet Propulsion Laboratory of Marine Design and Research Institute of China (MARIC) under project No. 61422230103162223005.

Conflicts of Interest: The authors declare no conflicts of interest.

References

1. Oberkampf, W.L.; Sindir, M.; Conlisk, A.T. *Guide for the Verification and Validation of Computational Fluid Dynamics Simulations;* Report No. AIAA G-077-1998; American Institute of Aeronautics and Astronautics: Reston, VA, USA, 1998.
2. Roache, P.J. *Verification and Validation in Computational Science and Engineering;* Hermosa: Albuquerque, NM, USA, 1998.
3. Roache, P.J. Verification of codes and calculations. *AIAA J.* **1998**, *36*, 696–702. [CrossRef]
4. ITTC. Uncertainty analysis in CFD, uncertainty assessment methodology. ITTC-Quality Manual, 4.9-04-01-01. In Proceedings of the International Towing Tank Conference, Shanghai, China, 5–11 September 1999.
5. ITTC. Uncertainty Analysis in CFD, Guidelines for RANS Codes. ITTC–Recommended Procedures and Guidelines, 7.5-03-01-02. In Proceedings of the International Towing Tank Conference, Seoul, Korea; Shanghai, China, 5–11 September 1999.
6. Stern, F.; Wilson, R.V.; Coleman, H.W.; Paterson, E.G. *Verification and Validation of CFD Simulations;* Report No. 407; Iowa Institute of Hydraulic Research: Iowa City, IA, USA, 1999.
7. Coleman, H.W.; Stern, F. Uncertainties and CFD Code Validation. *J. Fluids Eng.* **1997**, *119*, 795–803. [CrossRef]

8. Richardson, L.F. The approximate arithmetical solution by finite differences of physical problems involving differential equations, with an application to the stresses in a masonry dam. *J. Philos. Trans. R. Soc. Lond. Ser. A* **1911**, *210*, 307–357. [CrossRef]
9. ITTC. Uncertainty analysis in CFD, uncertainty assessment methodology and Procedures. ITTC-Quality Manual, 7.5-03-01-01. In Proceedings of the International Towing Tank Conference, Venice, Italy, 8–14 September 2002.
10. ITTC. Uncertainty Analysis in CFD, Verification and Validation Methodology and Procedures. ITTC-Recommended Procedures and Guidelines, 7.5-03-01-01. In Proceedings of the International Towing Tank Conference, Wuxi, China, 18 September 2017.
11. Eça, L.; Vaz, G.; Hoekstra, M. Code verification, solution verification and validation in RANS solvers. In Proceedings of the ASME 2010 29th International Conference on Ocean, Offshore and Arctic Engineering, Shanghai, China, 6–11 June 2010; pp. 597–605.
12. Larsson, L.; Stern, F.; Visonneau, M. *Numerical Ship Hydrodynamics: An Assessment of the 6th Gothenburg 2010 Workshop*; Springer Science and Business Media: Berlin, Germany, 2013.
13. Shen, H.C.; Yao, Z.Q.; Wu, B.S.; Zhang, N.; Yang, R.Y. A new method on uncertainty analysis and assessment in ship CFD. *J. Ship Mech.* **2010**, *14*, 1071–1083.
14. Simonsen, C.D.; Stern, F. Verification and validation of RANS maneuvering simulation of Esso Osaka: Effects of drift and rudder angle on forces and moments. *J. Comput. Fluids* **2003**, *32*, 1325–1356. [CrossRef]
15. Stern, F.; Wilson, R.V.; Coleman, H.; Paterson, E. Comprehensive approach to verification and validation of CFD simulations—Part 1: Methodology and procedures. *J. Fluids Eng.* **2001**, *123*, 793–802. [CrossRef]
16. Zhang, Z.R.; Zhao, F.; Wu, C.S. Research on uncertainty analysis of SUBOFF viscous flow field CFD simulation. In Proceedings of the 2007 Ship Mechanics Conference, Beijing, China, 30 July–1 August 2007.
17. Yang, Y.R.; Shen, H.C.; Yao, H.Z. Uncertain analysis of CFD simulation on the open-water performance of the propeller. *J. Ship Mech.* **2010**, *14*, 472–480.
18. Rosetti, G.F.; Vaz, G.; Fujarra, A.L.C. URANS calculations for smooth circular cylinder flow in a wide range of Reynolds numbers: Solution verification and validation. *J. Fluids Eng.* **2012**, *134*, 121103. [CrossRef]
19. Diskin, B.; Schwöppe, A. Grid-convergence of Reynolds-Averaged Navier–Stokes solutions for benchmark flows in two dimensions. *AIAA J.* **2016**, *54*, 2563–2588. [CrossRef]
20. Oberkampf, W.L.; Blottner, F.G. Issues in computational fluid dynamics code verification and validation. *AIAA J.* **1998**, *36*, 209–272. [CrossRef]
21. Wilson, R.; Shao, J.; Stern, F. Discussion: Criticisms of the "Correction Factor" Verification Method 1. *J. Fluids Eng.* **2004**, *126*, 704–706. [CrossRef]
22. Rodi, W. Experience with two-layer models combining the k-epsilon model with a one-equation model near the wall. In Proceedings of the 29th Aerospace Sciences Meeting, Reno, Nevada, 7–10 January 1991.
23. Shih, T.H.; Liou, W.W.; Shabbir, A.; Yang, Z.; Zhu, J. A new k-ε eddy viscosity model for high Reynolds number turbulent flows-model development and validation. *Comput. Fluids* **1995**, *24*, 227–238. [CrossRef]
24. Eça, L.; Hoekstra, M. Code verification of unsteady flow solvers with method of manufactured solutions. In Proceedings of the Seventeenth International Offshore and Polar Engineering Conference. International Society of Offshore and Polar Engineers, Lisbon, Portugal, 1–6 July 2007.
25. Wang, Z.L. *Measurement of the Hydraulic Performance of Axial-Flow Pump Model No.1413*; Report No. SM-2016-024; Marine Design and Research Institute of China: Shanghai, China, 2016.
26. Strasberg, M.; Breslin, J.P. Frequencies of the alternating forces due to interactions of contrarotating propellers. *J. Hydronaut.* **1976**, *10*, 62–64. [CrossRef]

MDPI
St. Alban-Anlage 66
4052 Basel
Switzerland
Tel. +41 61 683 77 34
Fax +41 61 302 89 18
www.mdpi.com

Journal of Marine Science and Engineering Editorial Office
E-mail: jmse@mdpi.com
www.mdpi.com/journal/jmse

www.ingramcontent.com/pod-product-compliance
Lightning Source LLC
Chambersburg PA
CBHW051711210326
41597CB00032B/5440
* 9 7 8 3 0 3 8 9 7 2 0 2 0 *